T0313750

FUNDAMENTALS OF MECHANICAL VIBRATIONS

Wiley-ASME Press Series List

FUNDAMENTALS OF MECHANICAL VIBRATIONS

Liang-Wu Cai

Kansas State University, USA

This Work is a co-publication between ASME Press and John Wiley & Sons, Ltd.

WILEY

Library of Congress Cataloging-in-Publication Data

Names: Cai, Liang-Wu author.
Title: Fundamentals of mechanical vibrations / Liang-Wu Cai.
Description: Hoboken : John Wiley & Sons, Inc., 2016. | Includes
  bibliographical references and index.
Identifiers: LCCN 2015044273 (print) | LCCN 2015045689 (ebook) | ISBN
  9781119050124 (cloth) | ISBN 9781119050230 (pdf) | ISBN 9781119050223
  (epub)
Subjects: LCSH: Machinery–Vibration. | Vibration–Mathematical models.
Classification: LCC TJ177 .C35 2016 (print) | LCC TJ177 (ebook) | DDC
  621.8/11–dc23
LC record available at http://lccn.loc.gov/2015044273

A catalogue record for this book is available from the British Library.

Cover image: Getty/odmeyer

Set in 10/12pt, TimesLTStd by SPi Global, Chennai, India.

1   2016

Contents

Series Preface

The Wiley-ASME Press Series in Mechanical Engineering brings together two established leaders in mechanical engineering publishing to deliver high-quality, peer-reviewed books covering topics of current interest to engineers and researchers worldwide.

The series publishes across the breadth of mechanical engineering, comprising research, design and development, and manufacturing. It includes monographs, references and course texts.

Prospective topics include emerging and advanced technologies in Engineering Design; Computer-Aided Design; Energy Conversion & Resources; Heat Transfer; Manufacturing & Processing; Systems & Devices; Renewable Energy; Robotics; and Biotechnology.

Preface

Why a new book on mechanical vibrations?

Mechanical vibration is a core subject in mechanical engineering that has been taught for many decades, with many old classics and many more excellent contemporary textbooks available in a rather crowded market. However, after teaching this subject for several years, I feel that there are two major challenges facing the current generation of engineering students studying vibrations, and, till this day, I am still unable to identify a textbook that satisfactorily addresses these challenges. This book represents my efforts in addressing these challenges.

A vibration analysis of a system starts from the equation(s) of motion for the system. The first major challenge facing the students is reaching this starting point. Although a prerequisite, the undergraduate Newtonian dynamics course does not adequately prepare students to derive the equation(s) of motion for a system that is slightly more complex than a mass–spring–dashpot system. Without teaching them on how to get from "here" (a given system) to "there" (the equations of motion), most textbooks parade through the analyses starting from "there."

In my view, Lagrangian dynamics is the ideal approach to reach this starting point. This view can be corroborated by the fact that almost all vibration textbooks contain a brief section, often in less than 10 pages, on Lagrangian dynamics, often just before vibration analyses of multi-degree-of-freedom systems. Lagrangian dynamics is typically a graduate-level topic that is rather mathematical and abstract for undergraduate students. The abstract nature compounded with the brevity in coverage does not prepare the students to reach the starting point with confidence.

I am partial to Lagrangian dynamics in part because of where I came from. When I was a doctoral student at Massachusetts Institute of Technology, for many years, I was a teaching assistant for the dynamics course in which Lagrangian dynamics was taught to sophomores, whose previous exposure to dynamics was college physics. Professor James H. Williams, Jr., who taught the course at that time, was also writing his own textbook on Lagrangian dynamics (*Fundamentals of Applied Dynamics*, John Wiley & Sons, 1996), which was the culmination of his years of award-winning teaching of Lagrangian dynamics to undergraduate students from a classic textbook (*Dynamics of Mechanical and Electromechanical Systems*, by Crandall, Karnopp, Kurtz, and Pridmore-Brown, McGraw-Hill, 1968). I was also helping him on the preparation for the solutions manual. I observed how such a difficult topic was taught to sophomores without losing its rigor. That experience gave me the belief that Lagrangian dynamics can be taught at undergraduate level, and Professor Williams had presented a successful paradigm in his textbook.

Another major challenge facing students is the symbolic analysis. In a vibration analysis, the solution is almost always in the form of an analytical expression. The current generation of engineering students, with sophisticated scientific calculators at their handy disposal, have grown comfortable with numbers: given a problem with a set of numbers, they apply a solution process to yield another number called the *answer*. The answer, right or wrong, represents a closure, as well as a sense of accomplishment. But in a symbolic analysis, the lack of numerical values given for the parameters in the problem appears to pull them away from the physical sense of the system, and the lack of a numerical answer makes them feel uncertain whether they have reached the final solution to the problem.

In my view, a graphical representation of an analytical solution could fill in as "the answer" to a symbolic problem. Graphing an analytical expression involves mundane details and some important steps such as nondimensionalizing the expression. The mundane details can be eliminated by incorporating mathematical software such as MatLab into the learning process, essentially giving the graphing capability in their handy disposal. This could reduce their uneasiness toward symbolic analysis and eventually learn this vitally important skill.

I feel that the introductory course on vibration is a perfect place to start training students for this skill. On the one hand, by junior or senior years, students have acquired a sense for the importance of the symbolic analysis: they have the motivation. On the other hand, many traditional vibration analysis topics can be enlivened by the computational capabilities brought by MatLab. Topics such as steady-state responses due to general periodic loadings by using Fourier expansions and the convolution integral as a general-purpose numerical method for an arbitrary transient loading can benefit from numerical computations.

Empowering students with these skills and capabilities does not require the commitment of a significant amount of time and effort, but requires the right instructional materials. After teaching the vibration course at Kansas State University for a couple of years using different textbooks, I started writing my own lecture notes.

Starting from a clean slate, I focused on the following set of learning objectives that I deem the most essential for prospective engineers:

- Being able to reach the starting point of a vibration analysis with confidence.
- Being able to establish a simple mathematical model for vibration analysis of real-world structures.
- Being able to handle the mathematics of vibration analyses, and, with the aid of a computer if necessary, to interpret the physical meanings of the results.
- Being able to tackle mathematical analysis symbolically.
- Being ready to use more powerful simulation tools when the needs call for.

The materials presented in this book are accumulated and refined over the past 10-plus years. To help the students achieving the above set of learning objectives, this textbook incorporates the following carefully designed pedagogical features:

- Staying true to the fundamentals. This book emphasizes providing thorough and clear explanations to the fundamental concepts and theories. It does not stray into advanced topics that are more suitable for advanced studies in graduate school. The time (and space) saved can be better spent on honing in the skills of symbolic analysis and the capabilities afforded by MatLab.

- Treating Lagrangian dynamics with rigor while maintaining accessibility. By limiting the consideration to Newtonian particles and rigid bodies, Lagrangian dynamics is tuned down its generality and the associated mathematical complexity. Students having the prerequisite of Newtonian dynamics are offered a fresh perspective. One of the recurring comments I received from students is that they "finally learned the dynamics!"
- Provoking higher level learning and thinking by enforcing symbolic analysis. MATLAB is used initially as merely a tool for visualizing analytical solutions. By limiting the use of MATLAB to this rudimentary and humble goal, students are greatly empowered in their analytical skills.
- Instilling the engineers' philosophy that having a solution to a problem is not the end to the problem; in contrary, it is the beginning of an exploration. In every vibration example, a section entitled *Exploring the Solution with MATLAB* is dedicated to curiosity-driven observations and explorations.
- Expanding the horizon by exploiting the capabilities of MATLAB. MATLAB is gradually used as a computational tool for analyzing more complicated problems such as modeling a fascinating phenomenon of levitating Slinky. In the end, a complete finite element analysis code is developed in MATLAB for vibration analysis of one-dimensional beams.
- Providing a consistent approach to modeling engineering systems. This book offers the following components to prepare the students for the industry:
  - A systematic procedure for establishing lumped-parameter models for simple engineering structures and systems. This gives students the skill to produce a quick and reasonably accurate estimate of practical problems.
  - The same lump-parameter modeling procedure is used for finite element formulation. This serves two purposes: it gives students an understanding of the working principle of the finite element method; and it boosts students the confidence on the effective lump-parameter modeling.
  - A set of illustrated tutorials for vibration analyses of a real structure using a commercial finite element analysis software package is also provided.

It is my sincere hope that this new textbook will bring a new breeze to the market that is crowded with many excellent textbooks over the past few decades. I wholeheartedly welcome all criticisms and suggestions to make it a better textbook.

In writing this book, first and foremost, I am greatly indebted to my former advisor at MIT, Professor James H. Williams, Jr., for offering me the opportunity as the teaching assistant for his class and as an assistant for preparing his textbook. His teaching of dynamics forms the basis of my firm belief that Lagrangian dynamics should be the starting point of the book, which inevitability bears the signature of his paradigm. Over the years, various draft versions of this book have been used in my class, and many students have provided valuable feedback. In particular, Ms. Congrui Jin and Mr. Ryan Cater offered extensive corrections to early versions. I would also like to thank Mr. Paul Petralia, senior acquisitions editor at Wiley, for his patience and encouragements, and Clive Lawson (UK), Preethi Belkese (India), and other editors at Wiley for their assistance. Finally, I would like to thank my wife, Huimin, for her unconditional support and love during this endeavor.

Liang-Wu Cai
September 2015

1

A Crash Course on Lagrangian Dynamics

1.1 Objectives

This chapter presents the fundamental concepts in Lagrangian dynamics and outlines a procedure for deriving the equation(s) of motion for holonomic mechanical systems using Lagrange's equation. Extensive examples are presented to cover a large variety of mechanical systems containing particles and rigid bodies. Finally, the procedures for finding the equilibrium position(s) and linearizing the equation(s) of motion in preparation for vibration analysis are also presented.

1.2 Concept of "Equation of Motion"

An *equation of motion* for a mechanical system is a differential equation that governs the changes in positions of components in the system with respect to time.

There are three key phrases in the above definition. Phrases *differential equation* and *with respect to time* specify a particular mathematical form for the equation and distinguish the equation of motion from its solution. That is, an equation of motion is a differential equation involving time derivatives. The phrase *positions* specifies a particular kinematic quantity, the most fundamental one from which other measures of motion, namely, displacements, velocities, and accelerations, can be obtained.

The concept of the *equation of motion* has suffered a degree of misuse in some dynamics textbooks. In these textbooks, Newton's second law, which many of us conveniently recite as $F = ma$, is called the equation of motion. In fact, Newton's second law is the fundamental physical law that governs the motion of any physical system. It is often a crucial tool to use in obtaining the equation(s) of motion for a system. But it is too primitive a form to be called an equation of motion. In our daily lives, we do not call a foundation as a structure. The same goes here.

Fundamentals of Mechanical Vibrations, First Edition. Liang-Wu Cai.
© 2016 John Wiley & Sons, Ltd. Published 2016 by John Wiley & Sons, Ltd.
Companion Website: www.wiley.com/go/cai/fundamentals_mechanical_vibrations

In the mean time, we must keep our minds open to the notion that there are other ways to establishing the equation of motion for a system. *Lagrangian dynamics* is one such alternative.

Before jumping into the details of Lagrangian dynamics, let us look at the way the equation of motion for a system can be obtained by using Newton's second law.

■ Example 1.1: Simple Mass–Spring–Dashpot System

Derive the equation of motion for the mass–spring–dashpot system as shown in Fig. 1.1.

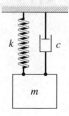

Figure 1.1   Mass–spring–dashpot system

□ Solution 1:

• Define x as the downward displacement of the mass, measured from its position when the spring is unstretched.

• *Kinematics of Mass m*: This example is simple enough to allow us to write directly:

$$\text{velocity} = \dot{x} \qquad \text{and} \qquad \text{acceleration} = \ddot{x} \tag{a}$$

where, as in dynamics, an overhead dot represents the time derivative, and overhead double-dot represents the second derivative with respect to time t.

• *Kinetics of Mass m*: A free-body diagram can be drawn, as in Fig. 1.2, to show all the forces acting on the mass in a generic instant in time when the system is in motion. In Fig. 1.2, F_s and F_d are the forces exerted by the spring and the dashpot, respectively. Applying Newton's second law based on the free-body diagram in Fig. 1.2 gives

$$mg - F_s - F_d = m\ddot{x} \tag{b}$$

Figure 1.2   Free-body diagram for mass m at a generic time instant

• *Constitutive Relations*: Besides the mass, the system contains a spring and a dashpot, whose forces are proportional to the deflection and the velocity, respectively. That is,

$$F_s = kx \qquad \text{and} \qquad F_d = c\dot{x} \tag{c}$$

- Substituting eqns. (c) into eqn. (b) gives, with a slight rearrangement,

$$m\ddot{x} + c\dot{x} + kx = mg \tag{d}$$

which is the equation of motion for this system.

□ Solution 2:

- Define y as the downward displacement of the mass, measured from its static equilibrium position.
- *Kinematics of Mass m*: Kinematics of mass m is unchanged, except replacing x by y.
- *Kinetic of Mass m*: The free-body diagram remains the same as in Fig. 1.2; and hence Newton's second law gives

$$mg - F_s - F_d = m\ddot{y} \tag{e}$$

- *Constitutive Relations* for system components are

$$F_s = k(y + \Delta) \qquad \text{and} \qquad F_d = c\dot{y} \tag{f}$$

where we note that the spring is already stretched at equilibrium, denoted as Δ, and that the spring force is proportional to the total amount of deformation in the spring.
- Substituting eqn. (f) into eqn. (e) gives

$$mg - ky - k\Delta - c\dot{y} = m\ddot{y} \tag{g}$$

- To eliminate Δ from eqn. (g), we look at the static equilibrium: the dashpot does not exert any force to the mass. The free-body diagram for the mass in its static equilibrium state is shown in Fig. 1.3. The equilibrium requires

$$k\Delta = mg \tag{h}$$

Figure 1.3   Free-body diagram for mass m at the static equilibrium

- Combining eqns. (g) and (h) gives, after a slight rearrangement,

$$m\ddot{y} + c\dot{y} + ky = 0 \tag{i}$$

which is the equation of motion for the system.

———— · ————

This example illustrates a typical process for deriving the equation(s) of motion for a mechanical system by using Newton's second law. Through this example, we can make the following observations:

- Before we can proceed to deriving the equation, some variables (such as x and y) must be defined so that the locations of the system's components can be described. In the end, the equations of motion for the system are differential equations of these variables. Such variables are called the *generalized coordinates*. We shall study this concept more thoroughly and rigorously in the next section.
- Different definitions of the generalized coordinates result in different equations of motion for the same system. That is, the equations of motion are not unique, depending on the choice of the generalized coordinates by which the system is described.
- The following three pieces of information are generally needed: (1) kinematics of the system components at a generic time instant during the motion; (2) the properties of system components, which are called the *constitutive relations*; and (3) Newton's second law that threads all these pieces of information together.

However, there are some shortcomings in this approach of deriving the equations of motion for a system. For example, how do we know how many equations should be obtained? For a complex system, which part or parts of the system should be isolated to draw the free-body diagrams? Finding the answer to these questions is *ad hoc*: we have to look into individual systems case by case. This means that we may be able to solve one problem effortlessly, but we might stumble on the next. We all have experienced the situations in which drawing a free-body diagram reveals more unknowns than desired and calls for drawing more free-body diagrams and writing out more equations. We may also recall that, in kinematics, finding the acceleration for a particle or a point in a rigid body involves substantially more work than finding the velocity.

Lagrangian dynamics avoids most of these issues and provides a structured way to analyze a system and to subsequently obtain its equation(s) of motion. It works in a way similar to the energy method: only positions and velocities are required in the formulation. The procedure is unchanged regardless of the system's complexity. Furthermore, it can be extended to handle systems in other physical domains, such as electrical, electromagnetic, and electromechanical systems.

Deriving the equation of motion is the first step toward understanding a system. Having the equation(s) of motion in hand, the first and foremost, we can find and subsequently analyze the solution to the equation(s) of motion. This plainly stated activity, in fact, encompasses almost all subjects of study in mechanical engineering, including vibration analyses. As we shall see, vibration analyses are to find, analyze, and study the solutions to a particular class of equations of motion.

Having the equation of motion, before putting all the efforts into finding the solution, a few less ambitious things having great engineering interests can be done:

- Determine the system's equilibrium configuration(s). In the equilibrium state, the system does not move at all. For the above example, setting the velocity and acceleration to zero, the equation of motion derived in *Solution 1* would become

$$kx_{eq} = mg \qquad \text{or} \qquad x_{eq} = \frac{mg}{k}$$

where x_{eq} denotes the position of the mass at equilibrium. The equation of motion derived in *Solution 2* would give $y_{eq} = 0$, as expected.
- Analyze the dynamic stability and other associated behaviors of the system.

- Introduce approximations to simplify the equation(s) of motion or to obtain approximate solutions.

We discuss these topics separately later in this chapter. In fact, these things must be explored before we can proceed with vibration analyses of a system.

1.3   Generalized Coordinates

Coordinates usually are associated with a particular coordinate system and usually appear in a set. For instance, in a Cartesian coordinate system, the coordinates for a point are (x, y, z). In a polar coordinate system, the coordinates are (r, θ). Adding the word *generalized* frees us from abiding to any particular coordinate system so that we can choose whatever parameter that is convenient to describe the position of a point in a system. Hence,

A *generalized coordinate* is a parameter that is used to locate a part of a system. A generalized coordinate is a scalar quantity.

A group of parameters that is used to locate a system is a *set* of generalized coordinates. To denote a set of generalized coordinates, we follow the mathematical notation for explicitly defining a set by listing all elements contained in the set. We write, for example, $\{x_1, x_2\}$.

A *complete set* of generalized coordinates is a set of generalized coordinates that completely locates all parts of a system in all geometrically admissible configurations. A *geometrically admissible configuration* is a configuration that is allowed by the geometrical constraints in the system.

An *independent set* of generalized coordinates is a set of generalized coordinates in which if all but one of them are fixed, there still exists a continuous range of values for that unfixed generalized coordinate.

A *complete and independent set* of generalized coordinates is, literally, a set of generalized coordinates that is both complete and independent.

Because generalized coordinates can be chosen or defined almost at will, generalized coordinates for a system are not unique. For this same reason, it is utterly important to define them clearly, aided by a schematic depiction if necessary. When defining a generalized coordinate, usually the following four vital aspects should be clearly indicated for each generalized coordinate: the exact physical meaning, where it is measured from, and relative to whom, and which direction is positive.

■ Example 1.2: Particle Moves Freely in Three-Dimensional Space

A particle can move freely in a three-dimensional space. Define a set of generalized coordinates and subsequently judge whether it is a complete and independent set of generalized coordinates.

□ Solution:

Before we start, as a preparatory setup, we define a Cartesian coordinate system to facilitate the discussions that follow. The Cartesian coordinate system is fixed in space.

We define $\{x, y, z\}$ as a set of generalized coordinates, where x, y, and z are the Cartesian coordinates of the particle. Note that the positive directions for the coordinates have already been defined in the Cartesian coordinate system.

- *Is It a Complete Set?* YES, because there is only one particle in the three-dimensional space, the three coordinates completely specified a point in the three-dimensional space.
- *Is It an Independent Set?* To answer this question, we need to conduct a series of tests: if all but one of the generalized coordinates are fixed, will there be a continuous range of values for the unfixed one to change. In this problem, if x and y are fixed, z can still be freely changed along a line parallel to the z-axis. Similar conclusion can be drawn for leaving any other coordinate unfixed. So, the answer to this question is also YES.

Therefore, we conclude that $\{x, y, z\}$ as defined is a complete and independent set of generalized coordinates.

□ **Discussion: Using a Coordinate System**

When using coordinates in a well-established coordinate system, such as the Cartesian coordinate system in this example, as the generalized coordinates, in general, the coordinate system has already included the specifications for the directions of positive coordinates. An important task in defining a coordinate system is to specify how its origin moves, such as relative to whom, and how the coordinate axes are oriented.

■ **Example 1.3: Simple Planar Pendulum**

A simple pendulum is made up by a particle hung to a pivot point through a massless string. A planar pendulum is a simple pendulum whose motion is restricted to within a plane, typically the plane of paper, as shown in Fig. 1.4. Assume that the string of length l remains taut at all times. Define a set of generalized coordinates for this pendulum and subsequently judge whether it is a complete and independent set of generalized coordinates.

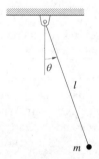

Figure 1.4   Simple planar pendulum

□ **Solution 1: Angular Displacement**

We define $\{\theta\}$ as a set of generalized coordinates, where θ is the angular displacement of the pendulum, measured in the counterclockwise direction from the vertical, as shown in Fig. 1.4. Note that, in the sketch, the arrow for θ goes only in one direction, indicating its positive direction.

- *Is It a Complete Set?* YES, because there is only one particle in the system and it can only move around a circle of radius l centered at the pivot point. Once the angle θ is determined, the location of the particle is uniquely determined.

- *Is It an Independent Set?* To answer this question, we need to conduct a test: if all but one generalized coordinates are fixed, will there be a continuous range of values for the unfixed one to vary? In this case, there is only one generalized coordinate. When "all but one are fixed," actually nothing is fixed, and the unfixed one is θ. As the particle is allowed to move along the circle of radius l, θ indeed has a continuous range to vary. So, the answer to this question is also YES.

Therefore, we conclude that $\{\theta\}$ as defined is a complete and independent set of generalized coordinates.

□ **Discussion: Geometrically Admissible Configurations**

The definition for the completeness of a set of generalized coordinate includes a qualifying term "geometrically admissible configurations." However, this phrase is not mentioned in the preceding discussions. A geometrically admissible configuration requires the pendulum remain in the plane of paper and the string being taut. These constraints have been implicitly invoked when we say "the pendulum can only move around a circle."

□ **Solution 2: Cartesian Coordinates**

In the preparatory setup, we define a Cartesian coordinate system Oxy such that its origin O is fixed at the pivot point, the x-axis points horizontally to the right, and the y-axis points vertically upward.

We then define $\{x, y\}$ as a set of generalized coordinates, where x and y are Cartesian coordinates of the mass of the pendulum.

- *Is It a Complete Set?* YES, because there is only one particle that can only move within the plane of paper: once the Cartesian coordinates x and y are specified, the location of the mass is completely determined.
- *Is It an Independent Set?* To answer this question, we need to conduct the following test: if all but one generalized coordinates are fixed, will there be a continuous range of values for the unfixed one to vary? If we fix x, since the particle can only move along the dotted circle shown in Fig. 1.5, y can be located at two possible positions as indicated: at the intersections of the vertical line and the circle. However, they are two *discrete* positions, and do not constitute a *continuous range*. This suffices to give NO as the answer to this question without conducting any further test.

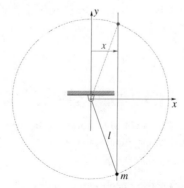

Figure 1.5 Two possible positions for the pendulum when x is fixed

Therefore, we conclude that $\{x, y\}$ as defined is not a complete and independent set of generalized coordinates. Specifically, it is a complete set but not an independent set.

□ **Solution 3: Cartesian Coordinates, Remedied**

Although the problem statement does not ask us to define a complete and independent set of generalized coordinates, we are still curious as how to remedy the set we just defined to make it a complete and independent set.

We define $\{x\}$ as a set of generalized coordinates, where x is the x-coordinate of the mass of the pendulum in the Cartesian coordinate system as defined in *Solution 2*.

- *Is It a Complete Set?* When x is fixed, as discussed earlier, there are two possible positions for the pendulum. In the Cartesian coordinate system, the y-coordinate can be found from the following relation:

$$x^2 + y^2 - l^2 \qquad \text{or} \qquad y = \pm\sqrt{l^2 - x^2} \qquad \text{(a)}$$

This identifies two possible y values in Fig. 1.5. The exact locations of the two possible positions are completely determined. We consider the answer to this question is YES.
- *Is It an Independent Set?* To answer this question, we need to conduct the following test: if all but one generalized coordinates are fixed, will there be a continuous range of values for the unfixed one to vary? In this case, if we fix all but x, nothing is actually fixed, and x can vary in the range $-l \leq x \leq l$. Thus, the answer to this question is YES.

Therefore, we conclude that $\{x\}$ as defined is a complete and independent set of generalized coordinates.

□ **Discussion**

The condition such as the one in eqn. (a) is called a *geometric constraint* of the system. The question about which of the two positions the particle is located at a given time will be answered or become apparent by other information of the system under consideration, such as the initial conditions of the system. A mechanical system moves continuously in space from one location into an adjacent one.

■ **Example 1.4: Rigid Slender Rod Moves in a Plane**

A rigid slender rod of length L is allowed to move freely on the plane of paper. Define a set of generalized coordinates and subsequently judge whether it is a complete and independent set of generalized coordinates.

□ **Solution:**

In the preparatory setup, we define a Cartesian coordinate system Oxy that is fixed in space, as shown in Fig. 1.6.

We then define $\{x_1, x_2, y_1, y_2\}$ as a set of generalized coordinate, where x_1 and y_1 are the Cartesian coordinates of the left end of the rod and x_2 and y_2 are the Cartesian coordinates of the other end of the rod.

- *Is It a Complete Set?* YES, because once the two ends of the rod are fixed, every point on the rod can be located.

- *Is It an Independent Set?* To answer this question, we need to conduct the following test: if all but one generalized coordinates are fixed, will there be a continuous range of values for the unfixed one to vary? We first fix x_1, y_1, and x_2 and leave y_2 unfixed. As a rigid body, the length of the rod, L, is fixed. This geometrical constraint gives y_2 as

$$y_2 = y_1 \pm \sqrt{L^2 - (x_1 - x_2)^2} \tag{a}$$

There are two possible values for y_2, as indicated by the $\pm$ sign, as shown in Fig. 1.7. They are two discrete values, but do not constitute a continuous range of values. Thus, the answer to this question is NO.

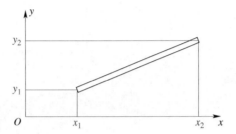

Figure 1.6   Rigid slender rod moves on the plane

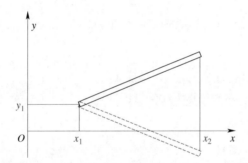

Figure 1.7   Two possible locations of the rod when x_1, y_1, and x_2 are fixed

Therefore, we conclude that $\{x_1, x_2, y_1, y_2\}$ as defined is not a complete and independent set of generalized coordinates. Specifically, it is a complete set but not an independent set.

☐ **Remedied Solution:**

To define a set of complete and independent set, we only need to pick out any three of the four generalized coordinates, knowing that the fourth can be obtained from the geometrical constraint in eqn. (a).

In this solution, we pick out $\{x_1, y_1, x_2\}$ as a set of generalized coordinates. Their definitions are exactly the same as before and would not be repeated here.

- *Is It a Complete Set?* YES. As analyzed before, when needed, the remaining coordinate y_2 can be found from eqn. (a). Then, the exact location of the rod is completely determined.

- *Is It an Independent Set?* We cannot take for granted that such a choice would automatically be a complete and independent set. We still need to conduct a series of tests.
  - First, we fix x_1 and y_1 and leave x_2 unfixed. The possible scenario is that the rod can rotate about its left end, while its right end moves along a circle, as shown in Fig. 1.8. Along this circle, x_2 certainly has a continuous range to vary. Furthermore, the range is between the extreme ends of the circle, that is, $x_1 - L \leq x_2 \leq x_1 + L$.

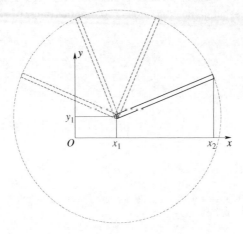

Figure 1.8   Possible locations of the rod when x_1 and y_1 are fixed

  - Next, we fix x_1 and x_2 and leave y_1 unfixed. This allows the rod to slide up and down along a vertical strip confined by x_1 and x_2, as shown in Fig. 1.9. Thus, y_2 has a continuous range to vary. Furthermore, the range is unlimited: $-\infty < y_1 < \infty$.

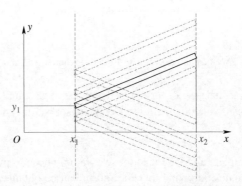

Figure 1.9   Possible locations of the rod when x_1 and x_2 are fixed

  - Lastly, we fix y_1 and x_2 and leave x_1 unfixed. The scenario is similar to a ladder sliding along a wall–floor corner while remaining in simultaneous contacts with both the wall and the floor. In this analogy, the "wall" is defined by the vertical lines $x = x_2$; while the "floor" is defined by the horizontal line $y = y_1$. The difference is that the image mirrored either by the "wall" or by the "floor" is also valid, as sketched in Fig. 1.10. Thus, x_1 has a continuous range to vary. Furthermore, the range is $x_2 - L \le x_1 \le x_2 + L$.

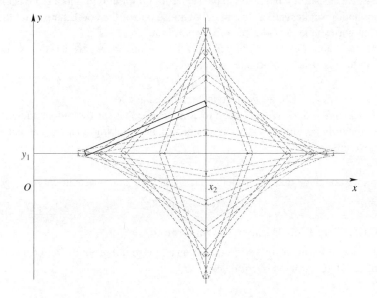

Figure 1.10   Possible locations of the rod when y_1 and x_2 are fixed

This completes the independence test. In all three tests, the unfixed coordinate always has a continuous range of values to vary. Thus, the answer to this question is YES.

We conclude that $\{x_1, y_1, x_2\}$ as defined is a complete and independent set of generalized coordinates.

■ **Example 1.5: Particle Moves Along a Wire of Known Shape**

A particle moves along a wire on the xy-plane of a known shape given by

$$y = a + bx^2$$

as shown in Fig. 1.11. Define a complete and independent set of generalized coordinates.

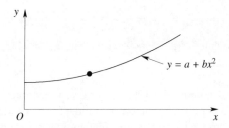

Figure 1.11   Particle moves on a wire of known shape

□ **Solution:**

Since a Cartesian coordinate system has already been defined, it is convenient to choose the Cartesian coordinates x and y of the particle as the generalized coordinates. However, Example 1.4 gives us sufficient reason to pause before jumping into making the declaration. A careful examination suggests that we can only choose one of the coordinates, and the function describing the wire shape provides the other coordinate.

Therefore, we define $\{x\}$ as a set of generalized coordinates, where x is the x-coordinate of the particle in the Cartesian coordinate system Oxy.

- *Is It a Complete Set?* YES, as we have just analyzed above.
- *Is It an Independent Set?* We still need to conduct the test for the independence. If all but one fixed, the only possibility is to leave x unfixed, and nothing is fixed. The particle is still free to move along the wire.

Therefore, we conclude that $\{x\}$ as defined is a complete and independent set of generalized coordinates.

■ **Example 1.6: Disk Rolls Without Slip on Ground**

A circular disk of radius R rolls without slip on a horizontal ground. Define a complete and independent set of generalized coordinates.

□ **Solution:**

In the preparatory setup, we paint a radius on the disk between its center and the contact point with ground in a reference configuration. We also define a Cartesian coordinate system Oxy such that its origin is fixed at the contact point in the reference configuration and its x-axis lies on the ground. Figure 1.12 shows both the reference and the displaced configurations of the disk. In the displaced configuration, C is the center of the disk, A is the contact point, and B is the contact at the reference configuration. Angle θ denotes the angular displacement of the painted radius.

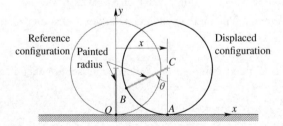

Figure 1.12 Disk rolls without slip on horizontal ground, showing reference and displaced configurations

We can now define the x-coordinate of the center C as a generalized coordinate. Once x is fixed, the location of the center C is completely determined. As a rigid body, locating the center is not sufficient. We need to fix its orientation. The angle θ seems to be a perfect candidate for another generalized coordinate. However, since the disk rolls without slip, it requires that the length OA equals to the arc length $\overset{\frown}{AB}$. Since length OA equals x, given the radius R of the disk, θ can be uniquely determined.

Therefore, we define $\{x\}$ as a set of generalized coordinate, where x is the x-coordinate of the center of the disk in the Cartesian coordinate system Oxy.

- *Is It a Complete Set?* YES. As we have just analyzed, this is a complete set.
- *Is It an Independent Set?* We need to conduct the following test: if all but one generalized coordinates are fixed, will there be a continuous range of values for the unfixed one to vary. If we fix all but x, actually nothing is fixed, the disk can roll on the ground, and hence x has a continuous range to vary. Thus, the answer to this question is YES.

Finally, we conclude that $\{x\}$ as defined is a complete and independent set of generalized coordinates.

■ **Example 1.7: Disk Rolls on Ground, Slipping Allowed**

A circular disk of radius R rolls on a horizontal ground. The disk is allowed to slip while rolling. Define a complete and independent set of generalized coordinates.

□ **Solution:**

The preparatory setup is the same as in Example 1.6.

We define $\{x, \theta\}$ as a set of generalized coordinate, where x is the x-coordinate of the center of the disk and θ is the clockwise angular displacement of the painted radius.

- *Is It a Complete Set?* Based on the analysis in Example 1.6, we can readily conclude that, YES, it is a complete set.
- *Is It an Independent Set?* We need to conduct the following series of tests: if all but one generalized coordinates are fixed, will there be a continuous range of values for the unfixed one to vary. We first fix x and allow θ to vary. Since slipping is allowed, when the center is fixed, the disk can still rotate just like a spinning wheel of a car on a jack stand. If we fix θ and allow x to vary, because slipping is allowed, the disk can slide without rotation, just like a locked wheel skidding on ice. So, YES, it is an independent set.

Therefore, we conclude that $\{x, \theta\}$ is a complete and independent set of generalized coordinates.

1.4   Admissible Variations

A *variation* is a hypothetical small change in a generalized coordinate.

A variation is a close cousin of a *virtual displacement*. Let us discuss the virtual displacement first. A virtual displacement differs from a real displacement in two aspects: (1) It occurs instantaneously without advancing the time. Or, put it differently: it is the difference between the real position and an alternative position, as if we say to ourselves: "what if at this particular moment this point is located there instead of here." (2) It is *small* in the sense of a differential change.

Now let us look at the relation between a virtual displacement and a variation. A real change in position is a *displacement* (vector). Since the system is described by the chosen complete and independent set of generalized coordinates, this displacement is expressible in terms of this set of generalized coordinates (scalars). A virtual displacement is a *hypothetical* and *small* displacement. It is expressible in *hypothetical* and *small* changes in generalized coordinates, which are defined as *variations*. Normally, displacements are vectors, so are the virtual displacements. Generalized coordinates are scalars, so are the variations.

An *admissible variation* is a hypothetical small change in a generalized coordinate that is allowed by the geometrical constraints of the system.

Since the admissible variations are associated with the generalized coordinates, once a set of generalized coordinates has been defined, a set of admissible variations can be naturally derived. In other words, it does not need to be defined. For a set of generalized coordinates, we usually write, for example, $\{\delta x_1, \delta x_2\}$ as the associated set of admissible variations. Mathematically, the variation operator δ follows the same rules as the differential operator d. But, what is important is to conduct similar tests to see whether the set of admissible variations is *complete and independent*. Passing both completeness and independence tests makes the set a *complete and independent set* of admissible variations.

■ Example 1.8: Rigid Slender Rod Moves on Plane, Variations

Determine whether the set of variations associated with the set of generalized coordinates defined in Example 1.4 is a complete and independent set of admissible variations.

□ Solution:

In Example 1.4, a complete and independent set of generalized coordinate has been defined as $\{x_1, y_1, x_2\}$, where x_1, x_2, and y_1 are all Cartesian coordinates. The corresponding set of admissible variations is $\{\delta x_1, \delta y_1, \delta x_2\}$. Being "admissible" means that the varied configuration remains entirely on the Oxy plane.

Assume a varied configuration of the system, shown as the dashed configuration in Fig. 1.13. The left end is shown in an enlarged view, in which δr_A is its virtual displacement; δx_1 and δy_1 are the admissible variations associated with x_1 and y_1, respectively. They, along with x_1 and y_1, locate the left end of rod in the varied configuration. Similarly, δx_2 and x_2 locate the x-coordinate of the right end of the rod in the varied configuration. With the geometrical constraint of a fixed length L, the right end of the rod in the varied configuration is thus located. Consequently, every point in the rod in the varied configuration can be located. Thus, this set of admissible variations is complete.

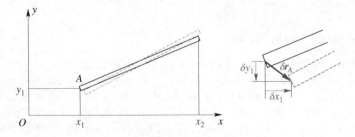

Figure 1.13 Varied configuration of slender rod on plane. The enlarged view of left end A on the right shows virtual displacement δr_A being represented by variations δx_1 and δy_1

Furthermore, we need to test when all but one admissible variations are fixed, whether the remaining unfixed one has a continuous range of values to vary. If δx_1 and δy_1 are fixed, the rod is allowed to rotate about the varied position of the left end while the other end moves around the circle of radius L, similar to Fig. 1.8. So, any small variation within the circle is the range in which δx_2 can vary. We can run through the other two tests and find that the situations are similar to the respective tests for the generalized coordinates in Example 1.4. In all tests, the unfixed one has a continuous range of values to vary.

Therefore, we conclude that $\{\delta x_1, \delta y_1, \delta x_2\}$ is a set of complete and independent admissible variations.

■ Example 1.9: Particle Moves Along a Wire of Known Shape, Variations

Determine whether the set of variations associated the set of generalized coordinates defined in Example 1.5 is a complete and independent set of admissible variations.

□ Solution:

In Example 1.5, a complete and independent set of generalized coordinates has been defined as $\{x\}$, where x is the Cartesian coordinates of the particle. The associated set of admissible variations is $\{\delta x\}$. We verify that a varied location can be completely located by δx along with x and that δx has a continuous range of values to vary. Therefore, we conclude that $\{\delta x\}$ is a set of complete and independent admissible variations for this set of generalized coordinates.

■ Example 1.10: Disk Rolls without Slip on Ground, Variations

Determine whether the set of variations associated the set of generalized coordinates defined in Example 1.6 is a complete and independent set of admissible variations.

□ Solution:

In Example 1.6, a complete and independent set of generalized coordinates has been defined as $\{x\}$, where x is the position of the center of the disk from its initial position and positive in the right direction. The associated set of admissible variations is $\{\delta x\}$. We verify that a varied location can be completely located by δx along with x; and that δx has a continuous range of values to vary. Therefore, we conclude that $\{\delta x\}$ is a set of complete and independent admissible variations for this set of generalized coordinates.

■ Example 1.11: Disk Rolls on Ground, Slipping Allowed, Variations

Determine whether the set of variations for the set of generalized coordinates defined in Example 1.7 is a set of complete and independent admissible variations.

□ Solution:

In Example 1.7, a complete and independent set of generalized coordinates has been defined as $\{x, \theta\}$, where x is the Cartesian coordinate of the particle and θ is the angle of the vertical makes with the line that is initially painted as vertical. The associated set of admissible variations is $\{\delta x, \delta\theta\}$. We verify that a varied location can be completely located by δx and $\delta\theta$, along with x and θ; and that when either one is fixed, the unfixed one has a continuous range of values to vary. Therefore, we conclude that $\{\delta x, \delta\theta\}$ is a set of complete and independent admissible variations for this set of generalized coordinates.

1.5   Degrees of Freedom

The number of variations in a set of complete and independent admissible variations for a system is called the number of *degrees of freedom* of the system.

A system is said to be *holonomic* if the number of degrees of freedom equals to the number of generalized coordinates in a set of complete and independent generalized coordinates. Otherwise, the system is *nonholonomic*. Most of the mechanical systems are holonomic and we will only discuss holonomic systems in this book. Treatments for nonholonomic system are beyond the scope of this book.

The number of degrees of freedom is a property of a system. No matter how we define the generalized coordinates, and there is probably an infinite number of ways of doing so, the number of degrees of freedom is always the same.

■ **Example 1.12: Revisiting Earlier Examples**

Determine the number of degrees of freedom and the holonomicity for systems in Examples 1.4 through 1.7.

☐ **Solution:**

• In Example 1.4: The number of degrees of freedom is 3. The system is holonomic.
• In Example 1.5: The number of degrees of freedom is 1. The system is holonomic.
• In Example 1.6: The number of degrees of freedom is 1. The system is holonomic.
• In Example 1.7: The number of degrees of freedom is 2. The system is holonomic.

■ **Example 1.13: Specified Motion**

The system shown in Fig. 1.14 consists of a cart of mass m_1 that moves on the horizontal rail and a pendulum of length l and mass m_2. The pendulum is pinned to the cart. The cart moves in accordance with $x = x_0(t)$, driven by an external mechanism (not shown), where x is measured from a fixed reference position. Determine the number of degrees of freedom of the system.

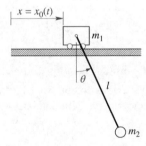

Figure 1.14   Mass–pendulum system with a specified motion

☐ **Solution:**

In this example, by intuition we are inclined to define x and θ as a set of generalized coordinates, where x is as defined in the problem statement and θ is the counterclockwise angular displacement of the pendulum with respect to the vertical. However, we must recognize that such a set is not independent, because $x(t)$ is known for all times, as a part of the geometrical constraint of the system. More importantly, it cannot be varied. In such a case, the cart is said to be undergoing a *specified motion*, and the condition $x = x_0(t)$ is an *auxiliary condition*, which can be considered as a generalized coordinate (but not included in the *complete and independent* set) when needed in deriving the equations of motion for the system.

We define $\{\theta\}$ as a complete and independent set of generalized coordinates for the system; and $\{\delta\theta\}$ is a complete and independent set of admissible variations in this set of generalized coordinates. The system is holonomic and has one degree of freedom.

———————— · ————————

So far, all examples include only holonomic systems. A curious reader might wonder: how a nonholonomic system looks like, or whether nonholonomic systems even exit. In fact, we deal with nonholonomic system almost daily. When we parallel park a car, if we park it too far from the curb in the first attempt and decide to "correct" it instead of starting over, we have to move the car back and forth in order to move the car laterally just a little bit. That little bit cannot be achieved directly by small variations. A careful reader shall examine this situation as an exercise, by modeling a wheel of the car as a disk that rolls without slip on the ground, and also examine the difference from the situation described in Example 1.6.

In general, a nonholonomic system contains nonholonomic constraints. A nonholonomic constraint is one that cannot be explicitly expressed as an equation involving only the generalized coordinates and time. The analysis in the remainder of this chapter is not applicable to nonholonomic systems.

1.6   Virtual Work and Generalized Forces

Loosely speaking, a *virtual work* is the work done by a force over a virtual displacement. However, in Lagrangian dynamics, we would like to give it a more articulate definition: the virtual work is the *total work done by all nonconservative forces acting on the system over a variation in the admissible configuration.*

There are two key points in this definition. First, the virtual work accounts only for work done by nonconservative forces. Typical nonconservative forces include those that introduce energy into the system, such as externally applied forces, or those that cause energy losses, such as dashpot forces and frictional forces. Second, the variation in the admissible configuration must be general enough to cause a virtual displacement at every location wherever a force is acting upon.

This definition of virtual for the virtual work results in the following expression:

$$\delta W^{\text{n.c.}} = \sum_{i=1}^{N} \boldsymbol{F}_i \cdot \delta \boldsymbol{R}_i \tag{1.1}$$

where N is the number of nonconservative forces acting on the system, $\boldsymbol{F}_i$'s are the individual forces, and $\delta \boldsymbol{R}_i$'s are the virtual displacements at the locations where the forces are acting.

On the other hand, having selected a complete and independent set of generalized coordinates for a holonomic system guarantees that any virtual displacement can be expressed in terms of admissible variations. Therefore, the virtual work can be expressed as

$$\delta W^{\text{n.c.}} = \sum_{j=1}^{M} \Xi_j \delta \xi_j \tag{1.2}$$

where M is the number of admissible variations in a complete and independent set, which is also the number of degrees of freedom of the system and Ξ_j is called the *generalized force* associated with the generalized coordinate ξ_j.

Equation (1.1) is used to calculate the virtual work; and eqn. (1.2) is used to identify the generalized forces. In calculating the virtual work, there are two methods that can be deployed. One method is to identify a varied configuration and determine the virtual displacement at every location where a force is applied, and use the vector dot product to calculate the virtual work. Another method is to vary one generalized coordinate at a time, which is mapped into a set of corresponding virtual displacements, and calculate the resulting virtual work. After we have varied all the generalized coordinates, summing the resulting virtual work gives the total virtual work. In practice, we need to keep the following in mind:

- Conservative forces should not be included in calculating the generalized forces. Common conservative forces include weights and spring forces. They are included in the system's potential energy.
- A generalized force is a scalar quantity.
- The dimension (unit) of a generalized force is not always the force. It depends on the corresponding generalized coordinate. But the product of a generalized force and the corresponding generalized coordinate always has a unit of work or energy. When a generalized coordinate is a displacement or a position measured in a linear length, the generalized force has the dimension of a force. When a generalized coordinate is an angular displacement or a position measured by an angle, the generalized force has the dimension of a moment or a torque.

■ **Example 1.14: Generalized Forces for Externally Applied Force**

Find the generalized forces for the force $F(t)$ acting on a particle moving in the plane, as shown in Fig. 1.15. The force forms an angle α with the x-axis.

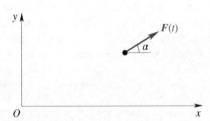

Figure 1.15   Force acting on a particle

□ **Solution 1: Using Cartesian Coordinates**

- *Generalized Coordinates*: We defined $\{x, y\}$ as a complete and independent set of generalized coordinates, where x and y are the Cartesian coordinates of the particle.
- *Admissible Variations*: We verify that $\{\delta x, \delta y\}$ is a complete and independent set of admissible variations in this set of generalized coordinates.
- *Virtual Work*: We use both methods to calculate the virtual work.

Method 1: We introduce a virtual displacement δr to the particle, as shown in Fig. 1.16a. The virtual work done by the force over this virtual displacement is, according to eqn. (1.1),

$$\delta W^{\text{n.c.}} = F(t) \cdot \delta r \qquad (a)$$

To carry out the dot product, we decompose both δr and $F(t)$ into the x- and y-components, as shown in Fig. 1.16b. Then,

$$\delta W^{\text{n.c.}} = \left[F(t) \cos \alpha i + F(t) \sin \alpha j \right] \cdot (\delta x i + \delta y j)$$
$$= F(t) \cos \alpha \delta x + F(t) \sin \alpha \delta y \qquad (b)$$

where we have used i and j to denote the unit vectors for the x- and y-axes, respectively.

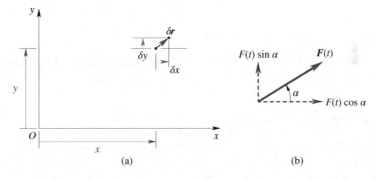

(a)   (b)

Figure 1.16   (a) Virtual displacement δr is decomposed into admissible variations δx and δy. (b) F is decomposed into the x- and y-components

Method 2: We first vary x by δx: the particle moves along the x-axis at a distance δx, as shown in Fig. 1.16a. Only the x-component of the force, $F(t) \cos \alpha$, does work over this virtual displacement, and the virtual work is $F(t) \cos \alpha \delta x$. We then vary y by δy. Similarly, the work done over this virtual displacement is $F(t) \sin \alpha \delta y$. Summing up, the total virtual work due to force $F(t)$ is

$$\delta W^{\text{n.c.}} = F(t) \cos \alpha \delta x + F(t) \sin \alpha \delta y \qquad (c)$$

which is the same as using the first method, as in eqn. (b).
- *Generalized Forces*: According to eqn. (1.2), the generalized forces are

$$\Xi_x = F(t) \cos \alpha \qquad (d)$$
$$\Xi_y = F(t) \sin \alpha. \qquad (e)$$

□ **Solution 2: Using Polar Coordinates**
- *Generalized Coordinates*: We alternatively defined $\{r, \theta\}$ as a complete and independent set of generalized coordinates, where r and θ are the polar coordinates of the particle.
- *Admissible Variations*: We verify that $\{\delta r, \delta \theta\}$ is a complete and independent set of admissible variations in this set of generalized coordinates.

- *Virtual Work*: To find the virtual work, we first vary r by δr. In the meantime, the force is decomposed into radial and tangential (azimuthal) components, as shown in Fig. 1.17. Note that the angle between F and the radial direction is $\alpha - \theta$. The virtual work due to this variation is $F(t)\cos(\alpha - \theta)\delta r$. We then vary θ by $\delta\theta$. At the location where the force is applied, this variation results in a linear virtual displacement of $r\delta\theta$, also shown in Fig. 1.17. Hence, the virtual work due to this variation is $F(t)\sin(\alpha - \theta)r\delta\theta$. Thus, the virtual work done by the force $F(t)$ is

$$\delta W^{\text{n.c.}} = F(t)\cos(\alpha - \theta)\delta r + F(t)\sin(\alpha - \theta)r\delta\theta \tag{f}$$

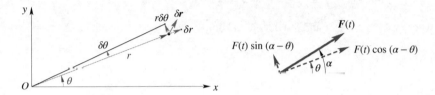

Figure 1.17 Variations in generalized coordinates and decomposition of force $F(t)$ into radial and tangential (azimuthal) components

- *Generalized Forces*: According to eqn. (1.2), the generalized forces are

$$\Xi_r = F(t)\cos(\alpha - \theta) \tag{g}$$

$$\Xi_\theta = F(t)r\sin(\alpha - \theta). \tag{h}$$

- Note that the generalized force Ξ_θ has the dimension of a moment or a torque.

■ Example 1.15: Buoyant Force and Drag on Pendulum

A planar pendulum of mass m and length l is submerged in water, which exerts a constant buoyant force F_0 on the bob vertically upward, as shown in Fig. 1.18. The water also exerts a drag force whose magnitude is proportional to the speed, cv, where c is a constant and v is the speed of the bob, and in the opposite direction of the velocity. The bob remains submerged during the motion. Find the generalized force for the system.

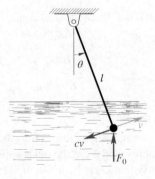

Figure 1.18 Planar pendulum with its bob submerged in water

□ **Solution:**

- *Generalized Coordinates*: We define $\{\theta\}$ as a complete and independent set of generalized coordinates, where θ is the angle the pendulum makes with the vertical, which is positive in the counterclockwise direction.
- *Admissible Variations*: We verify that $\{\delta\theta\}$ is a complete and independent set of admissible variations in this set of generalized coordinates.
- *Virtual Work*: The buoyant force is a constant value, which is very similar to the weight: the work it does depends only on the change in vertical elevation and is independent of the path. This is the signature for a conservative force. Thus, we do not include the buoyant force in the virtual work. Instead, we consider it in the potential energy.

   For the drag, its magnitude is $cl\dot{\theta}$. If we vary θ by $\delta\theta$, this variation causes the bob to travel a virtual displacement of an arc length $l\delta\theta$. Then, the virtual work done is

$$\delta W^{\text{n.c.}} = -(cl\dot{\theta})(l\delta\theta) = -cl^2\dot{\theta}\delta\theta \tag{a}$$

- *Generalized Forces*: According to eqn. (1.2), the generalized force is

$$\Xi_\theta = -cl^2\dot{\theta} \tag{b}$$

- *Potential Energy for Buoyant Force*: The potential energy due to the buoyant force is its hydrostatic pressure. If the mass density of the water is ρ_0, the potential energy can be written as

$$V_{\text{buoyant}} = \rho_0 V_{\text{obj}} gd \tag{c}$$

where V_{obj} is the volume of the object and d is the depth, which is positive in the downward direction, of its geometric center. Furthermore, $\rho_0 V_{\text{obj}} g$ gives the buoyant force F_0. Using the pivot point as the datum, the bob is located below the pivot point and the potential energy for the buoyant force is

$$V_{\text{buoyant}} = F_0 l \cos\theta \tag{d}$$

■ **Example 1.16: Generalized Force for Dashpot**

A dashpot of dashpot constant c has one of its ends connected to a wall while the other end can move freely, as shown in Fig. 1.19. Find the generalized force for the dashpot.

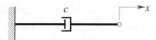

Figure 1.19   Dashpot with one end fixed to the wall

□ **Solution:**

- *Generalized Coordinates*: We define $\{x\}$ as a complete and independent set of generalized coordinates, where x is the rightward displacement of the movable end of the dashpot measured from a fixed reference position.
- *Admissible Variations*: We verify that $\{\delta x\}$ is a complete and independent set of admissible variations in this set of generalized coordinates.

- *Virtual Work*: The dashpot is a nonconservative element that produces a nonconservative force. To calculate the virtual work, we vary x by δx. Although the displacement is varied, the velocity $\dot{x}$ remains constant during the variation. In symbolic analysis, we only need to focus on the positive senses of parameters. Since x is positive when it moves to the right, both the velocity $\dot{x}$ and the variation δx are moving to the right. The force produced by the dashpot equals to $c\dot{x}$, in the direction opposite to $\dot{x}$ and δx. Thus, the virtual work done by the dashpot due to this variation is negative and is expressible as

$$\delta W^{\text{n.c.}} = -c\dot{x}\delta x \tag{a}$$

- *Generalized Forces*: According to eqn. (1.2), the generalized force corresponding to x is

$$\Xi_x = -c\dot{x} \tag{b}$$

■ **Example 1.17: Generalized Forces for "Floating Dashpot"**

A "floating dashpot" is a dashpot whose two ends move independently, such as the one shown in Fig. 1.20. Find the generalized forces for the "floating dashpot" in Fig. 1.20.

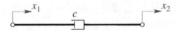

Figure 1.20   "Floating" dashpot: dashpot with independently moveable ends

□ **Solution 1: Absolute Displacements as Generalized Coordinates**

- *Generalized Coordinates*: We define $\{x_1, x_2\}$ as a complete and independent set of generalized coordinates, where x_1 and x_2 are the displacements of the left and the right ends, respectively, of the dashpot, with respect to the ground, measured to the right from their respective positions in a reference configuration (such as when $t = 0$).
- *Admissible Variations*: We verify that $\{\delta x_1, \delta x_2\}$ is a complete and independent set of admissible variations in this set of generalized coordinates.
- *Virtual Work*: At any instant, two ends of the dashpot have speeds of $\dot{x}_1$ and $\dot{x}_2$. They both are positive to the right. There are two possible scenarios, discussed separately below.
  Case 1: If $\dot{x}_1 > \dot{x}_2$, the dashpot is being compressed. The dashpot forces acting on the ends, which have the same magnitude and are denoted as $F_d = c(\dot{x}_1 - \dot{x}_2)$, are against the compression. Thus, they are in directions of stretching the dashpot, as shown in Fig. 1.21. If we vary x_1 by δx_1 and x_2 by δx_2, the dashpot force on the left end is in the opposite direction of δx_1; while the dashpot force on the right end is in the same direction as δx_2. Thus, the total virtual work in this scenario is

$$\delta W^{\text{n.c.}} = -c(\dot{x}_1 - \dot{x}_2)\delta x_1 + c(\dot{x}_1 - \dot{x}_2)\delta x_2 \tag{a}$$

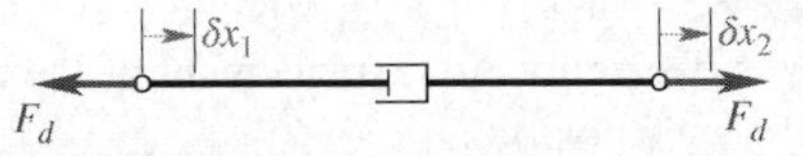

Figure 1.21   Dashpot force when the floating dashpot is compressed

Case 2: If $\dot{x}_1 < \dot{x}_2$, the dashpot is being stretched. The dashpot forces acting on the two ends, expressible as $F_d = c(\dot{x}_2 - \dot{x}_1)$, are against the stretching. Thus, they are in the directions of compressing the dashpot, as shown in Fig. 1.22.

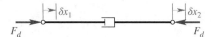

Figure 1.22 Dashpot force when the floating dashpot is stretched

If we vary x_1 by δx_1 and x_2 by δx_2, the dashpot force on the left end is in the same direction of δx_1 while the dashpot force on the right end is in the opposite direction as δx_2. Thus, the total virtual work in this scenario is

$$\delta W^{\text{n.c.}} = c(\dot{x}_2 - \dot{x}_1)\delta x_1 - c(\dot{x}_2 - \dot{x}_1)\delta x_2 \qquad (b)$$

Conclusion: Equations (a) and (b) are in fact identical and can be uniformly written as

$$\delta W^{\text{n.c.}} = -c(\dot{x}_1 - \dot{x}_2)(\delta x_1 - \delta x_2) = -c(\dot{x}_2 - \dot{x}_1)(\delta x_2 - \delta x_1) \qquad (c)$$

The ingredients in these two expressions are: the negative sign, the dashpot constant, the difference in the velocities, and the difference in the variations. The key is that the order for the subscripts in these two pairs of parentheses is the same. Also note that the two cases discussed earlier do not include the case $\dot{x}_1 = \dot{x}_2$. This is a special case where $\delta W^{\text{n.c.}} = 0$ and is included in the uniform expression in eqn. (c).

- *Generalized Forces*: Therefore, according to eqn. (1.2), the generalized forces are

$$\Xi_{x_1} = -c(\dot{x}_1 - \dot{x}_2) \qquad (d)$$

$$\Xi_{x_2} = -c(\dot{x}_2 - \dot{x}_1) \qquad (e)$$

□ **Solution 2: Relative Displacement as Generalized Coordinate**

- *Generalized Coordinates*: We define $\{x_1, y_2\}$ as a complete and independent set of generalized coordinates, where x_1 is the absolute displacement of m_1 and y_2 is the displacement of m_2 relative to m_1, both are positive to the right, measured from their respective positions in a reference configuration.
- *Admissible Variations*: We verify that $\{\delta x_1, \delta y_2\}$ is a complete and independent set of admissible variations in this set of generalized coordinates.
- *Virtual Work*: The dashpot force is determined by the relative velocities of its two ends, that is, $c\dot{y}_2$. If x_1 is varied by δx_1 but y_2 is fixed, both masses have the same virtual displacement of δx_1. But the dashpot forces on the two masses are in opposite directions. This means the virtual works done by the dashpot forces cancel each other out; hence, the total virtual work is zero. If x_1 is fixed while y_2 is varied by δy_2, there is no virtual displacement at m_1, but m_2 has a virtual displacement of δy_2. Hence, the virtual work is $-c\dot{y}_2\delta y_2$. Summing up, the total virtual work done by the dashpot is

$$\delta W^{\text{n.c.}} = -c\dot{y}_2\delta y_2 \qquad (f)$$

This is similar to the case when one end of the dashpot is fixed, as in Example 1.16.

- *Generalized Forces*: According to eqn. (1.2), the generalized forces are

$$\Xi_{x_1} = 0 \tag{g}$$

$$\Xi_{y_2} = -c\dot{y}_2 \tag{h}$$

1.7   Lagrangian

The *Lagrangian* of a system is given by

$$\mathcal{L} = T - V \tag{1.3}$$

where T is the total kinetic energy[1] of the system, and V is the total potential energy of the system. Since the system is completely described by the chosen set of generalized coordinates, denoted as ζ_j where $j = 1, 2, \ldots, M$, the system's potential energy, which involves positions, is expressible in terms of ξ_j; and the system's kinetic energy, which involves velocities and sometimes the positions as well, is expressible in terms of ξ_j and $\dot{\xi}_j$. Therefore, the Lagrangian, in general, is a function of both ξ_j and $\dot{\xi}_j$, that is, $\mathcal{L} = \mathcal{L}(\xi_j, \dot{\xi}_j)$.

1.8   Lagrange's Equation

For every generalized coordinate ξ_j, the *Lagrange's equation* is

$$\frac{d}{dt}\left(\frac{\partial \mathcal{L}}{\partial \dot{\xi}_j}\right) - \frac{\partial \mathcal{L}}{\partial \xi_j} = \Xi_j \tag{1.4}$$

This comprises the set of *equations of motion* for the system. There is exactly one equation of motion corresponding to every generalized coordinate. The number of equations of motion for a system equals to the number of generalized coordinates, which, in turn, equals to the number of degrees of freedom of the system.

Lagrange's equation is derived from *Hamilton's principle*, which is also known as the *principle of virtual work*. Historically, there are claims that Hamilton's principle is the most fundamental principle for dynamics from which Newton's second law can be derived. Not surprisingly, there are counter-claims that Newton's second law is more fundamental from which Hamilton's principle can be derived. No matter how science historians will settle their claims, it is safe to say that using Lagrange's equation is a valid alternative to Newton's second law.

1.9   Procedure for Deriving Equation(s) of Motion

Here, we outline the procedure for deriving the equation(s) of motion for a mechanical system. Note that, in the process, Newton's second law is never invoked.

---

[1] In a strict sense, the Lagrangian is defined as $\mathcal{L} = T^* - V$, where T^* is called the *kinetic coenergy*. For Newtonian particles and rigid bodies that move at speeds significantly below the speed of light, $T = T^*$. So here we use what most other textbooks use. Interested readers are referred to an excellent textbook on Lagrangian dynamics by J. H. Williams, Jr., *Fundamentals of Applied Dynamics* (John Wiley & Sons, 1995, New York) and a classic textbook by Stephen H. Crandall, D. C. Karnopp, E. F. Kurtz & D. C. Pridmore-Brown, *Dynamics of Mechanical and Electromechanical Systems* (McGraw-Hill, 1968, New York) for more comprehensive and rigorous discussions.

- Preparatory setup: This optional step establishes aids, such as coordinate systems, that facilitate mathematical description of the geometry of the problem.
- Define a set of generalized coordinates and verify its completeness and independence. The definition of a generalized coordinate should include four vital aspects: what, where, relative to whom, and in which direction. The verification can be a mental experiment, without being written out.
- Verify that the associate set of admissible variations is complete and independent.
- Check the holonomicity of the system. If holonomic, identify the number of degrees of freedom of the system. If not, Lagrange's equation is not applicable.
- Derive the expression for the total virtual work according to eqn. (1.1) and subsequently obtain the generalized forces through virtual work, using eqn. (1.2).
- Derive the Lagrangian for the system. This can be separated into three substeps: the kinetic energy, the potential energy, and then the Lagrangian via eqn. (1.3).
- Use Lagrange's equation in eqn. (1.4) to obtain the equation(s) of motion for the system: one equation for each generalized coordinate.

1.10 Worked Examples

In this section, we present a carefully selected collection of worked examples of deriving the equation(s) of motion for the systems. In many of these examples, equation(s) of motion are also derived for additional sets of generalized coordinates. The first several examples focus on systems consisting of only particles and then followed by examples on systems consisting of rigid bodies.

1.10.1 Systems Containing Only Particles

■ Example 1.18: Simple Mass–Spring–Dashpot System, Revisited

Use Lagrangian dynamics to rederive the equation of motion for the mass–spring–dashpot system as shown in Fig. 1.1.

□ Solution 1: Generalized Coordinate Defined from Spring's Unstretched Position

- *Generalized Coordinates*: We define $\{x\}$ as a complete and independent set of generalized coordinates, where x is the downward displacement of the mass, measured from its position when the spring is unstretched, as sketched in Fig. 1.23.

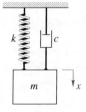

Figure 1.23 Mass–spring–dashpot system

- *Admissible Variations*: We verify that $\{\delta x\}$ is a complete and independent set of admissible variation in this set of generalized coordinates.
- *Holonomicity*: We observe that the number of generalized coordinate (one) equals to the number of admissible variation (one). Thus, we conclude that the system is holonomic. Furthermore, the system has one degree of freedom.
- *Generalized Forces*: The dashpot connects the mass to the wall. We can directly use the results from Example 1.16 for the virtual work done by the dashpot, as

$$\delta W^{\text{n.c.}} = -c\dot{x}\delta x \tag{a}$$

Thus, according to eqn. (1.2), the generalized force associated with x is

$$\Xi_x = -c\dot{x} \tag{b}$$

- *Kinetic Energy*: The velocity of the mass is $\dot{x}$. Thus,

$$T = \frac{1}{2}m\dot{x}^2 \tag{c}$$

- *Potential Energy*: Two elements in the system contribute to the potential energy: the mass and the spring. Note that the generalized coordinate is defined as positive downward and the gravitational energy decreases as x increases.

$$V = -mgx + \frac{1}{2}kx^2 \tag{d}$$

- *Lagrangian*: Combining eqns. (c) and (d) gives, according to eqn. (1.3),

$$\mathcal{L} = T - V = \frac{1}{2}m\dot{x}^2 + mgx - \frac{1}{2}kx^2 \tag{e}$$

- *Lagrange's Equation*: The Lagrange's equation for x is, according to eqn. (1.4),

$$\frac{d}{dt}\left(\frac{\partial \mathcal{L}}{\partial \dot{x}}\right) - \frac{\partial \mathcal{L}}{\partial x} = \Xi_x \tag{f}$$

When taking the partial derivative of $\mathcal{L}$ with respect to x or $\dot{x}$, we should treat x and $\dot{x}$ as independent variables. Therefore,

$$\frac{\partial \mathcal{L}}{\partial \dot{x}} = m\dot{x} \quad \text{and} \quad \frac{\partial \mathcal{L}}{\partial x} = mg - kx \tag{g}$$

Substituting eqns. (g) and (a) into eqn. (f) gives

$$m\ddot{x} - (mg - kx) = -c\dot{x}$$

or, moving the constant term mg to the right-hand side and $c\dot{x}$ term to the left-hand side, which are customarily done for differential equations, gives

$$m\ddot{x} + c\dot{x} + kx = mg \tag{h}$$

Equation (h) is the equation of motion for the system.

□ Solution 2: Generalized Coordinate Defined from Equilibrium Position

- *Generalized Coordinates*: We defined $\{y\}$ as a complete and independent set of generalized coordinates, where y is the downward displacement of the mass, measured from its static equilibrium position.
- *Admissible Variations*: We verify that $\{\delta y\}$ is a complete and independent set of admissible variations in this set of generalized coordinates.
- *Holonomicity*: We conclude that the system is holonomic and has one degree of freedom.
- *Generalized Forces*: With this new definition of the generalized coordinate, the relation between the dashpot force and the virtual displacement remains the same. That is,

$$\delta W^{\text{n.c.}} = -c\dot{y}\delta y \tag{i}$$

Thus, according to eqn. (1.2), the generalized force associated with y is

$$\Xi_y = -c\dot{y} \tag{j}$$

- *Kinetic Energy*: The velocity of the mass is $\dot{y}$. Thus,

$$T = \frac{1}{2}m\dot{y}^2 \tag{k}$$

- *Potential Energy*: At equilibrium, the spring is already stretched, and the amount is denoted as Δ. In calculating the potential energy stored in the spring, we need to add Δ into displacement y to account for the *total amount of stretch* in the spring. Thus,

$$V = -mgy + \frac{1}{2}k(y + \Delta)^2 \tag{l}$$

- *Lagrangian*: Combining eqns. (k) and (l) gives, according to eqn. (1.3),

$$\mathcal{L} = T - V = \frac{1}{2}m\dot{y}^2 + mgy - \frac{1}{2}k(y + \Delta)^2 \tag{m}$$

- *Lagrange's equation*: The Lagrange's equation for y, according to eqn. (1.4), is

$$\frac{d}{dt}\left(\frac{\partial\mathcal{L}}{\partial\dot{y}}\right) - \frac{\partial\mathcal{L}}{\partial y} = \Xi_y \tag{n}$$

Since

$$\frac{\partial\mathcal{L}}{\partial\dot{y}} = m\dot{y} \quad \text{and} \quad \frac{\partial\mathcal{L}}{\partial y} = mg - k(y + \Delta) \tag{o}$$

substituting eqns. (o) and (j) into eqn. (n) gives

$$m\ddot{y} - [mg - k(y + \Delta)] = -c\dot{y} \tag{p}$$

Moving the constant terms to the right-hand side,

$$m\ddot{y} + c\dot{y} + ky = mg - k\Delta \tag{q}$$

Equation (q) is the equation of motion for the system. But it is not the final form, since Δ is not a parameter given in the problem statement. Its expression needs to be determined. Recall that the equation of motion can be used to determine the equilibrium positions of the system. In the above solution process, eqn. (q) is already the equation of motion for the system. At the static equilibrium, $\ddot{y} = \dot{y} = 0$; and $y = 0$ since y is measured from the equilibrium position. Equation (q) itself can be used to find that, at equilibrium, $mg = k\Delta$. Substituting this relation into eqn. (q) gives

$$m\ddot{y} + c\dot{y} + ky = 0 \tag{r}$$

Equation (r) is the equation of motion for the system.

□ Discussion: Equilibrium Condition:

According to the analysis of the static equilibrium of the system in eqn. (h) in *Solution 2* of Example 1.1, $k\Delta = mg$. This is, not surprisingly, identical as the one found in Example 1.1.

In general, if the generalized coordinates are defined from the equilibrium configuration, we do not need to perform a separate analysis of the equilibrium state. This is especially convenient if we do not need to know the exact expression for Δ. We can simply state that, because $\ddot{y} = \dot{y} = y = 0$ at equilibrium, the constant terms (such as those gathered on the right-hand side of eqn. (q)) in the equation sum to zero as the *equilibrium condition*.

■ Example 1.19: Submerged Simple Pendulum

Derive the equation of motion for the submerged planar pendulum in Example 1.15.

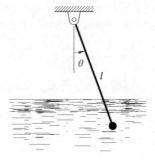

Figure 1.24   Simple pendulum submerged in water

□ Solution:

- *Generalized Coordinates*: We define $\{\theta\}$ as a complete and independent set of generalized coordinates, where θ is the angle the pendulum makes with the vertical, which is positive in the counterclockwise direction. This is the same definition used in Example 1.15.
- *Admissible Variations*: We verify that $\{\delta\theta\}$ is a complete and independent set of admissible variations in this set of generalized coordinates.
- *Holonomicity*: We conclude that the system is holonomic and has one degree of freedom.
- *Generalized Forces*: The generalized force has been found in Example 1.15 as

$$\Xi_\theta = -cl^2\dot{\theta} \tag{a}$$

- *Kinetic Energy*: The bob in the pendulum has a velocity of $l\dot{\theta}$. Thus,

$$T = \frac{1}{2}ml^2\dot{\theta}^2 \tag{b}$$

- *Potential Energy*: We choose the pivot point as the datum for both gravitational and buoyant potential energies. The bob is located below the datum by a distance of $l \cos \theta$. The potential energy for the buoyant force has been found in Example 1.15. Hence,

$$V = -mgl \cos \theta + F_0 l \cos \theta = (F_0 - mg)l \cos \theta \qquad \text{(c)}$$

- *Lagrangian*: Combining eqns. (b) and (c) gives, according to eqn. (1.3),

$$\mathcal{L} = T - V = \frac{1}{2}ml^2\dot{\theta}^2 - (F_0 - mg)l \cos \theta \qquad \text{(d)}$$

- *Lagrange's Equation*: The Lagrange's equation for θ is, according to eqn. (1.4),

$$\frac{d}{dt}\left(\frac{\partial \mathcal{L}}{\partial \dot{\theta}}\right) - \frac{\partial \mathcal{L}}{\partial \theta} = \Xi_\theta \qquad \text{(e)}$$

Since

$$\frac{\partial \mathcal{L}}{\partial \dot{\theta}} = ml^2\dot{\theta} \qquad \text{and} \qquad \frac{\partial \mathcal{L}}{\partial \theta} = (F_0 - mg)l \sin \theta \qquad \text{(f)}$$

substituting eqns. (f) and (a) into eqn. (e) gives

$$ml^2\ddot{\theta} - (F_0 - mg)l \sin \theta = -cl^2\dot{\theta} \qquad \text{(g)}$$

Rearranging and canceling an l give

$$ml\ddot{\theta} + cl\dot{\theta} + (mg - F_0)\sin \theta = 0 \qquad \text{(h)}$$

Equation (h) is the equation of motion for the system.

□ **Discussion: Small Motions**

A pendulum is a common device for studying vibration in the introductory engineering physics course. In most cases, we assume "small motions" and use the approximation $\sin \theta \approx \theta$ to simplify the equation. For the present system, doing so gives

$$ml\ddot{\theta} + cl\dot{\theta} + (mg - F_0)\theta = 0 \qquad \text{(i)}$$

where the left-hand side consists only of linear terms of $\ddot{\theta}$, $\dot{\theta}$, and θ. This process is known as linearization, in which we keep only up to the first-order (linear) small terms.

We can employ this "small motion" assumption much earlier in order to reduce the complexity in some expressions. We should keep in mind that, when using Lagrange's equation, both energies will be taken partial derivatives. This means that, when writing the expressions for energies, we need to keep up to the second-order small terms if we want to keep the first-order small terms in the resulting equation(s) of motion.

□ **Solution 2: Early Adoption of "Small Motion" Assumption**

We redo this example by adopting the small motion assumption early. We continue to use the same generalized coordinate. Hence, we start from the step of generalized forces.

- *Generalized Forces*: The generalized force is unchanged as in eqn. (a).
- *Kinetic Energy*: This is unchanged as in eqn. (b).

- *Potential Energy*: Using the "small motion" assumption: when θ is small, keeping up to the second order of θ gives

$$\sin \theta \approx \theta \qquad \text{and} \qquad \cos \theta \approx 1 - \frac{1}{2}\theta^2 \qquad \text{(j)}$$

The potential energy in eqn. (c) becomes

$$V = (F_0 - mg)l\left(1 - \frac{1}{2}\theta^2\right) \qquad \text{(k)}$$

- *Lagrangian*: Combining eqns. (b) and (k), according to eqn. (1.4), gives

$$\mathcal{L} = T - V = \frac{1}{2}ml^2\dot{\theta}^2 - (F_0 - mg)l\left(1 - \frac{1}{2}\theta^2\right) \qquad \text{(l)}$$

- *Lagrange's Equation*: The Lagrange's equation for θ is, according to eqn. (1.4),

$$\frac{d}{dt}\left(\frac{\partial \mathcal{L}}{\partial \dot{\theta}}\right) - \frac{\partial \mathcal{L}}{\partial \theta} = \Xi_\theta \qquad \text{(m)}$$

Since

$$\frac{\partial \mathcal{L}}{\partial \dot{\theta}} = ml^2\dot{\theta} \qquad \text{and} \qquad \frac{\partial \mathcal{L}}{\partial \theta} = (F_0 - mg)l\theta \qquad \text{(n)}$$

substituting eqns. (n) and (a) into eqn. (m) gives

$$ml^2\ddot{\theta} - (F_0 - mg)l\theta = -cl^2\dot{\theta} \qquad \text{(o)}$$

Rearranging and canceling out an l give

$$ml\ddot{\theta} + cl\dot{\theta} + (mg - F_0)\theta = 0 \qquad \text{(p)}$$

Equation (p) is the equation of motion for the system under the "small motion" assumption, which is the same as eqn. (i) discussed earlier.

■ **Example 1.20: Pendulum with Specified Base Motion**

The system shown in Fig. 1.25 consists of a cart of mass m_1 that moves on the horizontal rail, and a pendulum of mass m_2 and length l is pinned to the cart. The pendulum swings freely without friction. The cart moves in accordance with $x = x_0(t)$, driven by an external mechanism (not shown), where x is the location of the cart measured from a fixed reference position. Derive the equation(s) of motion for the system.

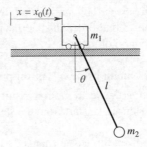

Figure 1.25 Pendulum with specified motion at its base

□ **Solution:**

- *Preparatory Setup*: We have discussed in Example 1.13 that $x_0(t)$ is a specified motion and shall not be considered as a generalized coordinate.
- *Generalized Coordinates*: We define $\{\theta\}$ as a complete and independent set of generalized coordinates, where θ is the counterclockwise angular displacement of the pendulum, measured from the vertical.
- *Admissible Variations*: We verify that $\{\delta\theta\}$ is a complete and independent set of admissible variations in this set of generalized coordinates.
- *Holonomicity*: We conclude that the system is holonomic and has one degree of freedom.
- *Generalized Forces*: Although there must be an external force acting on the cart to make it move, this force is specified through a specified motion and will be accounted for by the kinematics of the system. Besides this force, there is no other nonconservative force in the system. Thus, $\delta W^{\text{n.c.}} = 0$, and, according to eqn. (1.2),

$$\Xi_\theta = 0 \tag{a}$$

- *Kinetic Energy*: The pendulum swings relative to the translating cart. Recall the relative motions in the particle kinematics in eqn. (A.12),

$$v_{\text{bob}} = v_{\text{cart}} + v_{\text{bob/cart}} \tag{b}$$

where v_{bob} and v_{cart} are the absolute velocities of the pendulum bob and the cart, respectively, and $v_{\text{bob/cart}}$ is the relative velocity of the bob with respect to the cart. The word "absolute" is added to emphasize its difference from a relative velocity. This vector summation is sketched in Fig. 1.26, where $l\dot{\theta}$ is $v_{\text{bob/cart}}$ and $\dot{x}$ is v_{cart}.

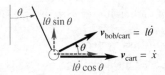

Figure 1.26 Velocity composition of the pendulum pivoted on moving cart

To calculate the kinetic energy, the velocity due to the swinging of the pendulum $v_{\text{bob/cart}}$ is decomposed into horizontal and vertical components, as shown in the dashed vectors in Fig. 1.26. Thus,

$$T = \frac{1}{2}m_1\dot{x}^2 + \frac{1}{2}m_2\left[\left(\dot{x} + l\dot{\theta}\cos\theta\right)^2 + \left(l\dot{\theta}\sin\theta\right)^2\right]$$

$$= \frac{1}{2}m_1\dot{x}^2 + \frac{1}{2}m_2\left(\dot{x}^2 + l^2\dot{\theta}^2 + 2\dot{x}l\dot{\theta}\cos\theta\right) \tag{c}$$

- *Potential Energy*: We choose the pivot point of the pendulum as the datum for gravitational potential energy. Although this is a moving point, its vertical position remains fixed and hence can be used as a datum point. Thus,

$$V = -m_2gl\cos\theta \tag{d}$$

- *Lagrangian*: Combining eqns. (c) and (d), according to eqn. (1.3),

$$\mathcal{L} = T - V = \frac{1}{2}m_1\dot{x}^2 + \frac{1}{2}m_2\left(\dot{x}^2 + l^2\dot{\theta}^2 + 2\dot{x}l\dot{\theta}\cos\theta\right) + m_2 g l\cos\theta \tag{e}$$

- *Lagrange's Equation*: The Lagrange's equation for θ is, according to eqn. (1.4),

$$\frac{d}{dt}\left(\frac{\partial\mathcal{L}}{\partial\dot{\theta}}\right) - \frac{\partial\mathcal{L}}{\partial\theta} = \Xi_\theta \tag{f}$$

Since

$$\frac{\partial\mathcal{L}}{\partial\dot{\theta}} = m_2 l^2\dot{\theta} + m_2 l\dot{x}\cos\theta \tag{g}$$

and

$$\frac{\partial\mathcal{L}}{\partial\theta} = -m_2 g l\sin\theta - m_2\dot{x}l\dot{\theta}\sin\theta \tag{h}$$

substituting eqns. (g), (h), and (a) into eqn. (f) gives

$$\frac{d}{dt}\left(m_2 l^2\dot{\theta} + m_2 l\dot{x}\cos\theta\right) + m_2 g l\sin\theta + m_2\dot{x}l\dot{\theta}\sin\theta = 0$$

or, expanding the derivative and canceling the common factor $m_2 l$,

$$l\ddot{\theta} + \ddot{x}\cos\theta - \dot{x}\dot{\theta}\sin\theta + g\sin\theta + \dot{x}\dot{\theta}\sin\theta = 0$$

Canceling out the two $\dot{x}\dot{\theta}\sin\theta$ terms gives

$$l\ddot{\theta} + \ddot{x}_0(t)\cos\theta + g\sin\theta = 0 \tag{i}$$

Equation (i) is the equation of motion for the system.

■ Example 1.21: Damped Two-Mass System

Two masses m_1 and m_2 are constrained between two walls by three springs k_1, k_2, and k_3 and three dashpots c_1, c_2, and c_3, as shown in Fig. 1.27. Assume masses move frictionlessly on the horizontal floor. Derive the equation(s) of motion for the system.

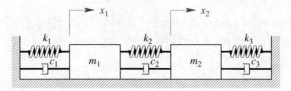

Figure 1.27 Two-mass mass–spring–dashpot system

□ Solution 1: Absolute Displacements as Generalized Coordinates

- *Generalized Coordinates*: We define $\{x_1, x_2\}$ as a complete and independent set of generalized coordinates, where x_1 and x_2 are the displacement of masses m_1 and m_2, respectively, with respect to the ground, measured to the right from their respective positions in equilibrium.

- *Admissible Variations*: We verify that $\{\delta x_1, \delta x_2\}$ is a set of complete and independent set of admissible variations in this set of generalized coordinates.
- *Holonomicity*: We conclude that the system is holonomic and has two degrees of freedom.
- *Generalized Forces*: There are three nonconservative force-producing dashpots in the system. Among them, c_1 and c_3 are simple dashpots as analyzed in Example 1.16, and c_2 is a "floating dashpot" as analyzed in Example 1.17. Thus, the total virtual work done by these dashpots is

$$\delta W^{n.c.} = -c_1 \dot{x}_1 \delta x_1 - c_2 (\dot{x}_2 - \dot{x}_1)(\delta x_2 - \delta x_1) - c_3 \dot{x}_2 \delta x_2$$

$$= \left[-c_1 \dot{x}_1 + c_2 (\dot{x}_2 - \dot{x}_1)\right] \delta x_1 + \left[-c_2 (\dot{x}_2 - \dot{x}_1) - c_3 \dot{x}_2\right] \delta x_2$$

Thus, according to eqn. (1.2),

$$\Xi_{x_1} = -(c_1 + c_2)\dot{x}_1 + c_2 \dot{x}_2 \tag{a}$$

$$\Xi_{x_2} = c_2 \dot{x}_1 - (c_2 + c_3)\dot{x}_2 \tag{b}$$

- *Kinetic Energy*: The velocities of the masses are $\dot{x}_1$ and $\dot{x}_2$. Thus,

$$T = \frac{1}{2}m_1 \dot{x}_1^2 + \frac{1}{2}m_2 \dot{x}_2^2$$

- *Potential Energy*: There is no change in the height of the masses, so the gravitational potential energies for both masses do not change. The springs may have already been stretched at equilibrium, denoted as Δ_1, Δ_2, and Δ_3, respectively. These are in addition to stretches due to displacements, which are, x_1, $x_2 - x_1$, and $-x_2$, respectively. Thus,

$$V = \frac{1}{2}k_1(x_1 + \Delta_1)^2 + \frac{1}{2}k_2(x_2 - x_1 + \Delta_2)^2 + \frac{1}{2}k_3(-x_2 + \Delta_3)^2$$

- *Lagrangian*: According to eqn. (1.3)

$$\mathcal{L} = T - V = \frac{1}{2}m_1 \dot{x}_1^2 + \frac{1}{2}m_2 \dot{x}_2^2 - \frac{1}{2}k_1(x_1 + \Delta_1)^2$$

$$- \frac{1}{2}k_2(x_2 - x_1 + \Delta_2)^2 - \frac{1}{2}k_3(-x_2 + \Delta_3)^2 \tag{c}$$

- *Lagrange's Equation*: The system has two equations of motion.
  - *The x_1-Equation*: The Lagrange's equation for x_1 is, according to eqn. (1.4),

$$\frac{d}{dt}\left(\frac{\partial \mathcal{L}}{\partial \dot{x}_1}\right) - \frac{\partial \mathcal{L}}{\partial x_1} = \Xi_{x_1} \tag{d}$$

Since

$$\frac{\partial \mathcal{L}}{\partial \dot{x}_1} = m_1 \dot{x}_1 \tag{e}$$

and

$$\frac{\partial \mathcal{L}}{\partial x_1} = -k_1\left(x_1 + \Delta_1\right) + k_2\left(x_2 - x_1 + \Delta_2\right) \tag{f}$$

substituting eqns. (e), (f), and (a) into eqn. (d) gives

$$m_1 \ddot{x}_1 + \left(c_1 + c_2\right)\dot{x}_1 - c_2 \dot{x}_2 + \left(k_1 + k_2\right)x_1 - k_2 x_2 = -k_1 \Delta_1 + k_2 \Delta_2 \tag{g}$$

– *The x_2-Equation*: The Lagrange's equation for x_2 is, according to eqn. (1.4),

$$\frac{d}{dt}\left(\frac{\partial \mathcal{L}}{\partial \dot{x}_2}\right) - \frac{\partial \mathcal{L}}{dx_2} = \Xi_{x_2} \tag{h}$$

Since

$$\frac{\partial \mathcal{L}}{\partial \dot{x}_2} = m_2 \dot{x}_2 \tag{i}$$

and

$$\frac{\partial \mathcal{L}}{\partial x_2} = -k_2\left(x_2 - x_1 + \Delta_2\right) + k_3\left(-x_2 + \Delta_3\right) \tag{j}$$

substituting eqns. (i), (j), and (b) into eqn. (h) gives

$$m_2 \ddot{x}_2 - c_2 \dot{x}_1 + \left(c_2 + c_3\right)\dot{x}_2 - k_2 x_1 + \left(k_2 + k_3\right)x_2 = -k_2 \Delta_2 + k_3 \Delta_3 \tag{k}$$

– *Equilibrium Conditions*: Since the generalized coordinates are defined from the respective equilibrium positions, we can simply require all constant terms in both equations to vanish as the equilibrium conditions. That is,

$$-k_1 \Delta_1 + k_2 \Delta_2 = 0$$

$$-k_2 \Delta_2 + k_3 \Delta_3 = 0$$

We have only two equations to solve for three Δ's: the system is statically indeterminate. However, if we do not need to know the explicit expressions for these "equilibrium stretches," they suffice to remove these Δ's from the equations of motion for the system, in eqns. (g) and (k), to give

$$m_1 \ddot{x}_1 + \left(c_1 + c_2\right)\dot{x}_1 - c_2 \dot{x}_2 + \left(k_1 + k_2\right)x_1 - k_2 x_2 = 0 \tag{l}$$

$$m_2 \ddot{x}_2 - c_2 \dot{x}_1 + \left(c_2 + c_3\right)\dot{x}_2 - k_2 x_1 + \left(k_2 + k_3\right)x_2 = 0 \tag{m}$$

Equations (l) and (m) comprise the equations of motion for the system.

□ Solution 2: Relative Displacement as Generalize Coordinate

- *Generalized Coordinates*: We define $\{x_1, y_2\}$ as a complete and independent set of generalized coordinates, where x_1 is the displacement of mass m_1 with respect to the ground, measured from its equilibrium position, and y_2 is the relative displacement of mass m_2 with respect to mass m_1, measured from its equilibrium position.
- *Admissible Variations*: We verify that $\{\delta x_1, \delta y_2\}$ is a complete and independent set of admissible variations in this set of generalized coordinates.
- *Holonomicity*: We conclude that the system is holonomic and has two degrees of freedom.
- *Generalized Forces*: There are three nonconservative elements. Dashpot c_1 is a simple dashpot as analyzed in Example 1.16, which gives the virtual work as $-c_1 \dot{x}_1 \delta x_1$. Dashpot c_2 is a "floating dashpot" as analyzed in *Solution 2* in Example 1.17, which gives the virtual work as $-c_2 \dot{y}_2 \delta y_2$.

For dashpot c_3, since the (absolute) velocity of m_2 is $\dot{x} + \dot{y}$, the dashpot force is $c_3(\dot{x}_1 + \dot{y}_2)$. If x_1 is varied by δx_1 but y_2 is fixed, m_2 has a virtual displacement of δx_1; the virtual work is $-c_3(\dot{x}_1 + \dot{y}_2)\delta x_1$. If x_1 is fixed but y_2 is varied by δy_2, m_2 has a virtual displacement of δy_2, and the virtual work is $-c_3(\dot{x}_1 + \dot{y}_2)\delta y_2$. Thus, the total virtual work done by dashpot c_3 is $-c_3(\dot{x}_1 + \dot{y}_2)(\delta x_1 + \delta y_2)$.

Summing up, the total virtual work done by all three dashpots is

$$\delta W^{\text{n.c.}} = -c_1\dot{x}_1\delta x_1 - c_2\dot{y}_2\delta y_2 - c_3(\dot{x}_1 + \dot{y}_2)(\delta x_1 + \delta y_2)$$
$$= \left[-c_1\dot{x}_1 - c_3(\dot{x}_1 + \dot{y}_2)\right]\delta x_1 + \left[-c_2\dot{y}_2 - c_3(\dot{y}_2 + \dot{x}_1)\right]\delta y_2$$

Then, according to eqn. (1.2),

$$\Xi_{x_1} = -(c_1 + c_3)\dot{x}_1 - c_3\dot{y}_2 \tag{n}$$

$$\Xi_{y_2} = -c_3\dot{x}_1 - (c_2 + c_3)\dot{y}_2 \tag{o}$$

- *Kinetic Energy*: The velocity of m_1 is $\dot{x}_1$, and the (absolute) velocity of m_2 is $(\dot{x}_1 + \dot{y}_2)$. Thus,

$$T = \frac{1}{2}m_1\dot{x}_1^2 + \frac{1}{2}m_2(\dot{x}_1 + \dot{y}_2)^2$$

- *Potential Energy*: Denote the amount of stretches in the springs at equilibrium as Δ_1, Δ_2, and Δ_3. The total amount of stretch in three springs are $x_1 + \Delta_1$, $y_2 + \Delta_2$, and $-x_1 - y_2 + \Delta_3$, respectively.

$$V = \frac{1}{2}k_1(x_1 + \Delta_1)^2 + \frac{1}{2}k_2(y_2 + \Delta_2)^2 + \frac{1}{2}k_3(-x_1 - y_2 + \Delta_3)^2$$

- *Lagrangian*: According to eqn. (1.3),

$$\mathcal{L} = T - V = \frac{1}{2}m_1\dot{x}_1^2 + \frac{1}{2}m_2(\dot{x}_1 + \dot{y}_2)^2 - \frac{1}{2}k_1(x_1 + \Delta_1)^2$$
$$- \frac{1}{2}k_2(y_2 + \Delta_2)^2 - \frac{1}{2}k_3(-x_1 - y_2 + \Delta_3)^2 \tag{p}$$

- *Lagrange's Equation*: The system has two equations of motion.
- *The x_1-Equation*: The Lagrange's equation for x_1 is, according to eqn. (1.4),

$$\frac{d}{dt}\left(\frac{\partial \mathcal{L}}{\partial \dot{x}_1}\right) - \frac{\partial \mathcal{L}}{dx_1} = \Xi_{x_1} \tag{q}$$

Since

$$\frac{\partial \mathcal{L}}{\partial \dot{x}_1} = m_1\dot{x}_1 + m_2(\dot{x}_1 + \dot{y}_2) = (m_1 + m_2)\dot{x}_1 + m_2\dot{y}_2 \tag{r}$$

and

$$\frac{\partial \mathcal{L}}{\partial x_1} = -k_1(x_1 + \Delta_1) - k_3(x_1 + y_2 - \Delta_3) \tag{s}$$

substituting eqns. (r), (s), and (n) into eqn. (q) gives

$$(m_1 + m_2)\ddot{x}_1 + m_2\ddot{y}_2 + (c_1 + c_3)\dot{x}_1 + c_3\dot{y}_2 + (k_1 + k_3)x_1 + k_3y_2 = -k_1\Delta_1 + k_3\Delta_3 \tag{t}$$

– The y_2-Equation: The Lagrange's equation for y_2 is, according to eqn. (1.4),

$$\frac{d}{dt}\left(\frac{\partial \mathcal{L}}{\partial \dot{y}_2}\right) - \frac{\partial \mathcal{L}}{\partial y_2} = \Xi_{y_2} \tag{u}$$

Since

$$\frac{\partial \mathcal{L}}{\partial \dot{y}_2} = m_2(\dot{y}_2 + \dot{x}_1) \tag{v}$$

and

$$\frac{\partial \mathcal{L}}{\partial y_2} = -k_2(y_2 + \Delta_2) - k_3(x_1 + y_2 - \Delta_3) \tag{w}$$

substituting eqns. (v), (w), and (o) into eqn. (u) gives

$$m_2\ddot{x}_1 + m_2\ddot{y}_2 + c_3\dot{x}_1 + (c_2 + c_3)\dot{y}_2 + k_3 x_1 + (k_2 + k_3)y_2 = -k_2\Delta_2 + k_3\Delta_3 \tag{x}$$

The constant terms in eqns. (t) and (x) have been moved to the right-hand sides. They vanish under the "equilibrium conditions." Thus,

$$(m_1 + m_2)\ddot{x}_1 + m_2\ddot{y}_2 + (c_1 + c_3)\dot{x}_1 + c_3\dot{y}_2 + (k_1 + k_3)x_1 + k_3 y_2 = 0 \tag{y}$$

$$m_2\ddot{x}_1 + m_2\ddot{y}_2 + c_3\dot{x}_1 + (c_2 + c_3)\dot{y}_2 + k_3 x_1 + (k_2 + k_3)y_2 = 0 \tag{z}$$

Equations (y) and (z) comprise the equations of motion for the system.

■ Example 1.22: Double Pendulum

A planar pendulum of mass m and length l is attached to another identical planar pendulum, which is connected to the ceiling, as shown in Fig. 1.28. Derive the equation(s) of the motion for the system.

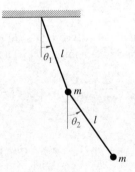

Figure 1.28   Double pendulum

□ **Solution:**

- *Generalized Coordinates*: We define $\{\theta_1, \theta_2\}$ as a set of complete and independent generalized coordinates, where θ_1 and θ_2 are the counterclockwise angular displacements of the two pendulums with respect to the vertical.

- *Admissible Variations*: We verify that $\{\delta\theta_1, \delta\theta_2\}$ is a complete and independent set of admissible variations in this set of generalized coordinates.
- *Holonomicity*: We conclude that the system is holonomic and has two degrees of freedom.
- *Generalized Forces*: There is no nonconservative element in the system. Thus, $\delta W^{n.c.} = 0$. Then, according to eqn. (1.2),

$$\Xi_{\theta_1} = 0 \quad \Xi_{\theta_2} = 0 \tag{a}$$

- *Kinetic Energy*: The mass of pendulum 1 has a velocity of $l\dot\theta_1$, which can be decomposed into the x- and y-components, as

$$v_1 = l\dot\theta_1(\cos\theta_1 i + \sin\theta_1 j) \tag{b}$$

The mass of pendulum 2 has a velocity of $l\dot\theta_2$ relative to a translating reference frame attached to the mass of pendulum 1 (but does not rotate with the pendulum), that is,

$$v_{2/1} = l\dot\theta_2(\cos\theta_2 i + \sin\theta_2 j) \tag{c}$$

Recall the relation for relative motions in eqn. (A.12), the absolute velocity of pendulum 2 is

$$v_2 = v_1 + v_{2/1} = l\left(\dot\theta_1\cos\theta_1 + \dot\theta_2\cos\theta_2\right)i + l\left(\dot\theta_1\sin\theta_1 + \dot\theta_2\sin\theta_2\right)j \tag{d}$$

Hence,

$$T = \frac{1}{2}ml^2\dot\theta_1^2 + \frac{1}{2}ml^2\left[\dot\theta_1^2 + \dot\theta_2^2 + 2\dot\theta_1\dot\theta_2\cos(\theta_2 - \theta_1)\right] \tag{e}$$

- *Potential Energy*: We choose the pivot point as the datum for the gravitational potential energy. Both masses are located below the datum. Thus,

$$V = mg(-l\cos\theta_1) + mg(-l\cos\theta_1 - l\cos\theta_2)$$
$$= -mgl(2\cos\theta_1 + \cos\theta_2) \tag{f}$$

- *Lagrangian*: Combining eqns. (e) and (f) gives, according to eqn. (1.3),

$$\mathcal{L} = T - V$$
$$= ml^2\left[\dot\theta_1^2 + \frac{1}{2}\dot\theta_2^2 + \dot\theta_1\dot\theta_2\cos(\theta_2 - \theta_1)\right] + mgl(2\cos\theta_1 + \cos\theta_2) \tag{g}$$

- *Lagrange's Equations*: The system has two equations of motion.
  - *The θ_1-Equation*: Lagrange's equation for θ_1 is

$$\frac{d}{dt}\left(\frac{\partial\mathcal{L}}{\partial\dot\theta_1}\right) - \frac{\partial\mathcal{L}}{\partial\theta_1} = \Xi_{\theta_1} \tag{h}$$

Since

$$\frac{\partial\mathcal{L}}{\partial\dot\theta_1} = 2ml^2\dot\theta_1 + ml^2\dot\theta_2\cos(\theta_2 - \theta_1) \tag{i}$$

and

$$\frac{\partial\mathcal{L}}{\partial\theta_1} = ml^2\dot\theta_1\dot\theta_2\sin(\theta_2 - \theta_1) - 2mgl\sin\theta_1 \tag{j}$$

substituting eqns. (i), (j), and (a) into eqn. (h) gives

$$2ml^2\ddot{\theta}_1 + ml^2\ddot{\theta}_2 \cos(\theta_2 - \theta_1) - ml^2\dot{\theta}_2(\dot{\theta}_2 - \dot{\theta}_1)\sin(\theta_2 - \theta_1)$$
$$-ml^2\dot{\theta}_1\dot{\theta}_2 \sin(\theta_2 - \theta_1) + 2mgl\,\sin\theta_1 = 0 \tag{k}$$

Canceling out the two terms of $ml^2\dot{\theta}_1\dot{\theta}_2 \sin(\theta_2 - \theta_1)$ gives

$$2ml^2\ddot{\theta}_1 + ml^2\ddot{\theta}_2 \cos(\theta_2 - \theta_1) - ml^2\dot{\theta}_2^2 \sin(\theta_2 - \theta_1) + 2mgl\,\sin\theta_1 = 0 \tag{l}$$

– *The θ_2-Equation*: Lagrange's equation for θ_2 is

$$\frac{d}{dt}\left(\frac{\partial \mathcal{L}}{\partial \dot{\theta}_2}\right) - \frac{\partial \mathcal{L}}{\partial \theta_2} = \Xi_{\theta_2} \tag{m}$$

Since

$$\frac{\partial \mathcal{L}}{\partial \dot{\theta}_2} = ml^2\dot{\theta}_2 + ml^2\dot{\theta}_1 \cos(\theta_2 - \theta_1) \tag{n}$$

and

$$\frac{\partial \mathcal{L}}{\partial \theta_2} = -ml^2\dot{\theta}_1\dot{\theta}_2 \sin(\theta_2 - \theta_1) - mgl\,\sin\theta_2 \tag{o}$$

substituting eqns. (n), (o), and (a) into eqn. (m) gives

$$ml^2\ddot{\theta}_2 + ml^2\ddot{\theta}_1 \cos(\theta_2 - \theta_1) - ml^2\dot{\theta}_1(\dot{\theta}_2 - \dot{\theta}_1)\sin(\theta_2 - \theta_1)$$
$$+ ml^2\dot{\theta}_1\dot{\theta}_2 \sin(\theta_2 - \theta_1) + mgl\,\sin\theta_2 = 0 \tag{p}$$

Canceling out the two terms of $ml^2\dot{\theta}_1\dot{\theta}_2 \sin(\theta_2 - \theta_1)$ gives

$$ml^2\ddot{\theta}_2 + ml^2\ddot{\theta}_1 \cos(\theta_2 - \theta_1) - ml^2\dot{\theta}_1^2 \sin(\theta_2 - \theta_1) + mgl\,\sin\theta_2 = 0 \tag{q}$$

Equations (l) and (q) are the equations of motion for the system.

1.10.2   Systems Containing Rigid Bodies

In this section, the systems under consideration contain rigid bodies. The expressions for kinetic and potential energies of a rigid body are slightly more complicated. The readers are advised to review Appendix A before proceeding. But the procedure for deriving the equation(s) of motion remains the same.

■ **Example 1.23: Rigid-Link Pendulum**

A uniform rigid slender rod of mass m and length L is pivoted at its top end and is allowed to move within the plane of paper, as shown in Fig. 1.29. Assume the pivot is frictionless. Derive the equation(s) of motion for the system.

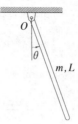

Figure 1.29   Rigid-link planar pendulum

□ Solution:

- *Generalized Coordinates*: We define $\{\theta\}$ as a complete and independent set of generalized coordinates, where θ is the counterclockwise angular displacement of the rod measured from the vertical.
- *Admissible Variations*: We verify that $\{\delta\theta\}$ is a complete and independent set of admissible variations in this set of generalized coordinates.
- *Holonomicity*: We conclude that the system is holonomic and has one degree of freedom.
- *Generalized Forces*: There is no nonconservative force in the system. Thus, $\delta W^{\text{n.c.}} = 0$. According to eqn. (1.2),

$$\Xi_\theta = 0 \tag{a}$$

- *Kinetic Energy*: The rod is a rigid body rotating about a fixed point O. We can use a simpler version of the equation for the kinetic energy of the rigid body:

$$T = \frac{1}{2}I_O\omega^2$$

where I_O is the moment of inertia of the rigid body about the fixed point O and ω is its angular velocity. Recall the moment of inertia for a uniform slender rod about its centroid

$$I_C = \frac{1}{12}mL^2$$

Using the parallel axes theorem (see eqn. (A.21)), the moment of inertia for the rigid slender rod about the pivot point O is

$$I_O = I_C + m\left(\frac{L}{2}\right)^2 = \frac{1}{12}mL^2 + \frac{1}{4}mL^2 = \frac{1}{3}mL^2$$

The angular velocity of the rod is $\dot{\theta}$. Thus,

$$T = \frac{1}{2}\left(\frac{1}{3}mL^2\right)\dot{\theta}^2 = \frac{1}{6}mL^2\dot{\theta}^2 \tag{b}$$

- *Potential Energy*: We choose the pivot point as the datum for gravitational potential energy. The centroid is located below the datum.

$$V = -mg\frac{L}{2}\cos\theta = -\frac{1}{2}mgL\cos\theta \tag{c}$$

- *Lagrangian*: Combining eqns. (b) and (c) gives, according to eqn. (1.3),

$$\mathcal{L} = T - V = \frac{1}{6}mL^2\dot{\theta}^2 + \frac{1}{2}mgL \cos \theta \qquad (d)$$

- *Lagrange's Equation*: The Lagrange's equation for θ is, according to eqn. (1.4),

$$\frac{d}{dt}\left(\frac{\partial \mathcal{L}}{\partial \dot{\theta}}\right) - \frac{\partial \mathcal{L}}{\partial \theta} = \Xi_\theta \qquad (e)$$

Since

$$\frac{\partial \mathcal{L}}{\partial \dot{\theta}} = \frac{1}{3}mL^2\dot{\theta} \qquad \text{and} \qquad \frac{\partial \mathcal{L}}{\partial \theta} = -\frac{1}{2}mgL \sin \theta \qquad (f)$$

substituting eqns. (f) and (a) into eqn. (e) gives

$$\frac{1}{3}mL^2\ddot{\theta} + \frac{1}{2}mgL \sin \theta = 0 \qquad (g)$$

Equation (g) is the equation of motion for the system.

■ Example 1.24: Inverted Rigid-Link Pendulum

A uniform rigid slender rod of mass m and length L is pivoted at its bottom end and is allowed to rotate within the plane of paper. Its top end is constrained by two identical linear springs, whose pivot points are a distance L from the pivot for the rod, as shown in Fig. 1.30. The springs are unstretched when the rod is vertical. Derive the equation(s) of motion for the system.

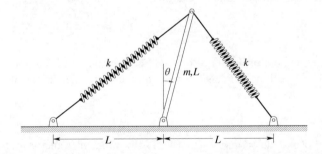

Figure 1.30   Inverted rigid-link planar pendulum

□ Solution:

- *Generalized Coordinates*: We define $\{\theta\}$ as a complete and independent set of generalized coordinates, where θ is clockwise angular displacement of the rod measured from the vertical.
- *Admissible Variations*: We verify that $\{\delta\theta\}$ is a complete and independent set of admissible variations in this set of generalized coordinates.
- *Holonomicity*: We conclude that the system is holonomic and has one degree of freedom.
- *Generalized Forces*: Since there is no nonconservative force in the system, $\delta W^{\text{n.c.}} = 0$. Thus,

$$\Xi_\theta = 0 \qquad (a)$$

- *Kinetic Energy*: The rod is rotating about a fixed point at an angular velocity of $\dot{\theta}$. The moment of inertia about the pivot point, as calculated in Example 1.23, is $I_O = \frac{1}{3}mL^2$. Thus,

$$T = \frac{1}{2}\left(\frac{1}{3}mL^2\right)\dot{\theta}^2 = \frac{1}{6}mL^2\dot{\theta}^2 \tag{b}$$

- *Potential Energy*: We choose the pivot point as the datum for the gravitational potential energy. The gravitational potential energy is

$$V_{\text{gravity}} = mg\frac{L}{2}\cos\theta = \frac{1}{2}mgL\cos\theta \tag{c}$$

For the potential energy stored in the springs, consider first the spring on the left. The current length of this spring, L_L, can be found by the law of cosine as

$$L_L^2 = L^2 + L^2 - 2L^2\cos\left(\frac{\pi}{2} + \theta\right) = 2L^2(1 + \sin\theta) \tag{d}$$

Similarly, for the spring on the right,

$$L_R^2 = L^2 + L^2 - 2L^2\cos\left(\frac{\pi}{2} - \theta\right) = 2L^2(1 - \sin\theta) \tag{e}$$

The unstretched length of both springs is $\sqrt{2}L$. Thus, the potential energy stored in the two springs is

$$
\begin{aligned}
V_{\text{springs}} &= \frac{1}{2}k\left(\sqrt{2}L\sqrt{1-\sin\theta} - \sqrt{2}L\right)^2 + \frac{1}{2}k\left(\sqrt{2}L\sqrt{1+\sin\theta} - \sqrt{2}L\right)^2 \\
&= kL^2\left[\left(\sqrt{1-\sin\theta} - 1\right)^2 + \left(\sqrt{1+\sin\theta} - 1\right)^2\right] \\
&= 2kL^2\left[2 - \left(\sqrt{1-\sin\theta} + \sqrt{1+\sin\theta}\right)\right]
\end{aligned}
\tag{f}
$$

- *Lagrangian*: Combining eqns. (b) and (f), according to eqn. (1.3), gives

$$\mathcal{L} = T - V = \frac{1}{6}mL^2\dot{\theta}^2 - \frac{1}{2}mgL\cos\theta - 2kL^2\left[2 - \left(\sqrt{1-\sin\theta} + \sqrt{1+\sin\theta}\right)\right] \tag{g}$$

- *Lagrange's Equation*: The Lagrange's equation for θ is, according to eqn. (1.4),

$$\frac{d}{dt}\left(\frac{\partial\mathcal{L}}{\partial\dot{\theta}}\right) - \frac{\partial\mathcal{L}}{\partial\theta} = \Xi_\theta \tag{h}$$

Note that

$$\frac{\partial\mathcal{L}}{\partial\dot{\theta}} = \frac{1}{3}mL^2\dot{\theta} \tag{i}$$

and

$$
\begin{aligned}
\frac{\partial\mathcal{L}}{\partial\theta} &= \frac{1}{2}mgL\sin\theta + 2kL^2\left(\frac{1}{2}\frac{-\cos\theta}{\sqrt{1-\sin\theta}} + \frac{1}{2}\frac{\cos\theta}{\sqrt{1+\sin\theta}}\right) \\
&= \frac{1}{2}mgL\sin\theta - kL^2\left(\sqrt{1+\sin\theta} - \sqrt{1-\sin\theta}\right)
\end{aligned}
\tag{j}
$$

where, considering the geometrically admissible range of $-\pi/2 \le \theta \le \pi/2$, the following relation has been used:

$$\sqrt{(1 - \sin \theta)(1 + \sin \theta)} = \cos \theta$$

Substituting eqns. (i), (j), and (a) into eqn. (h) gives

$$\frac{1}{3}mL^2\ddot{\theta} - \frac{1}{2}mgL \sin \theta + kL^2 \left(\sqrt{1 + \sin \theta} - \sqrt{1 - \sin \theta} \right) = 0 \qquad \text{(k)}$$

Equation (k) is the equation of motion for the system. Noting that

$$\left(\sqrt{1 + \sin \theta} - \sqrt{1 - \sin \theta} \right)^2 = 2 - 2\sqrt{(1 - \sin \theta)(1 + \sin \theta)} = 2(1 - \cos \theta) \qquad \text{(l)}$$

Equation (k) can be further simplified into

$$\frac{1}{3}mL^2\ddot{\theta} - \frac{1}{2}mgL \sin \theta + \sqrt{2}kL^2\sqrt{1 - \cos \theta} = 0 \qquad \text{(m)}$$

■ Example 1.25: Restrained Double Pulley

A double pulley comprises two pulleys, modeled as uniform circular disks, wielded together such that they rotate about their common center as one piece. The larger pulley has mass m_1 and radius R_1, and the smaller pulley has mass m_2 and radius R_2. A mass m_0 is attached to the double pulley via a massless inextensible rope that is wrapped around the perimeter of the smaller pulley without slippage. The larger pulley is restrained by a spring k and a dashpot c, as shown in Fig. 1.31. Derive the equation(s) of motion for the system.

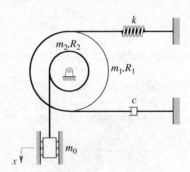

Figure 1.31   A double pulley constrained by spring and dashpot

□ Solution 1: Linear Displacement as Generalized Coordinate

- *Generalized Coordinates*: We define $\{x\}$ as a complete and independent set of generalized coordinates, where x is the downward displacement of mass m_0 with respect to the ground measured from the equilibrium position.
- *Admissible Variations*: We verify that $\{\delta x\}$ is a complete and independent set of admissible variations in this set of generalized coordinates.
- *Holonomicity*: We conclude that the system is holonomic and has one degree of freedom.

- *Generalized Forces*: The dashpot is the nonconservative element in the system. A velocity of $\dot{x}$ at the perimeter of the smaller pulley of radius R_2 produces an angular velocity of $\dot{x}/R_2$ in the double pulley and, in turn, a velocity of $\dot{x}R_1/R_2$ at the dashpot. Similarly, a variation in x of δx at radius R_2 produces a virtual angular displacement of $\delta x/R_2$ in the double pulley and a virtual displacement of $\delta x R_1/R_2$ at the dashpot. Thus,

$$\delta W^{\text{n.c}} = -c\left(\dot{x}\frac{R_1}{R_2}\right)\left(\delta x\frac{R_1}{R_2}\right) = -c\left(\frac{R_1}{R_2}\right)^2 \dot{x}\delta x \tag{a}$$

According to eqn. (1.2),

$$\Xi_x = -c\left(\frac{R_1}{R_2}\right)^2 \dot{x} \tag{b}$$

- *Kinetic Energy*: As mass m_0 moves at a speed of $\dot{x}$, the double pulley rotates at an angular velocity of $\dot{x}/R_2$. Recall that the moment of inertia for a circular disk of mass m and radius R about its centroid is $\frac{1}{2}mR^2$. Hence,

$$T = \frac{1}{2}m_0\dot{x}^2 + \frac{1}{2}\left(\frac{1}{2}m_1 R_1^2\right)\left(\frac{\dot{x}}{R_2}\right)^2 + \frac{1}{2}\left(\frac{1}{2}m_2 R_2^2\right)\left(\frac{\dot{x}}{R_2}\right)^2$$

$$= \frac{1}{2}\left(m_0 + \frac{R_1^2}{2R_2^2}m_1 + \frac{1}{2}m_2\right)\dot{x}^2 \tag{c}$$

- *Potential Energy*: When mass m_0 moves a distance x, the pulley rotates by an angle of x/R_2, and, on the outer rim, the spring is stretched by an additional xR_1/R_2. Thus,

$$V = -mgx + \frac{1}{2}k\left(x\frac{R_1}{R_2} + \Delta\right)^2 \tag{d}$$

where Δ is the amount of stretch in the spring at equilibrium.
- *Lagrangian*: Combining eqns. (c) and (d) gives, according to eqn. (1.3),

$$\mathcal{L} = T - V = \frac{1}{2}\left(m_0 + \frac{R_1^2}{2R_2^2}m_1 + \frac{1}{2}m_2\right)\dot{x}^2 + mgx - \frac{1}{2}k\left(x\frac{R_1}{R_2} + \Delta\right)^2 \tag{e}$$

- *Lagrange's Equation*: The Lagrange's equation for x is, according to eqn. (1.4),

$$\frac{d}{dt}\left(\frac{\partial \mathcal{L}}{\partial \dot{x}}\right) - \frac{\partial \mathcal{L}}{\partial x} = \Xi_x \tag{f}$$

Since

$$\frac{\partial \mathcal{L}}{\partial \dot{x}} = \left(m_0 + \frac{R_1^2}{2R_2^2}m_1 + \frac{1}{2}m_2\right)\dot{x} \tag{g}$$

and

$$\frac{\partial \mathcal{L}}{\partial x} = mg - k\left(x\frac{R_1}{R_2} + \Delta\right)\frac{R_1}{R_2} \tag{h}$$

substituting eqns. (g), (h), and (b) into eqn. (f) gives

$$\left(m_0 + \frac{R_1^2}{2R_2^2}m_1 + \frac{1}{2}m_2\right)\ddot{x} - mg + k\left(x\frac{R_1}{R_2} + \Delta\right)\frac{R_1}{R_2} = -c\left(\frac{R_1}{R_2}\right)^2\dot{x}$$

or

$$\left(m_0 + \frac{R_1^2}{2R_2^2}m_1 + \frac{1}{2}m_2\right)\ddot{x} + c\left(\frac{R_1}{R_2}\right)^2\dot{x} + k\left(\frac{R_1}{R_2}\right)^2 x + k\frac{R_1}{R_2}\Delta - mg = 0 \qquad \text{(i)}$$

At equilibrium, all constant terms (Δ and mg) cancel out as the equilibrium condition, and eqn. (i) becomes

$$\left[(2m_0 + m_2)R_2^2 + m_1 R_1^2\right]\ddot{x} + 2cR_1^2\dot{x} + 2kR_1^2 x = 0 \qquad \text{(j)}$$

Equation (j) is the equation of motion for the system.

□ **Solution 2: Angular Displacement as Generalized Coordinate**

- *Generalized Coordinates*: We define $\{\theta\}$ as a complete and independent set of generalized coordinates, where θ is the counterclockwise angular displacement of the pulleys measured from the equilibrium configuration, as shown in Fig. 1.32.

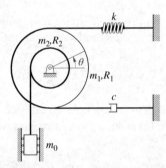

Figure 1.32   Alternative generalized coordinate for the double pulley constrained by spring and dashpot

- *Admissible Variations*: We verify that $\{\delta\theta\}$ is a complete and independent set of admissible variations in this set of generalized coordinates.
- *Holonomicity*: We conclude that the system is holonomic and has one degree of freedom.
- *Generalized Forces*: The dashpot is the nonconservative element in the system. An angular velocity of $\dot{\theta}$ in the pulleys produces a velocity of $\dot{\theta}R_1$ at the dashpot. A variation $\delta\theta$ produces a virtual displacement of $R_1\delta\theta$ at the end of the dashpot. Then,

$$\delta W^{\text{n.c}} = -c\left(R_1\dot{\theta}\right)\left(R_1\delta\theta\right) = -cR_1^2\dot{\theta}\delta\theta \qquad \text{(k)}$$

Thus, according to eqn. (1.2),

$$\Xi_\theta = -cR_1^2\dot{\theta} \qquad \text{(l)}$$

- *Kinetic Energy*: When the pulley rotates at an angular velocity $\dot{\theta}$, mass m_0 moves at the same velocity as the outer rim of the smaller pulley, at $\dot{\theta}R_2$. Hence,

$$T = \frac{1}{2}m_0\left(\dot{\theta}R_2\right)^2 + \frac{1}{2}\left(\frac{1}{2}m_1R_1^2\right)\dot{\theta}^2 + \frac{1}{2}\left(\frac{1}{2}m_2R_2^2\right)\dot{\theta}^2$$

$$= \frac{1}{2}\left(m_0R_2^2 + \frac{1}{2}m_1R_1^2 + \frac{1}{2}m_2R_2^2\right)\dot{\theta}^2 \tag{m}$$

- *Potential Energy*: When the pulley rotates by an angle θ, the spring is additionally stretched by θR_1 while mass m_0 is lowered by θR_2. Thus,

$$V = -mgR_2\theta + \frac{1}{2}k\left(\theta R_1 + \Delta\right)^2 \tag{n}$$

where Δ is the amount of stretch in the spring at equilibrium, and the equilibrium position of the mass is used as the datum for its gravitational potential energy.

- *Lagrangian*: Combining eqns. (m) and (n) gives, according to eqn. (1.3)

$$\mathcal{L} = T - V = \frac{1}{2}\left(m_0R_2^2 + \frac{1}{2}m_1R_1^2 + \frac{1}{2}m_2R_2^2\right)\dot{\theta}^2 + mgR_2\theta - \frac{1}{2}k\left(\theta R_1 + \Delta\right)^2 \tag{o}$$

- *Lagrange's Equation*: The Lagrange's equation for θ is, according to eqn. (1.4),

$$\frac{d}{dt}\left(\frac{\partial\mathcal{L}}{\partial\dot{\theta}}\right) - \frac{\partial\mathcal{L}}{\partial\theta} = \Xi_\theta \tag{p}$$

Since

$$\frac{\partial\mathcal{L}}{\partial\dot{\theta}} = \left(m_0R_2^2 + \frac{1}{2}m_1R_1^2 + \frac{1}{2}m_2R_2^2\right)\dot{\theta} \tag{q}$$

and

$$\frac{\partial\mathcal{L}}{\partial\theta} = mgR_2 - kR_1\left(\theta R_1 + \Delta\right) \tag{r}$$

substituting eqns. (q), (r), and (l) into eqn. (p) gives

$$\left(m_0R_2^2 + \frac{1}{2}m_1R_1^2 + \frac{1}{2}m_2R_2^2\right)\ddot{\theta} - mgR_2 + kR_1\left(\theta R_1 + \Delta\right) = -cR_1^2\dot{\theta} \tag{s}$$

At equilibrium, constant terms (Δ and mg) form the equilibrium condition and cancel out, and eqn. (s) becomes

$$\left(m_0R_2^2 + \frac{1}{2}m_1R_1^2 + \frac{1}{2}m_2R_2^2\right)\ddot{\theta} + cR_1^2\dot{\theta} + kR_1^2\theta = 0 \tag{t}$$

Equation (t) is the equation of motion for the system.

■ **Example 1.26: Platform on Two Supports**

A platform, which is modeled as a uniform rigid slender rod of a mass m and length L, is supported by two springs k_1 and k_2, and two dashpots c_1 and c_2 at its two ends, as shown in Fig. 1.33. Assume small motions. Derive the equation(s) of motion for the system.

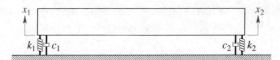

Figure 1.33   Platform supported by springs and dashpots at two ends

□ Solution 1: Using Two End Displacements as Generalized Coordinates

- *Generalized Coordinates*: We define $\{x_1, x_2\}$ as a complete and independent set of generalized coordinates, where x_1 and x_2 are the upward displacement of the left and right ends, respectively, of the platform with respect to ground and measured from their respective equilibrium positions.
- *Admissible Variations*: We verify that $\{\delta x_1, \delta x_2\}$ is a complete and independent set of admissible variations in this set of generalized coordinates.
- *Holonomicity*: We conclude that the system is holonomic and has two degrees of freedom.
- *Generalized Forces*: There are two nonconservative elements (dashpots) in the system. Each dashpot is connected in the same way as the one analyzed in Example 1.16. Thus,

$$\delta W^{\text{n.c.}} = -c_1 \dot{x}_1 \delta x_1 - c_2 \dot{x}_2 \delta x_2$$

Thus, according to eqn. (1.2),

$$\Xi_{x_1} = -c_1 \dot{x}_1 \tag{a}$$

$$\Xi_{x_2} = -c_2 \dot{x}_2 \tag{b}$$

- *Kinetic Energy*: There is no identifiable fixed point in the rigid body, we use the general expression for the kinetic energy in eqn. (A.27) and recall that the centroidal moment of inertia for a slender rod is $I_C = \frac{1}{12} mL^2$.

  To find the velocity at the center of the platform, we need to analyze the geometry, as sketched in Fig. 1.34. In Fig. 1.34, the bottom horizontal line represents the ground; l's are the unstretched length of the springs, Δ's are the compressions in the springs at equilibrium, and x's are the current positions. From the geometry, the y-coordinate of the centroid is expressible as

$$y_C = \frac{(l_1 + x_1 - \Delta_1) + (l_2 + x_2 - \Delta_2)}{2} \tag{c}$$

and the inclined angle of the platform, denoted as θ, is expressible as

$$\theta \approx \sin\theta = \frac{(l_1 + x_1 - \Delta_1) - (l_2 + x_2 - \Delta_2)}{L} \tag{d}$$

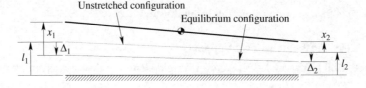

Figure 1.34   Geometry of the platform at a generic moment

In writing these equations, we have used the small motion assumption such that all displacements are in vertical direction, and the horizontal distance between two ends of the platform remains as L.

Taking a time derivative, the velocity of the centroid and the angular velocity of the platform are

$$v_C = \frac{\dot{x}_1 + \dot{x}_2}{2} \quad \text{and} \quad \dot{\theta} = \frac{\dot{x}_2 - \dot{x}_1}{L} \tag{e}$$

Thus, the kinetic energy of the platform is

$$T = \frac{1}{2} v_C^2 + \frac{1}{2} I_C \dot{\theta}^2$$

$$= \frac{1}{2} m \left(\frac{\dot{x}_1 + \dot{x}_2}{2} \right)^2 + \frac{1}{2} \left(\frac{1}{12} mL^2 \right) \left(\frac{\dot{x}_2 - \dot{x}_1}{L} \right)^2$$

$$= \frac{1}{6} m \left(\dot{x}_1^2 + \dot{x}_2^2 + \dot{x}_1 \dot{x}_2 \right) \tag{f}$$

- *Potential Energy*: Again, referring to the geometry in Fig. 1.34, the amounts of stretches in springs are $x_1 - \Delta_1$ and $x_2 - \Delta_2$, respectively. For the gravitational potential energy, we can use the y-coordinate for the centroid y_C in eqn. (c). Thus,

$$V = \frac{1}{2} k_1 (x_1 - \Delta_1)^2 + \frac{1}{2} k_2 (x_2 - \Delta_2)^2 + \frac{1}{2} mg \left[(l_1 + x_1 - \Delta_1) + (l_2 + x_2 - \Delta_2) \right] \tag{g}$$

- *Lagrangian*: Combining eqns. (f) and (g) gives, according to eqn. (1.3),

$$\mathcal{L} = T - V = \frac{1}{6} m \left(\dot{x}_1^2 + \dot{x}_2^2 + \dot{x}_1 \dot{x}_2 \right) - \frac{1}{2} k_1 (x_1 - \Delta_1)^2 - \frac{1}{2} k_2 (x_2 - \Delta_2)^2$$

$$- \frac{1}{2} mg \left[(l_1 + x_1 - \Delta_1) + (l_2 + x_2 - \Delta_2) \right] \tag{h}$$

- *Lagrange's Equation*: The system has two equations of motion.
  - The x_1-Equation: The Lagrange's equation for x_1 is, according to eqn. (1.4),

$$\frac{d}{dt} \left(\frac{\partial \mathcal{L}}{\partial \dot{x}_1} \right) - \frac{\partial \mathcal{L}}{\partial x_1} = \Xi_{x_1} \tag{i}$$

Since

$$\frac{\partial \mathcal{L}}{\partial \dot{x}_1} = \frac{1}{3} m \dot{x}_1 + \frac{1}{6} m \dot{x}_2 \tag{j}$$

and

$$\frac{\partial \mathcal{L}}{\partial x_1} = -k_1 (x_1 - \Delta_1) - \frac{1}{2} mg \tag{k}$$

substituting eqns. (j), (k), and (a) into eqn. (i) gives

$$\frac{1}{3} m \ddot{x}_1 + \frac{1}{6} m \ddot{x}_2 + k_1 (x_1 - \Delta_1) + \frac{1}{2} mg = -c_1 \dot{x}_1$$

Rearranging and canceling the constant terms using the equilibrium condition give

$$\frac{1}{3} m \ddot{x}_1 + \frac{1}{6} m \ddot{x}_2 + c_1 \dot{x}_1 + k_1 x_1 = 0 \tag{l}$$

– *The x_2-Equation*: The Lagrange's equation for x_2 is, according to eqn. (1.4),

$$\frac{d}{dt}\left(\frac{\partial \mathcal{L}}{\partial \dot{x}_2}\right) - \frac{\partial \mathcal{L}}{dx_2} = \Xi_{x_2} \tag{m}$$

Since

$$\frac{\partial \mathcal{L}}{\partial \dot{x}_2} = \frac{1}{6}m\dot{x}_1 + \frac{1}{3}m\dot{x}_2 \tag{n}$$

and

$$\frac{\partial \mathcal{L}}{\partial x_2} = -k_2(x_2 - \Delta_2) - \frac{1}{2}mg \tag{o}$$

substituting eqns. (n), (o), and (b) into eqn. (m) gives

$$\frac{1}{6}m\ddot{x}_1 + \frac{1}{3}m\ddot{x}_2 + k_2(x_2 - \Delta_2) + \frac{1}{2}mg = -c_2\dot{x}_2$$

Rearranging and canceling the constant terms using the equilibrium condition give

$$\frac{1}{6}m\ddot{x}_1 + \frac{1}{3}m\ddot{x}_2 + c_2\dot{x}_2 + k_2 x_2 = 0 \tag{p}$$

Equations (l) and (p) comprise the equations of motion for the system.

□ **Solution 2: Using Displacement and Rotation as Generalized Coordinates**

- *Generalized Coordinates*: We define $\{y, \theta\}$ as a complete and independent set of generalized coordinates, where y is the upward displacement of the centroid of the platform with respect to the ground and θ is the counterclockwise angular displacement of the platform; both are measured from the equilibrium configuration, as shown in Fig. 1.35.

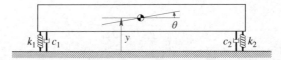

Figure 1.35   Linear and angular displacements as generalized coordinates for platform

- *Admissible Variations*: We verify that $\{\delta y, \delta \theta\}$ is a complete and independent set of admissible variations in this set of generalized coordinates.
- *Holonomicity*: We conclude that the system is holonomic and has two degrees of freedom.
- *Generalized Forces*: There are two nonconservative elements (dashpots). In determining the virtual work, we look at the velocities first. The platform is rotating about the centroid at angular velocity of $\dot{\theta}$ while being carried at the velocity of the centroid $\dot{y}$. Assuming small motions, the relative velocities at both ends due to the rotation are in the vertical direction. Thus, the left end has a velocity of $\dot{y} - \frac{1}{2}L\dot{\theta}$, and the right end has a velocity of $\dot{y} + \frac{1}{2}L\dot{\theta}$.

    Now consider the virtual displacements. If y is varied by δy but θ is fixed, both ends have an upward virtual displacement of δy. If y is fixed but θ is varied by $\delta \theta$, the left end has a downward virtual displacement of $\frac{1}{2}L\delta\theta$ and the right end has an upward virtual displacement of the same amount, as sketched in Fig. 1.36. Thus, the combined total virtual displacement at the left end is $\delta y - \frac{1}{2}L\delta\theta$ and at the right end is $\delta y + \frac{1}{2}L\delta\theta$.

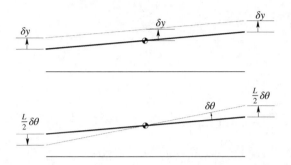

Figure 1.36   Variations and virtual displacements in the platform

Thus, the total virtual work is

$$\delta W^{\text{n.c.}} = -c_1\left(\dot{y} - \frac{1}{2}L\dot{\theta}\right)\left(\delta y - \frac{1}{2}L\delta\theta\right) - c_2\left(\dot{y} + \frac{1}{2}L\dot{\theta}\right)\left(\delta y + \frac{1}{2}L\delta\theta\right)$$

$$= -\left[(c_1 + c_2)\dot{y} + \frac{1}{2}(c_2 - c_1)L\dot{\theta}\right]\delta y + \frac{1}{2}L\left[(c_1 - c_2)\dot{y} - \frac{1}{2}(c_1 + c_2)L\dot{\theta}\right]\delta\theta$$

Then, according to eqn. (1.2),

$$\Xi_y = -(c_1 + c_2)\dot{y} - \frac{1}{2}(c_2 - c_1)L\dot{\theta} \tag{q}$$

$$\Xi_\theta = \frac{1}{2}L\left[(c_1 - c_2)\dot{y} - \frac{1}{2}(c_1 + c_2)L\dot{\theta}\right] \tag{r}$$

- *Kinetic Energy*: The centroid of the platform has a velocity of $\dot{y}$ and the platform has an angular velocity of $\dot{\theta}$. Thus,

$$T = \frac{1}{2}m\dot{y}^2 + \frac{1}{24}mL^2\dot{\theta}^2 \tag{s}$$

- *Potential Energy*: Assume that the spring compressions at equilibrium are Δ_1 and Δ_2, respectively. The displacements at the two ends are $y \pm \frac{1}{2}L\theta$. For the gravitational potential energy, we choose the location of the centroid at equilibrium as the datum. Then,

$$V = \frac{1}{2}k_1\left(y - \frac{1}{2}\theta L - \Delta_1\right)^2 + \frac{1}{2}k_2\left(y + \frac{1}{2}\theta L - \Delta_2\right)^2 + mgy \tag{t}$$

- *Lagrangian*: Combining eqns. (s) and (t) gives, according to eqn. (1.3),

$$\mathcal{L} = \frac{1}{2}m\dot{y}^2 + \frac{1}{24}mL^2\dot{\theta}^2 - \frac{1}{2}k_1\left(y - \frac{1}{2}\theta L - \Delta_1\right)^2$$

$$- \frac{1}{2}k_2\left(y + \frac{1}{2}\theta L - \Delta_2\right)^2 - mgy \tag{u}$$

- *Lagrange's Equation*: The system has two equations of motion.

– *The y-Equation*: The Lagrange's equation for y is, according to eqn. (1.4),

$$\frac{d}{dt}\left(\frac{\partial \mathcal{L}}{\partial \dot{y}}\right) - \frac{\partial \mathcal{L}}{\partial y} = \Xi_y \tag{v}$$

Since

$$\frac{\partial \mathcal{L}}{\partial \dot{y}} = m\dot{y} \tag{w}$$

and

$$\frac{\partial \mathcal{L}}{\partial y} = -k_1\left(y - \frac{1}{2}\theta L - \Delta_1\right) - k_2\left(y + \frac{1}{2}\theta L - \Delta_2\right) - mg \tag{x}$$

substituting eqns. (w), (x), and (q) into eqn. (v) gives

$$m\ddot{y} + k_1\left(y - \frac{1}{2}\theta L - \Delta_1\right) + k_2\left(y + \frac{1}{2}\theta L + \Delta_2\right) - mg$$

$$= -(c_1 + c_2)\dot{y} - \frac{1}{2}(c_2 - c_1)L\dot{\theta}$$

Rearranging and canceling out the constant terms using the equilibrium condition give

$$m\ddot{y} + (c_1 + c_2)\dot{y} - \frac{1}{2}(c_1 - c_2)L\dot{\theta} + (k_1 + k_2)y + \frac{1}{2}(k_2 - k_1)L\theta = 0 \tag{y}$$

– *The θ-Equation*: The Lagrange's equation for θ is, according to eqn. (1.4),

$$\frac{d}{dt}\left(\frac{\partial \mathcal{L}}{\partial \dot{\theta}}\right) - \frac{\partial \mathcal{L}}{\partial \theta} = \Xi_\theta \tag{z}$$

Since

$$\frac{\partial \mathcal{L}}{\partial \dot{\theta}} = \frac{1}{12}mL^2\dot{\theta} \tag{aa}$$

and

$$\frac{\partial \mathcal{L}}{\partial \theta} = k_1\left(y - \frac{1}{2}\theta L - \Delta_1\right)\frac{1}{2}L - k_2\left(y + \frac{1}{2}\theta L - \Delta_2\right)\frac{1}{2}L$$

$$= -\frac{L}{2}(k_2 - k_1)y - \frac{L^2}{4}(k_1 + k_2)\theta - \frac{L}{2}(k_1\Delta_1 - k_2\Delta_2) \tag{ab}$$

substituting eqns. (aa), (ab), and (r) into eqn. (z) gives

$$\frac{1}{12}mL^2\ddot{\theta} + \frac{L}{2}(k_2 - k_1)y + \frac{L^2}{4}(k_1 + k_2)\theta + \frac{L}{2}(k_1\Delta_1 - k_2\Delta_2)$$

$$= \frac{L}{2}\left[(c_1 - c_2)\dot{y} - (c_1 + c_2)\frac{L}{2}\dot{\theta}\right]$$

Rearranging and canceling out the constant terms using the equilibrium condition give

$$mL\ddot{\theta} + 6(c_2 - c_1)\dot{y} + 3(c_1 + c_2)L\dot{\theta} + 6(k_2 - k_1)y + 3(k_1 + k_2)L\theta = 0 \tag{ac}$$

Equations (y) and (ac) comprise the equations of motion for the system.

■ Example 1.27: Washing Machine with Load Imbalance

A front-loading washing machine is confined to move only in the vertical direction. Its rubber feet are modeled as two springs of $\frac{1}{2}k$ and a dashpot c, as sketched in Fig. 1.37. The drum and the clothes are modeled together as one piece of mass m and centroidal moment of inertia I_C, and its centroid is located at a distance e from its geometric center. The machine body has mass M. Assume the drum rotates at a constant angular velocity of Ω. Derive the equation(s) of motion for the system.

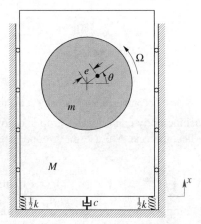

Figure 1.37   Model for front-loading washing machine

□ Solution:

- *Preparatory Setup*: The drum is undergoing a specified motion. We define θ as the angle the line connecting the mass center to geometric center makes with the horizontal. This angle can be described as

$$\theta = \Omega t + \theta_0 \tag{a}$$

where θ_0 is the angle at $t = 0$. We shall keep in mind that θ is completely specified at all times, and hence shall not be included as a generalized coordinate.
- *Generalized Coordinates*: We define $\{x\}$ as a set of complete and independent generalized coordinates, where x is the upward displacement of the machine body measured from its equilibrium position with respect to the ground.
- *Admissible Variations*: We verify that $\{\delta x\}$ is a complete and independent set of admissible variations in this set of generalized coordinates.
- *Holonomicity*: We conclude that the system is holonomic and has one degree of freedom.
- *Generalized Forces*: There is one nonconservative element: a simple dashpot. Thus,

$$\delta W^{\text{n.c.}} = -c\dot{x}\delta x \tag{b}$$

Then, according to eqn. (1.2),

$$\Xi_x = -c\dot{x} \tag{c}$$

- *Kinetic Energy*: The machine body simply moves up and down at a velocity of $\dot{x}$. The drum rotates relative to the moving body. The velocity composition at the mass center of the drum is sketched in Fig. 1.38. Thus, the kinetic energy of the washing machine is

$$T = \frac{1}{2}M\dot{x}^2 + \frac{1}{2}m[(\dot{x} + e\Omega\cos\theta)^2 + (e\Omega\sin\theta)^2] + \frac{1}{2}I_C\Omega^2$$

$$= \frac{1}{2}M\dot{x}^2 + \frac{1}{2}m\left[\dot{x}^2 + (e\Omega)^2 + 2e\dot{x}\Omega\cos\theta\right] + \frac{1}{2}I_C\Omega^2 \tag{d}$$

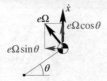

Figure 1.38 Composition and decomposition of velocity at the mass center of the drum

- *Potential Energy*: The machine body can only move up and down. The two springs move in unison and effectively become one. We use the ground as the datum for gravitational potential energy. Then,

$$V = \frac{1}{2}k(x - \Delta)^2 + Mgx + mg(x + e\sin\theta) \tag{e}$$

where Δ is the amount of compression in the spring at equilibrium.

- *Lagrangian*: Combining eqns. (d) and (e) gives, according to eqn. (1.3),

$$\mathcal{L} = T - V = \frac{1}{2}M\dot{x}^2 + \frac{1}{2}m\left[\dot{x}^2 + (e\Omega)^2 + 2e\dot{x}\Omega\cos\theta\right] + \frac{1}{2}I_C\Omega^2$$

$$- \frac{1}{2}k(x - \Delta)^2 - Mgx - mg(x + e\sin\theta) \tag{f}$$

- *Lagrange's Equation*: Lagrange's equation for x is

$$\frac{d}{dt}\left(\frac{\partial\mathcal{L}}{\partial\dot{x}}\right) - \frac{\partial\mathcal{L}}{\partial x} = \Xi_x \tag{g}$$

Since

$$\frac{\partial\mathcal{L}}{\partial\dot{x}} = M\dot{x} + m(\dot{x} + e\Omega\cos\theta) \tag{h}$$

and

$$\frac{\partial\mathcal{L}}{\partial x} = -k(x - \Delta) - Mg - mg \tag{i}$$

Substituting eqns. (h), (i), and (c) into eqn. (g) gives

$$(M + m)\ddot{x} - me\Omega^2\sin\theta + kx - k\Delta + (M + m)g = -c\dot{x} \tag{j}$$

Using the equilibrium condition to eliminate the constant terms gives the following equation of motion for the system:

$$(M + m)\ddot{x} + c\dot{x} + kx = me\Omega^2\sin(\Omega t + \theta_0) \tag{k}$$

where eqn. (a) has been used.

■ Example 1.28: Disk Rolling Inside Circular Track

A circular disk of mass m and radius r can roll without slip inside a fixed circular track of radius R $(R > r)$, as shown in Fig. 1.39. Derive the equation(s) of motion for the disk.

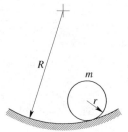

Figure 1.39   Small disk rolling without slip on a large circular track

□ Solution:

- *Preparatory Setup*: We paint a radius on the disk that connects its center to the contact point when the disk is located at its lowest position, which is used as a reference configuration. The geometry of the disk in a displaced configuration is depicted in Fig. 1.40. In this figure, point B on the disk is in contact with point A on the track in the reference configuration. BC is the painted radius and CD is the radius that connects the contact point D to the center of the disk C in the displaced configuration, whose extension passes through the center of the track O.

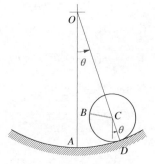

Figure 1.40   Geometry of disk rolling on a circular track

- *Generalized Coordinates*: We define $\{\theta\}$ as a complete and independent set of generalized coordinates, where θ is the angle that the line connecting the two centers forms with the vertical, which is positive in the counterclockwise direction, as shown in Fig. 1.40.
- *Admissible Variations*: We verify that $\{\delta\theta\}$ is a complete and independent set of admissible variations in this set of generalized coordinates.
- *Holonomicity*: We conclude that the system is holonomic and has one degree of freedom.
- *Generalized Forces*: There is no externally applied force, nor nonconservative elements. Thus, $\delta W^{\text{n.c.}} = 0$. According to eqn. (1.2),

$$\Xi_\theta = 0 \tag{a}$$

- *Kinetic Energy*: There is no apparent fixed point in the disk. Thus, we use the general formula in eqn. (A.27) for the kinetic energy.

  For the velocity of the center of the disk v_C, note that line OC rotates at $\dot{\theta}$,

$$v_C = (R - r)\dot{\theta}$$

For the angular velocity of the disk, we look at the angular displacement first. Because of the nonslip condition, the arc length of $\overparen{AD}$ (of radius R) must equal to the arc length of $\overparen{BD}$ (of radius r). On the other hand, the angular displacement of the disk is the angle between BC line and the vertical, which is

$$\text{Angular displacement of disk} = \frac{\overparen{BD}}{r} - \theta = \frac{\overparen{AD}}{r} - \theta = \left(\frac{R}{r} - 1\right)\theta$$

Taking a time derivative of the above relation gives the angular velocity of the disk as

$$\omega_{\text{disk}} = \left(\frac{R}{r} - 1\right)\dot{\theta}$$

Therefore, the kinetic energy of the disk is

$$T = \frac{1}{2}mv_C^2 + \frac{1}{2}I_C\omega_{\text{disk}}^2$$

$$= \frac{1}{2}m[(R-r)\dot{\theta}]^2 + \frac{1}{2}\left(\frac{1}{2}mr^2\right)\left[\left(\frac{R}{r}-1\right)\dot{\theta}\right]^2 = \frac{3}{4}m(R-r)^2\dot{\theta}^2 \qquad \text{(b)}$$

- *Potential Energy*: We use the center of the track O as the datum for the gravitational potential energy. Then,

$$V = -mg(R - r)\cos\theta \qquad \text{(c)}$$

- *Lagrangian*: Combining eqns. (b) and (c) gives, according to eqn. (1.3),

$$\mathcal{L} = T - V = \frac{3}{4}m(R-r)^2\dot{\theta}^2 + mg(R-r)\cos\theta \qquad \text{(d)}$$

- *Lagrange's Equation*: The Lagrange's equation for θ is, according to eqn. (1.4),

$$\frac{d}{dt}\left(\frac{\partial\mathcal{L}}{\partial\dot{\theta}}\right) - \frac{\partial\mathcal{L}}{\partial\theta} = \Xi_\theta \qquad \text{(e)}$$

Since

$$\frac{\partial\mathcal{L}}{\partial\dot{\theta}} = \frac{3}{2}m(R-r)^2\dot{\theta} \qquad \text{(f)}$$

and

$$\frac{\partial\mathcal{L}}{\partial\theta} = -mg(R-r)\sin\theta \qquad \text{(g)}$$

substituting eqns. (f), (g), and (a) into eqn. (e) gives

$$\frac{3}{2}m(R-r)^2\ddot{\theta} + mg(R-r)\sin\theta = 0$$

Canceling a common factor $m(R-r)$ gives

$$3(R-r)\ddot{\theta} + 2g\sin\theta = 0 \qquad \text{(h)}$$

Equation (h) is the equation of motion for the system.

■ Example 1.29: Rocking Cylinder Segment

A segment of a circular cylinder of radius R and mass m can roll without slip on a horizontal ground. The total height of the segment is H, and when it rests in upward equilibrium config-uration, as shown in Fig. 1.41, the center of mass is located at a distance h above the ground $(h < H)$. A force $F(t)$ acts on the edge of the segment and maintains in the horizontal direction at all times, and yet not large enough to overcome the static friction between the cylinder and the ground. The centroidal moment of inertia of the cylinder segment is I. Find the equation(s) of motion for the cylinder segment.

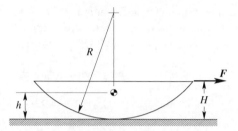

Figure 1.41   Rocking cylinder segment and its geometry

□ Solution:

- *Preparatory Setup*: Fig. 1.42 illustrates the geometry of the cylinder segment in a displaced configuration. In this figure, O is the center of the cylindrical surface, A is the contact point on the ground, A' is the matching contact point on the cylinder, B is the contact point on the ground in the equilibrium configuration, B' is the matching contact point on the cylinder, C is the centroid, D is the edge of the segment where the external force acts, CE and GD are horizontal projection lines, and θ_0 is constant angle describing the angular size of the segment, as

$$\cos\theta_0 = \frac{R - H}{R} \quad \text{or} \quad \theta_0 = \cos^{-1}\left(1 - \frac{H}{R}\right) \tag{a}$$

We also define a Cartesian coordinate system Bxy such that the x-axis lies on the ground.

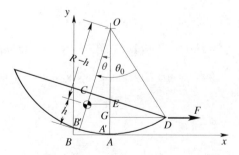

Figure 1.42   Generalized coordinate and geometry of a cylinder segment

- *Generalized Coordinates*: We define $\{\theta\}$ as a complete and independent set of generalized coordinates, where θ is the angle measured from the vertical to line OC, as shown in Fig. 1.42, positive in the clockwise direction.
- *Admissible Variations*: We verify that $\{\delta\theta\}$ is a complete and independent set of admissible variations in this set of generalized coordinates.
- *Holonomicity*: We conclude that the system is holonomic and has one degree of freedom.
- *Generalized Forces*: The virtual work by the applied force is expressible as

$$\delta W^{\text{n.c.}} = F\delta x_D \tag{b}$$

where δx_D is the horizontal virtual displacement at point D. To find δx_D, we first write the x-coordinate of point D in terms of the generalized coordinate θ. From the geometry in Fig. 1.42, because of the nonslip condition, $\overline{BA} = \overparen{B'A} = R\theta$. Then,

$$x_D - \overline{BA} + \overline{GD} = R\theta + R\sin(\theta_0 - \theta) \tag{c}$$

where θ_0 is given in eqn. (a). Then, δx_D can be determined by treating δ like a differential operator as

$$\delta x_D = R\delta\theta - R\cos(\theta_0 - \theta)\delta\theta = R\left[1 - \cos(\theta_0 - \theta)\right]\delta\theta \tag{d}$$

and, substituting eqn. (d) into eqn. (b) gives

$$\delta W^{\text{n.c.}} = FR\left[1 - \cos(\theta_0 - \theta)\right]\delta\theta \tag{e}$$

Then, according to eqn. (1.2),

$$\Xi_\theta = FR\left[1 - \cos(\theta_0 - \theta)\right] \tag{f}$$

- *Kinetic Energy*: There is no apparent fixed point in the cylinder segment, we will use the general expression for the kinetic energy as in eqn. (A.27). The angular velocity of the cylinder segment is $\dot{\theta}$. To find the velocity of the centroid C, a convenient approach is to find the coordinates for the centroid C first. According to Fig. 1.42,

$$x_C = \overline{BA} - \overline{EC} = R\theta - (R - h)\sin\theta \tag{g}$$

$$y_C = \overline{AE} = R - \overline{OE} = R - (R - h)\cos\theta \tag{h}$$

Taking a time derivative of the above expressions gives

$$v_C = [R\dot{\theta} - (R - h)\cos\theta\dot{\theta}]i + (R - h)\sin\theta\dot{\theta}j \tag{i}$$

where i and j are unit vectors for the Bxy coordinate system. Then, the kinetic energy for the cylinder segment is

$$\begin{aligned} T &= \frac{1}{2}mv_C^2 + \frac{1}{2}I_C\omega^2 \\ &= \frac{1}{2}m\left\{\left[R\dot{\theta} - (R - h)\cos\theta\dot{\theta}\right]^2 + \left[(R - h)\sin\theta\dot{\theta}\right]^2\right\} + \frac{1}{2}I\dot{\theta}^2 \\ &= \frac{1}{2}\left\{m\left[R^2 + (R - h)^2 - 2R(R - h)\cos\theta\right] + I\right\}\dot{\theta}^2 \end{aligned} \tag{j}$$

- *Potential Energy*: Since we already have the expression for y_C, we can simply choose the ground as the datum for the gravitational potential energy, and

$$V = mgy_C = mg\,[R - (R - h)\cos\,\theta] \tag{k}$$

- *Lagrangian*: Combining eqns. (j) and (k) gives, according to eqn. (1.3),

$$\mathcal{L} = \frac{1}{2}\left\{m\left[R^2 + (R - h)^2 - 2R(R - h)\cos\,\theta\right] + I\right\}\dot{\theta}^2 - mg\,[R - (R - h)\cos\,\theta] \tag{l}$$

- *Lagrange's Equation*: The Lagrange's equation for θ is, according to eqn. (1.4),

$$\frac{d}{dt}\left(\frac{\partial\mathcal{L}}{\partial\dot{\theta}}\right) - \frac{\partial\mathcal{L}}{\partial\theta} = \Xi_\theta \tag{m}$$

Since

$$\frac{\partial\mathcal{L}}{\partial\dot{\theta}} = \left\{m\left[R^2 + (R - h)^2 - 2R(R - h)\cos\,\theta\right] + I\right\}\dot{\theta} \tag{n}$$

and

$$\frac{\partial\mathcal{L}}{\partial\theta} = -mg(R - h)\sin\,\theta, \tag{o}$$

substituting eqns. (n), (o), and (f) into eqn. (m) gives

$$\left\{m\left[R^2 + (R - h)^2 - 2R(R - h)\cos\,\theta\right] + I\right\}\ddot{\theta}$$
$$+ mg(R - h)\sin\,\theta = FR[1 - \cos(\theta_0 - \theta)] \tag{p}$$

Equation (p) is the equation of motion for the system.

1.11   Linearization of Equations of Motion

Mechanical vibration studies the small motions of a system around its equilibrium configuration. Meanwhile we have observed that many equations of motion for different systems are often nonlinear; and solving such nonlinear equations could be extremely difficult, if possible at all. Fortunately, the key qualifying term *small motions* saves our day. By assuming small motions, we could reduce a nonlinear equation into a linear one, for which we could solve rather comfortably. Another key qualifying term *equilibrium configuration* indicates that the small motion assumption should be applied to the system near an equilibrium position. Before linearizing an equation of motion, we need to find the equilibrium position(s) first.

At the beginning of this chapter, we have briefly discussed the equilibrium position and the stability of a simple mass–spring–dashpot system in Example 1.1. In this section, we use another example to illustrate the procedure of finding the equilibrium position(s) and then linearizing the equation(s) of motion.

Consider the following example of an inverted rigid-link pendulum, constrained by two springs as shown in Fig. 1.43. The equation of motion for this system has been derived in Example 1.24.

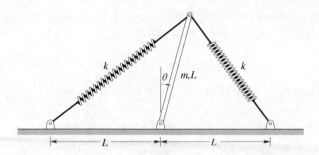

Figure 1.43   Inverted rigid-link planar pendulum

1.11.1   Equilibrium Position(s)

In the equilibrium state, the system does not move, and hence all velocities (including angular velocities) and accelerations (including angular accelerations) vanish. To find the equilibrium position, we set all time derivatives to zero. The resulting equation is called the *equilibrium equation*. Roots of this equation give the *equilibrium positions* of the system.

■ **Example 1.30:   Equilibrium Positions of Inverted Rigid-Link Pendulum**

Find all the equilibrium positions for the inverted rigid-link pendulum in Fig. 1.43.

□ **Solution:**

Using θ, the clockwise angular displacement of the pendulum measured from the vertical, as the generalized coordinate, the equation of motion for the inverted rigid-link pendulum, has been found in Example 1.24 as

$$\frac{1}{3}mL\ddot{\theta} - \frac{1}{2}mg\sin\theta + \sqrt{2}kL\sqrt{1-\cos\theta} = 0 \tag{a}$$

Setting $\ddot{\theta} = 0$ (and $\dot{\theta} = 0$ if this term also presents), eqn. (a) becomes the following *equilibrium equation*:

$$-mg\sin\theta + 2\sqrt{2}kL\sqrt{1-\cos\theta} = 0 \tag{b}$$

or

$$\sin\theta = \frac{2\sqrt{2}kL}{mg}\sqrt{1-\cos\theta} \tag{c}$$

To find the root(s) of this equation, we square both sides of eqn. (c) and make use of the following trigonometry identity $\sin^2\theta = 1 - \cos^2\theta = (1+\cos\theta)(1-\cos\theta)$,

$$(1+\cos\theta)(1-\cos\theta) = 8\left(\frac{kL}{mg}\right)^2(1-\cos\theta) \tag{d}$$

We shall exercise caution not to hastily cancel out the common factor $(1-\cos\theta)$ without a reason. Instead, we rewrite eqn. (d) as

$$(1-\cos\theta)\left[1+\cos\theta - 8\left(\frac{kL}{mg}\right)^2\right] = 0 \tag{e}$$

A root to this equation would make either of the two factors to vanish. That is,

$$\cos \theta = 1 \qquad \text{or} \qquad \cos \theta = 8\left(\frac{kL}{mg}\right)^2 - 1$$

Furthermore, when taking the inverse of the cosine function, we need to be cautious again not to miss the negative solution. In all, the system has three equilibrium positions:

$$\theta_1^{eq} = 0 \qquad \text{and} \qquad \theta_{2,3}^{eq} = \pm\cos^{-1}\left[8\left(\frac{KL}{mg}\right)^2 - 1\right] \tag{f}$$

□ **Discussion:**

We now can see that, had we carelessly cancel the common factor $(1 - \cos \theta)$, we would have missed one important (and actually obvious) equilibrium position $\theta = 0$.

The equilibrium positions 2 and 3 exist only when

$$0 \le 8\left(\frac{kL}{mg}\right)^2 - 1 \le 1 \tag{g}$$

This requirement comes from the geometrically admissible configuration of $-\pi/2 \le \theta \le \pi/2$, which gives $0 \le \cos \theta \le 1$. Rearranged, the above condition can be written as

$$\frac{\sqrt{2}}{2} \le \frac{2kL}{mg} \le 1 \tag{h}$$

Physically, this condition means: if the spring is either too strong or too weak, the system has only one equilibrium position that is vertically inverted. We will gain further insight as what would happen if the spring is too weak or too strong once we have obtained a linearized the equation of motion.

1.11.2   Linearization

Example 1.30 shows that a system may have multiple equilibrium positions. In such cases, linearization must be performed for a specific equilibrium position, and the resulting linearized equations of motion are generally different for different equilibrium positions. In the following, we linearize the equation of motion for each equilibrium position.

When we consider small motions around an equilibrium position, we can make the following variable substitution:

$$x = x^{eq} + y \tag{1.5}$$

where x is the generalized coordinate used in deriving the equation of motion, x^{eq} is the equilibrium position, and y represents the small motions. From eqn. (1.5), since x^{eq} is a constant, it is obvious that

$$\dot{x} = \dot{y} \qquad \ddot{x} = \ddot{y} \tag{1.6}$$

"Small" is a comparative term that should be restricted to nondimensional or normalized parameters. For example, a linear displacement can be normalized by a representative length in the system. This way, the displacement is said to be small compared to that length.

Mathematically, y being small is written as $y \ll 1$. It is generally also implied that the associated velocity and the acceleration are small, that is, $\dot{y} \ll 1$ and $\ddot{y} \ll 1$.

The fundamental idea in a linearization process is to keep only up to the first order small terms. Terms that are explicitly higher order small, such as y^2 or $y\dot{y}$, are simply dropped; and terms that are nonlinear but not explicitly higher order small are expanded in power series using Taylor expansions first and then keep only up to the first order small terms. Recall that the Taylor expansion for a general function $f(x)$ near the point $x = x^{eq}$ can be written as

$$f(x) = f(x^{eq}) + f'(x)|_{x=x^{eq}} y + \cdots \tag{1.7}$$

In the following, we illustrate the linearization process with two examples.

■ Example 1.31: Linearize the Equation of Motion for Equilibrium Position 1

Linearize the equation of motion for the inverted rigid-link pendulum for the equilibrium position $\theta_1^{eq} = 0$.

□ Solution:

We use ϕ to represent the "small motions" near this equilibrium position. The generalized coordinate θ is an angle, thus "small" means a small angle, which is a nondimensional parameter of the unit of radians. The change of variable is as the following:

$$\theta = \theta_1^{eq} + \phi \tag{a}$$

where ϕ is small. Since $\theta_1^{eq} = 0$, this quickly leads to $\phi = \theta$, $\dot{\phi} = \dot{\theta}$, and $\ddot{\phi} = \ddot{\theta}$.

To linearize the equation of motion in eqn. (a) in Example 1.30, we use Taylor expansion to expand the nonlinear terms in the equation of motion near $\theta = 0$. These terms include $\sin \theta$ and $\sqrt{1 - \cos \theta}$. We write their Taylor expansions as

$$\sin \theta \approx \sin \theta|_{\theta=0} + \cos \theta|_{\theta=0} \phi = \phi$$

$$\sqrt{1 - \cos \theta} \approx \sqrt{1 - \cos \theta}\Big|_{\theta=0} + \left(\sqrt{1 - \cos \theta}\right)'\Big|_{\theta=0} \phi = \frac{1}{2}\frac{\sin \theta}{\sqrt{1 - \cos \theta}}\Big|_{\theta=0} \phi$$

Note that $\sin \theta / \sqrt{1 - \cos \theta}$ would become $0/0$ at $\theta = 0$. This can be evaluated using the L'Hôpital's rule or replacing $\sin \theta$ by $\sqrt{(1 - \cos \theta)(1 + \cos \theta)}$. Using the latter,

$$\sqrt{1 - \cos \theta} \approx \frac{1}{2}\sqrt{1 + \cos \theta}\Big|_{\theta=0} \phi = \frac{\sqrt{2}}{2}\phi$$

Then, substituting the above expansions into the equation of motion in eqn. (a) in Example 1.30 while keeping up to the first-order terms gives

$$2mL\ddot{\phi} - 3mg\phi + 6\sqrt{2}kL\frac{\sqrt{2}}{2}\phi = 0 \tag{b}$$

or

$$\ddot{\phi} + \frac{3g}{2L}\left(\frac{2kL}{mg} - 1\right)\phi = 0 \tag{c}$$

which is the linearized equation of motion for small motions of the inverted rigid-link pendulum near its vertical equilibrium position.

◻ **Discussion:**

In the cases where $\theta^{eq} = 0$, we can directly assume θ is small and proceed with the analysis, without introducing a new variable ϕ. In fact, this is a very common scenario as we often define the generalized coordinates to be measure from an equilibrium position.

■ **Example 1.32: Linearize Equation of Motion for Equilibrium Positions 2 and 3**

Linearize the equation of motion for the inverted rigid-link pendulum about its equilibrium positions at

$$\theta_{2,3}^{eq} = \pm \cos^{-1}\left[8\left(\frac{kL}{mg}\right)^2 - 1\right]$$

◻ **Solution:**

We again use ϕ to represent the small motions near the equilibrium position, as

$$\theta = \theta^{eq} + \phi \tag{a}$$

where ϕ is small and θ^{eq} denotes either θ_2^{eq} or θ_3^{eq}, and

$$\cos \theta^{eq} = 8\left(\frac{kL}{mg}\right)^2 - 1 \tag{b}$$

Taylor expansions for $\sin \theta$ and $\sqrt{1 - \cos \theta}$ about θ^{eq} are

$$\sin \theta \approx \sin \theta^{eq} + \cos \theta^{eq} \phi \tag{c}$$

$$\sqrt{1 - \cos \theta} \approx \sqrt{1 - \cos \theta^{eq}} + \frac{1}{2}\sqrt{1 + \cos \theta^{eq}} \phi \tag{d}$$

Substituting eqns. (c) and (d) into the equation of motion in eqn. (a) in Example 1.30 gives

$$2mL\ddot{\phi} - 3mg\,(\sin \theta^{eq} + \cos \theta^{eq}\phi) + 6\sqrt{2}kL\left[\sqrt{1 - \cos \theta^{eq}} + \frac{1}{2}\sqrt{1 + \cos \theta^{eq}}\phi\right] = 0 \tag{e}$$

Since the constant terms (the first terms in the Taylor expansions in eqns. (c) and (d)) cancel out according to the equilibrium equation as given in eqn. (b) in Example 1.30, eqn. (e) simplifies to

$$2mL\ddot{\phi} + \left[-3mg\cos \theta^{eq} + 6\sqrt{2}kL\frac{1}{2}\sqrt{1 + \cos \theta^{eq}}\right]\phi = 0 \tag{f}$$

Substituting the expression for $\cos \theta^{eq}$ in eqn. (b) into eqn. (f) gives

$$2mL\ddot{\phi} + \left\{-3mg\left[8\left(\frac{kL}{mg}\right)^2 - 1\right] + 3\sqrt{2}kL\sqrt{8\frac{kL}{mg}}\right\}\phi = 0 \tag{g}$$

or

$$\ddot{\phi} + \frac{3g}{2L}\left[1 - \left(\frac{2kL}{mg}\right)^2\right]\phi = 0 \tag{h}$$

which is the linearized equation of motion for small motions of the inverted rigid-link pendulum near its two slanted equilibrium positions. Note that this linearized equation is different from the one for the vertically inverted equilibrium position in Example 1.31.

1.11.3   Observations and Further Discussions

Through this series of examples, we make the following observations:

- A system may have multiple equilibrium positions, and the linearized equation of motion is generally different for each equilibrium position. When finding the equilibrium position(s) for a system, make sure to find all possible roots to the equilibrium equation.
- The key for the linearization is using Taylor expansions while keeping only up to linear terms to approximate nonlinear terms. All higher-order terms are omitted.
- When the Taylor expansions are substituted into the nonlinear equation of motion, the zeroth-order terms satisfy the equilibrium equation and cancel out.
- The word *small* in "small motions" is a relative term. In order to assume a parameter being small, that parameter must be nondimensionalized or normalized.

1.11.3.1   Stability of an Equilibrium Position

At each equilibrium position, if the system is slightly disturbed, will the system come back to the equilibrium? This is the test for the stability of the equilibrium. If it comes back, the equilibrium is said to be *stable* and a vibration ensues, which is what we will study. If it does not come back, the equilibrium is said to be *unstable*.

A linearized equation of motion presents a clear indicator about the stability of the equilibrium. For many systems, like Examples 1.31 and 1.32, the linearized equation of motion is of the following form:

$$\ddot{x} + \lambda x = 0 \tag{1.8}$$

From differential equation theories, the solution for eqn. (1.8) is, in complex parameters, $e^{\pm i\sqrt{\lambda}t}$. If λ is positive, the solution can be alternatively written as $\sin\sqrt{\lambda}t$ or $\cos\sqrt{\lambda}t$: such a solution is oscillatory. However, if $\lambda < 0$, the solution would be written as $e^{\pm\sqrt{|\lambda|}t}$. In such a case, the component of $e^{-\sqrt{|\lambda|}t}$ will decade with time, but the component of $e^{\sqrt{|\lambda|}t}$ will grow exponentially with time. This means that the equilibrium is unstable. Vibration only occurs around a stable equilibrium position. Thus, we can conclude that if the resulting linearized equation of motion is of the form in eqn. (1.8), the *equilibrium is stable only when $\lambda > 0$*.

In a more general case, a linear second-order differential equation is of the form:

$$\ddot{x} + \gamma\dot{x} + \lambda x = 0 \tag{1.9}$$

It can be shown that the equilibrium is stable if and only if $\gamma \geq 0$ and $\lambda > 0$.

■ Example 1.33: Stability of Inverted Rigid-Link Pendulum

Determine the stabilities of the three equilibrium positions of the inverted rigid-link pendulum found in Example 1.30.

□ Solution:

For equilibrium position 1, the linearized equation of motion found in Example 1.31 is

$$\ddot{\phi} + \frac{3g}{2L}\left(\frac{2kL}{mg} - 1\right)\phi = 0 \tag{a}$$

The equilibrium is stable only when

$$\frac{2kL}{mg} - 1 > 0 \qquad \text{or} \qquad 2kL > mg \tag{b}$$

For equilibrium positions 2 and 3, the linearized equation of motion found is

$$\ddot{\phi} + \frac{3g}{2L}\left[1 - \left(\frac{2kL}{mg}\right)^2\right]\phi = 0 \tag{c}$$

The equilibrium is stable only when

$$1 - \left(\frac{2kL}{mg}\right)^2 > 0 \qquad \text{or} \qquad 2kL < mg \tag{d}$$

Thus, when the spring is sufficiently strong such that $2kL > mg$, equilibrium position 1 is stable and equilibrium positions 2 and 3 do not exist. For a weaker spring, equilibrium position 1 become unstable and equilibrium positions 2 and 3 appear and are stable. According to eqn. (h) in Example 1.30, the weakest spring for equilibrium positions 2 and 3 to exist is $2kL > mg/\sqrt{2}$.

1.11.3.2 Steady Position versus Equilibrium Position

Some systems may undergo steady motions, especially steady rotations, as its regular operation mode while vibration problems arise. There are a few such systems described in the end-of-chapter problems, for example, a pendulum encased in a container and placed on a steady-rotating platform in Problem 1.14. When the rotation of the platform steadies, the pendulum stays in an inclined position if there is no disturbance. Such a position is called a *steady position*.

The process of finding the steady positions is exactly the same as finding the equilibrium position, except that we need to be observant that angular velocity of the rotating platform is constant.

In this book, steady positions are categorically referred to as equilibrium positions, in which equilibrium is a relative sense: relative to a rotating base.

1.12 Chapter Summary

The procedure for deriving the equation(s) of motion for mechanical systems has been summarized in Section 1.9. The step of crucial importance is defining a set of generalized coordinates and subsequently conducting tests to ensure that it is a complete and independent set. If this is not carefully tested, we might end up with a wrong number of degrees of freedom for the system. Then, everything that follows that conclusion would be wrong.

The following are a few other important observations we have made:

- For a holonomic system, the number of equations of motion equals to the number of generalized coordinates and, in turn, equals to the number of degrees of freedom.
- When generalized coordinates are defined from an equilibrium configuration, the constants in Lagrange's equation satisfy the equilibrium condition and cancel out.

- For small motions, the mathematics could be simplified if the "small motion" assumption is incorporated earlier. Specifically, in the Lagrangian, we need to keep up to second-order small terms. In generalized forces, we only need to keep up to the first-order terms.
- A system may have multiple equilibrium positions. These positions can be found from the equation(s) of motion by setting all time derivatives (velocities and accelerations) of the generalized coordinates to zero.
- Linearization of the equation(s) of motion is specific to an equilibrium position.
- When the linearized equation of motion is normalized to the following form:

$$\ddot{x} + \gamma \dot{x} + \lambda x = 0$$

The equilibrium position is stable if and only if $\gamma \geq 0$ and $\lambda > 0$.

Problems

Problem 1.1: A planar pendulum is made of a single massless rigid link with two attached masses. The first mass m_1 is located at a distance a from the pivot point, and the second mass m_2 is at a distance b from the first mass, as shown in Fig. P1.1. Derive the equation(s) of motion for the system.

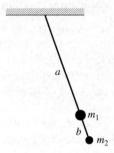

Figure P1.1 Two masses attached to a massless rigid-link pendulum

Problem 1.2: A particle of mass m can slide frictionlessly along a rigid fixed wire, whose shape is given by the equation $y = a + bx^2$, where both a and b are positive constants, as shown in Fig. P1.2. Derive the equation(s) of motion for the system.

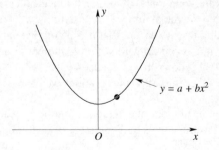

Figure P1.2 A particle slides along a wire of known shape

Problem 1.3: The rigid wire in Problem 1.2 is rotating about the *y*-axis at a constant angular velocity of Ω. Derive the equation(s) of motion for the system.

Problem 1.4: A particle of mass *m* can slide frictionlessly along a circular ring of radius *R* that is rotating about its vertical diameter at an angular velocity of Ω, as shown in Fig. P1.4. Note that a vertical stem holding the ring would prevent the particle from passing through it. Derive the equation(s) of motion for the system.

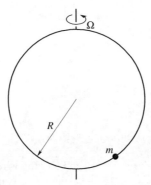

Figure P1.4   A particle slides along a circular ring that rotates about vertical axis

Problem 1.5: A particle of mass *m* slides frictionlessly inside a circular track of radius *R*, as shown in Fig. P1.5. Derive the equation(s) of motion for the system.

Figure P1.5   Particle slides frictionlessly inside a circular track

Problem 1.6: A pendulum of mass *m* and length *l* is attached to a mass *M*, which is confined to move within a vertical slot. The mass *M* is restrained by a spring *k* and a dashpot *c*, as shown in Fig. P1.6. Derive the equation(s) of motion for the system.

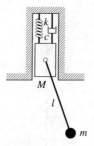

Figure P1.6   Pendulum attached to mass confined in a vertical slot

Problem 1.7: A *metronome*, also called a *rhythm timer* or a *music timer*, is modeled as comprising a massless hand (rigid link) pivoted at its root and a mass m near its tip. The hand is restrained by a torsional spring of spring constant k_t. The mass is located at a distance h from the pivot point, as shown in Fig. P1.7. When the hand is in its vertically upright position, the torsional spring is undeformed. Derive the equation(s) of motion for the system.

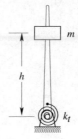

Figure P1.7   Metronome with mass attached to its hand

Problem 1.8: A planar pendulum of mass m and length l is restrained by a dashpot as shown in Fig. P1.8. The dashpot connected to a vertical wall is maintained in the horizontal orientation at all times. Derive the equation(s) of motion for the system.

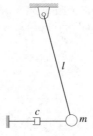

Figure P1.8   Pendulum constrained by a horizontal dashpot

Problem 1.9: A massless rigid link of length $2L$ is pivoted vertically at its center, with two masses, $2m$ and m, mounted at its two ends. The link can only move within the plane of paper and is further restrained by spring k and a dashpot c, which are attached at a distance a from the pivot, as shown in Fig. P1.9. When the link is in the vertical orientation, the spring and the dashpot are horizontal, the spring is undeformed, and of length l_0. Derive the equation(s) of motion for the system.

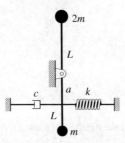

Figure P1.9   Masses mounted on rigid link and restrained by spring and dashpot

Problem 1.10: A mass m can move frictionlessly on the inclined surface of a cart, of mass M, which can move frictionlessly on the horizontal floor. Mass m is restrained by spring k and dashpot c, as shown in Fig. P1.10. The surface's incline angle is θ. Derive the equation(s) of motion for the system.

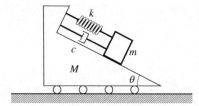

Figure P1.10   Mass on the inclined surface of a cart

Problem 1.11: A planar pendulum of mass m and length l is mounted on an L-shaped frame of mass M, which can move frictionlessly on the horizontal floor, as shown in Fig. P1.11. Assume that the pendulum never touches the frame, but its motion is not necessarily small. Derive the equation(s) of motion for the system.

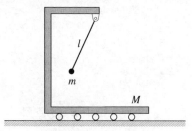

Figure P1.11   Planar pendulum mounted on a moving frame

Problem 1.12: The inextensible massless cable of a pendulum slides frictionlessly through a hole. Its upper end is held by a force $F(t)$, and the lower end attaches to a mass m, as shown in Fig. P1.12. Assume the pendulum is confined to move within the plane of paper, and the cable remains taut at all times. Derive the equation(s) of motion for the system.

Figure P1.12   Planar pendulum with force exerted at the end of its cable

Problem 1.13: The inextensible massless cable of a planar pendulum of mass m passes through a pivoting hole, wraps around a massless pulley, and then connects to a spring k, which in turn is attached to a wall, as shown in Fig. P1.13. At the equilibrium configuration, the length of the cable below pivoting hole is l. The unstretched length of the spring is l_0. Derive the equation(s) of motion for the system.

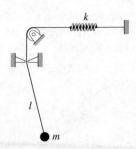

Figure P1.13   Planar pendulum with an inextensible cable and a spring

Problem 1.14: A planar pendulum of mass m and length l is mounted on an L-shaped frame that rests on a rotating platform, which rotates at a constant angular velocity of Ω. The pivot point is located at a distance e from the axis of rotation, as shown in Fig. P1.14. Assume that the pendulum never touches the frame, but its motion is not necessarily small. Derive the equation(s) of motion for the system.

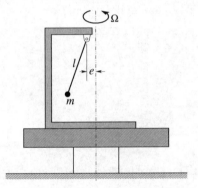

Figure P1.14   Planar pendulum mounted on a rotating frame

Problem 1.15: A mass m moves within a slot of length L located along a diameter of a horizontal rotating platform. It is further restrained by spring k and dashpot c, as shown in Fig. P1.15. The spring is unstretched when the mass is located at the center of the slot. The platform rotates at a constant angular velocity Ω. Derive the equation(s) of motion for the system.

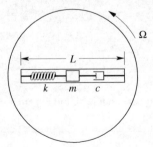

Figure P1.15   Mass–spring–dashpot system inside diametrical slot in a rotating platform

Problem 1.16: A mass m moves within a slot, of length L, in a horizontal rotating platform. It is further restrained by spring k and dashpot c, as shown in Fig. P1.16. The center line of the slot is at a distance a from the center of the platform, which rotates at a constant angular velocity Ω. The spring is unstretched when the mass is located at the center of the slot. Derive the equation(s) of motion for the system.

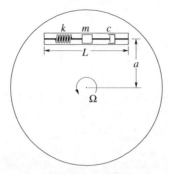

Figure P1.16   Mass–spring–dashpot system inside off-center slot in a rotating platform

Problem 1.17: Two masses m_1 and m_2 are attached to the ends of a rod that is rotating about the pivot at a constant angular velocity Ω. The collar of mass M, which houses the pivot, can slide frictionlessly along the vertical column but is restrained by two identical springs of spring constant k, as shown in Fig. P1.17. The masses are a distance r from the rotation center. Both springs have unstretched length of l_0, which is larger than the length of the entire column such that they are always compressed. Derive the equation(s) of motion for the system.

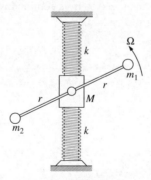

Figure P1.17   Masses attached to the ends of a rotating arm

Problem 1.18: A mass m is restrained by four identical springs, each of a spring constant k, between two perpendicular pairs of walls facing each other, as shown in Fig. P1.18. Assume that the distance between each pair of facing walls is $2L$ and the unstretched length of the springs is l_0 ($l_0 < L$). Consider only small motions of the mass within the horizontal plane of paper. Derive the equation(s) of motion for the system.

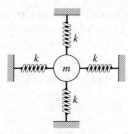

Figure P1.18   Mass restrained by four identical springs

Problem 1.19: A uniform rigid slender rod of mass m and length L is suspended from the ceiling by two identical inextensible cables of length l. It is free to swing within the plane of paper, as shown in Fig. P1.19. The distance between two suspension points is also L. Both cables remain taut at all times. Derive the equation(s) of motion for the system.

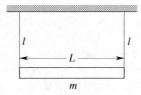

Figure P1.19   A rigid slender rod as a planar pendulum

Problem 1.20: A composite rigid-body pendulum is made of a uniform rigid slender rod of mass m and length L rigidly attached to a uniform circular disk, of mass M and radius r, as shown in Fig. P1.20. Assume the system is confined within the plane of paper. Derive the equation(s) of motion for the system.

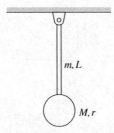

Figure P1.20   Composite rigid-body pendulum

Problem 1.21: A uniform rigid slender rod of mass m and length l has one of its ends attached to an inextensible massless string, also of length l, which, in turn, is attached to the ceiling, as shown in Fig. P1.21. Assume the system is confined to swing in the plane of paper. Derive the equation(s) of motion for the system.

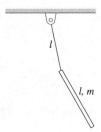

Figure P1.21   Another rigid slender rod as a planar pendulum

Problem 1.22: A planar double rigid-link pendulum comprises two identical uniform rigid links of mass m and length L joined in series and pivoted at one end, as shown in Fig. P1.22. They are confined within the plane of paper. Both the joint and the pivot are frictionless. A horizontal force $F(t)$ acts on the joint. Derive the equation(s) of motion for the system.

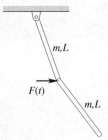

Figure P1.22   Planar double rigid-link pendulum

Problem 1.23: A U-shaped tube is filled with an inviscid and incompressible fluid of mass density ρ, as shown in Fig. P1.23. Its both ends are open to the atmosphere. At equilibrium, the fluid columns in both branches are of a height h, measured from its semicircular bottom of radius R, $h > R$. The tube has a uniform cross-sectional area A throughout. Derive the equation(s) of motion for the system.

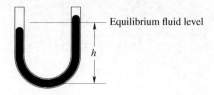

Figure P1.23   Fluid-filled U-shaped tube

Problem 1.24: A uniform rigid slender rod of mass M and length L is supported at its two ends by two springs k_1 and k_2, respectively. A mass m is attached to the center of the rod via spring k_3 and dashpot c, as shown in Fig. P1.24. Consider only small motions in the vertical direction. Derive the equation(s) of motion for the system.

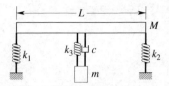

Figure P1.24   A mass attached to the center of a rod

Problem 1.25: A uniform rigid slender rod of mass m and length L is supported at its two ends by four springs and four dashpots, as shown in Fig. P1.25. Consider only small motions in the vertical direction. Derive the equation(s) of motion for the system.

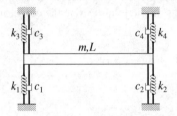

Figure P1.25   Rod supported by springs and dashpots at the ends

Problem 1.26: A uniform rigid slender rod of mass M is pivoted at its left end. Its right end rests on a spring k_1 and dashpot c_1. It is also attached to a mass m through a pulley and another spring k_2 and dashpot c_2, as shown in Fig. P1.26. The pulley can be treated as a uniform circular disk of mass m_0 and radius r. Assume small motions of the right end of the rod. Derive the equation(s) of motion for the system.

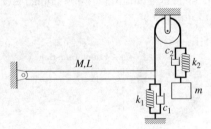

Figure P1.26   Mass attached to the end of pivoted rod via pulley

Problem 1.27: A uniform rigid slender rod of mass m_1 is pivoted at its left end. It is further supported by a spring k_1 and a dashpot c_1 at a distance a from the pivot point. A mass m_2 is attached to its right end, which is at a distance b from the supporting spring and dashpot, through another spring k_2 and another dashpot c_2, as shown in Fig. P1.27. Assume small motions. Derive the equation(s) of motion for the system.

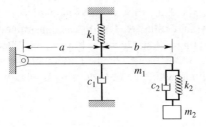

Figure P1.27 Mass attached to the end of pivoted rod

Problem 1.28: A circular rigid cylinder of mass m and radius r_1 can roll without slip on an inclined surface, of inclining angle θ. It is restrained by spring k whose end is wrapped around the core of the cylinder at radius r_2 and a dashpot is connected to its center, as shown in Fig. P1.28. The cylinder has a centroidal moment of inertia I. Derive the equation(s) of motion for the system.

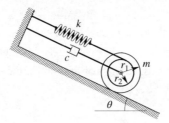

Figure P1.28 Cylinder restrained by spring and dashpot on an inclined surface

Problem 1.29: A uniform circular rigid cylinder of mass m and radius r is confined by spring k and dashpot c at its center on the inclined surface of a cart of mass M, as shown in Fig. P1.29. The cart can move frictionlessly on the horizontal floor, while the cylinder can roll without slip on the inclined surface. The surface's inclined angle is θ. Derive the equation(s) of motion for the system for the following two scenarios:

(a) A horizontal force $F(t) = F_0 \sin \Omega t$ acts on the cart, pointing to the right.
(b) The cart moves according to $x(t) = X_0 \sin \Omega t$, where x is the location of the left side of the cart measured from a reference configuration, which is positive to the right.

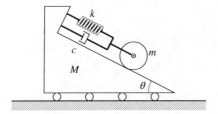

Figure P1.29 Mass restrained by spring and dashpot on the inclined surface of a cart

Problem 1.30: A uniform rigid slender rod of mass m and length L is pinned to the rim of a circular disk of mass M and radius R. The disk is constrained by spring k that connects the center of the disk to the wall, as shown in Fig. P1.30. When the spring is unstretched, the pin that connects the rod is located on the horizontal diameter. The disk rolls without slip on the horizontal floor, while the rod slides frictionlessly. Derive the equation(s) of motion for the system.

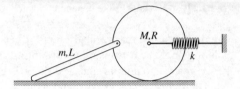

Figure P1.30   Slender rod pinned to the rim of a circular disk

Problem 1.31: A uniform semicircular rigid cylinder of mass m and radius r rolls without slip on the horizontal ground, as shown in Fig. P1.31. The centroid is located at a distance $a = \frac{4r}{3\pi}$ below the flat surface of the cylinder, and the centroidal moment of inertia is $I_C = m\left(\frac{r^2}{2} - a^2\right)$. Derive the equation(s) of motion for the system.

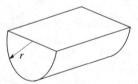

Figure P1.31   Semicircular cylinder rolls without slip on the horizontal ground

Problem 1.32: A uniform rigid slender rod of mass m and length L can slide frictionlessly inside a circular track of radius R, as shown in Fig. P1.32. The rod is confined to move within the plane of paper. Derive the equation(s) of motion for the system.

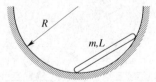

Figure P1.32   Slender rod slides inside a circular track

Problem 1.33: A uniform rigid slender rod of mass m and length L can roll without slip on top of a fixed circular cylindrical surface of radius R. At equilibrium, the rod is horizontal, as shown in Fig. P1.33. Derive the equation(s) of motion for the system.

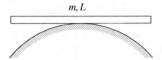

Figure P1.33   Rigid slender rod rolls without slip on a circular cylindrical surface

Problem 1.34: A uniform rigid circular disk of radius r and mass m_2 can roll without slip inside of a circular track of radius R. The disk is restrained by a rigid slender rod of mass m_1 to the center of the track, as shown in Fig. P1.34. All pin-connections are frictionless. Derive the equation(s) of motion for the system.

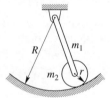

Figure P1.34   Restrained disk rolls without slip inside a circular track

Problem 1.35: Two identical circular disks of mass m and radius r are connected by a uniform rigid link of mass m_0 and length l. The disks roll without slip on a circular track of radius R, as shown in Fig. P1.35. Assume that the entire assembly moves in a vertical plane and remains below the center of the track; and all pin-connections are frictionless. Derive the equation(s) of motion for the system.

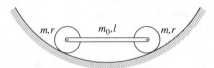

Figure P1.35   Circular disks connected by rod roll without slip on a circular track

Problem 1.36: Two identical circular disks of mass m and radius r are connected by a spring k and a dashpot c. The disks roll without slip on a circular track of radius R, as shown in Fig. P1.36. Assume that both disks move in a vertical plane and remain entirely below the center of the track at all times. The unstretched length of the spring is l_0. Derive the equation(s) of motion for the system.

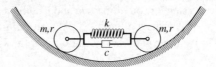

Figure P1.36   Disks connected by spring and dashpot roll without slip on a circular track

Problem 1.37: Two identical circular disks of mass m and radius r are pin-connected by a uniform rigid link of mass m_0 and length l. The assembly straddles on a fixed circular cylindrical surface of radius R, and is confined to move within the plane of paper, as shown in Fig. P1.37. The disks can roll without slip on the cylindrical surface. Derive the equation(s) of motion for the system.

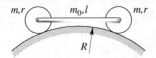

Figure P1.37   Disks connected by rod straddle on a circular cylindrical surface

Problem 1.38: Two identical circular disks of mass m and radius r are connected by a spring k. The assembly straddles on a fixed circular cylindrical surface of radius R and is confined to move within the plane of paper, as shown in Fig. P1.38. The disks can roll without slip on the cylindrical surface. The unstretched length of the spring is l_0. Derive the equation(s) of motion for the system.

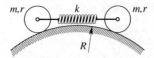

Figure P1.38   Disks connected by spring straddle on a circular cylindrical surface

Problem 1.39: Two uniform rigid slender beams of masses M and m_0, respectively, are pin-connected via two rigid links at their ends, as shown in Fig. P1.39. Both beams have a length of L; and both links have a length of l and mass m_1. The beam of mass M can slide frictionlessly on the horizontal surface. All pin-connections are frictionless. Derive the equation(s) of motion for the system.

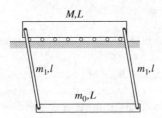

Figure P1.39   Two beams connected by two identical rigid links

Problem 1.40: A uniform rigid slender beam of mass M and length L is attached to two carts via rigid links at its ends. Two links are identical, and both are of mass m_1 and length l. The two carts are identical, and both are of mass m and can slide frictionlessly on the horizontal floor, as shown in Fig. P1.40. The carts are connected by spring k and dashpot c. The cart on the left is subjected to a prescribed motion $x_0(t)$ with respect to the ground. Assume small motions. Derive the equation(s) of motion for the system.

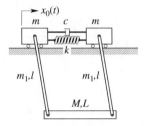

Figure P1.40   Uniform rigid slender beam connected to two carts via rigid links

Problem 1.41: Two identical circular rigid disks of mass M and radius R are pin-connected by a rigid link of mass m and length l. At equilibrium, the link is horizontal, and the pins are located at a distance a vertically below the centers, as shown in Fig. P1.41. A horizontal force $F(t)$ acts on the rim of a disk at the same level as the centers of the disks. Both pin-connections are frictionless; and both disks roll without slip on the horizontal ground. Derive the equation(s) of motion for the system.

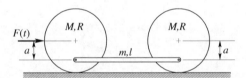

Figure P1.41   Two disks linked by rigid link roll without slip on the ground

Problem 1.42: A roller comprises a circular cylindrical core of mass m_2 and radius r_2, which is shrink-fitted into a circular annulus of outer radius r_1 and mass m_1. The protruded core rolls without slip on a circular track of radius R, as shown in Fig. P1.42. Derive the equation(s) of motion for the system.

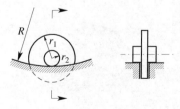

Figure P1.42   Roller's cylindrical core rolls without slip on a circular track

Problem 1.43: A rigid rocker having a concave circular cylindrical surface of radius R can roll without slip on a circular cylindrical support, of radius $r < R$, as shown in Fig. P1.43. The rocker has a mass m and a centroidal moment of inertia I. At equilibrium, the center of mass of the rocker is located vertically at a distance a above the contact point. Derive the equation(s) of motion for the system.

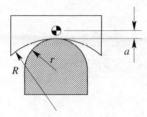

Figure P1.43 Rocker rolls without slip on a circular support

Problem 1.44: A uniform circular rigid cylinder of mass m_1 and radius r_1 rolls without slip on a horizontal surface. A pendulum comprises a circular disk of mass m_2 and radius r_2 and a rigid connection rod of mass m_3 and length L, rigidly joined together. The pendulum is attached to the cylinder as shown in Fig. P1.44. Derive the equation(s) of motion for the system for the following two scenarios:

(a) The connecting rod is welded to the cylinder m_1 rigidly.
(b) The connecting rod is pinned to the cylinder m_1 frictionlessly.

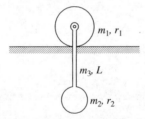

Figure P1.44 Pendulum connected to a circular cylinder that rolls on surface

Problem 1.45: A circular cylinder of radius r rolls without slip on the horizontal surface. The cylinder is made of two uniform semicircular cylinders welded together, as shown in Fig. P1.45. One has an areal mass density of ρ_1 and the other ρ_2 ($\rho_2 > \rho_1$). Derive the equation(s) of motion for the system. (*Hint: Refer to Problem 1.31 for the location of the centroid and the centroidal moment of inertia for a semicircular cylinder.*)

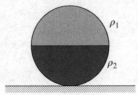

Figure P1.45 Cylinder made of two halves of uniform materials

Problem 1.46: A rotor blade of a helicopter is modeled as a uniform rigid slender rod of mass m and length L. It is pin-joined to a base of diameter a, which is rotating at a constant angular velocity Ω, as shown in Fig. P1.46. The rotor blade is free to "flap" about the pin-joint. Derive the equation(s) of motion for the system.

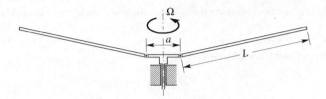

Figure P1.46 Rotor blades on a helicopter

Problem 1.47: A slender rod of mass m and length L is pin-joined to a rotating axle, as shown in Fig. P1.47. The pin-joint is planar, meaning that the rod can only rotate about the pin within a plane that is being rotated, at a given angular velocity Ω. Derive the equation(s) of motion for the system.

Figure P1.47 Rigid slender rod being rotated at an angular velocity Ω

Problem 1.48: Find all the equilibrium positions for the system described in Problem 1.3. Linearize the equation(s) of motion for small motions about each equilibrium position.

Problem 1.49: Find all the equilibrium positions for the system described in Problem 1.4. Linearize the equation(s) of motion for small motions about each equilibrium position.

Problem 1.50: Find all the equilibrium positions for the system described in Problem 1.7. Linearize the equation(s) of motion for small motions about each equilibrium position.

Problem 1.51: Find all the equilibrium positions for the system described in Problem 1.9. Linearize the equation(s) of motion for small motions about each equilibrium position.

Problem 1.52: Find all the equilibrium positions for the system described in Problem 1.14 when the offset $e = 0$. Linearize the equation(s) of motion for small motions about each equilibrium position.

Problem 1.53: Find all the equilibrium positions for the system described in Problem 1.16. Linearize the equation(s) of motion for small motions about each equilibrium position.

Problem 1.54: Find all the equilibrium positions for the system described in Problem 1.46 for the case $a = 0$. Linearize the equation(s) of motion for small motions about each equilibrium position.

Problem 1.55: Find all the equilibrium positions for the system described in Problem 1.47. Linearize the equation(s) of motion for small motions about each equilibrium position.

Problem 1.56: If the linearized equation of motion for a single degree-of-freedom system about an equilibrium position can be written

$$\ddot{x} + \gamma\dot{x} + \lambda x = 0$$

show that the equilibrium position is stable if and only if $\gamma \geq 0$ and $\lambda > 0$.

2

Vibrations of Single-DOF Systems

2.1 Objectives

This chapter presents the basic concepts and theories of vibrations of single-degree-of-freedom (single-DOF) systems. Vibration systems are classified into four types: undamped, underdamped, critically damped, and overdamped. The underdamped systems are the primary focus. Three types of vibrations will be analyzed: (1) the free vibration when the system is subjected to an initial disturbance but otherwise free from any loading; (2) the steady-state response, which is the long-term response, when the system is subjected to a persistent periodic loading; and (3) the transient response when the system is subjected to a time-varying loading.

Also, starting from this chapter, MATLAB, a mathematical software, will be used to explore the solutions for all examples. In engineering practice, having obtained the solution to a problem is not the end to the problem. In the contrary, it is just the beginning of an exploration. Through illustrative examples, the solutions to vibration problems will be further explored by nondimensionalizing the solutions, plotting and validating the response curves, observing the system's behaviors, and following our curiosity to examine other features rising from our observations.

2.2 Types of Vibration Analyses

After having obtained the equation(s) of motion for a system, we can study the motion of the system by finding the solution to the equation(s) of motion and examining the physical meanings of the solution. For a large class of mechanical systems, the motions are oscillatory about an equilibrium configuration of the system. Such oscillatory motions are called vibrations. In this course, we shall limit our study of vibration to small motions of systems about equilibrium configurations. Such small motions are called the *linear vibrations*.

In general, we will be dealing with a class of ordinary differential equations of the following form:

$$a\ddot{x} + b\dot{x} + cx = f(t) \tag{2.1}$$

Fundamentals of Mechanical Vibrations, First Edition. Liang-Wu Cai.
© 2016 John Wiley & Sons, Ltd. Published 2016 by John Wiley & Sons, Ltd.
Companion Website: www.wiley.com/go/cai/fundamentals_mechanical_vibrations

where x is the unknown function of time t; $f(t)$ is a known function of t; and a, b, and c are constant coefficients. In the theory of ordinary differential equations, such an equation is called a *linear second-order constant-coefficient ordinary differential equation*. This is quite a mouthful, but each term has a very specific meaning. *Second order* refers to the highest order of differentiation in the equation. *Constant coefficient* means that all the coefficients in front of the terms containing the unknown x and its derivatives, which include a, b, and c in eqn. (2.1), are constant. Such a system is also called a *time-invariant system*, meaning that the system's characteristics do not change with time. *Linear* means that the left-hand side contains a linear combination of the unknown and its derivatives. *Ordinary* means that the unknown x has only one independent variable. In our study for vibrations, this independent variable is always the time t. Furthermore, if the right-hand side of eqn. (2.1) vanishes, the equation contains only terms involving the unknown x and its derivatives. Such an equation is said to be *homogeneous*. Otherwise, it is *non-homogeneous*.

According to the theory of ordinary differential equations, the total solution to eqn. (2.1) consists of two parts and can be written as:

$$x(t) = x_h(t) + x_p(t) \tag{2.2}$$

where $x_h(t)$ is the *homogeneous solution*, which is the general solution to the corresponding *homogeneous equation*:

$$a\ddot{x} + b\dot{x} + cx = 0 \tag{2.3}$$

and $x_p(t)$ is the *particular solution*, which is a function that satisfies the differential equation in eqn. (2.1).

Our study of the vibrations generally entails the following three types of motions, which we will study in detail one by one in this chapter:

- *Free vibration* refers to the motion of the system when there is no externally applied loading. In such cases, the system moves at its own "will," as opposed to being "forced." The motion is initiated by other means, such as stretching or compressing the spring to a certain length and then releasing it. Such conditions are called the *initial conditions*. Mathematically, a free vibration constitutes a homogeneous solution $x_h(t)$.
- *Steady-state response* refers to the system's long-term motion under a persistent periodic loading. In this type of motion, an external agent forces a motion upon the system, and the system has reached a stage of "submission." Since in this case only the long-term behaviors of the system are of concern, the initial conditions are irrelevant. Mathematically, a steady-state response constitutes a particular solution $x_p(t)$, with an additional implication that the loading is periodic. When the loading is harmonic, that is, the temporal factor is either sine or cosine function, or a linear combination of sine and cosine functions of the same frequency, the analysis of the steady-state response is also called the *harmonic analysis*.
- *Transient response* refers to the system's motion immediately following the application of a load, which often varies with time, to the system. Mathematically, this is the total solution $x(t) = x_h(t) + x_p(t)$.

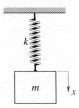

Figure 2.1   Mass–spring undamped single-degree-of-freedom system

2.3   Free Vibrations of Undamped System

Let us first consider the simplest vibration system: a spring–mass system, as shown in Fig. 2.1. This simple system gives us a chance to review the solution procedure for an ordinary differential equation, as well as some basic terminologies of vibration.

This system is similar to the one in Example 1.1 but without the dashpot. Such a system is called an undamped system. We directly adapt the equation of motion obtained in Examples 1.1 and 1.18, where the downward displacement of the mass measured from its equilibrium position, denoted as y, is used as the generalized coordinate. Here we will use x to denote this generalized coordinate. Without an externally applied force ($F = 0$), the equation of motion is

$$m\ddot{x} + kx = 0 \tag{2.4}$$

2.3.1   General Solution for Homogeneous Differential Equation

Here, as a refresher, we review how the general solution for eqn. (2.4) can be found in accordance with the theory of ordinary differential equations.

We assume a solution of the form $x(t) = e^{\lambda t}$, where λ is a constant that may be real or complex. Substituting this solution into the differential equation in eqn. (2.4) gives

$$m\lambda^2 e^{\lambda t} + k e^{\lambda t} = 0 \tag{2.5}$$

Since $e^{\lambda t}$ is not always zero, eliminating $e^{\lambda t}$ from both sides of eqn. (2.5) gives

$$\lambda^2 + \frac{k}{m} = 0 \tag{2.6}$$

which is called the *characteristic equation* of the system. This quadratic equation has two roots as

$$\lambda = \pm i\sqrt{\frac{k}{m}} \tag{2.7}$$

where $i = \sqrt{-1}$ is the unit of imaginary numbers. This means that there are two possible solutions for the differential equation in eqn. (2.4):

$$x_1(t) = e^{i\sqrt{\frac{k}{m}}\,t} \quad \text{and} \quad x_2(t) = e^{-i\sqrt{\frac{k}{m}}\,t}$$

Furthermore, any linear combination of these two possible solutions will still be a possible solution to the equation. Thus, the general solution for eqn. (2.4) is

$$x(t) = C_1 e^{i\omega_n t} + C_2 e^{-i\omega_n t} \tag{2.8}$$

where C_1 and C_2 are arbitrary complex constants, and, to simplify the notation, we have used

$$\omega_n = \sqrt{\frac{k}{m}} \tag{2.9}$$

In eqn. (2.8), the right-hand side comprises complex variables. But we started out from a physical system of a real mass and a real spring, that is, the left-hand side of the equation is a real quantity. What does it mean, when we equate a real quantity to a complex expression? Recall the well-known *Euler formula*:

$$e^{i\theta} = \cos\theta + i\sin\theta \tag{2.10}$$

We can expand the right-hand side of eqn. (2.8) as the following:

$$
\begin{aligned}
x(t) &= \left(\Re\{C_1\} + i\Im\{C_1\}\right)\left(\cos\omega_n t + i\sin\omega_n t\right) \\
&\quad + \left(\Re\{C_2\} + i\Im\{C_2\}\right)\left(\cos\omega_n t - i\sin\omega_n t\right) \\
&= \left(\Re\{C_1\} + \Re\{C_2\}\right)\cos\omega_n t + \left(-\Im\{C_1\} + \Im\{C_2\}\right)\sin\omega_n t \\
&\quad + i\left[\left(\Im\{C_1\} + \Im\{C_2\}\right)\cos\omega_n t + \left(\Re\{C_1\} - \Re\{C_2\}\right)\sin\omega_n t\right] \tag{2.11}
\end{aligned}
$$

where $\Re$ and $\Im$ denote the real and imaginary parts, respectively, of a complex number. Being real on the left-hand side means that the imaginary part of the right-hand side must vanish for all times, that is,

$$\Re\{C_1\} - \Re\{C_2\} = 0 \quad \text{and} \quad \Im\{C_1\} + \Im\{C_2\} = 0 \tag{2.12}$$

or

$$\Re\{C_1\} = \Re\{C_2\} \quad \text{and} \quad \Im\{C_1\} = -\Im\{C_2\} \tag{2.13}$$

Equation (2.13) suggests that C_1 and C_2 are a pair of *complex conjugates*. Consequently, eqn. (2.8) can be reexpressed as

$$x(t) = A\cos\omega_n t + B\sin\omega_n t \tag{2.14}$$

where, from its lineage as being derived from eqn. (2.11),

$$A = \Re\{C_1\} + \Re\{C_2\} = 2\Re\{C_1\} \tag{2.15}$$

$$B = -\Im\{C_1\} + \Im\{C_2\} = -2\Im\{C_1\} \tag{2.16}$$

But, for most cases, we shall simply state that A and B are arbitrary real constants.

The solution in the form of eqn. (2.8) is often called the *complex notation*, and the solution in the form of eqn. (2.14) is called the *real notation*. Sometimes, the complex notation is more convenient. Sometimes, the complex notation is more general if we allow the left-hand side to be complex. In such cases, C_1 and C_2 do not need to be conjugates, and the real and imaginary parts of the solution can be used to represent two similar but separate problems. We generally prefer to stay within the realm of real variables. Hence, eqn. (2.14) is our preferred form of solution.

2.3.2 Basic Vibration Terminologies

Now, let us take a closer look at the general solution in eqn. (2.14).

First, the expression in eqn. (2.14) has other two equally general alternative expressions as

$$x(t) = X \sin(\omega_n t + \phi_s) \tag{2.17}$$

and

$$x(t) = X \cos(\omega_n t - \phi_c) \tag{2.18}$$

These are the most general forms of the *harmonic functions*, in which X is a real and positive number called the *amplitude* and ϕ's are called the *phase angle*. We can trace their relations to the solution in eqn. (2.14) by recalling the following trigonometric identities:

$$\sin(\alpha + \beta) = \sin \alpha \cos \beta + \cos \alpha \sin \beta$$

$$\cos(\alpha + \beta) = \cos \alpha \cos \beta - \sin \alpha \sin \beta$$

If we expand eqns. (2.17) and (2.18) using these identities,

$$x(t) = X \sin \omega_n t \cos \phi_s + X \cos \omega_n t \sin \phi_s \tag{2.19}$$

$$x(t) = X \cos \omega_n t \cos \phi_c + X \sin \omega_n t \sin \phi_c \tag{2.20}$$

We see that they all go back to eqn. (2.14) if we set, for eqn. (2.17),

$$X \sin \phi_s = A \qquad X \cos \phi_s = B$$

and, for eqn. (2.18),

$$X \cos \phi_c = A \qquad X \sin \phi_c = B$$

Conversely,

$$X = \sqrt{A^2 + B^2} \quad \tan \phi_s = \frac{A}{B} \quad \tan \phi_c = \frac{B}{A} \tag{2.21}$$

Second, the two solutions in eqns. (2.17) and (2.18) can be combined to write another complex form as

$$x(t) = X e^{i(\omega_n t + \phi)} \tag{2.22}$$

Note that this complex solution is merely a mathematical construct to unify the two similar solutions in a complex function. It differs from the solution in complex notation in eqn. (2.8) for which the left-hand side is real. In eqn. (2.22), the left-hand side is complex, and we would take either the real or the imaginary part as the solution to a real physical problem.

The relation among the real solutions in eqn. (2.14) and eqns. (2.17) and (2.18) and complex expression in eqn. (2.22) is illustrated in Figure 2.2. On the complex plane, a particle moves along a circle of radius X at a constant angular velocity ω_n. Initially, the particle is located at an angle ϕ. At any time t, the particle is project to the real axis at $X \cos(\omega_n t + \phi)$ and to the imaginary axis at $X \sin(\omega_n t + \phi)$.

We now examine the physical meanings of these terms. As we can see, $\omega_n t$ is—and must be, as required by the exponential function[1]—a dimensionless parameter. This means ω_n has

---

[1] In fact, any mathematical function requires its argument be dimensionless.

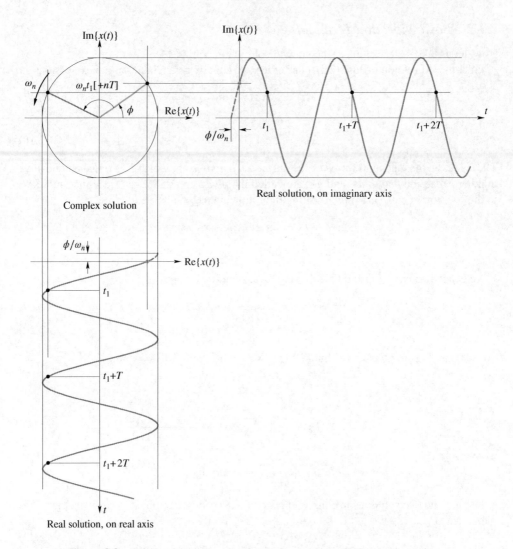

Figure 2.2   Relation between real and complex notations for harmonic motions

a unit of t^{-1}, which has the same unit as the "frequency," a quantity that is often associated with periodic motions. Also, because there is no external force that might have imposed a timing effect onto the system, this is the parameter that controls the timing of the events for the system. For this reason, ω_n is called the *natural frequency* of the system; "natural" means something inherent of the system. For this reason, eqn. (2.6) is also called the *frequency equation*.

The time for the particle to complete a revolution is a *natural period*, denoted as T_n:

$$T_n = \frac{2\pi}{\omega_n} \tag{2.23}$$

The reciprocal of the natural period is called the *natural frequency*, denoted as f_n:

$$f_n = \frac{1}{T_n} \tag{2.24}$$

When it is necessary to distinguish the two frequencies we have defined so far, f_n is called the *Hertzian frequency* as hertz (revolutions per second) is its standard unit; and ω_n is called the *radian frequency* as radians per second is its standard unit (radian is actually dimensionless). Because of the circular motion in the complex plane as depicted in Fig. 2.2, ω_n is also called the *circular frequency*.

2.3.3 Determining Constants via Initial Conditions

The unknown constants C_1 and C_2 in the solution in complex notation in eqn. (2.8) or A and B in the solution in real notation in eqn. (2.14) are determined by the initial conditions. In general, the number of initial conditions that are needed to obtain a solution equals to the order of the differential equation, which is two, which, by no coincidence, equals to the number of unknown constants.

Typically, for a mechanical system, the initial conditions are given as the initial displacement and the initial velocity of the mass, that is,

$$x(0) = x_0 \quad \text{and} \quad \dot{x}(0) = v_0 \tag{2.25}$$

We now use our preferred solution form, the real notation, to determine constants A and B. Taking a time derivative of the solution in eqn. (2.14) gives

$$\dot{x}(t) = -\omega_n A \sin \omega_n t + \omega_n B \cos \omega_n t \tag{2.26}$$

Substituting the initial conditions in eqn. (2.25) into eqns. (2.14) and (2.26) gives

$$x_0 = A \quad \text{and} \quad v_0 = \omega_n B \tag{2.27}$$

Solving the above two equations gives

$$A = x_0 \quad \text{and} \quad B = \frac{v_0}{\omega_n} \tag{2.28}$$

Thus, the response of the system due to the initial conditions in eqn. (2.25) is

$$x(t) = x_0 \cos \omega_n t + \frac{v_0}{\omega_n} \sin \omega_n t \tag{2.29}$$

which can also be written as

$$x(t) = X \cos (\omega_n t - \phi_c) \quad \text{or} \quad x(t) = X \sin (\omega_n t + \phi_s) \tag{2.30}$$

where, according to eqn. (2.21),

$$X = \sqrt{x_0^2 + \left(\frac{v_0}{\omega_n}\right)^2} = \sqrt{x_0^2 + \frac{m v_0^2}{k}} \tag{2.31}$$

and

$$\phi_c = \tan^{-1} \frac{v_0}{x_0 \omega_n} \qquad \phi_s = \tan^{-1} \frac{x_0 \omega_n}{v_0} \tag{2.32}$$

■ **Example 2.1: Initial Compression of Spring in Mass–Spring System**

For the system shown in Fig. 2.1, assume that the mass is held above its equilibrium position by the amount Δ and then released from rest, as shown in Fig. 2.3. Determine the subsequent motion of the mass.

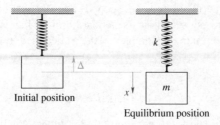

Figure 2.3   Initial spring compression in mass–spring system

□ **Solution:**

Since the generalized coordinate x is the downward displacement measured form the equilibrium position, the initial displacement is negative; and "from rest" means zero initial velocity. Thus, the initial conditions can be written mathematically as

$$x(0) = -\Delta \quad \text{and} \quad \dot{x}(0) = 0 \tag{a}$$

We can simply substitute the initial conditions above to eqn. (2.29). However, in this example, we take an alternative approach by using the general solution in complex notation. According to eqn. (2.8), the general solution and its first-order time derivative (velocity) are

$$x(t) = C_1 e^{i\omega_n t} + C_2 e^{-i\omega_n t} \tag{b}$$

$$\dot{x}(t) = i C_1 \omega_n e^{i\omega_n t} - i C_2 \omega_n e^{-i\omega_n t} \tag{c}$$

Substituting the initial conditions into the above expressions gives

$$-\Delta = C_1 + C_2 \tag{d}$$

$$0 = i(C_1 - C_2) \tag{e}$$

Solving these equations gives

$$C_1 = C_2 = -\frac{\Delta}{2} \tag{f}$$

Hence, the solution is

$$x(t) = -\frac{\Delta}{2} \left[e^{i\omega_n t} + e^{-i\omega_n t} \right] = -\Delta \cos(\omega_n t) \tag{g}$$

or, in the amplitude–phase angle form as

$$x(t) = \Delta \cos(\omega_n t + \pi) \tag{h}$$

We note that, in the end, the response actually turns out to be a real quantity, although in the solution process, we did not particularly specify that C_1 and C_2 be complex conjugate. In fact, eqns. (d) and (e) should be viewed as complex equations. The left-hand sides of both equations being real in fact require that the imaginary parts of right-hand sides to vanish.

□ Caution on Using Amplitude–Phase Solution

We might be tempted to use the amplitude–phase type of solution in eqns. (2.17) and (2.18) from the very beginning. In this example, we might have used the cosine form as

$$x(t) = X \cos(\omega_n t + \phi_c) \tag{i}$$

where, according to eqns. (2.31) and (2.32),

$$X = \Delta \qquad \phi_c = -\tan^{-1} 0 = 0 \tag{j}$$

and the solution might have been written as

$$x(t) = \Delta \cos \omega_n t$$

which, compared to eqn. (g), is missing the negative sign! The problem comes from the multivalued nature of $\tan^{-1}$ function. In the hindsight, we see that $\phi_c = \pi$ also gives $\tan \phi_c = 0$, but we would not have seen this error had we not solved the problem first using the complex solution. The lesson here is that, in general, we should not use the amplitude–phase form of the solution from the beginning. Instead, we would start with a general solution either in complex notation as eqn. (2.8) or in real notation as eqn. (2.14), and only after having determined the constants, convert the solution into the amplitude–phase form.

□ Exploring the Solution with MATLAB

To an engineer, having obtained a solution to a problem does not mean the end to the problem. In the contrary, it marks the beginning of an exploration. Throughout this textbook, we will use MATLAB to explore the solution that we will obtain for every example. MATLAB is a mathematical software that is particularly useful in many engineering fields. For a brief tutorial on using MATLAB please read Appendix B.

As the first example using MATLAB, we shall limit ourselves to a humble goal of plotting the response curve in eqn. (g). The simplicity of eqn. (g) is a particularly good place to start. How shall we plot the response curve $x(t)$ when it contains only symbols Δ and ω_n without any numerical value? We first introduce a nondimensionalized displacement $\bar{x}$ and a nondimensionalized time $\bar{t}$ as follows:

$$\bar{x} = \frac{x}{\Delta} \quad \text{and} \quad \bar{t} = \omega_n t \tag{k}$$

where we are reminded that ω_n has the unit of t^{-1}. The physical meanings of displacement and time have now become comparative, rather than absolute. For example, when $\bar{x} = 2$, it means x is twice of Δ, regardless what the value of Δ might be. When $\bar{t} = 2\pi$, $t = 2\pi/\omega_n = T_n$, which is one natural period of the system, regardless of the natural frequency of the system. Such comparative meanings make the curves that we will plot much more useful. Using these nondimensionalized parameters, the response in eqn. (g) can be rewritten as

$$\bar{x}(\bar{t}) = -\cos \bar{t} \tag{l}$$

This process is called the *nondimensionalization*. In a nondimensionalized expression such as eqn. (l), all parameters are dimensionless.

The following MATLAB code plots this response up to $\bar{t} = 20$.

```
t=[0:0.01:20];              % Range of nondimensionalized time
x = -cos(t);                % System's response
plot(t,x);                  % Plot the curve
xlabel('\omega_n t');       % Labeling the x-axis
ylabel('x(t)/\Delta');      % labeling the y-axis
```

In the above code, t represents $\bar{t}$ and x represents $\bar{x}$. In the first line, a row matrix t is defined, whose elements contain values of $\bar{t}$, sequentially from 0 to 20 at an increment of 0.01. The second line calls MATLAB's function cos() to calculate the response, producing a matrix x that has the same size as t, with each matrix element storing the system's response at the corresponding $\bar{t}$ value stored in t. The third line simply plots the curve, using t as the abscissa and x as the ordinate. The last two lines produce the labels for the two axes. The plot is shown in Fig. 2.4.

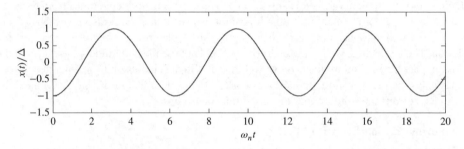

Figure 2.4   Free vibration of mass–spring system due to initial compression in spring

Note that we literally write out the definitions for the nondimensionalized parameters on the labels. This greatly improves the readability of the figure and is thus preferred.

Although we are familiar with the cosine curve, we still must verify that the curve is plotted correctly because producing such a curve is a computational process. Note that one natural period corresponds to $\bar{t} = 2\pi$; and the amplitude is 1. The plotted curve agrees with these. We thus conclude that the curve has indeed been plotted correctly.

□ **Discussion: What Makes a System Vibrate?**

Let us pause for a moment and take another look at the plotted response and the original system. It is mind-boggling to imagine the implication of the solution for the simple mass–spring system, if we allow ourselves to be placed in the ages without electricity and those modern electronic necessities, even without calculus, not to mention differential equations. In those "dark ages," how do we measure time? With a simple spring and a mass, we have a precision timer! This is the fundamental principle behind a major kind of mechanical watches and clocks that has been in use for centuries and is still in use. The spring-loading mechanism for watches and clocks was invented more than 500 years ago,[2] which is about 100 years before Galileo's time (who was the first to study the pendulum to be used as a clock), about 150 years before

---

[2] According to A. P. Usher (*A History of Mechanical Inventions, Revised Edition*, McGraw-Hill, New York, 1929, p. 271), the invention is believed to have been made during the last decade of the fifteenth century.

Hooke devised his Hooke's law describing the relation between spring force and displacement, and about 200 years before Newton found his laws.

This simplest vibration system gives us a reason and a chance to ponder on a very elementary question: what makes a system vibrate? The system comprises of a spring and a mass. That is all. To help this inquisition, we note another physically simpler but mathematically a little more complicated system: a pendulum, as shown in Fig. 1.4. What do these two primitive systems have in common? They both have a mass. If we put two masses together, they will impact with each other but otherwise each moves in its own way. There must be something more. Recall that, when deriving the equation of motion, we consider kinetic energy for the mass and potential energy for the spring. Bingo! The other crucial element in a vibration system is a device or a mechanism to store the potential energy: the spring in the system. For the pendulum, it is the gravitational field. Vibration occurs in a system in which there is a channel to allow the conversion between kinetic energy and potential energy back and forth. For the mass–spring system, the connection between the mass and the spring establishes the channel. For the pendulum, it is the geometric arrangement.

The rate at which the channel is capable of converting the energy gives the system an inherent timer that is called the natural frequency. In this example, the system starts out with a given potential energy (compression in the spring) but no kinetic energy. During the motion, all the potential energy is converted into the kinetic energy when the system reaches its equilibrium position. Then the system moves on to convert the kinetic energy back into potential energy by continuing to stretch the spring.

We shall check this using MATLAB. Based on the solution, the kinetic energy in the mass is

$$T = \frac{1}{2}m\dot{x}^2 = \frac{1}{2}m\omega_n^2\Delta^2\sin^2\omega_n t = \frac{1}{2}k\Delta^2\sin^2\omega_n t \tag{m}$$

where the relation $\omega_n^2 = k/m$ has been used. The potential energy in the spring is

$$V_{spring} = \frac{1}{2}k(x+\Delta_0)^2 = \frac{1}{2}k(\Delta_0 - \Delta\cos\omega_n t)^2 \tag{n}$$

where Δ_0 is the amount of spring stretch at equilibrium, which satisfies the relation $k\Delta_0 = mg$. The potential energy due to the gravity is

$$V_{gravity} = -mgx = k\Delta_0\Delta\cos\omega_n t \tag{o}$$

The total potential energy is, using eqns. (n) and (o),

$$V = V_{spring} + V_{gravity} = \frac{1}{2}k\left(\Delta_0^2 + \Delta^2\cos^2\omega_n t\right) \tag{p}$$

To nondimensionalize the energies of the system, we chose the potential energy stored in the spring at equilibrium $V_0 = \frac{1}{2}k\Delta_0^2$ as the normalization factor. We also introduce a ratio $r = \Delta/\Delta_0$ between the amount of compression in comparison to the amount of spring stretch at equilibrium. Then, the nondimensionalized energies are

$$\overline{T} = \frac{T}{V_0} = r^2\sin^2\bar{t}$$

$$\overline{V}_{spring} = \frac{V_{spring}}{V_0} = \left(1 - r\cos\bar{t}\right)^2 \tag{q}$$

$$\overline{V}_{gravity} = \frac{V_{gravity}}{V_0} = 2r\cos\bar{t} \tag{r}$$

We also noticed that the energy conservation holds

$$\overline{T} + \overline{V} = 1 + r^2 \tag{s}$$

The following MATLAB code plots the kinetic energy and two potential energies.

```
r = 0.5;                        % Static-deflection to Initial Disp Ratio
t = [0:0.01:20];                % Time range
T = r^2 * sin(t) .^2;           % Kinetic energy
Vs = (1 - r * cos(t) ).^2;      % Potential energy in spring
Vg = 2*r*cos(t);                % Gravitational potential energy
plot(t,T,'r-.', t,Vs,'--g', t,Vg,':b', t,T+Vs+Vg,'m');
legend('Kinetic energy', 'Spring potential energy', ...
   'Gravitational potential energy','Total mechanical energy');
xlabel('\omega n t');
ylabel('T/V_0, V/V_0');
```

In the above code, .^ is a special operator that performs element-by-element power. The dot operator can be added to the front of most MATLAB arithmetic operators. The plot() command takes many more parameters. Every three parameters form a set for one curve, with the third (optional) parameter being the color for the curve. The legend() command places a legend box in the figure. At the end of this line, ... signals that the following line is a continuation of the present line.

The curves are shown in Figs. 2.5 and 2.6 for the cases $r = 0.5$ and 1, respectively.

It is easy to verify that both sets of curves are correct. We observe the following: (1) The curves for the two individual potential energies have the same period as the displacement. (2) Both kinetic energy and potential energy have periods that are half of that of the motion. This means the system completes two cycles of energy conversion in one period. (3) The individual potential energies have amplitudes much greater than that of the kinetic energy. (4) The energy stored in the spring is not a sinusoidal function.

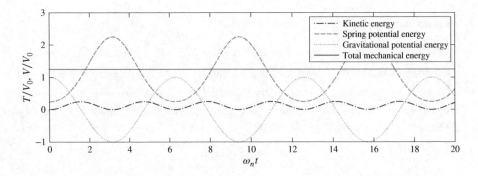

Figure 2.5   Energies in the spring–mass system when $\Delta/\Delta_0 = 0.5$

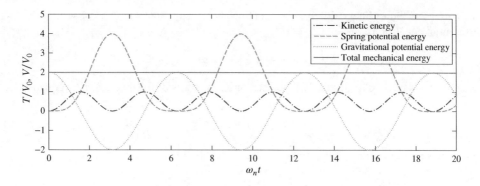

Figure 2.6 Energy in the spring–mass system when $\Delta/\Delta_0 = 1$

2.4 Free Vibrations of Damped Systems

By adding the dashpot between the mass and the ceiling for the system in Fig. 2.1, the equation of motion for the free vibration of the single-DOF system becomes, as having been derived in Examples 1.1 and 1.18,

$$m\ddot{x} + c\dot{x} + kx = 0 \tag{2.33}$$

The solution process remains the same: by assuming the solution of the form $x(t) = e^{\lambda t}$, we obtain the following characteristic equation:

$$m\lambda^2 + c\lambda + k = 0 \tag{2.34}$$

which is a quadratic equation with two roots:

$$\lambda_{1,2} = \frac{-c \pm \sqrt{c^2 - 4km}}{2m} \tag{2.35}$$

There are some complications coming from taking the square root: in a physical system where all m, k, and c are real and positive, there is no guarantee that the quantity under the square root is positive. We have to consider all the possibilities.

Case 1: $c^2 - 4km > 0$ This case is most straightforward. The general solution for the differential equation can be written as

$$x(t) = e^{-\frac{ct}{2m}} \left(A e^{\frac{\sqrt{c^2 - 4km}}{2m}t} + B e^{-\frac{\sqrt{c^2 - 4km}}{2m}t} \right) \tag{2.36}$$

Case 2: $c^2 - 4km = 0$ In this case, the quadratic characteristic equation has only one root, called the *repeated root*:

$$\lambda_1 = \lambda_2 = -\frac{c}{2m} \tag{2.37}$$

In such cases, the ordinary differential equation theory says that the general solution to the differential equation is

$$x(t) = Ae^{-\frac{c}{2m}t} + Bte^{-\frac{c}{2m}t} = e^{-\frac{c}{2m}t}(A + Bt) \tag{2.38}$$

Case 3: $c^2 - 4km < 0$ In this case, we can write the roots of the characteristic equation as

$$\lambda_1 = \frac{-c + i\sqrt{4km - c^2}}{2m} \quad \text{and} \quad \lambda_2 = \frac{-c - i\sqrt{4km - c^2}}{2m} \tag{2.39}$$

Thus, the solution for the differential equation in eqn. (2.33) can be written as

$$x(t) = e^{-\frac{ct}{2m}} \left(C_1 e^{i\frac{\sqrt{4km-c^2}}{2m}t} + C_2 e^{-i\frac{\sqrt{4km-c^2}}{2m}t} \right) \tag{2.40}$$

We are familiar with the terms in the parenthesis: they are complex notation for harmonic motions as those in eqn. (2.8), and we can rewrite them in our preferred real notation as

$$x(t) = e^{-\frac{c}{2m}t} \left[A\cos\left(\frac{\sqrt{4km - c^2}}{2m}t\right) + B\sin\left(\frac{\sqrt{4km - c^2}}{2m}t\right) \right] \tag{2.41}$$

The above solutions are rather cumbersome and sometimes make it difficult to fully comprehend the nature of the vibration in such form of expressions. We would like to seek an alternative way to express our solutions.

2.5   Using Normalized Equation of Motion

2.5.1   Normalization of Equation of Motion

To normalize the equation of motion, we divide both sides of eqn. (2.33) by the mass m, giving

$$\ddot{x} + \frac{c}{m}\dot{x} + \frac{k}{m}x = 0 \tag{2.42}$$

We see that, according to eqn. (2.9), the coefficient for the x-term is ω_n^2 where ω_n is the natural frequency of the mass–spring system. We introduce another parameter ζ such that

$$\frac{c}{m} = 2\zeta\omega_n \tag{2.43}$$

Note that ζ is a dimensionless parameter. We shall see shortly why we introduce this parameter in this strange way. Here, we proceed to write the normalized equation of motion as

$$\ddot{x} + 2\zeta\omega_n\dot{x} + \omega_n^2 x = 0 \tag{2.44}$$

The following advantages of using this normalized form of the equation of motion immediately become apparent when compared with the original equation of motion: (1) The normalized equation completely hides the physical parameters of the original system, which makes the equation independent of the physical construction of the system. (2) The parameters in this normalized equation of motion (ζ and ω_n) have very specific meanings that characterize the system. This gives us a glimpse of the system's vibration characteristics even before actually solving the equation. (3) When compared with the original equation of motion, this normalized equation gives us hints about what physical parameter controls what vibration characteristics. (4) It consolidates the number of parameters. In the original physical system, there are three parameters: m, k, and c. In the normalized equation, only two parameters appear, namely ω_n and ζ. This suppresses a superfluous parameter and gives a better picture of who actually controls the system's behaviors.

Another advantage, although not "immediately apparent," is that the normalized equation can be viewed as a mathematical form, which is the most general form of a *homogeneous linear second-order constant-coefficient ordinary differential equation*. This enables the solution obtained for one system be easily adapted for another system. Of course, when we derive the equation of motion, the thus-obtained equation may not appear in this form. Very often, nonlinear terms appear. However, after the linearization, all single-DOF second-order linear systems will have the equations of motion of the form of eqn. (2.44) on the left-hand side.

We now proceed to finding the general solution for this new equation of motion in the same way as Section 2.4. Following the same procedure, we obtain the characteristic equation as

$$\lambda^2 + 2\zeta\omega_n\lambda + \omega_n^2 = 0 \tag{2.45}$$

whose two roots can be written as

$$\lambda_{1,2} = -\zeta\omega_n \pm \omega_n\sqrt{\zeta^2 - 1} \tag{2.46}$$

2.5.2   Classification of Vibration Systems

Equation (2.46) indicates that the deciding factor that divides the solution into three different cases is ζ. Specifically, the division is based on the value of ζ compared to unity. Since ζ is related to damping effects of the dashpot, it is thus called the *damping ratio* of the system. We have the following classification of vibration systems:

$$\zeta = 0 : \quad \textit{undamped} \text{ system;}$$
$$0 < \zeta < 1 : \quad \textit{underdamped} \text{ system;}$$
$$\zeta = 1 : \quad \textit{critically damped} \text{ system;}$$
$$\zeta > 1 : \quad \textit{overdamped} \text{ system.}$$

An undamped system has $\zeta = 0$, and an underdamped system has a damping ratio $0 < \zeta < 1$. Since almost all solutions and observations about underdamped systems remain valid when ζ is set to zero and identical to those of the undamped systems, for convenience, an undamped system is often considered as a special case of an underdamped system. We shall collectively refer both undamped and underdamped systems as underdamped systems, with damping ratio $0 \leq \zeta < 1$.

For the mass–spring–dashpot system, if we denote the dashpot constant that would produce a critical damping as c_{critical}, since

$$\zeta = \frac{c}{2m\omega_n} = \frac{c}{2\sqrt{mk}} \tag{2.47}$$

we find that

$$c_{\text{critical}} = 2\sqrt{mk} \tag{2.48}$$

and the dashpot constant can be written as

$$c = \zeta c_{\text{critical}} \quad \text{or} \quad \zeta = \frac{c}{c_{\text{critical}}} \tag{2.49}$$

which gives a further endorsement for calling ζ as the damping ratio.

We now rework our free-vibration solutions for this system separately for each type of system.

2.5.3   Free Vibration of Underdamped Systems

In this case, $\zeta^2 - 1 < 0$, the roots in eqn. (2.46) can be written as

$$\lambda_{1,2} = -\zeta\omega_n \pm i\omega_d \tag{2.50}$$

where

$$\omega_d = \omega_n\sqrt{1 - \zeta^2} \tag{2.51}$$

The general solution can thus be written as, in complex notation,

$$x(t) = e^{-\zeta\omega_n t}\left[C_1 e^{i\omega_d t} + C_2 e^{-i\omega_d t}\right] \tag{2.52}$$

or, equivalently, in real notation,

$$x(t) = e^{-\zeta\omega_n t}\left[A\cos\omega_d t + B\sin\omega_d t\right] \tag{2.53}$$

We can see that the terms in square brackets are almost identical to those of the undamped motion in eqn. (2.14), except that the frequency is ω_d. For this reason, ω_d is called the *damped natural frequency*.

The response of the system comprises two factors: one is harmonics with a frequency that is slightly lower than the natural frequency; and the other is an exponential decay factor that eventually suppresses all the motions. We shall explore this in the next example using MATLAB.

In the following, we use our preferred solution form in eqn. (2.53) to finalize the solution for a general set of initial conditions

$$x(0) = x_0 \quad \text{and} \quad \dot{x}(0) = v_0 \tag{2.54}$$

Taking the time derivative of eqn. (2.53) gives the velocity of the mass as

$$\dot{x}(t) = -\zeta\omega_n e^{-\zeta\omega_n t}\left[A\cos\omega_d t + B\sin\omega_d t\right]$$
$$+ \omega_d e^{-\zeta\omega_n t}\left[-A\sin\omega_d t + B\cos\omega_d t\right] \tag{2.55}$$

Substituting the initial conditions in eqns. (2.54) into eqns. (2.53) and (2.55) gives

$$x_0 = A$$
$$v_0 = -\zeta\omega_n A + \omega_d B$$

which can be solved to give

$$A = x_0 \quad \text{and} \quad B = \frac{v_0 + \zeta\omega_n x_0}{\omega_d}$$

Thus, the system's response due to the general initial conditions in eqn. (2.54) is

$$x(t) = e^{-\zeta\omega_n t}\left[x_0 \cos \omega_d t + \frac{v_0 + \zeta\omega_n x_0}{\omega_d} \sin \omega_d t\right] \qquad (2.56)$$

■ Example 2.2: Initial Compression of Spring in Underdamped Systems

Consider an underdamped mass–spring–dashpot system. Assume that initially the mass is held above its equilibrium position by the amount Δ and then released from rest, as shown in Fig. 2.7. Determine the subsequent motion of the mass.

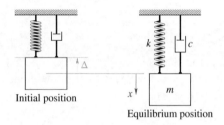

Figure 2.7 Initial spring compression in mass–spring–dashpot system

□ **Solution:**

The initial conditions can be written as

$$x_0 = -\Delta \quad \text{and} \quad v_0 = 0 \qquad (a)$$

The solution is, according to eqn. (2.56),

$$x(t) = -\Delta e^{-\zeta\omega_n t}\left(\cos \omega_d t + \frac{\zeta\omega_n}{\omega_d} \sin \omega_d t\right) \qquad (b)$$

This solution can be further simplified into the amplitude–phase angle form of eqn. (2.18)

$$x(t) = Xe^{-\zeta\omega_n t} \cos(\omega_d t - \phi) \qquad (c)$$

and, according to eqn. (2.21),

$$X = \Delta\sqrt{1 + \frac{\zeta^2\omega_n^2}{\omega_d^2}} = \Delta\sqrt{1 + \frac{\zeta^2}{1-\zeta^2}} = \frac{\Delta}{\sqrt{1-\zeta^2}} \qquad (d)$$

and

$$\tan \phi = \frac{\zeta\omega_n}{\omega_d} = \frac{\zeta}{\sqrt{1-\zeta^2}} \quad \text{or} \quad \phi = \tan^{-1}\frac{\zeta}{\sqrt{1-\zeta^2}} \qquad (e)$$

Thus, the solution can be written as

$$x(t) = -\frac{\Delta}{\sqrt{1-\zeta^2}} e^{-\zeta\omega_n t} \cos(\omega_d t - \phi) \tag{f}$$

□ Discussion: Which of Equations (b) and (f) Is the Final Solution?

In problems solved in symbolic form, a question is often being asked: can we stop at eqn. (b), or do we have to go all the way to eqn. (f)? The answer is "up to you!" Generally, this is a matter of personal preference. Sometimes, one may have reasons to prefer one over the other. The important thing here is to recognize that the two expressions are identical.

□ Exploring the Solution with MATLAB

We define the following nondimensionalization parameters:

$$\bar{x} = \frac{x}{\Delta} \quad \bar{t} = \omega_n t \quad \text{and} \quad \bar{\omega}_d = \frac{\omega_d}{\omega_n} = \sqrt{1-\zeta^2} \tag{g}$$

Note that $\bar{x}$ and $\bar{t}$ are defined in the same way as in Example 2.1; and that ζ is already a nondimensionalized parameter and does not need to be further nondimensionalized. Then, the solution in eqn. (b) can be rewritten as

$$\bar{x}(\bar{t}) = -e^{-\zeta\bar{t}} \left(\cos\bar{\omega}_d\bar{t} + \frac{\zeta}{\bar{\omega}_d} \sin\bar{\omega}_d\bar{t} \right) \tag{h}$$

Comparing the nondimensionalized eqn. (h) with the original expression in eqn. (b), they are strikingly similar. Specifically, if we set the natural frequency ω_n and initial displacement Δ to unity in eqn. (b), we would get the same expression as in eqn. (h) except the overbars. With careful choice of nondimensionalization parameters, this would be true for most cases.

The following MATLAB code implements the above solution:

```
zeta = .25;                        % Damping ratio
omega_d = sqrt( 1-zeta^2 );        % Damped natural frequency
t=[0:0.01:20];                     % Time range
x = -exp(-zeta*t) .* ( cos(omega_d*t) + zeta/omega_d * sin(omega_d*t) );
plot(t,x);
xlabel('\omega_n t');
ylabel('x(t)/\Delta');
```

We can examine the system's responses by varying the system's various parameters. Interestingly enough, ζ (zeta in the code) is the only parameter that can be varied. Recall that, in the original system, we have three parameters m, k, and c. After using the normalized equation in eqn. (2.33), we are left with two parameters ζ and ω_n. Now, after the nondimensionalization, we are left with only one parameter ζ. This shows another advantage of going through the nondimensionalization process: it consolidates the number of parameters that affect the system's responses. Varying the damping ratio ζ, we obtain the response curves as shown in Fig. 2.8 for damping ratios $\zeta = 0$ (undamped), 0.1, 0.25, 0.5, 0.75, 0.99 (near critically damped).

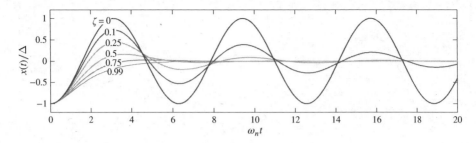

Figure 2.8   Responses of mass–spring–dashpot system to initial spring compression at different damping ratios, including cases of undamped and near critically damped

Having plotted the response curves, the first thing we must do is to verify that the curves are plotted correctly. We check the initial conditions. Given the initial condition in eqn. (a), we verify that every curve starts from $\bar{x} = -1$, with a slope of zero. Additionally, we also verify that, in the undamped case, the curve has a period of 2π.

We can now examine the curves and make the following observations. (1) In the presence of damping, the period of the oscillatory motion is slightly longer than 2π. This agrees with the definition in eqn. (2.51). (2) The higher the damping ratio, the longer the period. (3) The motion becomes more "tamed" by increasing the damping ratio. This is exactly what the damping is doing to the system. (4) We can hardly see any oscillatory motion in cases $\zeta = 0.75$ and 0.99.

As mentioned earlier, the solution consists of two factors. Here, we explore the roles of these two factors using the following MATLAB code:

```
zeta = .1;                              % Damping ratio
omega_d = sqrt( 1-zeta^2 );             % Damped natural frequency
t=[0:0.01:20];                          % Time range
f1 = exp( -zeta*t );                    % f1 and f2 factors
f2 = -( cos( omega_d*t ) + zeta/omega_d * sin( omega_d*t ) );
x  = f1 .* f2;
plot(t, f1, '--r', t, -f1, '--r', t, f2, ':g', t, x, 'b');
legend('f_1(t)','-f_1(t)','f_2(t)','x(t)')
xlabel('\omega_n t');
ylabel('x(t)/\Delta, f_1(t)/\Delta, f_2(t)/\Delta');
```

In the above code, f1 and -f1 (plotted as dashed curves) represent the decaying factor $\pm e^{-\zeta t}$ and f2 (the dotted curve) represents the harmonics factor enclosed in the parenthesis in eqn. (h) and thus the negative sign. The response (the solid curve) is simply the product of these two factors. The plotted curves are shown in Fig. 2.9 for the case $\zeta = 0.1$.

We verify that the curve for the response is identical to the one in Fig. 2.8. We observe that the decay factor controls the overall amplitude, and the harmonics factor provides the oscillation. We might be tempted to say that the decay factor curves form the envelope for the system's response. However, this is only approximately true. The reason is that, although the expression has been normalized and nondimensionalized, the amplitude of the harmonics, which equals to $1/\sqrt{1 - \zeta^2}$ as shown in eqn. (f), exceeds unity. The envelop can be formed if the decay factor is multiplied by the amplitude.

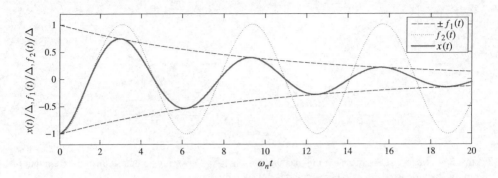

Figure 2.9 Two factors in the response for underdamped system

2.5.4 Free Vibration of Critically Damped System

In this case, $\zeta = 1$. The content under the square root in eqn. (2.46) is zero, and the characteristic equation has one repeated root, as

$$\lambda_1 = \lambda_2 = -\omega_n$$

According to the theory of ordinary differential equations, the general solution in such a case is

$$x(t) = e^{-\omega_n t}(A + Bt) \tag{2.57}$$

Correspondingly, the velocity of the mass is

$$\dot{x}(t) = -\omega_n e^{-\omega_n t}(A + Bt) + Be^{-\omega_n t} \tag{2.58}$$

The motion is no longer oscillatory. When such a system is given the same general initial conditions in eqn. (2.54), substituting eqns. (2.57) and (2.58) into the initial conditions gives

$$x_0 = A$$
$$v_0 = -\omega_n A + B$$

Solving the above two equations gives

$$A = x_0 \quad \text{and} \quad B = v_0 + \omega_n x_0$$

Finally, the system's response to the general initial conditions in eqn. (2.54) is

$$x(t) = e^{-\omega_n t} \left[x_0 + (v_0 + \omega_n x_0)t \right] \tag{2.59}$$

■ **Example 2.3: Initial Compression of Spring in Critically Damped Systems**

Determine the response of a critically damped mass–spring–dashpot system for the same initial conditions as in Example 2.2.

□ **Solution:**

The initial conditions can be written as

$$x_0 = -\Delta \quad \text{and} \quad v_0 = 0 \tag{a}$$

The solution is, according to eqn. (2.59),

$$x(t) = -\Delta e^{-\omega_n t}(1 + \omega_n t) \tag{b}$$

□ **Exploring the Solution with MATLAB**

We define the following nondimensionalization parameters:

$$\bar{x} = \frac{x}{\Delta} \quad \text{and} \quad \bar{t} = \omega_n t \tag{c}$$

Then, the solution can be written as

$$\bar{x}\left(\bar{t}\right) = -e^{-\bar{t}}(1 + \bar{t}) \tag{d}$$

The following MATLAB code implements the above solution.

```
t = [0:0.01:20];               % Time range
x = -exp(-t) .* (1+t);         % System's response
plot(t,x,'r');                 % Plotting
xlabel('\omega_n t');          % Labeling the x-axis
ylabel('x(t)/\Delta');         % Labeling the y-axis
```

There is no parameter that we can vary. The response is plotted in Fig. 2.10. We verify that the curve starts from $\bar{x} = -1$, with a slope of zero, and settles to zero with a sufficiently large time $\bar{t}$. We observe that, as expected, there is no oscillatory motion, and this curve is extremely close to the case of $\zeta = 0.99$ in Fig. 2.8. The motion of the system has essentially been suppressed by $\bar{t} = 2\pi$, one natural period.

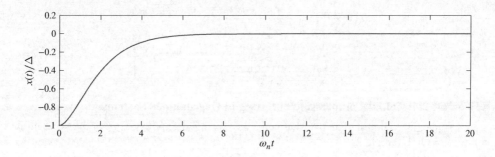

Figure 2.10   Response of critically damped mass–spring–dashpot system to initial spring compression

2.5.5   Free Vibration of Overdamped System

In this case, $\zeta > 1$, the content under the square root in eqn. (2.46) is positive and the characteristic equation has the following two roots:

$$\lambda_1 = -\left(\zeta - \sqrt{\zeta^2 - 1}\right)\omega_n \qquad \lambda_2 = -\left(\zeta + \sqrt{\zeta^2 - 1}\right)\omega_n \qquad (2.60)$$

Both roots are real and negative. The resulting general solution is expressible as

$$x(t) = Ae^{-\left(\zeta - \sqrt{\zeta^2 - 1}\right)\omega_n t} + Be^{-\left(\zeta + \sqrt{\zeta^2 - 1}\right)\omega_n t} \qquad (2.61)$$

Again, the motion is not oscillatory. It simply involves two exponentially decaying functions. Taking a time derivative, the velocity can be written as

$$\dot{x}(t) = -\left(\zeta + \sqrt{\zeta^2 - 1}\right)\omega_n Ae^{-\left(\zeta + \sqrt{\zeta^2 - 1}\right)\omega_n t}$$

$$-\left(\zeta - \sqrt{\zeta^2 - 1}\right)\omega_n Be^{-\left(\zeta - \sqrt{\zeta^2 - 1}\right)\omega_n t} \qquad (2.62)$$

When the system is subjected to the general initial conditions in eqn. (2.54), substituting eqns. (2.61) and (2.62) into eqn. (2.54) gives

$$x_0 = A + B \qquad (2.63)$$

$$v_0 = -\left(\zeta + \sqrt{\zeta^2 - 1}\right)\omega_n A - \left(\zeta - \sqrt{\zeta^2 - 1}\right)\omega_n B \qquad (2.64)$$

Solving the above two equations gives

$$A = \frac{1}{2}\left(x_0 - \frac{v_0 + \zeta\omega_n x_0}{\omega_n\sqrt{\zeta^2 - 1}}\right)$$

$$B = \frac{1}{2}\left(x_0 + \frac{v_0 + \zeta\omega_n x_0}{\omega_n\sqrt{\zeta^2 - 1}}\right)$$

Thus, the system's response to the general initial conditions in eqn. (2.54) is

$$x(t) = \frac{1}{2}\left[\left(x_0 - \frac{v_0 + \zeta\omega_n x_0}{\omega_n\sqrt{\zeta^2 - 1}}\right)e^{-\left(\zeta + \sqrt{\zeta^2 - 1}\right)\omega_n t}\right.$$

$$\left. + \left(x_0 + \frac{v_0 + \zeta\omega_n x_0}{\omega_n\sqrt{\zeta^2 - 1}}\right)e^{-\left(\zeta - \sqrt{\zeta^2 - 1}\right)\omega_n t}\right] \qquad (2.65)$$

■ **Example 2.4: Initial Compression of Spring in Overdamped Systems**

Determine the response of an overdamped mass–spring–dashpot system for the same initial condition as in Example 2.2.

□ **Solution:**

The initial conditions can be written as

$$x_0 = -\Delta \quad \text{and} \quad v_0 = 0 \qquad (a)$$

The solution is, according to eqn. (2.65),

$$x(t) = -\frac{\Delta}{2}\left[\left(1 - \frac{\zeta}{\sqrt{\zeta^2 - 1}}\right)e^{-\left(\zeta+\sqrt{\zeta^2-1}\right)\omega_n t}\right.$$

$$\left.+ \left(1 + \frac{\zeta}{\sqrt{\zeta^2 - 1}}\right)e^{-\left(\zeta-\sqrt{\zeta^2-1}\right)\omega_n t}\right]$$

$$= \frac{\Delta}{2\sqrt{\zeta^2 - 1}}\left[\left(\sqrt{\zeta^2 - 1} - \zeta\right)e^{-\left(\zeta+\sqrt{\zeta^2-1}\right)\omega_n t}\right.$$

$$\left.+ \left(\sqrt{\zeta^2 - 1} + \zeta\right)e^{-\left(\zeta-\sqrt{\zeta^2-1}\right)\omega_n t}\right] \tag{b}$$

Denote

$$d_1 = \zeta + \sqrt{\zeta^2 - 1} \quad \text{and} \quad d_2 = \zeta - \sqrt{\zeta^2 - 1} \tag{c}$$

Then,

$$x(t) = -\frac{\Delta}{2\sqrt{\zeta^2 - 1}}\left[d_1 e^{-d_2\omega_n t} - d_2 e^{-d_1\omega_n t}\right] \tag{d}$$

□ **Exploring the Solution with MATLAB**

We define the following nondimensionalized parameters:

$$\bar{x} = \frac{x}{\Delta} \quad \text{and} \quad \bar{t} = \omega_n t \tag{e}$$

Then, the solution can be rewritten as

$$\bar{x}\left(\bar{t}\right) = -\frac{1}{2\sqrt{\zeta^2 - 1}}\left[d_1 e^{-d_2\bar{t}} - d_2 e^{-d_1\bar{t}}\right] \tag{f}$$

where it is noted that d_1 and d_2 in eqn. (c) are already dimensionless. The following MATLAB code implements the above solution.

```
zeta = 2;                           % Damping ratio
t=[0:0.01:20];                      % Time range
d1 = zeta + sqrt(zeta^2-1);         % Two roots: d1 and d2
d2 = zeta - sqrt(zeta^2-1);
x = -.5 / sqrt(zeta^2-1) * ( d1 .* exp(-d2*t) - d2 .* exp(-d1*t) );
plot(t,x);
xlabel('\omega_n t');
ylabel('x(t)/\Delta');
```

The responses for damping ratios $\zeta = 1.5$, 2, and 5 are plotted in Fig. 2.11. We verify that every curve starts from $\bar{x} = -1$, with a slope of zero, and settles to zero with a sufficiently large time $\bar{t}$.

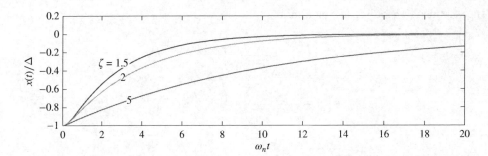

Figure 2.11 Response of overdamped mass–spring–dashpot system to initial spring compression at different damping ratios

We observed that the higher the damping ratio, the slower the system returns to its equilibrium position. This does not seem to be very intuitive to us. Since we have MATLAB in hand, we shall explore this issue. Based on the solution in eqn. (f), we decide to plot the following two components in the solution separately:

$$f_1\left(\bar{t}\right) = d_1 e^{-d_2 \bar{t}} \quad \text{and} \quad f_2\left(\bar{t}\right) = d_2 e^{-d_1 \bar{t}} \tag{g}$$

These two factors allow us to observe which is dominating and how they change when the damping ratio is varied. The following code is used to generate the plots in Fig. 2.12.

```
zeta = 1.5;                        % Damping ratio
t=[0:0.01:10];                     % Time range
d1 = zeta + sqrt(zeta^2-1);        % Two roots: d1 and d2
d2 = zeta - sqrt(zeta^2-1);
f1 = d1 * exp( -d2*t );            % Two factors: f1 and f2
f2 = d2 * exp( -d1*t );
subplot(1,2,1);                    % Pick the left subplot in 1x2 division
plot(t,f1);                        % Plotting the f1 factor
xlabel('\omega_n t');
ylabel('f_1(t)');
subplot(1,2,2);                    % Pick the right subplot in 1x2 division
plot(t,f2);                        % Plotting the f2 factor
xlabel('\omega_n t');
ylabel('f_2(t)');
```

In the above code, we use a new MATLAB command subplot(m,n,p) that accomplishes two tasks: the plotting area is divided into an m row × n column grid; and all subsequent plotting actions are directed to subplot p. Consequently, the axis labels need to be repeated for each subplot. Plots for $\zeta = 1.5$, 2, and 5 are shown in Fig. 2.12: f_1 on Fig. 2.12a and f_2 on Fig. 2.12b.

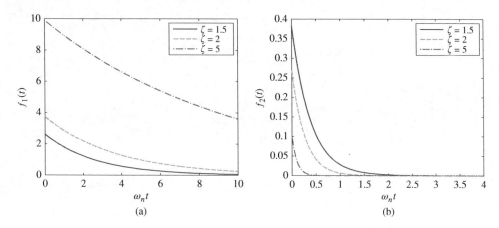

Figure 2.12 Comparison of contributions of two factors f_1 and f_2 defined in eqn. (g) in overdamped system. (a) f_1; (b) f_2

Based on Fig. 2.12, we can make the following observations: (1) The contribution by factor f_1 is much more significant than by f_2, as the vertical axes in the two subplots differ by one order of magnitude. (2) The higher the damping ratio ζ, factor f_1 decays the slower, but f_2 decays the faster. (3) The higher the damping ratio ζ, the larger the amplitude of f_1, and the smaller the amplitude of f_2. Combining these observations, we can conclude that f_1 is the factor that dominates the system's response.

We can further observe how d_1 and d_2, as defined in eqn. (c), vary with ζ. First, we note that, when ζ is large,

$$\lim_{\zeta \to \infty} d_1 = 2\zeta \qquad \lim_{\zeta \to \infty} d_2 = 0 \tag{h}$$

The simplified expression for a function when the variable approaches infinity, such as those in eqn. (h), is called the *asymptote* for the function. We can observe the behaviors of d_1 and d_2 with the following MATLAB code:

```
zeta = [1:0.01:10];               % Range of damping ratio
d1 = zeta + sqrt( zeta .^2 -1);   % d1 and d2 factors
d2 = zeta - sqrt( zeta .^ 2-1);
plot(zeta,d1,'r',zeta,d2,':b');   % Plotting d1 and d2
hold on;                          % Turn on "hold" for next plot command
plot([0 10],[0,20],'--g');        % Plotting asymptote for d1
xlabel('\zeta');
ylabel('d_1, d_2');
```

In the above code, a new MATLAB command hold on is used, which allows multiple plot commands to plot on the same figure. The second plot command plots a dashed line that connects two points $(\zeta, d_1) = (0, 0)$ and $(10, 20)$, which represent the asymptote for $d_1(\zeta)$ in eqn. (h). The variations of d_1 and d_2 with the damping ratio ζ are shown in Fig. 2.13. We verify that these curves are correct as they coincide at $\zeta = 1$ with $d_1 = d_2 = 1$, and the asymptotes in eqn. (h) are followed closely.

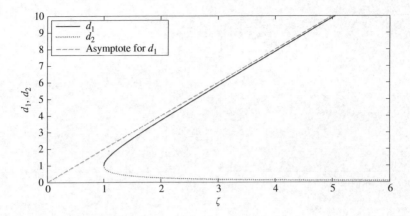

Figure 2.13   Variations of d_1 and d_2 as defined in eqn. (c) with damping ratio

From Fig. 2.13, we observe that d_1 and d_2 quickly depart from their juncture point at $\zeta = 1$. As ζ increases, d_1 quickly increases and d_2 decreases. If we are inclined, we can continue to investigate how the ratios d_1 and d_2 vary with ζ. However, we would like stop here by concluding that, when $\zeta > 1.5$, the effects of d_2, and in turn f_2, are negligible.

■ Example 2.5: Initial Velocity in Damped Systems

Suppose that, in the mass–spring–dashpot system, the mass is given an initial velocity V_0. Determine the subsequent response of the system. In this example, we want to compare all three possible damping cases.

□ Solution:

The initial condition can be written as

$$x_0 = 0 \quad \text{and} \quad v_0 = V_0 \tag{a}$$

For the **underdamped system**, the solution is, according to eqn. (2.56),

$$x(t) = \frac{V_0}{\omega_d} e^{-\zeta \omega_n t} \sin \omega_d t \tag{b}$$

For the **critically damped system**, the solution is, according to eqn. (2.59),

$$x(t) = V_0 t e^{-\zeta \omega_n t} \tag{c}$$

For the **overdamped system**, the solution is, according to eqn. (2.65),

$$x(t) = \frac{V_0}{2\omega_n \sqrt{\zeta^2 - 1}} \left[-e^{-d_1 \omega_n t} + e^{-d_2 \omega_n t} \right] \tag{d}$$

where, as in the previous example,

$$d_1 = \zeta + \sqrt{\zeta^2 - 1} \quad \text{and} \quad d_2 = \zeta - \sqrt{\zeta^2 - 1} \tag{e}$$

□ Exploring the Solution with MATLAB

In selecting the nondimensionalization parameter for the displacement, we want the response to be compared to the initial conditions; so ideally, the displacement would be nondimensionalized by V_0. However, x and V_0 are of different dimensions (units). We find, taking the hint from eqns. (b) and (d), that V_0/ω_n can be used for such a purpose. We introduce the following nondimensionalization parameters:

$$\bar{x} = \frac{x\omega_n}{V_0} \qquad \bar{t} = \omega_n t \quad \text{and} \quad \bar{\omega}_d = \frac{\omega_d}{\omega_n} \tag{f}$$

Thus, the nondimensionalized responses can be rewritten as

$$\bar{x}_{\text{under}}(\bar{t}) = \frac{1}{\bar{\omega}_d} e^{-\zeta\bar{t}} \sin \bar{\omega}_d \bar{t} \tag{g}$$

$$\bar{x}_{\text{critical}}(\bar{t}) = \bar{t} e^{-\zeta\bar{t}} \tag{h}$$

$$\bar{x}_{\text{over}}(\bar{t}) = \frac{1}{2\sqrt{\zeta^2 - 1}} \left[-e^{-d_1\bar{t}} + e^{-d_2\bar{t}} \right] \tag{i}$$

The following MATLAB code implements these three solutions:

```
zeta = 1;                        % Damping ratio
t=[0:0.01:20];                   % Time range
if (zeta<1.)                     % Case of underdamped
    omega_d = sqrt( 1-zeta^2 );
    x = exp( -zeta*t ) .* sin( omega_d*t )/omega_d;
elseif (zeta==1)                 % Case of critically damped
    x = t .* exp( -zeta*t );
else                             % Case of overdamped
    x = .5 * ( -exp( -(zeta+sqrt(zeta^2-1))*t ) + ...
         exp( -(zeta-sqrt(zeta^2-1))*t ) ) /sqrt(zeta^2-1);
end
plot(t,x);
xlabel('\omega_nt');
ylabel('x(t)/(V_0/\omega_n)');
```

In the above code, a new feature is introduced: the `if...elseif...end` logic structure, which is a flow control mechanisms existing in almost every programming language.

Again, the damping ratio is the only parameter that we can vary, thanks to the nondimensionalization process. Varying the damping ratio ζ (`zeta` in the code), we obtain the responses as illustrated in Fig. 2.14 for damping ratios $\zeta = 0, 0.1, 0.25, 0.5, 0.75, 1$ (critically damped) and 2 (overdamped).

We verify that, based on the initial conditions, all curves start from $\bar{x} = 0$ and all curves have the same slope at $\bar{t} = 0$. During the period immediately after the initial release, the displacement in all cases is positive because the mass is given a positive velocity. We observe once again that the overdamped system returns to the equilibrium position slower than the critically damped system; all other characteristics of the motion are similar to what we observed in the earlier examples when the system is subjected to an initial displacement.

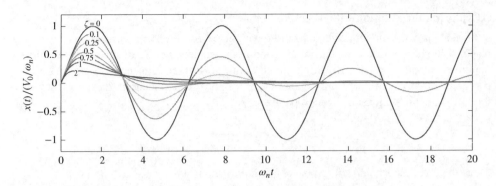

Figure 2.14 Responses to initial velocity at different damping ratio ζ, including undamped, under-damped, critically damped, and overdamped

2.6   Forced Vibrations I: Steady-State Responses

We will study two types of responses of the system to forced excitations. In the first, we want to investigate the long-term effects of the system when it is subjected to a persistent loading. We further limit our consideration to periodic loadings. Such long-term solutions are generally called the *steady-state responses* of the system. In such cases, the equation of motion is

$$m\ddot{x} + c\dot{x} + kx = F(t) \tag{2.66}$$

We have by now seen that the normalized equation of motion brings not only notational simplicity but also benefits of easier interpretation of the solutions. Thus, we continue to use the normalized equation of motion. Dividing both sides by m gives

$$\ddot{x} + 2\zeta\omega_n\dot{x} + \omega_n^2 x = f(t) \tag{2.67}$$

where the normalized force $f(t) = F(t)/m$ is called the *forcing function*, or the *forcing* for short.

2.6.1   Harmonic Loading

Assume that the forcing function is of the following harmonic form

$$f(t) = f_0 \cos \Omega t \tag{2.68}$$

where Ω is the called *driving frequency* or the *forcing frequency*.

As we are looking for the long-term solution under a persistent harmonic loading at a given frequency, we expect that, after an initial stage of "hand-shakes," the system eventually becomes submissive and exhibits a degree of "harmony," said to have reached the *steady state*, by moving at the same frequency as the forcing. Therefore, we assume the solution to be of the form

$$x(t) = A \cos \Omega t + B \sin \Omega t \tag{2.69}$$

and consequently,

$$\dot{x}(t) = \Omega(-A \sin \Omega t + B \cos \Omega t) \tag{2.70}$$

$$\ddot{x}(t) = -\Omega^2(A \cos \Omega t + B \sin \Omega t) \tag{2.71}$$

Substituting this solution and its derivatives, along with the forcing function in eqn. (2.68), into the equation of motion in eqn. (2.67) gives

$$-\Omega^2(A \cos \Omega t + B \sin \Omega t) + 2\zeta \omega_n \Omega(-A \sin \Omega t + B \cos \Omega t)$$

$$+\omega_n^2(A \cos \Omega t + B \sin \Omega t) = f_0 \cos \Omega t$$

We expect the above equation to hold for all time t in the steady state. This requires the coefficients for $\cos \Omega t$ on both sides to be equal, and the coefficients for $\sin \Omega t$ to be equal, too, for which the coefficient on the right-hand side is actually 0. Therefore,

$$(\omega_n^2 - \Omega^2)A + 2\zeta \omega_n \Omega B = f_0 \tag{2.72}$$

$$-2\zeta \omega_n \Omega A + (\omega_n^2 - \Omega^2)B = 0 \tag{2.73}$$

These two equations can be solved for A and B. Here, we write the equations in matrix form as

$$\begin{bmatrix} \omega_n^2 - \Omega^2 & 2\zeta \omega_n \Omega \\ -2\zeta \omega_n \Omega & \omega_n^2 - \Omega^2 \end{bmatrix} \begin{Bmatrix} A \\ B \end{Bmatrix} = \begin{Bmatrix} f_0 \\ 0 \end{Bmatrix} \tag{2.74}$$

and solve the above equation using *Cramer's rule* as

$$A = \frac{\begin{vmatrix} f_0 & 2\zeta \omega_n \Omega \\ 0 & \omega_n^2 - \Omega^2 \end{vmatrix}}{\begin{vmatrix} \omega_n^2 - \Omega^2 & 2\zeta \omega_n \Omega \\ -2\zeta \omega_n \Omega & \omega_n^2 - \Omega^2 \end{vmatrix}} = \frac{f_0(\omega_n^2 - \Omega^2)}{(\omega_n^2 - \Omega^2)^2 + (2\zeta \omega_n \Omega)^2} \tag{2.75}$$

$$B = \frac{\begin{vmatrix} \omega_n^2 - \Omega^2 & f_0 \\ -2\zeta \omega_n \Omega & 0 \end{vmatrix}}{\begin{vmatrix} \omega_n^2 - \Omega^2 & 2\zeta \omega_n \Omega \\ -2\zeta \omega_n \Omega & \omega_n^2 - \Omega^2 \end{vmatrix}} = \frac{f_0(2\zeta \omega_n \Omega)}{(\omega_n^2 - \Omega^2)^2 + (2\zeta \omega_n \Omega)^2} \tag{2.76}$$

These two expressions look rather complicated. To simplify, we introduce one more parameter called the *frequency ratio*:

$$r = \frac{\Omega}{\omega_n} \tag{2.77}$$

which is the ratio between the driving frequency Ω and the system's natural frequency ω_n. Then, dividing both the numerator and the denominator on the right-hand side of eqn. (2.75) by ω_n^4 gives

$$A = \frac{\frac{f_0}{\omega_n^2}\left(1 - \frac{\Omega^2}{\omega_n^2}\right)}{\left(1 - \frac{\Omega^2}{\omega_n^2}\right)^2 + \left(2\zeta \frac{\Omega}{\omega_n}\right)^2} = \frac{f_0}{\omega_n^2} \frac{1 - r^2}{(1 - r^2)^2 + (2\zeta r)^2} \tag{2.78}$$

Similarly, eqn. (2.76) becomes

$$B = \frac{f_0}{\omega_n^2} \frac{2\zeta r}{(1 - r^2)^2 + (2\zeta r)^2} \tag{2.79}$$

Before we put the expressions for A and B back into eqn. (2.69) to write the final expression for the system's steady-state response, let us take a second look at the solution: we like to express the solution in the form of

$$x(t) = X \cos(\Omega t - \phi) \tag{2.80}$$

In this form, the physical meanings of individual components are apparent: X is the *amplitude* of the steady-state response, and ϕ is called the *phase lag*: the phase angle the response lagging behind (because of the negative sign) that of the driving force. As we have done this conversion before, substituting the expressions for A and B in eqns. (2.78) and (2.79) into eqn. (2.21) gives

$$X = \frac{f_0}{\omega_n^2} \sqrt{\frac{(1 - r^2)^2 + (2\zeta r)^2}{\left[(1 - r^2)^2 + (2\zeta r)^2\right]^2}} = \frac{f_0}{\omega_n^2} \frac{1}{\sqrt{(1 - r^2)^2 + (2\zeta r)^2}} \tag{2.81}$$

and

$$\phi = \tan^{-1} \frac{2\zeta r}{1 - r^2} \tag{2.82}$$

Finally, the steady-state response of the system can be written as

$$x(t) = \frac{f_0}{\omega_n^2} \cdot \frac{1}{\sqrt{(1 - r^2)^2 + (2\zeta r)^2}} \cdot \cos(\Omega t - \phi) \tag{2.83}$$

2.6.2 Mechanical Significance of Steady-State Solution

A careful reader might have already noticed that, when the final expression for the steady-state response is written in eqn. (2.83), two dots have been explicitly inserted into the expression. They separate the steady-state solution into three factors, and their meanings are as the following:

$\dfrac{f_0}{\omega_n^2}$:   *Static Deflection*: The deflection of the mass if the load F_0 is applied as a static force.

$\dfrac{1}{\sqrt{(1 - r^2)^2 + (2\zeta r)^2}}$:   *Dynamic Amplification*: The amplification of the static deflection due to dynamical effects of the system.

$\cos(\Omega t - \phi)$:   *Response's Functional Form*: It takes the same cosinusoidal form as the driving function; Ω is the driving frequency; and ϕ represents the *phase lag*.

2.6.2.1 Static Deflection

Recall that

$$\omega_n^2 = \frac{k}{m} \quad \text{and} \quad f_0 = \frac{F_0}{m}$$

Hence,

$$\frac{f_0}{\omega_n^2} = \frac{F_0}{m}\frac{m}{k} = \frac{F_0}{k} \tag{2.84}$$

which gives the amount of stretch in the spring (the *deflection* of the system) at static equilibrium under a constant force F_0. In general, we prefer the notation of f_0/ω_n^2 for the same reason we prefer the normalized equation of motion.

2.6.2.2 Dynamic Amplification

The dynamic amplification, also called the *magnification factor*, is often written as the amplitude of dynamic response nondimensionalized by the static deflection, that is,

$$\frac{X}{f_0/\omega_n^2} = \frac{1}{\sqrt{(1 - r^2)^2 + (2\zeta r)^2}} \tag{2.85}$$

Variations of dynamic amplification versus frequency ratio r are plotted in Fig. 2.15 for various values of ζ. Each curve in Fig. 2.15 is called a *frequency response spectrum* of the system at the given ζ.

Three regions in Fig. 2.15 are of great engineering interest. The first is in the vicinity of $\Omega/\omega_n \approx 1$, that is, when the forcing frequency is close to the system's natural frequency. When $\zeta = 0$, $X/(f_0/\omega_n^2)$ approaches ∞ when $\Omega = \omega_n$; otherwise, for a wide range of ζ values, $X/(f_0/\omega_n^2)$ is near its peak when $\Omega = \omega_n$. This phenomenon is generally known as the *resonance*, and ω_n is also called the *resonant frequency*. At the resonance, the forcing matches

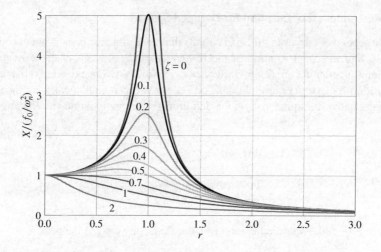

Figure 2.15   Dynamic amplification of mass–spring–dashpot system

the system inherent frequency and keeps injecting energy into the system, resulting in a disproportionally large amplitude in the response. Mathematically, as $r \to 1$,

$$\lim_{r \to 1} \frac{X}{f_0/\omega_n^2} = \frac{1}{2\zeta} \tag{2.86}$$

Equation (2.86) shows that, in this region, the dynamic amplification is controlled by the damping ratio. Thus, this region is called the *damping-dominated region*.

The second region is when Ω is very small compared to ω_n. This is the low-frequency limit, when the forcing varies slowly, and the system essentially follows the forcing much similar to the static deflection. Dynamical effects essentially disappear. This is sometimes called a *quasi-static* process. Mathematically, as $r \to 0$,

$$\lim_{r \to 0} \frac{X}{f_0/\omega_n^2} = 1 \quad \text{or} \quad \lim_{r \to 0} X = \frac{f_0}{\omega_n^2} = \frac{F_0}{k} \tag{2.87}$$

Equation (2.87) shows that, in this region, the stiffness k controls the response amplitude. Thus, this region is called the *stiffness-dominated region*.

The third region is when Ω is very large compared to ω_n. This is the high-frequency limit when the forcing varies rapidly. The system, having its own timing, cannot keep up with the forcing, and in effect shields the motion such that the wall does not feel the forcing. Mathematically, as $r \to \infty$,

$$\lim_{r \to \infty} \frac{X}{f_0/\omega_n^2} = \frac{1}{r^2} \quad \text{or} \quad \lim_{r \to \infty} X = \frac{f_0}{\omega_n^2}\frac{\omega_n^2}{\Omega^2} = \frac{F_0}{m\Omega^2} \tag{2.88}$$

Equation (2.88) shows that the mass (inertia) of the system controls the response amplitude. Thus, this region is called the *inertia-dominated region*.

2.6.2.3 Resonance Peak Location for Damped System

A closer inspection of curves in Fig. 2.15 reveals that the resonance peak is not exactly located at $r = 1$. To find the exact location of the resonance peak, we can set the derivative of the dynamic amplification in eqn. (2.85) with respect to r to zero. Since the derivative being zero is the condition for extremes, either peaks or valleys, we can equivalently set the derivative of the content under the square root in the denominator of the dynamic amplification to zero, that is,

$$\frac{\partial}{\partial r}\left[(1 - r^2)^2 + (2\zeta r)^2\right] = 0$$

which gives

$$4r(-1 + r^2 + 2\zeta^2) = 0$$

The nonzero root of the above equation gives the frequency ratio at the peak at

$$r_{\text{peak}} = \sqrt{1 - 2\zeta^2} \tag{2.89}$$

Equation (2.89) also indicates that, when $\zeta > 1/\sqrt{2}$, there will be no peak in the spectrum. Furthermore, at the frequency ratio r_{peak}, we find that

$$\left.\frac{X}{f_0/\omega_n^2}\right|_{peak} = \frac{1}{\sqrt{[1 - (1 - 2\zeta^2)]^2 + 4\zeta^2(1 - 2\zeta^2)}} = \frac{1}{2\zeta\sqrt{1 - \zeta^2}} \tag{2.90}$$

2.6.2.4 The Q-Factor

The *Q-factor* is defined as the dynamic amplification at $r = 1$, that is,

$$Q = \left.\frac{X}{f_0/\omega_n^2}\right|_{r=1} = \frac{1}{2\zeta} \tag{2.91}$$

For most mechanical systems, the damping ratio ζ is rather small and the dynamic amplification at $r = 1$ and at $r = r_{peak}$ are very close to each other. Very often this distinction is ignored, and in such cases, the *Q*-factor approximately represents the amplification at the peak. That is, eqn. (2.90) can be approximated as

$$\left.\frac{X}{f_0/\omega_n^2}\right|_{peak} \approx \frac{1}{2\zeta} = Q \tag{2.92}$$

2.6.2.5 The Peak Width

Another important parameter that can be measured by the Q-factor is the *peak width*, which is also sometimes called the *bandwidth* of the resonance. Since the peak generally is of a smooth form, the peak width is defined as the frequency range between the points when the dynamic amplification reaches $1/\sqrt{2}$ of the peak value. The reason for the square root is that, as we have seen in an earlier example, the energy stored in the system is proportional to the square of the amplitude. Hence, $1/\sqrt{2}$ of the peak amplitude corresponds to a half of the peak power. The definition of the peak width is illustrated in Fig. 2.16: a horizontal line at $1/\sqrt{2}$ of the peak height intersects with the spectrum at two points, denoted as r_1 and r_2. Those two intersection points are called the *half-peak-power points*, and the distance between these points is defined as the peak width.

The peak width can be determined as the difference between the two roots of the following equation:

$$\frac{1}{\sqrt{2}}\frac{1}{2\zeta} = \frac{1}{\sqrt{(1 - r^2)^2 + (2\zeta r)^2}} \tag{2.93}$$

where the approximate peak height in eqn. (2.92) has been used. The roots can be found by equating the squares of the denominators of both sides

$$8\zeta^2 = 1 - 2r^2 + r^4 + 4\zeta^2 r^2$$

or

$$r^4 - 2(1 - 2\zeta^2)r^2 + 1 - 8\zeta^2 = 0$$

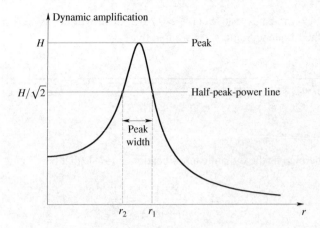

Figure 2.16 Peak width defined by half-peak-power of spectrum

This is a quadratic equation for r^2, which has two roots as

$$r_{1,2}^2 = \frac{2(1 - 2\zeta^2) \pm \sqrt{4(1 - 2\zeta^2)^2 - 4(1 - 8\zeta^2)}}{2}$$

For the quadratic equation, assuming $r_1 > r_2$,

$$r_1^2 - r_2^2 = \sqrt{4(1 - 2\zeta^2)^2 - 4(1 - 8\zeta^2)} = 4\zeta\sqrt{1 + \zeta^2}$$

or

$$(r_1 - r_2)(r_1 + r_2) = 4\zeta\sqrt{1 + \zeta^2}$$

When ζ is small, which is true for most systems and certainly is when we talk about the Q-factor, the two points r_1 and r_2 are located on the opposite sides of the peak and their average $(r_1 + r_2)/2$ is very close to 1, that is, $r_1 + r_2 \approx 2$. Then,

$$r_1 - r_2 \approx 2\zeta\sqrt{1 + \zeta^2} \approx 2\zeta = \frac{1}{Q} \tag{2.94}$$

which means that the peak width is the reciprocal of the Q-factor.

As we can see, a single Q-factor represents three things in a damped system: (1) the damping ratio, in the form $Q = 1/(2\zeta)$; (2) the reciprocal of the bandwidth of the resonance peak; and (3) the peak height. For this reason, the Q-factor is widely used in engineering practice, even much more so than the damping ratio. Sometimes, the Q-factor is also called the *quality factor*, as someone tries hard to figure out what the letter Q stands for. In general, the higher the Q-factor, the lower the damping ratio, and the higher and the narrower resonance peak.

2.6.2.6 Phase Lag

The phase lag in eqn. (2.82) is shown in Fig. 2.17 as it varies with the frequency ratio at different damping ratios. We have mentioned the three frequency regions. It helps to look at the phase lag to understand the characteristics in these regions.

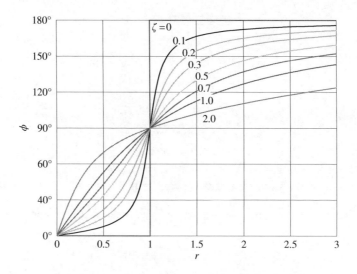

Figure 2.17   Phase lag for mass–spring–dashpot system

In the low-frequency region, the phase lag is almost zero, for which the response is said to be *in phase* with the driving force, and the system's response follows the forcing closely.

In the resonance region, the phase lag is about $\pi/2$. If we assume that the driving force is of the form of a cosine function, the phase lag makes the response a sine function, whose time derivative is a cosine function. In other words, the velocity of the response is in phase with the driving force. This means that the driving force is doing positive work and keeps injecting energy into the system.

In the high-frequency limit, the phase lag is near π. For sinusoidal functions, this flips the sign for the response, that is, the response is the opposite of the driving force. This means that the system response is opposing the driving force.

For most single-DOF systems, the phase lag ranges between 0 and π. However, as a phase angle in a harmonics, ϕ has a period of 2π and its range is typically chosen as $-\pi \le \phi \le \pi$. It is called phase lag as we have, due to the negative sign preceding it, when $\phi \ge 0$, and it is called the *phase lead* when $\phi < 0$.

2.6.3   Other Examples of Harmonic Loading

In the following, we use examples to study two other important types of harmonic excitations to a system.

■ Example 2.6: System Excited at Base

Assume that a mass–spring–dashpot system is enclosed in a casing, which is then placed on a vibration table, as shown in Fig. 2.18. The vibration table provides a specified motion to the casing $y_0(t) = H + Y_0 \sin \Omega t$, where y_0 is measured upward from the ground, and H and Y_0 are constants. Determine the steady-state response of the enclosed system.

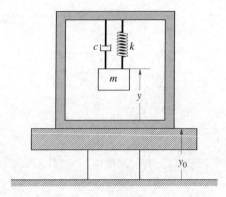

Figure 2.18   Enclosed mass–spring–dashpot system on the vibration table

□ **Solution:**

Equation of Motion

- *Generalized Coordinates*: We define $\{\, y \,\}$ as a set of complete and independent generalized coordinates, where y is the location of the mass from its equilibrium position measured upward with respect to the casing, as sketched in Fig. 2.18.
- *Admissible Variations*: We verify that $\{\, \delta y \,\}$ is a complete and independent set of admissible variations in this set of generalized coordinates.
- *Holonomicity*: We verify that the system is holonomic and has one degree of freedom.
- *Generalized Forces*: The dashpot is a floating dashpot but the generalized coordinate y measures a relative motion. Using the results from Example 1.17,

$$\delta W^{\text{n.c.}} = -c\dot{y}\delta y \tag{a}$$

Thus, according to eqn. (1.2),

$$\Xi_y = -c\dot{y} \tag{b}$$

- *Kinetic Energy*: We do not need to be concerned with the mass of the casing, as it moves with the table. The absolute velocity of the mass is $\dot{y} + \dot{y}_0$. Thus,

$$T = \frac{1}{2}m(\dot{y} + \dot{y}_0)^2 \tag{c}$$

- *Potential Energy*: As y is a relative displacement, the potential energy in the spring is simplified. But for the gravitational potential energy, we need the absolute location of the mass. Thus,

$$V = \frac{1}{2}k(y - \Delta)^2 + mg(y + y_0) \tag{d}$$

where Δ is the amount of stretch in the spring at equilibrium.

- *Lagrangian*: Combining eqns. (c) and (d) gives, according to eqn. (1.3),

$$\mathcal{L} = \frac{1}{2}m(\dot{y} + \dot{y}_0)^2 - \frac{1}{2}k(y - \Delta)^2 - mg(y + y_0) \tag{e}$$

- *Lagrange's Equation*: Lagrange's equation for y is

$$\frac{d}{dt}\left(\frac{\partial \mathcal{L}}{\partial \dot{y}}\right) - \frac{\partial \mathcal{L}}{\partial y} = \Xi_y \tag{f}$$

Since

$$\frac{\partial \mathcal{L}}{\partial \dot{y}} = m(\dot{y} + \dot{y}_0) \qquad \frac{\partial \mathcal{L}}{\partial y} = -k(y - \Delta) - mg \tag{g}$$

Lagrange's equation gives

$$m(\ddot{y} + \ddot{y}_0) + k(y - \Delta) + mg = -c\dot{y} \tag{h}$$

Using the equilibrium condition $k\Delta - mg = 0$ gives

$$m\ddot{y} + c\dot{y} + ky = -m\ddot{y}_0 \tag{i}$$

Finally, substituting the expression for $y_0(t)$ into the above equation gives

$$m\ddot{y} + c\dot{y} + ky = mY_0\Omega^2 \sin \Omega t \tag{j}$$

Steady-State Vibration Analysis

We now can start the vibration analysis, as eqn. (j) is already linearized. We continue to use the normalized equation of motion, which is obtained by dividing both sides of eqn. (j) by m, as

$$\ddot{y} + 2\zeta\omega_n\dot{y} + \omega_n^2 y = Y_0\Omega^2 \sin \Omega t \tag{k}$$

The steady-state response is assumed to be of the form

$$y(t) = A \cos \Omega t + B \sin \Omega t$$

and substituting this solution into the equation of motion in eqn. (k) gives

$$(\omega_n^2 - \Omega^2)(A \cos \Omega t + B \sin \Omega t)$$
$$+ 2\zeta\omega_n\Omega(-A \sin \Omega t + B \cos \Omega t) = Y_0\Omega^2 \sin \Omega t \tag{l}$$

Dividing both sides by ω_n^2 and denoting $r = \Omega/\omega_n$ give

$$(1 - r^2)(A \cos \Omega t + B \sin \Omega t)$$
$$+ 2\zeta r(-A \sin \Omega t + B \cos \Omega t) = r^2 Y_0 \sin \Omega t \tag{m}$$

Then, requiring the coefficients for sine and cosine terms on both sides to equal gives the following set of linear equation system:

$$\begin{bmatrix} 1 - r^2 & 2\zeta r \\ -2\zeta r & 1 - r^2 \end{bmatrix} \begin{Bmatrix} A \\ B \end{Bmatrix} = Y_0 \begin{Bmatrix} 0 \\ r^2 \end{Bmatrix} \tag{n}$$

which can be solved via Cramer's rule to give

$$A = \frac{-2\zeta r^3}{(1 - r^2)^2 + (2\zeta r)^2} Y_0 \tag{o}$$

$$B = \frac{r^2(1 - r^2)}{(1 - r^2)^2 + (2\zeta r)^2} Y_0 \tag{p}$$

To express the final system's steady-state response in the amplitude–phase lag form, we note that the forcing is a sine function; hence, the response is to be written in the same functional form such that the term *phase lag* could retain its meaning. That is, we seek to express the system's response in the following form

$$y(t) = Y \sin (\Omega t - \phi) \tag{q}$$

Recall such conversions in eqn. (2.21),

$$\frac{Y}{Y_0} = \frac{\sqrt{A^2 + B^2}}{Y_0} = \frac{r^2}{\sqrt{(1 - r^2)^2 + (2\zeta r)^2}} \tag{r}$$

which is still called the dynamics amplification although the nondimensionalization parameter is difference, and

$$\phi = -\tan^{-1}\frac{A}{B} = \tan^{-1}\frac{2\zeta r}{1 - r^2} \tag{s}$$

We note that the phase lag ϕ has the same expression as in eqn. (2.82).

□ Exploring the Solution with MATLAB

The expressions for the dynamic amplification and the phase lag are already dimensionless. The dynamic amplification can be plotted by the following MATLAB code:

```
zeta=0.1;                      % Damping ratio
r = [0.:0.01:3];               % Range of frequency ratio
M = r .^ 2 ./ sqrt( (1-r.^2) .^ 2  + (2*zeta*r) .^2 );   % Magnification
plot(r,M, 'Color', [1 0 1]);   % Plotting with user-specified color
xlabel('r');
ylabel('X/Y_0');
```

In the above code, the plot() function takes a new form that allows a direct specification of the color with a red–green–blue triplet, each valued between 0 and 1. The spectra are shown in Fig. 2.19 for a series of values of damping ratio ranging from $\zeta = 0$ to 2.

To verify, we check the limits when r is small and large, at which, according to eqn. (r), the dynamic amplification approaches 0 and 1, respectively. This is the most observable feature of Fig. 2.19, in comparison with Fig. 2.15.

The expression for the phase lag is the same as eqn. (2.82). Thus, Fig. 2.17 is still applicable.

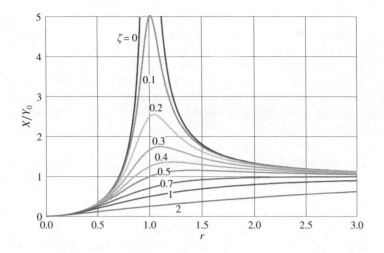

Figure 2.19   Dynamic amplification for mass–spring–dashpot system subject to base excitation

□ **Discussion: Engineering Applications**

This example has two significant engineering applications. In the low-frequency limit,

$$\lim_{r\to 0} \frac{Y}{Y_0} = r^2 \quad \text{and} \quad \lim_{r\to 0} \phi = 0 \tag{t}$$

which give

$$\lim_{r\to 0} y(t) = \frac{Y_0}{\omega_n^2} \Omega^2 \sin \Omega t = \frac{1}{\omega_n^2} \ddot{y}_0(t) \tag{u}$$

The right-hand side of eqn. (u) is the acceleration of the vibration table, multiplied by a characteristic constant of the system. This relation provides the basic principle for the design of the *accelerometer*, which measures the acceleration of the surface to which it is attached (the "vibration table" in the this example). Although eqn. (u) is only valid for small r, Fig. 2.19 suggests that, for almost any not-too-large damping ratio, the approximation is valid up to $r \approx 0.4$ and becomes exact for up to the resonant frequency if $\zeta = 0$. In any case, an accelerometer is designed to operate in a frequency range that is significantly lower than its own natural frequency.

In the other extreme, in the high-frequency limit,

$$\lim_{r\to \infty} \frac{Y}{Y_0} = 1 \quad \text{and} \quad \lim_{r\to \infty} \phi = \pi \tag{v}$$

which give

$$\lim_{r\to \infty} y(t) = Y_0 \sin (\Omega t - \pi) = H - y_0(t) \tag{w}$$

The right-hand side of eqn. (w) reflects the displacement of the surface. This is the basic principle for *vibrometers* (or *seismometers* when used as a seismic device) that measure the displacement of the surface to which it is attached. Typically, if we want to measure a displacement, we need a fixed point as the datum. In the event of an impending earthquake, no place can be reliably assumed as "fixed," this device provides a measurement. Generally, seismometer are heavy and bulky, such that the seismic wave has a significantly higher frequency (in the range of 0.1–20 Hz) than the device's natural frequency.

It is noted that the generalized coordinate is defined with respect to the casing. The relative motion of the mass can be indicated by a needle pointing to a scale painted on the casing to provide a reading. Had we defined the generalized coordinates differently (such as measured from the ground), we would not have seen the potential applications of this setup.

■ Example 2.7: Load Imbalance in Washing Machine

We probably have experienced first hand the noise that an unbalanced washing machine or dryer can make. Similar problems exist for many other rotating machinery. We have obtained the equation of motion for a front-loading washing machine in Example 1.27, whose figure is repeated here in Fig. 2.20. In this example, assume that the drum of the washing machine rotates at a constant angular velocity of Ω during a spin cycle. Find the steady-state vibration of the washing machine during the spin circle.

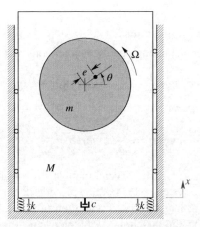

Figure 2.20   Model for front-loading washing machine

□ Solution:

The equation for the motion has been obtained in Example 1.27 as

$$(M + m)\ddot{x} + c\dot{x} + kx = me\Omega^2 \sin \Omega t \tag{a}$$

where x is the upward displacement of the machine body with respect to ground, measured from its equilibrium position. We can normalize the equation of motion by dividing both sides of eqn. (a) by $(M + m)$,

$$\ddot{x} + 2\zeta\omega_n\dot{x} + \omega_n^2 x = \frac{me}{M + m}\Omega^2 \sin \Omega t \tag{b}$$

We see that the equation of motion is almost the same as the mass–spring–dashpot system on a vibration table in Example 2.6, with only two differences: (1) the total mass is $(M + m)$, which also affects the natural frequency and the damping ratio; and (2) Y_0 is replaced by $me/(M + m)$. We can simply adapt the result in that example and write the system's steady-state response in the following form

$$x(t) = X \sin(\Omega t - \phi) \tag{c}$$

where

$$\frac{X}{me/(M+m)} = \frac{r^2}{\sqrt{(1-r^2)^2 + (2\zeta r)^2}} \quad \text{and} \quad \phi = \tan^{-1}\frac{2\zeta r}{1 - r^2} \tag{d}$$

The figures for the dynamic amplification and the phase lag will not be repeated here because they are identical to those in Example 2.6. However, we can still observe the plot in that example, Fig. 2.19, while reflecting upon the current system. We can make the following observations: (1) As the washing machine spins up to the desired speed, the driving frequency increases continuously from 0 until the working frequency is reached. To avoid any possible resonance, the machine must operate below the resonant frequency, that is, the fastest spin speed must be noticeably lower than the system's natural frequency. (2) In order to minimize the amplitude of the vibration due to the imbalance, we can increase the system's natural frequency. One way is to bolt the machine to the ground instead of letting it sit on its rubber feet. When bolted, the metal bolts effectively become the spring in the system, which are much stiffer than the rubber feet. (3) In general, the vibration is worst in the spin cycle when the drum is rotating at the highest speed.

□ **Exploring the Solution with MATLAB**

We might wonder, if we live in an apartment, what our neighbor downstairs would feel about our washing machine. Here, we use MATLAB to explore this issue.

As the drum runs, it exerts a force into the floor. What is the magnitude of this force? The total force includes the force exerted by the spring, which equals to kx, and the force exerted by the dashpot, which equals to $c\dot{x}$. Thus,

$$F(t) = X\left[k \sin(\Omega t - \phi) + c\Omega \cos(\Omega t - \phi)\right]$$

$$= \frac{kme}{M+m} \frac{r^2}{\sqrt{(1-r^2)^2 + (2\zeta r)^2}} \left[\sin(\Omega t - \phi) + 2\zeta r \cos(\Omega t - \phi)\right] \tag{e}$$

where

$$\omega_n^2 = \frac{k}{M+m} \qquad r = \frac{\Omega}{\omega_n} \quad \text{and} \quad \frac{kme r^2}{M+m} = me\Omega^2$$

Note that $me\Omega^2$ is the amplitude of the centrifugal force produced by the imbalance. Then, the ratio of the amplitude of the transmitted force to that of the applied force is called the *transmissibility* of a system, denoted as TR, which is

$$\text{TR} = \frac{|F(t)|}{me\Omega^2} = \sqrt{\frac{1 + (2\zeta r)^2}{(1-r^2)^2 + (2\zeta r)^2}} \tag{f}$$

The following MATLAB code plots the transmissibility curve for the system:

```
zeta = .1;                              % Damping ratio
r = [0.:0.01:3];                        % Range of frequency ratio
TR = sqrt( (1+ (2*zeta*r) .^2) ./ ((1-r.^2) .^ 2 + (2*zeta*r) .^2 ));
plot(r,TR,'Color', [1 .3 1]);           % Plotting with user-defined color
xlabel('r');
ylabel('|F|/(me\Omega^2)');
```

The transmissibility of the washing machine at different damping ratios ranging from $\zeta = 0$ to 2 is show in Fig. 2.21.

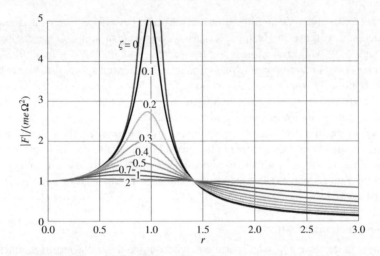

Figure 2.21   Transmissibility of front-loading washing machine

We verify these curves by checking the limits as r is either very small or very large, where, based on eqn. (f), the transmissibility approaches to 1 and $2\zeta/r$, respectively.

We observe that, at low frequencies, the force exerted to the floor will not be significantly magnified if the machine is operated away from the resonance, such as $r < 0.5$. At very high frequencies, the transmissibility is not amplified either. Within the resonance range, the amplification could be significant at low damping ratio, but much smaller if the damping is relatively large. This implies that increasing the damping could be an effective way of reducing the transmitted force.

Another interesting observation is that all curves for different damping ratios pass through the same point. Since the location of this point is independent of ζ, the derivative of the transmissibility with respect to ζ would be zero at this point, that is,

$$\frac{d\text{TR}}{d\zeta} = 0 = \frac{8\zeta r^2[(1 - r^2)^2 + (2\zeta r)^2] - [1 + (2\zeta r)^2]8\zeta r^2}{[(1 - r^2)^2 + (2\zeta r)^2]^2}$$

$$= \frac{8\zeta r^4(r^2 - 2)}{[(1 - r^2)^2 + (2\zeta r)^2]^2} \tag{g}$$

which gives $r = 0$ or $r = \sqrt{2}$. Furthermore, at $r = \sqrt{2}$, TR = 1.

However, the above transmissibility curves might be misleading for our present problem. The reason is that the nondimensionalization factor for the transmissibility varies with the driving frequency. A more meaningful comparison would be using a constant force to nondimensionalize the transmitted force. Thus, we replace Ω by ω_n, which is a characteristic constant of the system. Then, the nondimensionalized force is

$$\overline{F} = \frac{F}{me\omega_n^2} = \frac{F}{\dfrac{kme}{M+m}} \tag{h}$$

and

$$\left|\overline{F}(t)\right| = r^2 \sqrt{\frac{1 + (2\zeta r)^2}{(1 - r^2)^2 + (2\zeta r)^2}} \tag{i}$$

which has a factor r^2 compared to eqn. (f). This curve can be plotted using the following MATLAB code:

```
zeta=0.1;                          % Damping ratio
r = [0.:0.01:3];                   % Range of frequency ratio
M = r .^2 .* sqrt( (1+(2*zeta*r).^2)./((1-r.^2).^ 2 + (2*zeta*r).^2 ));
plot(r,M,'Color', [1 .2 .0]);      % Plotting with user-defined color
xlabel('r');
ylabel('|F|/[kme/(M+m)]');
```

The amplitude of the transmitted force is shown in Fig. 2.22 for the same set of damping ratios as in Fig. 2.21. We verify those curves by checking the limits when r is very small and very large, where $|\overline{F}|$ approaches to 0 and $2\zeta r$, respectively. We also verify that all curves pass through the same point at $r = \sqrt{2}$, as the multiplication of r^2 would not affect the derivative with respect to ζ. We can calculate that at the common point, $|\overline{F}| = 2$, which is also shown in the curves.

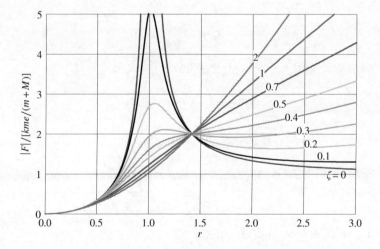

Figure 2.22   Amplitude of transmitted force due to washing machine

We observe that, at low frequencies, the force transmitted to the floor is still very small. However, at higher frequencies, the force becomes large and increases with the frequency. This is because of the r^2 factor: as the frequency increases, the centrifugal force increases in the order of r^2. The system has the capability to suppress it to r. Nevertheless, it still grows as the frequency increases. Hence, a washing machine should never be operated above its resonant frequency.

This example reflects the typical situation of harmonic excitation caused by rotating machinery, such as wheels on a vehicle: an imbalance of tires can cause essentially the same effects. The resonance frequency in such a system is often called the *critical speed*.

2.6.4   General Periodic Loading

The examples in the previous section have loadings that are simply a sine or a cosine function. If the loading is a combination of both sine and cosine functions of the same driving frequency, it can be analyzed exactly the same way.

If the loading consists of two different frequencies, it can be split into two problems, each dealing with only one frequency. This is possible because of the *principle of superposition*, which states that, for a linear system, the response of the system to multiple loadings applied simultaneously equals to the sum of all responses to individual loadings applied individually.

We can take this idea one step further to analyze system's responses to general periodic loadings. A general periodic function can be viewed as comprising an infinite number of single-frequency harmonics. This is the view of the Fourier serial expansion for a periodic function. We can analyze the response due to each individual frequency components and then sum the responses up.

To make this to work, the forcing function is expanded into Fourier series first. Assuming a forcing function $f(t)$ has a period of T, its *Fourier expansion* is

$$f(t) = \frac{a_0}{2} + \sum_{n=1}^{\infty} \left(a_n \cos \Omega_n t + b_n \sin \Omega_n t \right) \tag{2.95}$$

where

$$\Omega_n = \frac{2n\pi}{T} \tag{2.96}$$

and the *Fourier coefficients* a_n and b_n are determined by, for $n = 0, 1, 2, \cdots$,

$$a_n = \frac{2}{T} \int_0^T f(t) \cos \Omega_n t \, dt \tag{2.97}$$

$$b_n = \frac{2}{T} \int_0^T f(t) \sin \Omega_n t \, dt \tag{2.98}$$

Then, each component of the forcing is applied to the system, resulting in the following equation of motion

$$\ddot{x} + 2\zeta \omega_n \dot{x} + \omega_n^2 x = a_n \cos \Omega_n t + b_n \sin \Omega_n t \tag{2.99}$$

We assume the solution to be of the form

$$x_n(t) = A_n \cos \Omega_n t + B_n \sin \Omega_n t \tag{2.100}$$

and solve for the system's response as before.

The case $n = 0$ needs special care. At $n = 0$, $\cos \Omega_n t = 1$ while $\sin \Omega_n t = 0$. The coefficient a_0 obtained from eqn. (2.97) gives twice the average of the forcing over one period, while eqn. (2.98) always gives $b_0 = 0$. Since a_0 is a constant, the steady-state response to a constant loading is in fact the static deflection. Thus,

$$A_0 = \frac{a_0}{\omega_n^2} \tag{2.101}$$

Finally, the system's response due to the original forcing function $f(t)$ is the sum of these responses, as

$$x(t) = \frac{A_0}{2} + \sum_{n=1}^{\infty} (A_n \cos \Omega_n t + B_n \sin \Omega_n t) \tag{2.102}$$

which is the Fourier serial expansion for the system's response.

■ Example 2.8: Square-Form Periodic Loading on Mass–Spring–Dashpot System

Consider a mass–spring–dashpot system subjected to a periodic loading of a period T. Within each period, the force is a constant F_0 for the first half-period and vanishes in the second half-period. Find the system's steady-state response to this periodic loading.

□ **Solution:**

The normalized equation of motion for the system can be written as

$$\ddot{x} + 2\zeta \omega_n \dot{x} + \omega_n^2 x = f(t) \tag{a}$$

The forcing function $f(t)$ can be written as

$$f(t) = \begin{cases} f_0 & \text{when } 0 < t \leq T/2 \\ 0 & \text{when } t \geq T/2 \end{cases} \tag{b}$$

where $f_0 = F_0/m$. The Fourier coefficients for the forcing function can be calculated according to eqns. (2.97) and (2.98) as, for $n = 0$,

$$a_0 = \frac{2}{T} \int_0^{\frac{T}{2}} f_0 \, dt = f_0 \tag{c}$$

and for $n \geq 1$,

$$a_n = \frac{2f_0}{T} \int_0^{\frac{T}{2}} \cos \frac{2n\pi t}{T} dt = \frac{2f_0}{T} \frac{T}{2n\pi} \sin \frac{2n\pi t}{T} \Big|_0^{\frac{T}{2}} = 0$$

$$b_n = \frac{2f_0}{T} \int_0^{\frac{T}{2}} \sin \frac{2n\pi t}{T} dt = \frac{2f_0}{T} \frac{T}{2n\pi} \left(-\cos \frac{2n\pi t}{T} \Big|_0^{\frac{T}{2}} \right) \tag{d}$$

$$= \frac{f_0}{n\pi} (1 - \cos n\pi) = \begin{cases} \dfrac{2f_0}{n\pi} & \text{when } n \text{ is odd,} \\ 0 & \text{when } n \text{ is even.} \end{cases} \tag{e}$$

The Fourier serial expansion for the forcing function can be written as

$$f(t) = \frac{f_0}{2} + \frac{2f_0}{\pi} \sum_{n=1,3,5,\ldots}^{\infty} \frac{1}{n} \sin \frac{2n\pi t}{T} \tag{f}$$

For each forcing component $n \geq 1$, the equation of motion is

$$\ddot{x}_n + 2\zeta \omega_n \dot{x}_n + \omega_n^2 x_n = b_n \sin \Omega_n t \tag{g}$$

Assuming the solution to be of the form

$$x_n(t) = A_n \cos \Omega_n t + B_n \sin \Omega_n t \tag{h}$$

Substituting eqn. (h) into eqn. (g) gives

$$(\omega_n^2 - \Omega_n^2)(A_n \cos \Omega_n t + B_n \sin \Omega_n t)$$
$$+ 2\zeta \omega_n \Omega_n (-A_n \sin \Omega_n t + B_n \cos \Omega_n t) = b_n \sin \Omega_n t \tag{i}$$

We introduce a normalized *frequency ratio* similar as earlier. In a multifrequency loading, the Ω_1 is the *fundamental frequency*. We use this frequency to define the frequency ratio as

$$r = \frac{\Omega_1}{\omega_n} = \frac{2\pi}{\omega_n T} \tag{j}$$

This way, $\Omega_n / \omega_n = nr$, eqn. (i) can be written as

$$(1 - n^2 r^2)(A_n \cos \Omega_n t + B_n \sin \Omega_n t)$$
$$+ 2\zeta nr(-A_n \sin \Omega_n t + B_n \cos \Omega_n t) = \frac{b_n}{\omega_n^2} \sin \Omega_n t \tag{k}$$

which gives

$$\begin{bmatrix} 1 - n^2 r^2 & 2n\zeta r \\ -2n\zeta r & 1 - n^2 r^2 \end{bmatrix} \begin{Bmatrix} A_n \\ B_n \end{Bmatrix} = \begin{Bmatrix} 0 \\ \frac{b_n}{\omega_n^2} \end{Bmatrix} \tag{l}$$

Solving the above equation using Cramer's rule gives

$$A_n = -\frac{b_n}{\omega_n^2} \frac{2n\zeta r}{(1 - n^2 r^2)^2 + (2n\zeta r)^2} \tag{m}$$

$$B_n = \frac{b_n}{\omega_n^2} \frac{1 - n^2 r^2}{(1 - n^2 r^2)^2 + (2n\zeta r)^2} \tag{n}$$

Thus, the nth component of the system's response is

$$x_n(t) = \frac{b_n}{\omega_n^2} \frac{-2n\zeta r \cos \Omega_n t + (1 - n^2 r^2) \sin \Omega_n t}{[(1 - n^2 r^2)^2 + (2n\zeta r)^2]}$$
$$= \frac{b_n}{\omega_n^2} \frac{\sin (\Omega_n t - \phi_n)}{\sqrt{(1 - n^2 r^2)^2 + (2n\zeta r)^2}} \tag{o}$$

where

$$\phi_n = \tan^{-1} \frac{2n\zeta r}{1 - n^2 r^2} \tag{p}$$

The case $n = 0$ needs to be treated separately. Substituting eqn. (c) into eqn. (2.101) gives $A_0 = f_0 / \omega_n^2$.

Finally, assembling the responses to all components of forcing gives the system's steady-state response as

$$x(t) = \frac{f_0}{\omega_n^2} \left\{ \frac{1}{2} + \frac{2}{\pi} \sum_{n=1,3,\ldots}^{\infty} \frac{\sin(nr\omega_n t - \phi_n)}{n\sqrt{(1 - n^2 r^2)^2 + (2\zeta n r)^2}} \right\} \tag{q}$$

Note that the summation is over all odd-valued n. In general, we prefer a summation to be over every integer, because in computations such as using MATLAB, the summation can be conveniently performed as a matrix product, placing 0's into a matrix not only unnecessarily bloats the size of the matrix; but also causes some useless computations to be performed (as 0 multiply with something). For this reason, we introduce a new subscript i such that $n = 2i - 1$. This way, when i varies from 1 to ∞ in steps of 1, n jumps in odd values in steps of 2. Thus, replacing n by $2i - 1$ gives

$$\frac{x(t)}{f_0 / \omega_n^2} = \frac{1}{2} + \frac{2}{\pi} \sum_{i=1}^{\infty} \frac{\sin\left[(2i - 1)r\omega_n t - \phi_{2i-1}\right]}{(2i - 1)\sqrt{\left[1 - (2i - 1)^2 r^2\right]^2 + \left[2(2i - 1)\zeta r\right]^2}} \tag{r}$$

□ **Exploring the Solution with MATLAB**

We define the following nondimensionalized parameters:

$$\bar{x} = \frac{x}{f_0 / \omega_n^2} \quad \text{and} \quad \bar{t} = \omega_n t \tag{s}$$

Then, the system's response in eqn. (r) can be nondimensionalized as

$$\bar{x}(\bar{t}) = \frac{1}{2} + \frac{2}{\pi} \sum_{i=1}^{\infty} \frac{\sin\left[(2i - 1)r\bar{t} - \phi_{2i-1}\right]}{(2i - 1)\sqrt{\left[1 - (2i - 1)^2 r^2\right]^2 + \left[2\zeta r(2i - 1)\right]^2}} \tag{t}$$

We would like to plot the Fourier series for the forcing function $f(t)$ as a comparison and also as an additional means of verifying the computation. We further introduce

$$\bar{f} = \frac{f}{f_0} \tag{u}$$

Then, the Fourier serial expansion for the nondimensionalized forcing function is

$$\bar{f}(\bar{t}) = \frac{1}{2} + \frac{2}{\pi} \sum_{i=1}^{\infty} \frac{\sin\left[(2i - 1)r\bar{t}\right]}{2i - 1} \tag{v}$$

In order to plot these functions, another issue must be resolved: how to perform the summation of infinity series, as in eqns. (t) and (v)? In numerical computations, typically we only include the first few, say N, terms in the computation. That is, replacing the upper limit of ∞

in the summations in eqns. (t) and (v) by N, and N is called the *truncation size*. It is possible to determine an appropriate truncation size based on the desired level of accuracy. Here, we would simply select a truncation size *a priori*, and we will explore the effects of different truncation sizes should the curiosity take us.

The following MATLAB code plots the Fourier serial expansions for both the forcing function and the system's steady-state response:

```
zeta=0.2;                % Damping ratio
r = .1;                  % Frequency ratio: t0/Tn
N = 200;                 % Truncation size in Fourier expansion
t = [0.:0.01:50];        % Time array
i = [1:N];               % Index i in Fourier serial expansions
n = 2*i-1;               % Converting i into n for Fourier expansions
freq = n*r;              % Frequency components of forcing
phi = atan2( 2*zeta*freq, 1-freq .^2 );   % Phase lag for all components
fcoef = 2/pi ./ n;       % Fourier coefficient for forcing
xcoef = fcoef ./sqrt( (1-freq .^2) .^2 + (2*zeta*freq).^2 ); % for response
f = .5 + fcoef * sin ( freq' *t );      % Fourier series for forcing
x = .5 + xcoef * sin ( freq' *t - phi' * ones(1,length(t))); % for response
plot(t,x, 'r', t, f, '--g');
legend('x(t)','F(t)');
xlabel('\omega_nt');
ylabel('x(t)/(F_0/k), F(t)/F_0');
```

In the above code, the phase lag `phi` is computed using a different MATLAB function `atan2(y,x)`. This function calculates the phase angle according to the definition $\tan^{-1}(y/x)$ but avoids two issues with the function `atan()` we used in an earlier example: (1) the result ranges between $-\pi$ and π, which matches the conventional range for the phase lag, and (2) it avoids the problematic "division by zero" error when x=0.

The computations for the two Fourier series need some strategic planning. In the code, `freq`, `phi`, `fcoef`, and `xcoef` store $\overline{\Omega}_n$, ϕ_n, $\overline{b}_n$, and $\overline{X}_n$, respectively. They are all $1{\times}N$ row matrices, where N is the truncation size for the Fourier expansions. On the other hand, `t` stores the time series, which is a $1 \times N_t$ row matrix, where N_t is the number of time instants for each curve. The summation part of the Fourier expansion for the forcing function is computed by the code segment `fcoef*sin(freq'*t)`. The symbol `'` following `freq` means a matrix transpose. Thus, `freq'*t` works as following:

$$\{\overline{\boldsymbol{\Omega}}\}^T \{\bar{t}\} = \left\{ \begin{matrix} \overline{\Omega}_1 \\ \overline{\Omega}_2 \\ \overline{\Omega}_2 \\ \vdots \end{matrix} \right\}_{N \times 1} \left\{ \bar{t}_1 \ \bar{t}_2 \ \bar{t}_3 \ \cdots \right\}_{1 \times N_t}$$

$$= \begin{bmatrix} \overline{\Omega}_1\bar{t}_1 & \overline{\Omega}_1\bar{t}_2 & \overline{\Omega}_1\bar{t}_3 & \cdots \\ \overline{\Omega}_2\bar{t}_1 & \overline{\Omega}_2\bar{t}_2 & \overline{\Omega}_2\bar{t}_3 & \cdots \\ \overline{\Omega}_3\bar{t}_1 & \overline{\Omega}_3\bar{t}_2 & \overline{\Omega}_3\bar{t}_3 & \cdots \\ & \cdots & & \end{bmatrix}_{N \times N_t} \tag{w}$$

The `sin()` function simply performs an element-by-element evaluation. The result is then left multiplied by `fcoef`. The code segment `fcoef * sin(freq' *t)` means the following mathematical operations:

$$\{\bar{b}\}\sin\left(\{\overline{\boldsymbol{\Omega}}\}^T\{\bar{t}\}\right) = \{\bar{b}_1\ \bar{b}_2\ \bar{b}_3\ \cdots\}_{1\times N}\begin{bmatrix}\sin\overline{\Omega}_1\bar{t}_1 & \sin\overline{\Omega}_1\bar{t}_2 & \sin\overline{\Omega}_1\bar{t}_3 & \cdots \\ \sin\overline{\Omega}_2\bar{t}_1 & \sin\overline{\Omega}_2\bar{t}_2 & \sin\overline{\Omega}_2\bar{t}_3 & \cdots \\ \sin\overline{\Omega}_3\bar{t}_1 & \sin\overline{\Omega}_3\bar{t}_2 & \sin\overline{\Omega}_3\bar{t}_3 & \cdots \\ & & \cdots \end{bmatrix}_{N\times N_t}$$

$$= \begin{bmatrix}\bar{b}_1\sin\overline{\Omega}_1\bar{t}_1 + \bar{b}_2\sin\overline{\Omega}_2\bar{t}_1 + \bar{b}_3\sin\overline{\Omega}_3\bar{t}_1 + \cdots \\ \bar{b}_1\sin\overline{\Omega}_1\bar{t}_2 + \bar{b}_2\sin\overline{\Omega}_2\bar{t}_2 + \bar{b}_3\sin\overline{\Omega}_3\bar{t}_2 + \cdots \\ \bar{b}_1\sin\overline{\Omega}_1\bar{t}_3 + \bar{b}_2\sin\overline{\Omega}_2\bar{t}_3 + \bar{b}_3\sin\overline{\Omega}_3\bar{t}_3 + \cdots \\ \vdots \end{bmatrix}_{1\times N_t} \quad (x)$$

That is, this code segment produces the serial summation part of the Fourier expansions for all time instants contained in t, one row for each time instant.

The computation for the system's response is similar. The main difference is that the code segment phi'*ones(1,length(t)) expands the phase angle to a matrix of size N $\times N_t$, compatible with freq'*t, where MATLAB function ones(n,m) produces a matrix of size n $\times$ m with all elements being 1.

Figures 2.23 through 2.27 show the system's steady-state responses for frequency ratios of $r = 0.1, 0.2, 0.4, 1$, and 2 for a fixed damping ratio of $\zeta = 0.1$.

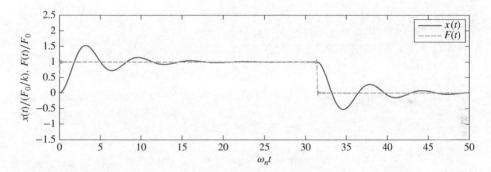

Figure 2.23 Steady-state response for system of damping ratio $\zeta = 0.2$ to square-form forcing at frequency ratio $r = 0.1$

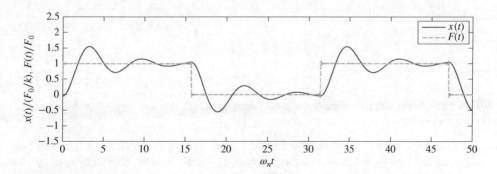

Figure 2.24 Steady-state response for system of damping ratio $\zeta = 0.2$ to square-form forcing at frequency ratio $r = 0.2$

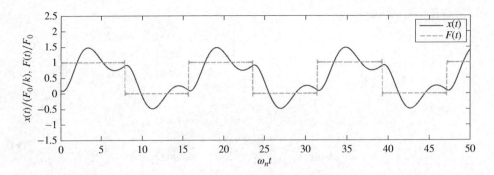

Figure 2.25   Steady-state response for system of damping ratio $\zeta = 0.2$ to square-form forcing at frequency ratio $r = 0.4$

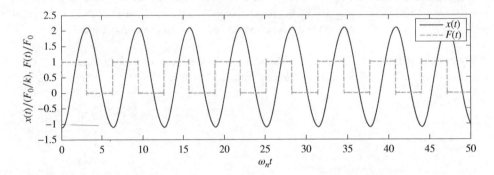

Figure 2.26   Steady-state response for system of damping ratio $\zeta = 0.2$ to square-form forcing at frequency ratio $r = 1$

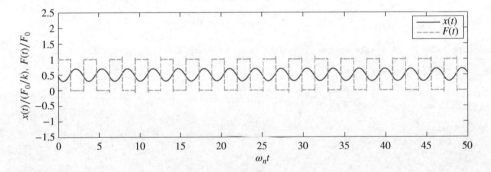

Figure 2.27   Steady-state response for system of damping ratio $\zeta = 0.2$ to square-form forcing at frequency ratio $r = 2$

When there is no obvious way of checking the correctness of the plotted curves, we include the forcing function to make sure that this function is plotted correctly. This way, we know that the Fourier serial summations are constructed correctly. We verify that the forcing functions are indeed plotted correctly.

From the responses in Figs. 2.24 through 2.27, we observe that, when the forcing frequency is low, such as $r < 0.4$, the system's response is very similar to the free vibration with a displacement initial condition, initiated one after another. At $r = 0.4$, one free vibration does not fully developed before the next one occurs. However, at higher forcing frequencies, the responses are very different: the responses are essentially harmonic. This is counter-intuitive because the input has a square shape. We want to explore this further.

Fourier Spectrum

The forcing function has been decomposed into different frequency components. Probably only a few frequency components of the forcing gain significant dynamic amplifications while the remaining components are suppressed. If we compare forcing frequency components with the system's response spectrum, we can identify which components are amplified and which components are suppressed. If we plot the Fourier coefficient for individual components in an amplitude versus frequency plot, we obtain a so-called *Fourier spectrum* of the forcing function. Typically, such a spectrum is not continuous, as a Fourier serial expansion contains only discrete frequencies Ω_n. A component is represented as a vertical bar whose height represents the value. The comparison between the system's response spectrum and the forcing function's Fourier spectrum for the cases $r = 0.1, 0.4, 1$, and 2 is shown in Fig. 2.28, which is produced by the following MATLAB code:

```
zeta=0.2;                      % Damping ratio
omega_d = sqrt(1-zeta^2);      % Damped natural frequency
N = 30;                        % Truncation size for Fourier expansions

raxis = [0.:0.01:6];           % Range of r
spect = 1./sqrt((1-raxis.^2).^ 2+(2*zeta*raxis).^2 ); % Response spectrum

r=[2; 1; .4; .1];              % Ratio r for four forcing cases
i=[1:N];                       % Index i in Fourier expansion
n = 2*i-1;                     % Index n in Fourier expansion
freq = r*n;                    % Frequencies in Fourier expansion
coef = 2/pi ./ n;              % Fourier coefficients

plot(raxis,spect,'m');         % Plotting response spectrum
hold on;
stem( freq(1,:), coef, 'Color','r','Marker','s'); % Plot Fourier spectra
stem( freq(2,:), coef, 'Color','y','Marker','d'); % for 4 loading cases
stem( freq(3,:), coef, 'Color','g','Marker','o');
stem( freq(4,:), coef, 'Color','b','Marker','^');
axis([0 5 -.2 3]);
legend('System response spectrum','Fourier spectrum for r=2',...
    'Fourier spectrum for r=1', 'Fourier spectrum for r=0.4', ...
    'Fourier spectrum for r=0.2');
xlabel('r');
ylabel('Dynamic amplification and Fourier coefficient');
```

In the above code, `raxis` and `spect` are a pair of arrays that holds the system's response spectrum. The Fourier spectrum for the forcing functions is stored in `freq` and `coef` pair. The command `stem()` plots discrete data using a stem-like marker.

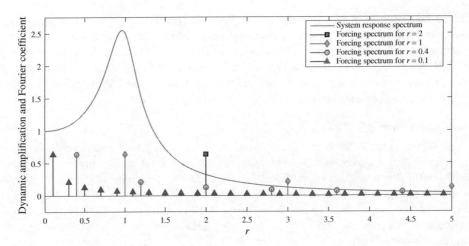

Figure 2.28 Fourier spectra of four forcing frequency ratios of $r = 2, 1, 0.4$, and 0.1 plotted against system's response spectrum at $\zeta = 0.2$

We verify that the system's response spectrum is identical to the one in Fig. 2.15, and that the Fourier spectra agree with eqn. (e) for the magnitudes. For the frequencies of Fourier components, the first component matches the r-coordinate value and the nth component is located at nr, which is present only when n is odd.

In the system's response spectrum, only when $r < \sqrt{2(1 - 2\zeta^2)} \approx 1.4$, the dynamic amplification is larger than unity; at higher frequencies, the motion is suppressed. In the amplified region, only forcing components whose frequencies fall within the system's resonance peak will be noticeably amplified. According to eqn. (2.94), the peak width is $2\zeta = 0.4$, the resonance peak is located between $r \approx 0.8$ and $r \approx 1.2$.

Comparing the forcing's Fourier spectra with the system's response spectrum, we observe the following: (1) For the case $r = 2$, none of its Fourier components is located within the range of being amplified, hence the overall amplitude is small. In such cases, only the first component makes a significant contribution; the second and higher frequency components have smaller forcing amplitudes and are more severely suppressed. Thus, the response in Fig. 2.27 has a small amplitude and is harmonic. (2) For the case $r = 1$, the first component is located right at the resonance peak, thus the overall amplitude is high. Other components again have comparatively much smaller amplitudes and all being suppressed. The response is again harmonic. (3) For the cases $r \leq 0.4$, the first component is located in the "slightly amplified" region, there are other components located in amplified range, some even at the resonance peak. Overall, the system's response exhibits characteristics of having multiple frequency components. We can expect that as r becomes smaller, that is, the loading duration becomes longer, more components will contribute to the response.

Gibbs Phenomenon

Another interesting feature in Figs. 2.23 through 2.27 is the wiggles at every corner of the forcing functions. This is the famous *Gibbs phenomenon* in Fourier expansion: at a discontinuity point, no matter how many terms we include in the Fourier serial expansion, there will always be such wiggles. We simplify the earlier MATLAB code to plot only the Fourier expansion for the forcing function as the following:

```
N = 200;                    % Truncation size
r = 1;                      % Frequency ratio
t = [0:.001:4];             % Time range
i = [1:N];                  % Index i in Fourier expansion
n = 2*i-1;                  % Index n in Fourier expansion
freq = n*r;                 % Component frequencies in Fourier expansion
coef = 2/pi ./ n;           % Fourier coefficients
forcing = .5 + coef * sin ( freq' *t );    % Forcing in Fourier expansion
plot ( t, forcing );
axis([2.5 3.5 -.5 1.5]);    % Limiting the range to be shown
xlabel('\omega_n t');
ylabel('F(t)/F_0');
```

Fourier serial expansions for the forcing function using 2, 20, 200, and 2000 terms are plotted in Fig. 2.29 for a frequency ratio of $r = 1$. At $r = 1$, the forcing drops from F_0 to 0 at $\bar{t} = \pi$, as seen in Fig. 2.26. To take a closer look at the jump, Fig. 2.29 shows only the range $2.5 \le \bar{t} \le 3.5$. We can see that, even with 2000 terms, the wiggle still exists. In Figs. 2.23 through 2.27, 200 terms are used in the computation for the Fourier expansion, precisely because of this reason. If we only need to compute the system's response, as demonstrated by the Fourier spectra, a much smaller truncation size suffices.

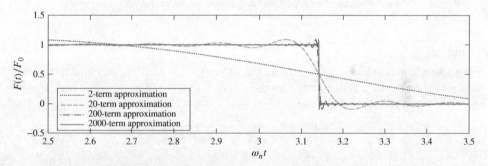

Figure 2.29   Gibbs phenomenon in Fourier expansion near the jump at $\bar{t} = \pi$

2.7   Forced Vibrations II: Transient Responses

The second type of system's responses due to forced excitations is immediately after the start of the excitation. This is referred to as the *transient response* of the system. Unless otherwise noted, we shall focus on underdamped systems.

The equation of motion in such cases is still the following:

$$\ddot{x} + 2\zeta\omega_n\dot{x} + \omega_n^2 x = f(t) \tag{2.103}$$

where $f(t) = F(t)/m$. But $f(t)$ can be any function. As mentioned at the beginning of this chapter, according to the theory of ordinary differential equations, the total solution to eqn. (2.103), which is the mathematical term for the transient response, consists of a particular solution $x_p(t)$ and the homogeneous solution $x_h(t)$.

Obviously, the free-vibration analysis has helped us find the homogeneous solution as in eqn. (2.53), which is repeated here for convenience:

$$x_h(t) = e^{-\zeta\omega_n t}\left(A\cos\omega_d t + B\sin\omega_d t\right) \tag{2.104}$$

The study of steady-state responses also gives us a procedure to find the particular solution when the loading is periodic. But the focus of this section is on general nonperiodic loadings.

Since we are fresh off studying the steady-state responses due to periodic loadings, we shall start with an example of periodic loading to observe how the solution evolves from the transient phase into the steady state.

2.7.1 Transient Response to Periodic Loading

For periodic loadings, combining the two solution components, namely the steady-state and the free-vibration responses, would conveniently give us the total solution. We use one example to illustrate the general procedure for finding the transient response of the system.

■ Example 2.9: Example of Periodic Loading

Assume that the mass–spring–dashpot system is initially at rest. At $t = 0$, it is subjected to a harmonic loading $F_0\cos\Omega t$. Find the system's response from the very beginning.

□ Solution:

We note that the loading is the same as the one in Section 2.6.1, and the steady-state response, which is the particular solution in the terminology of differential equations, has been found in eqn. (2.83) as

$$x_p(t) = \frac{f_0}{\omega_n^2}\frac{\cos(\Omega t - \phi)}{\sqrt{(1 - r^2)^2 + (2\zeta r)^2}} \tag{a}$$

where

$$r = \frac{\Omega}{\omega_n} \quad\text{and}\quad \phi = \tan^{-1}\frac{2\zeta r}{1 - r^2} \tag{b}$$

Combined with the homogeneous solution in eqn. (2.104), the total solution is assumed as

$$x(t) = e^{-\zeta\omega_n t}(A\cos\omega_d t + B\sin\omega_d t) + \frac{f_0}{\omega_n^2}\frac{\cos(\Omega t - \phi)}{\sqrt{(1 - r^2)^2 + (2\zeta r)^2}} \tag{c}$$

Before evaluating the initial conditions, we need to take a time derivative of the solution in eqn. (c) to obtain the expression for the velocity. Thus,

$$\dot{x}(t) = -\zeta\omega_n e^{-\zeta\omega_n t}(A\cos\omega_d t + B\sin\omega_d t)$$

$$+ \omega_d e^{-\zeta\omega_n t}(-A\sin\omega_d t + B\cos\omega_d t) - \Omega\frac{f_0}{\omega_n^2}\frac{\sin(\Omega t - \phi)}{\sqrt{(1 - r^2)^2 + (2\zeta r)^2}} \tag{d}$$

The initial conditions are that the system is "at rest," meaning

$$x(0) = 0 \quad \text{and} \quad \dot{x}(0) = 0. \tag{e}$$

Substituting eqns. (c) and (d) into the initial conditions in eqn. (e) gives

$$0 = A + \frac{f_0}{\omega_n^2} \frac{\cos\phi}{\sqrt{(1 - r^2)^2 + (2\zeta r)^2}} \tag{f}$$

$$0 = -\zeta\omega_n A + \omega_d B + \Omega \frac{f_0}{\omega_n^2} \frac{\sin\phi}{\sqrt{(1 - r^2)^2 + (2\zeta r)^2}} \tag{g}$$

Solving the above two equations gives

$$A = -\frac{f_0}{\omega_n^2} \frac{\cos\phi}{\sqrt{(1 - r^2)^2 + (2\zeta r)^2}} \tag{h}$$

$$B = -\frac{f_0}{\omega_n^2} \frac{\zeta\omega_n \cos\phi + \Omega \sin\phi}{\sqrt{(1 - r^2)^2 + (2\zeta r)^2}} \tag{i}$$

Thus, the total solution, or the system's transient response, is

$$x(t) = -\frac{f_0}{\omega_n^2} \frac{e^{-\zeta\omega_n t}}{\sqrt{(1 - r^2)^2 + (2\zeta r)^2}} \left[\cos\phi \cos\omega_d t + \left(\zeta\frac{\omega_n}{\omega_d} \cos\phi \right. \right.$$
$$\left. \left. + \frac{\Omega}{\omega_d}\sin\phi \right) \sin\omega_d t \right] + \frac{f_0}{\omega_n^2} \frac{\cos(\Omega t - \phi)}{\sqrt{(1 - r^2)^2 + (2\zeta r)^2}} \tag{j}$$

This might be the end of finding the total solution. But for those perfectionists among us, we can simplify it still further. First, we express the terms in the square brackets in the form of $C\cos(\omega_d t - \phi_0)$, as we have done many times earlier. This gives

$$C^2 = \cos^2\phi + \left(\zeta\frac{\omega_n}{\omega_d}\cos\phi + \frac{\Omega}{\omega_d}\sin\phi \right)^2$$

$$= \cos^2\phi \left[1 + \zeta^2 \left(\frac{\omega_n}{\omega_d} \right)^2 + 2\zeta\frac{\omega_n}{\omega_d}\frac{\Omega}{\omega_d}\tan\phi + \left(\frac{\Omega}{\omega_d} \right)^2 \tan^2\phi \right]$$

$$= \cos^2\phi \left[1 + \frac{\zeta^2}{1 - \zeta^2} + \frac{2\zeta r}{1 - \zeta^2}\frac{2\zeta r}{1 - r^2} + \frac{r^2}{1 - \zeta^2}\frac{(2\zeta r)^2}{(1 - r^2)^2} \right]$$

$$= \frac{(1 - r^2)^2}{(1 - r^2)^2 + (2\zeta r)^2} \frac{(1 - r^2)^2 + (2\zeta r)^2(1 - r^2) + r^2(2\zeta r)^2}{(1 - \zeta^2)(1 - r^2)^2} = \frac{1}{1 - \zeta^2} \tag{k}$$

where $\tan\phi$ as given in eqn. (b) has been used, and

$$\tan\phi_0 = \frac{\zeta\frac{\omega_n}{\omega_d}\cos\phi + \frac{\Omega}{\omega_d}\sin\phi}{\cos\phi} = \zeta\frac{\omega_n}{\omega_d} + \frac{\Omega}{\omega_d}\tan\phi$$

$$= \frac{\zeta}{\sqrt{1 - \zeta^2}} + \frac{r}{\sqrt{1 - \zeta^2}}\frac{2\zeta r}{1 - r^2} = \frac{\zeta(1 + r^2)}{\sqrt{1 - \zeta^2}(1 - r^2)} \tag{l}$$

Finally, the total solution can be written as

$$x(t) = \frac{f_0/\omega_n^2}{\sqrt{(1-r^2)^2 + (2\zeta r)^2}} \left[\cos(\Omega t - \phi) - \frac{e^{-\zeta\omega_n t}}{1-\zeta^2} \cos(\omega_d t - \phi_0) \right] \tag{m}$$

□ **Exploring the Solution with MATLAB**

We introduce the following nondimensionalization parameters:

$$\bar{x}(t) = \frac{x(t)}{f_0/\omega_n^2} \qquad \bar{t} = \omega_n t \quad \text{and} \quad \bar{\omega}_d = \frac{\omega_d}{\omega_n} = \sqrt{1-\zeta^2} \tag{n}$$

The system's nondimensionalized transient response can be written as

$$\bar{x}(t) = \frac{1}{\sqrt{(1-r^2)^2 + (2\zeta r)^2}} \left[\cos(r\bar{t} - \phi) - \frac{e^{-\zeta\bar{t}}}{1-\zeta^2} \cos(\bar{\omega}_d \bar{t} - \phi_0) \right] \tag{o}$$

and the phase angles are already expressed in a nondimensionalized form. The MATLAB code is as the following:

```
r = .2;                              % Frequency ratio
zeta = 0.05;                         % Damping ratio
omega_d = sqrt( 1 - zeta^2 );        % Damped natural frequency
phi = atan2( 2*r*zeta, 1-r^2 );      % Phase angle phi and phi0
phi0 = atan2( zeta*(1+r^2), (1-r^2)*sqrt(1-zeta^2) );

t=[0:0.02:100];                      % Time range
x = ( cos(r*t-phi) - exp(-zeta*t)/(1-zeta^2) .* cos(omega_d*t-phi0 ))...
      / sqrt( (1 - r^2)^2 + (2*zeta*r)^2 );   % Transient response
f = cos ( r*t );                     % Forcing
plot(t,x,'r');                       % Plotting the response
hold on;                             % Turn on "hold" for next plot command
plot(t,f,'--g');                     % Plotting the forcing
legend('x(t)','F(t)');               % Legend box
xlabel('\omega_n t');
ylabel('x(t)/(F_0/k)');
```

The system's responses for four different frequency ratios $r = 0.1, 0.5, 1$, and 2 are shown (solid curve) in Figs. 2.30 through 2.33 for a damping ratio of $\zeta = 0.1$. In each of these figures, the forcing function (dashed curve) is also plotted for comparison.

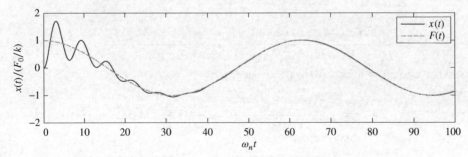

Figure 2.30 System's transient response for $r = 0.1$ at damping ratio $\zeta = 0.1$

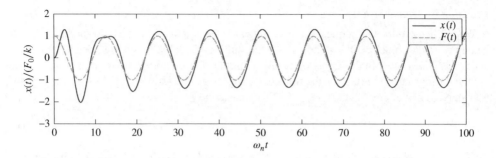

Figure 2.31 System's transient response for $r = 0.5$ at damping ratio $\zeta = 0.1$

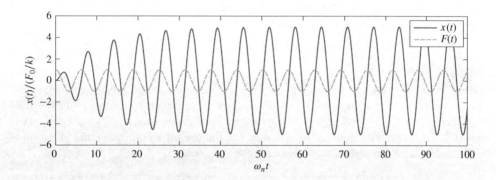

Figure 2.32 System's transient response for $r = 1$ at damping ratio $\zeta = 0.1$

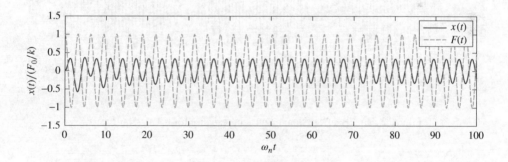

Figure 2.33 System's transient response for $r = 2$ at damping ratio $\zeta = 0.1$

We verify that each curve starts from $\bar{x} = 0$ with a zero slope. We also verify that the periods of the forcing functions are correct, and at a large time t, the system's response assumes a steady amplitude at the same period of the forcing function.

We observe that, at a low driving frequency ($r = 0.1$), the system's response is rather different from the other cases: it consists of a high-frequency component (at the system's natural frequency) riding on the top of a lower-frequency harmonic motion. In other cases, the system's natural frequency is not readily observable. The responses are quickly dominated by the steady-state responses.

Duration of Transient Response

We might ask ourselves a question: how long after the application of a periodic loading will be considered as reaching the steady state? Or, put it differently, how long does the transient state last? The answer to this question is rather simple. The solution has two components: one for the steady state, and one for the transient state. The latter is modulated by a decay factor $e^{-\zeta \omega_n t}$. Note that $e^{-\pi} \approx 0.04321 < 5\%$. This means, when

$$\zeta \omega_n t > \pi \quad \text{or} \quad \frac{t}{T} > \frac{\pi}{2\pi\zeta} = \frac{1}{2\zeta} = Q \tag{p}$$

the transient component of the response has been reduced to less than 5% of its initial amplitude. Interestingly, we have collected one more meanings for the Q-factor! For a system with a damping ratio of $\zeta = 0.1$, this only takes about five periods. We can see from curves in Figs. 2.30 through 2.33 that the transient effects vanish by the time $\omega_n t \approx 31$. The case of $r = 0.1$ is the clearest among the four cases.

Beat Phenomenon

Another interesting phenomenon that can be observed in this example is the so-called *beat*. This occurs when the damping ratio is small and the forcing frequency is close to the resonant frequency. Figures 2.34 through 2.36 show the beat phenomenon for $r = 0.95$ at $\zeta = 0, 0.01$, and 0.02. The main characteristic of a beat is that the overall amplitude of the vibration is modulated by another oscillation, at a much slower frequency.

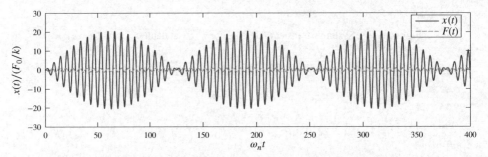

Figure 2.34 "Beat": system's transient response for $r = 0.95$ at $\zeta = 0$

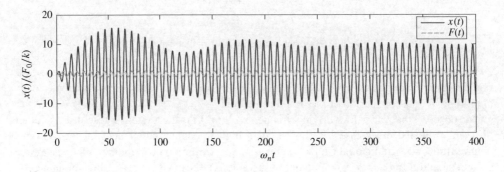

Figure 2.35 "Beat": system's transient response for $r = 0.95$ at $\zeta = 0.01$

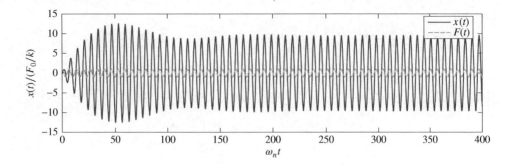

Figure 2.36   "Beat": system's transient response for $r = 0.95$ at $\zeta = 0.02$

The beat phenomenon is the result of transient response interferencing with the steady-state response. When the two frequencies are very close, they construct and destruct with each other periodically. We can analyze this using an illustrative case when the two components are of the same amplitude and without damping:

$$A \sin \Omega t + A \sin \omega_d t = 2A \sin \frac{\Omega + \omega_d}{2} t \cos \frac{\Omega - \omega_d}{2} t \qquad (q)$$

The higher frequency factor has a frequency of $(\Omega + \omega_d)/2 \approx \Omega \approx \omega_d$, which is the oscillatory part, and the lower-frequency component has a frequency of $(\Omega - \omega_d)/2$, which forms the beat. Physically, since the system is excited near its natural frequency, the system responds strongly at its natural frequency and in the mean time adjusts to the driving frequency. The beat dies down when the transient response dissipates.

2.7.2   General Loading: Direct Analytical Method

We now turn our attention to general nonperiodic loadings.

As mentioned earlier, the total solution for eqn. (2.103) consists of a particular solution plus the homogeneous solution. The latter, for underdamped systems, has been given in eqn. (2.104). Thus, the remaining challenge is to find the particular solution: any function of t that satisfies the differential equation. Of course, we might also encounter a practical engineering difficulty of expressing a general loading in a function form. Here, we assume that this difficulty has been overcome.

The first method to find the transient responses of systems to general loadings is called the *direct analytical method*. In this method, we try to solve the differential equation directly if the forcing function $f(t)$ is given in analytical form. In the solution process, we assume a function form for the particular solution. The general "rule of thumb" is to assume the particular solution to be a linear combination of functions that constitute the forcing function and its first and second derivatives.

■ **Example 2.10: Transient Response to Step Loading**

Assume that a mass–spring–dashpot system is initially at rest and is subject to a constant force F_0 starting from $t = 0$, that is,

$$F(t) = \begin{cases} F_0 & \text{when } t \geq 0 \\ 0 & \text{when } t < 0 \end{cases}$$

Find the subsequent transient response of the system.

□ **Solution:**

Incorporating the forcing $F(t)$, the normalized equation of motion is, for $t \geq 0$,

$$\ddot{x} + 2\zeta\omega_n\dot{x} + \omega_n^2 x = f_0 \tag{a}$$

where $f_0 = F_0/m$. We make a list of functions contained in the forcing and its derivatives:

$$\text{Forcing function itself} \quad f(t) \rightarrow 1 \text{ (constant)}$$
$$\text{First derivative} \quad \dot{f}(t) \rightarrow 0$$
$$\text{Second derivative} \quad \ddot{f}(t) \rightarrow 0$$

Note that we are only concerned with function forms here, not their coefficients. Thus, the constant function form is represented by 1. In the above tabulation, the only nonvanishing function form is a constant. Thus, we assume the particular solution as

$$x_p(t) = C \tag{b}$$

Substituting this particular solution into the equation of motion in eqn. (a) gives

$$C = \frac{f_0}{\omega_n^2}$$

Thus, combining the particular solution with the homogeneous solution in eqn. (2.104), the total solution can be assumed as

$$x(t) = \frac{f_0}{\omega_n^2} + e^{-\zeta\omega_n t}(A\cos\omega_d t + B\sin\omega_d t) \tag{c}$$

Taking a time derivative, the corresponding velocity is expressible as

$$\dot{x}(t) = -\zeta\omega_n e^{-\zeta\omega_n t}(A\cos\omega_d t + B\sin\omega_d t)$$
$$+ \omega_d e^{-\zeta\omega_n t}(-A\sin\omega_d t + B\cos\omega_d t) \tag{d}$$

The initial conditions are "at rest," meaning

$$x(0) = 0 \quad \text{and} \quad \dot{x}(0) = 0 \tag{e}$$

Substituting eqns. (c) and (d) into the initial conditions in eqn. (e) gives

$$0 = \frac{f_0}{\omega_n^2} + A \tag{f}$$

$$0 = -\zeta\omega_n A + \omega_d B \tag{g}$$

Solving the above two equations gives

$$A = -\frac{f_0}{\omega_n^2} \quad \text{and} \quad B = -\frac{\zeta\omega_n}{\omega_d}\frac{f_0}{\omega_n^2} = -\frac{\zeta}{\sqrt{1-\zeta^2}}\frac{f_0}{\omega_n^2} \tag{h}$$

Then, the system's response can be written as

$$x(t) = \frac{f_0}{\omega_n^2} \left[1 - e^{-\zeta\omega_n t} \left(\cos\omega_d t + \frac{\zeta}{\sqrt{1-\zeta^2}} \sin\omega_d t \right) \right] \tag{i}$$

We can further simplify the terms in the parentheses to the form of $X \cos(\omega_d t - \phi)$. According to eqn. (2.21),

$$X = \sqrt{1 + \left(\frac{\zeta}{\sqrt{1-\zeta^2}} \right)^2} = \frac{1}{\sqrt{1-\zeta^2}} \quad \text{and} \quad \phi = \tan^{-1}\frac{\zeta}{\sqrt{1-\zeta^2}} \tag{j}$$

Finally, the transient response of the system can be written as

$$x(t) = \frac{f_0}{\omega_n^2} \left[1 - \frac{e^{-\zeta\omega_n t}}{\sqrt{1-\zeta^2}} \cos(\omega_d t - \phi) \right] \tag{k}$$

Note that $f_0/\omega_n^2 = F_0/k$, which is the static deflection of the system due to F_0.

□ **Exploring the Solution with MATLAB**

We define the following nondimensionalized parameters:

$$\bar{x}(t) = \frac{x(t)}{f_0/\omega_n^2} = \frac{x(t)}{F_0/k} \qquad \bar{t} = \omega_n t \quad \text{and} \quad \bar{\omega}_d = \frac{\omega_d}{\omega_n} = \sqrt{1-\zeta^2}$$

Then the solution can be written as

$$\bar{x}(t) = 1 - \frac{e^{-\zeta\bar{t}}}{\bar{\omega}_d} \cos\left(\bar{\omega}_d \bar{t} - \phi\right) \tag{l}$$

The following MATLAB code plots the system's response:

```
zeta=0.1;                          % Damping ratio
omega_d = sqrt(1-zeta^2);          % Damped natural frequency
t = [0.:0.01:50];                  % Time range
phi = atan2( zeta/omega_d, 1);     % Phase angle
x = 1 - exp( -zeta*t ) .*  cos(omega_d *t - phi ) / omega_d;
plot(t,x);
xlabel('\omega_nt');
ylabel('x(t)/(F_0/k)');
```

The system's responses for damping ratios $\zeta = 0.1$, 0.25, 0.5, 0.75, and 0.99 are plotted in Fig. 2.37. We verify that all curves start from $\bar{x} = 0$ with a slope of zero and approach to unity for large time $\bar{t}$, which is more evident in the cases with higher damping ratios.

We observe that overall the system behaves just like free vibrations we observed earlier. In fact, comparing Fig. 2.37 with Fig. 2.8 in Example 2.2, they are identical except that the vertical coordinate is shifted. This is not a coincidence. Recall that, in Examples 1.1 and 1.18, the equation of motion for the mass–spring–dashpot system was derived twice in each example.

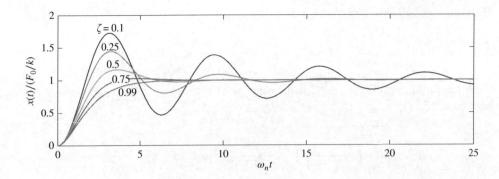

Figure 2.37   Transient responses to step loading at different damping ratios

In one case, when the displacement is measured from the equilibrium position, the resulting equation of motion is homogeneous as in Example 2.2. In another case, when the displacement is measured from the position when the spring is unstretched, the resulting equation of motion is exactly the same as eqn. (a), with F_0 being mg, as in Solution 1 in Examples 1.1 and 1.18.

———————— . ————————

Figure 2.37 is an important plot for many mechanical systems and, in fact, for any dynamical system. Many important terminologies often used in describing a dynamical system are illustrated by this figure. It is reproduced with just one curve and annotated in Fig. 2.38, and then discussed outside the context of the example.

The steady-state value represents the *target value* that the system is tasked to achieve. A portion of the response exceeding the target value is called an *overshoot*; and the *maximum overshoot* occurs at the first peak. The overshoot decreases when the damping is increased. When the damping ratio is sufficiently large, there will be no overshoot. However, suppressing the overshoot comes with a price in the system's responsiveness.

The responsiveness of the system is best observed along the time axis. The time when the system reaches the peak of the overshoot is called the *peak time*. However, a more meaningful measure for the system's responsiveness is the *rise time*, which is the time duration for the

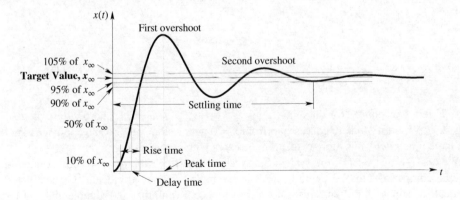

Figure 2.38   Key dynamical response characteristics of single-DOF system

system's response to rise from 10% to 90% of the target value. The *delay time* is the time when the response reaches 50% of the target value. The *settling time* is the time when the overall response reaches within a $\pm 5\%$ (or $\pm 2\%$ in some applications) bracket of the target value.

From this example, we summarize the procedure for finding the transient response of the system as the following:

- Construct a particular solution from a linear combination of the functions contained in the forcing function and its first and second derivatives.
- Substitute the particular solution into the equation of motion and determine the constants in the particular solution by matching coefficients of individual functions.
- Combine the particular solution and the homogeneous solution with yet-to-be-determined constants to form the total solution.
- Use the initial conditions to determine the constants in the total solution.

A common scenario when forcing a system into motion is to start when the system is at rest. Mathematically, such initial conditions are written as $x(0) = 0$ and $\dot{x}(0) = 0$. Such initial conditions are often called the *trivial initial conditions*.

In the following, we follow this procedure to work out another example.

■ Example 2.11: Transient Response to Ramp Loading

Assume that a mass–spring–dashpot system is initially at rest. It is subjected to a linearly increasing force $F(t) = at$ starting from $t = 0$. Find the system's transient response.

□ **Solution:**

Incorporating the force $F(t)$, the normalized equation of motion can be written as

$$\ddot{x} + 2\zeta\omega_n\dot{x} + \omega_n^2 x = \alpha t \tag{a}$$

where $\alpha = a/m$. We list the functions contained in the forcing function and its derivatives:

$$\text{Forcing function itself } f(t) \rightarrow t$$

$$\text{First derivative } \dot{f}(t) \rightarrow 1 \text{ (constant)}$$

$$\text{Second derivative } \ddot{f}(t) \rightarrow 0$$

Hence, we assume the particular solution as a linear combination of the functions above, as

$$x_p(t) = C_1 t + C_2 \tag{b}$$

Substituting this particular solution into the equation of motion in eqn. (a) gives

$$2\zeta\omega_n C_1 + \omega_n^2(C_1 t + C_2) = \alpha t \tag{c}$$

Since this equation must hold for all times, coefficients for respective powers of t must be equal on both sides of the equation, that is,

$$\omega_n^2 C_1 = \alpha \tag{d}$$

$$2\zeta\omega_n C_1 + \omega_n^2 C_2 = 0 \tag{e}$$

Solving the above two equations gives

$$C_1 = \frac{\alpha}{\omega_n^2} \quad \text{and} \quad C_2 = -\frac{2\zeta\alpha}{\omega_n^3} \tag{f}$$

Thus, the particular solution is

$$x_p(t) = \frac{\alpha}{\omega_n^3}\left(\omega_n t - 2\zeta\right) \tag{g}$$

Combining the particular solution with homogeneous solution in eqn. (2.104), the total solution can be assumed as

$$x(t) = \frac{\alpha}{\omega_n^3}\left(\omega_n t - 2\zeta\right) + e^{-\zeta\omega_n t}(A\cos\omega_d t + B\sin\omega_d t) \tag{h}$$

Taking a time derivative gives the corresponding velocity as

$$\dot{x}(t) = \frac{\alpha}{\omega_n^2} - \zeta\omega_n e^{-\zeta\omega_n t}(A\cos\omega_d t + B\sin\omega_d t)$$

$$+ \omega_d e^{-\zeta\omega_n t}(-A\sin\omega_d t + B\cos\omega_d t) \tag{i}$$

The initial conditions are the so-called trivial initial conditions. Evaluating eqns. (h) and (i) for the trivial initial conditions gives

$$0 = -\frac{2\zeta\alpha}{\omega_n^3} + A \tag{j}$$

$$0 = \frac{\alpha}{\omega_n^2} - \zeta\omega_n A + \omega_d B \tag{k}$$

which can be solved for

$$A = \frac{2\zeta\alpha}{\omega_n^3} \quad \text{and} \quad B = -\frac{(1 - 2\zeta^2)\alpha}{\omega_n^2\omega_d} \tag{l}$$

Then, the transient response of the system is

$$x(t) = \frac{\alpha}{\omega_n^3}\left(\omega_n t - 2\zeta\right) - e^{-\zeta\omega_n t}\frac{\alpha}{\omega_n^3}\left[-2\zeta\cos\omega_d t + \frac{(1 - 2\zeta^2)\omega_n}{\omega_d}\sin\omega_d t\right] \tag{m}$$

The terms inside the square brackets can be converted into an amplitude–phase form as

$$\frac{(1 - 2\zeta^2)\omega_n}{\omega_d}\sin\omega_d t - 2\zeta\cos\omega_d t = \frac{1}{\sqrt{1 - \zeta^2}}\sin(\omega_d t - \phi)$$

where the relation $\omega_d = \sqrt{1 - \zeta^2}\omega_n$ has been used and

$$\phi = \tan^{-1}\frac{2\zeta\sqrt{1 - \zeta^2}}{1 - 2\zeta^2} \tag{n}$$

Finally, the system's transient response can be written as

$$x(t) = \frac{\alpha}{\omega_n^3}\left[\omega_n t - 2\zeta - \frac{e^{-\zeta\omega_n t}}{\sqrt{1 - \zeta^2}}\sin(\omega_d t - \phi)\right] \tag{o}$$

□ Exploring the Solution with MATLAB

We introduce the following nondimensionalized parameters:

$$\bar{x}(t) = \frac{x(t)}{\alpha/\omega_n^3} \qquad \bar{t} = \omega_n t \quad \text{and} \quad \bar{\omega}_d = \frac{\omega_d}{\omega_n} = \sqrt{1 - \zeta^2} \tag{p}$$

Then, the nondimensionalized solution can be written as

$$\bar{x}(\bar{t}) = \bar{t} - 2\zeta - \frac{e^{-\zeta\bar{t}}}{\bar{\omega}_d} \sin\left(\bar{\omega}_d\bar{t} - \phi\right) \tag{q}$$

The following MATLAB code is used to plot the system's responses in Fig. 2.39 for damping ratios $\zeta = 0, 0.1, 0.25$, and 0.5:

```
zeta=0.1;                                        % Damping ratio
omega_d = sqrt(1-zeta^2);                         % Damped natural frequency
phi = atan2( 2*zeta*omega_d, 1 -2*zeta^2);        % Phase angle
t = [0.:0.01:25];                                 % Time range
x = t - 2*zeta - exp(-zeta*t)/omega_d .* sin(omega_d*t-phi);   % Response
plot(t,x,'Color', [0 0 1]);
xlabel('\omega_n t');
ylabel('x(t)/(\alpha/{\omega_n^3})');
```

We verify that each curve starts from $\bar{x} = 0$, with a zero slope, and quickly approaches its asymptote $\bar{x} = \bar{t} - 2\zeta$, as determined by eqn. (o).

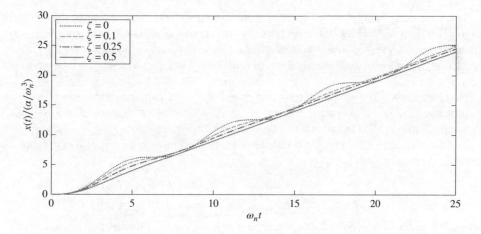

Figure 2.39 Transient responses to ramp loading at different damping ratios

We observe that the dominant feature is the particular solution, which is a straight line (linear function of time), having a shift that is proportional to the damping ratio. The transient effects are only noticeable at very low damping ratios of $\zeta = 0$ and 0.1. At a damping ratio of 0.25 and higher, the transient effects have largely decayed after a half-period, when the response has overcome the zero slope at the beginning.

2.7.3   Laplace Transform Method

We can also use the Laplace transform method to solve the transient response problem. The Laplace transform is typically applied to functions having time as its argument, transforming the real-valued argument time t, which is naturally defined only in the domain $t \geq 0$, into a function of a complex argument s. Furthermore, the transform process has already incorporated the initial conditions,

$$\mathscr{L}[x(t)] = \tilde{x}(s) \tag{2.105}$$

$$\mathscr{L}[\dot{x}(t)] = s\tilde{x}(s) - x(0) \tag{2.106}$$

$$\mathscr{L}[\ddot{x}(t)] = s^2\tilde{x}(s) - sx(0) - \dot{x}(0) \tag{2.107}$$

Applying the Laplace transform to the equation of motion, in conjunction with the initial conditions, gives

$$s^2\tilde{x}(s) - sx(0) - \dot{x}(0) + 2\zeta\omega_n[s\tilde{x}(s) - x(0)] + \omega_n^2\tilde{x}(s) = \tilde{f}(s) \tag{2.108}$$

Therefore, the transformed solution is

$$\tilde{x}(s) = \frac{\tilde{f}(s) + (s + 2\zeta\omega_n)x(0) + \dot{x}(0)}{s^2 + 2\zeta\omega_n s + \omega_n^2} \tag{2.109}$$

The only remaining task is to perform the inverse transform. Once the inverse transform is found, we have the solution. This sounds too easy! But, it is true! Just not as easy as it sounds. In fact, finding the inverse is the most difficult part, which seriously limits the wide applicability of this method.

There are many readily tabulated Laplace transform pairs. One way of finding the inverse is to look up from such tables, and Appendix C of this textbook obligatorily lists some Laplace transform pairs. There is also a general *Laplace inverse transform* to pair with the Laplace transform. But it usually involves analysis of functions of complex variables, which we would prefer to avoid.

Here, we introduce a simple method that does not require us to memorize too many techniques and tabulated transform pairs, and yet has a wide range of applicability to vibration problems. This is called the *partial fraction method*.

For many mechanical vibration problems, eqn. (2.109) takes the form of one polynomial divided by another, such as

$$\frac{a_m s^m + a_{m-1}s^{m-1} + a_{m-1}s^{m-1} + \cdots + a_1 s + a_0}{b_n s^n + b_{n-1}s^{n-1} + b_{n-1}s^{n-1} + \cdots + b_1 s + b_0} = \frac{P(s)}{Q(s)}$$

where the order of polynomial $Q(s)$, n, is generally higher than that of polynomial $P(s)$, m. Generally, a polynomial $Q(s)$ of order n has n *roots*, denoted as r_i where $i = 1, 2, \ldots, n$, such that $Q(r_i) = 0$, and complex roots come in conjugate pairs. Thus, $Q(s)$ can be factorized into

$$Q(s) = (s - r_1) \cdot (s - r_2) \cdot (s - r_3) \cdots \cdots (s - r_n) \tag{2.110}$$

Subsequently, $P(s)/Q(s)$ can be expressed as the following *partial fractions* as

$$\frac{P(s)}{Q(s)} = \frac{c_1}{s - r_1} + \frac{c_2}{s - r_2} + \cdots + \frac{c_i}{s - r_i} = \sum_{i=1}^{n} \frac{c_i}{s - r_i} \tag{2.111}$$

where c_i's can be found using the following formula:

$$c_i = \frac{P(s)}{(s - r_1) \cdot (s - r_2) \cdot \cdots \cdot (s - r_{i-1}) \cdot (s - r_{i+1}) \cdot \cdots \cdot (s - r_1)}\bigg|_{s=r_i} \qquad (2.112)$$

Note that the factor $(s - r_i)$ is absent from the denominator.

Having found the r_i's and c_i's, then, using the following Laplace transform pair

$$\mathscr{L}[e^{at}H(t)] = \frac{1}{s - a} \qquad (2.113)$$

where $H(t)$ is called the *Heaviside step function* defined as

$$H(t) = \begin{cases} 1 & \text{when } t \geq 0 \\ 0 & \text{when } t < 0 \end{cases} \qquad (2.114)$$

the inverse Laplace transform can be written as

$$\mathscr{L}^{-1}\left[\frac{P(s)}{Q(s)}\right] = \sum_{i=1}^{n} c_i e^{r_i t} \qquad (2.115)$$

We note that, in eqn. (2.115), the Heaviside step function $H(t)$ has been omitted. In general, the behavior of the Heaviside step function is implied in the solution, as the response cannot precede the loading.

■ **Example 2.12: Step Loading via Laplace Transform**

In this example, we rework the step-loading problem in Example 2.10 using the Laplace transform method. The step loading can be conveniently described by the Heaviside step function as $F(t) = F_0 H(t)$.

□ **Solution:**

Note that the Laplace transform for the Heaviside step function is

$$\mathscr{L}[H(t)] = \frac{1}{s} \qquad (a)$$

With the trivial initial conditions, the transformed solution is, according to eqn. (2.109),

$$\tilde{x}(s) = \frac{f_0}{s(s^2 + 2\zeta\omega_n s + \omega_n^2)} \qquad (b)$$

To find the inverse, we first find the roots of the denominator. Note that the setting second factor of the denominator to zero gives system's characteristic equation in eqn. (2.45), with roots $s = -\zeta\omega_n \pm i\omega_d$. Therefore, the roots of the denominator are

$$r_1 = 0 \qquad r_2 = -\zeta\omega_n - i\omega_d \qquad r_3 = -\zeta\omega_n + i\omega_d \qquad (c)$$

Subsequently, the transformed solution can be written as the following partial fractions:

$$\tilde{x}(s) = \frac{c_1}{s} + \frac{c_2}{s + \zeta\omega_n + i\omega_d} + \frac{c_3}{s + \zeta\omega_n - i\omega_d}$$

According to eqn. (2.112), the coefficients c_i's are

$$c_1 = \frac{f_0}{s^2 + 2\zeta\omega_n s + \omega_n^2}\bigg|_{s=0} = \frac{f_0}{\omega_n^2} \tag{d}$$

$$c_2 = \frac{f_0}{s(s + \zeta\omega_n - i\omega_d)}\bigg|_{s=-\zeta\omega_n - i\omega_d} = \frac{f_0}{2i\omega_d(\zeta\omega_n + i\omega_d)} \tag{e}$$

$$c_3 = \frac{f_0}{s(s + \zeta\omega_n + i\omega_d)}\bigg|_{s=-\zeta\omega_n + i\omega_d} = -\frac{f_0}{2i\omega_d(\zeta\omega_n - i\omega_d)} \tag{f}$$

Therefore, the system's response, according to eqn. (2.115), is

$$x(t) = \frac{f_0}{\omega_n^2} + \frac{f_0}{2i\omega_d(\zeta\omega_n + i\omega_d)}e^{(-\zeta\omega_n - i\omega_d)t} - \frac{f_0}{2i\omega_d(\zeta\omega_n - i\omega_d)}e^{(-\zeta\omega_n + i\omega_d)t}$$

$$= \frac{f_0}{\omega_n^2} + \frac{f_0 e^{-\zeta\omega_n t}}{2i\omega_d[(\zeta\omega_n)^2 + \omega_d^2]}\left[(\zeta\omega_n - i\omega_d)e^{-i\omega_d t} - (\zeta\omega_n + i\omega_d)e^{i\omega_d t}\right] \tag{g}$$

Using Euler's formula

$$e^{\pm i\omega_d t} = \cos\omega_d t \pm i\sin\omega_d t \tag{h}$$

The solution can be further simplified as

$$x(t) = \frac{f_0}{\omega_n^2} + \frac{f_0 e^{-\zeta\omega_n t}}{\omega_d[(\zeta\omega_n)^2 + \omega_d^2]}\left(-\omega_d\cos\omega_d t - \zeta\omega_n\sin\omega_d t\right)$$

$$= \frac{f_0}{\omega_n^2}\left[1 - e^{-\zeta\omega_n t}\left(\cos\omega_d t + \frac{\zeta}{\sqrt{1-\zeta^2}}\sin\omega_d t\right)\right] \tag{i}$$

which is the same as eqn. (i) in Example 2.10.

_____ · _____

Through the above example, we shall acknowledge that, in general, this is probably not the fastest approach to find the inverse transform. However, the reason we emphasize this approach, or rather, this procedure, is that it requires memorizing a minimal number of Laplace transform pairs. It is simple, just slightly tedious.

There are cases when a root for polynomial $Q(s)$ is a repeated root, such as

$$Q(s) = (s - r_1) \cdot (s - r_2) \cdots\cdots (s - r_i)^p \cdots\cdots (s - r_{n-p})$$

In this case, r_i is said to be a p-fold repeated root, the partial fraction $P(s)/Q(s)$ can be written as the following:

$$\frac{P(s)}{Q(s)} = \frac{c_1}{s - r_1} + \frac{c_2}{s - r_2} + \cdots$$

$$+ \frac{c_{ip}}{(s - r_i)^p} + \frac{c_{i(p-1)}}{(s - r_i)^{p-1}} + \cdots + \frac{c_{i1}}{s - r_i}$$

$$+ \cdots + \frac{c_{n-p}}{s - r_{n-p}} \tag{2.116}$$

and the coefficients c_{ip} to c_{i1} are determined by the following:

$$c_{ip} = \left. \frac{(s - r_i)^p P(s)}{Q(s)} \right|_{s=r_n} \tag{2.117}$$

$$c_{i(p-1)} = \left. \frac{1}{1!} \frac{d}{ds} \left[\frac{(s - r_i)^p P(s)}{Q(s)} \right] \right|_{s=r_n} \tag{2.118}$$

. . .

$$c_{i1} = \left. \frac{1}{(p - 1)!} \frac{d^{(p-1)}}{ds^{(p-1)}} \left[\frac{(s - r_i)^p P(s)}{Q(s)} \right] \right|_{s=r_i} \tag{2.119}$$

Then, the following Laplace transform pair is useful for finding the inverse

$$\mathcal{L}^{-1} \left[\frac{1}{(s - a)^n} \right] = \frac{1}{(n - 1)!} t^{n-1} e^{at} \tag{2.120}$$

■ **Example 2.13: Ramp Loading via Laplace Transform**

In this example, we reexamine the response of the mass–spring–dashpot system due to a ramp loading: $F(t) = at$, with trivial initial conditions.

□ **Solution:**

The normalized forcing function is

$$f(t) = \alpha t \tag{a}$$

where $\alpha = a/m$. The Laplace transform for the normalized forcing function is

$$\mathcal{L}\{\alpha t\} = \frac{\alpha}{s^2} \tag{b}$$

With the trivial initial conditions, according to eqn. (2.109), the transformed solution is

$$\tilde{x}(s) = \frac{1}{s^2(s^2 + 2\zeta\omega_n s + \omega_n^2)} \tag{c}$$

In this case, $s = 0$ is a twofold repeated root. The other two roots are still $s = -\zeta\omega_n \pm i\omega_d$, the same as in the previous example. (We should have noticed by now that this pair of roots is the characteristic of the system, regardless of the loading.) The transformed solution can be written as the following partial fraction:

$$\tilde{x}(s) = \frac{c_{12}}{s^2} + \frac{c_{11}}{s} + \frac{c_2}{s + \zeta\omega_n + i\omega_d} + \frac{c_3}{s + \zeta\omega_n - i\omega_d}$$

Coefficients c_{12} and c_{11} can be found in accordance with eqns. (2.117) through (2.119) as

$$c_{12} = \left. \frac{\alpha(s - 0)^2}{s^2(s^2 + 2\zeta\omega_n s + \omega_n^2)} \right|_{s=0} = \left. \frac{\alpha}{s^2 + 2\zeta\omega_n s + \omega_n^2} \right|_{s=0} = \frac{\alpha}{\omega_n^2} \tag{d}$$

$$c_{11} = \left. \frac{1}{1!} \frac{d}{ds} \left[\frac{\alpha(s - 0)^2}{s^2(s^2 + 2\zeta\omega_n s + \omega_n^2)} \right] \right|_{s=0} = \left. -\frac{\alpha(2s + 2\zeta\omega_n)}{(s^2 + 2\zeta\omega_n s + \omega_n^2)^2} \right|_{s=0} = -\frac{2\alpha\zeta}{\omega_n^3} \tag{e}$$

and coefficients c_2 and c_3 can be found in accordance with eqn. (2.112) as

$$c_2 = \left.\frac{\alpha}{s^2(s + \zeta\omega_n - i\omega_d)}\right|_{s=-(\zeta\omega_n+i\omega_d)} = -\frac{\alpha}{2i\omega_d(\zeta\omega_n + i\omega_d)^2} \tag{f}$$

$$c_3 = \left.\frac{\alpha}{s^2(s + \zeta\omega_n + i\omega_d)}\right|_{s=(-\zeta\omega_n+i\omega_d)} = \frac{\alpha}{2i\omega_d(\zeta\omega_n - i\omega_d)^2} \tag{g}$$

Therefore, the system's response can be written as, according to eqn. (2.115),

$$
\begin{aligned}
x(t) &= \frac{\alpha}{\omega_n^2}\mathscr{L}^{-1}\left[\frac{1}{s^2}\right] - \frac{2\alpha\zeta}{\omega_n^3}\mathscr{L}^{-1}\left[\frac{1}{s}\right] - \frac{\alpha}{2i\omega_d(\zeta\omega_n + i\omega_d)^2}\mathscr{L}^{-1}\left[\frac{1}{s + \zeta\omega_n + i\omega_d}\right] \\
&\quad + \frac{\alpha}{2i\omega_d(\zeta\omega_n - i\omega_d)^2}\mathscr{L}^{-1}\left[\frac{1}{s + \zeta\omega_n - i\omega_d}\right] \\
&= \frac{\alpha}{\omega_n^2}t - \frac{2\alpha\zeta}{\omega_n^3} + \frac{\alpha}{2i\omega_d}\left[-\frac{e^{-(\zeta\omega_n+i\omega_d)t}}{(\zeta\omega_n + i\omega_d)^2} + \frac{e^{-(\zeta\omega_n-i\omega_d)t}}{(\zeta\omega_n - i\omega_d)^2}\right] \\
&= \frac{\alpha}{\omega_n^2}t - \frac{2\alpha\zeta}{\omega_n^3} + \frac{\alpha e^{-\zeta\omega_n t}}{\omega_d}\left[\frac{2\zeta\omega_n\omega_d\cos\omega_d t + [(\zeta\omega_n)^2 - \omega_d^2]\sin\omega_d t}{[(\zeta\omega_n)^2 - \omega_d^2]^2 + (2\zeta\omega_n\omega_d)^2]}\right] \\
&= \frac{\alpha}{\omega_n^2}t - \frac{2\alpha\zeta}{\omega_n^3} + \frac{\alpha e^{-\zeta\omega_n t}}{\omega_n^2}\left[\frac{2\zeta}{\omega_n}\cos\omega_d t + \frac{2\zeta^2 - 1}{\omega_d}\sin\omega_d t\right] \tag{h}
\end{aligned}
$$

which can be verified to be the same as eqn. (m) in Example 2.11.

2.7.4 Decomposition Method

There are many cases when the loadings are readily divided into segments. A straightforward method is to write the analytical expression for each segment and repeat the procedures we have learn so far to solve for each segment. In this process, we need to remember that the end conditions (displacement and velocity) for one segment become the initial conditions for the next segment. This often becomes the pain point of this method: often, the initial conditions are very complicated.

Very often, the forcing function for each segment is rather simple, such as step or ramp forms, for which we have already solved using the direct analytical method. It would be great if we can reuse those solutions. Following this line of thought, the *decomposition method* is developed. In this method, we decompose the forcing into several components such that each component of forcing starts from a certain time t_0 but lasts forever. Then, according to the principle of superposition, the solution to the original problem is the sum of the solutions to all these problems. This way, the problem is split into several problems, each with a complete loading history.

2.7.4.1 Decomposition Method with Trivial Initial Conditions

This method is rather straightforward if the system is initially at rest, which is often called the *trivial initial conditions*, that is, both the initial displacement and the initial velocity vanish. We shall illustrate this method with examples.

■ **Example 2.14: Square Loading via Decomposition Method**

Consider the response of the mass–spring–dashpot system initially at rest and is subjected to a *square loading* for a duration of t_0, that is,

$$F(t) = \begin{cases} F_0 & \text{when } 0 \le t \le t_0 \\ 0 & \text{when } t > t_0 \end{cases} \tag{a}$$

where F_0 is a constant. Determine the transient response of the system.

□ **Solution:**

We decompose the forcing function into two components. The first, denoted as $f_1(t)$, is a constant $f_0 = F_0/m$ that starts at time $t = 0$ and lasts forever. The second, denoted as $f_2(t)$, is a constant $-f_0$ that starts at time $t = t_0$ and also lasts forever. This decomposition is depicted in Fig. 2.40. Mathematically, the forcing function can be written as

$$f(t) = \begin{cases} f_1(t) & \text{when } 0 \le t \le t_0 \\ f_1(t) + f_2(t) & \text{when } t > t_0 \end{cases} \tag{b}$$

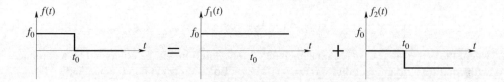

Figure 2.40   Decomposition of a square loading into two step loadings

We denote the corresponding responses of the system due to these individual loadings as $x_1(t)$ and $x_2(t)$, respectively. According to the principle of superposition, the system's transient response is given by

$$x(t) = \begin{cases} x_1(t) & \text{when } 0 \le t \le t_0 \\ x_1(t) + x_2(t) & \text{when } t > t_0 \end{cases} \tag{c}$$

Forcing $f_1(t)$ is a step loading, to which the system's transient response has been found in Examples 2.10 and 2.12. Here, we simply copy the result from those examples, as

$$x_1(t) = \frac{f_0}{\omega_n^2} \left[1 - \frac{e^{-\zeta \omega_n t}}{\sqrt{1 - \zeta^2}} \cos(\omega_d t - \phi) \right] \tag{d}$$

where

$$\phi = \tan^{-1} \frac{\zeta}{\sqrt{1 - \zeta^2}} \tag{e}$$

Forcing $f_2(t)$ is in fact another step loading, a delayed step loading, expressible as $f_2(t) = -f_1(t - t_0)$. Consequently, the system's transient response to $f_2(t)$ can be written as $x_2(t) = -x_1(t - t_0)$. So, without resolving the same problem over, we can write

$$x_2(t) = -\frac{f_0}{\omega_n^2} \left\{ 1 - \frac{e^{-\zeta\omega_n(t-t_0)}}{\sqrt{1-\zeta^2}} \cos\left[\omega_d(t-t_0) - \phi\right] \right\} \tag{f}$$

Putting these responses together, the system's transient response is as the following: When $0 \le t \le t_0$, $x(t) = x_1(t)$ as given in eqn. (d). When $t > t_0$, the system's response is the sum of responses $x_1(t) + x_2(t)$ as

$$x(t) = \frac{f_0}{\omega_n^2} \left[1 - \frac{e^{-\zeta\omega_n t}}{\sqrt{1-\zeta^2}} \cos\left(\omega_d t - \phi\right) \right] - \frac{f_0}{\omega_n^2} \left\{ 1 - \frac{e^{-\zeta\omega_n(t-t_0)}}{\sqrt{1-\zeta^2}} \cos\left[\omega_d(t-t_0) - \phi\right] \right\}$$

$$= \frac{f_0}{\omega_n^2\sqrt{1-\zeta^2}} \{ e^{-\zeta\omega_n(t-t_0)} \cos\left[\omega_d(t-t_0) - \phi\right] - e^{-\zeta\omega_n t} \cos\left(\omega_d t - \phi\right) \} \tag{g}$$

□ **Exploring the Solution with MatLab**

We define the following nondimensionalized parameters:

$$\bar{x} = \frac{x}{f_0/\omega_n^2} \qquad \bar{t} = \omega_n t \qquad r = \frac{t_0}{T_n} = \frac{\omega_n t_0}{2\pi} \qquad \bar{\omega}_d = \frac{\omega_d}{\omega_n} = \sqrt{1-\zeta^2} \tag{h}$$

In numerical computations, we shall exploit the structural simplicity of this method and only implement the solution $x_1(t)$, which can be nondimensionalized as the following:

$$\bar{x}_1(\bar{t}) = 1 - \frac{e^{-\zeta\bar{t}}}{\bar{\omega}_d} \cos\left(\bar{\omega}_d - \phi\right) \tag{i}$$

where the expression for ϕ remains unchanged. The solution will be computed in accordance with eqn. (c), with the fact that $x_2(t) = -x_1(t - t_0)$. Equation (c) is nondimensionalized to

$$\bar{x}(\bar{t}) = \begin{cases} \bar{x}_1(\bar{t}) & \text{when } 0 \le \bar{t} \le 2\pi r \\ \bar{x}_1(\bar{t}) - \bar{x}_1(\bar{t} - 2\pi r) & \text{when } \bar{t} \ge 2\pi r \end{cases} \tag{j}$$

The following MatLab code computes and plots the response as well as the forcing:

```
zeta=0.75;                          % Damping ratio
r=2.5;                              % Duration of loading
tmax = 40;                          % Maximum time to plot
dt = .1;                            % Time step size
omega_d = sqrt(1-zeta^2);           % Damped natural frequency
scoef = zeta/sqrt(1-zeta^2);        % Coefficient in front of sin term
phi = atan2( zeta, omega_d );       % Phase angle in step response

% Anonymous functions for step response and step loading
xstep = @(t) 1-exp(-zeta*t)/omega_d .* cos(omega_d*t - phi);
fstep = @(t) ones(1,length(t));

t = [0:dt:tmax];                    % Time array
x1 = xstep(t);                      % System's response for first segment
x2 = x1 - xstep(t-2*pi*r);          % System's response for second segment
```

```
f1 = fstep(t);                    % Forcing, first segment
f2 = f1 - fstep(t-2*pi*r);        % Forcing, second segment

% Repacking to form system's response and forcing
x = [x1(1:floor(2*pi*r/dt)) x2(floor(2*pi*r/dt)+1:length(t))];
f = [f1(1:floor(2*pi*r/dt)) f2(floor(2*pi*r/dt)+1:length(t))];
plot(t,x,'r', t,f,':m');
xlabel('\omega_n t');
ylabel('x(t)/(F_0/k), F(t)/F_0');
```

This above code involves some new features of MATLAB, which are explained as follows.

A new feature of the so-called *anonymous function* of MATLAB is used in the second part: the two statements containing @(t): one is xstep for the response to a step loading, and the other is fstep for the forcing of step loading. The code fragment @(t) declares a no-name (thus anonymous) function that takes t as its argument. It is then followed by the expression of the function. The left-hand side of such a statement, such as xstep, is called the *function handle* that can be used as the function's name. This mechanism allows us to define a function on the fly and to be invoked multiple times with different arguments.

The third part of the code calculates the two responses: x1 for $x_1(t)$ and x2 for $x_1(t) - x_1(t - t_0)$, as well as the forcing f1 and f2. They are computed for all time instances in t without regard to the time limits of these expressions. Then, appropriate segments from x1 and x2 are picked out and packed into a new array x that represents the system's response. The new array is created by the pair of square brackets, just like defining a new matrix, and two segments are concatenated to form a new row matrix. The x1 array is picked out from the first 2*pi*r/dt entries, where dt is the time step size, and MATLAB function floor, which returns the integer part of its argument, is used to ensure the index be an integer. The remaining entries of the solution x are picked out from x2, from where x1 is left, to the entire length of the time array t. The forcing f is constructed in the same way.

Figures 2.41 through 2.43 show the system's responses for loading durations of $r = t_0/T_n = 0.2, 1$, and 2.5. Each figure shows the responses at three damping ratios $\zeta = 0.1, 0.25$, and 0.75.

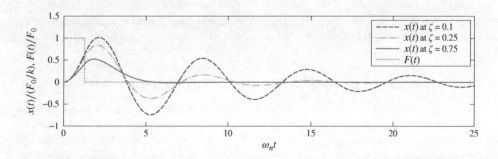

Figure 2.41 Transient responses to square loading of duration $t_0 = 0.2T_n$ at different damping ratios

We verify that the initial conditions are satisfied: all curves start from $\bar{x} = 0$ with a slope of zero, and that the transition at the end of loading is smooth, meaning that both the displacement and the velocity are continuous.

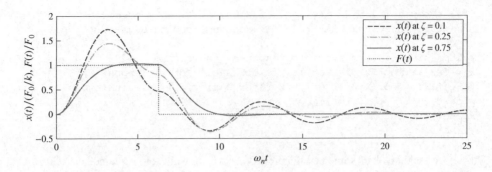

Figure 2.42   Transient responses to square loading of duration $t_0 = T_n$ at different damping ratios

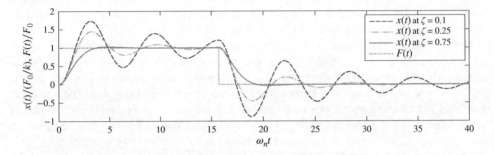

Figure 2.43   Transient responses to square loading of duration $t_0 = 2.5T_n$ at different damping ratios

We observe that the behavior of the system is rather similar to that of a periodic loading of the square wave form. The main difference is that the responses diminish as time progresses without the repeating pattern. When the loading time is very short (such as $r < 0.1$), there is no slightest trace of the square shape of loading. When the loading duration is long, the response is similar to that of a step loading.

2.7.4.2   Decomposition Method with Nontrivial Initial Conditions

When using the decomposition method, an important detail must be kept in mind: the base solution must be for the trivial initial conditions. For instance, in the above example, the solution $x_1(t)$ must be for trivial initial conditions. This way, when $x_2(t) = -x_1(t - t_0)$ is added to form the solution for $x > t_0$, the resulting response will have a smooth transition. Otherwise, the nontrivial initial conditions will be carried into $x_2(t)$ and added into the solution. This would be incorrect.

How do we deal with problems of nontrivial initial conditions? In such cases, the system's response in the first segment of forcing is further decomposed into two components: the first, still denoted as $x_1(t)$, is the response to $f_1(t)$ with the trivial initial conditions; and the second is the system's free-vibration response to the specified initial conditions, denoted as $x_{IC}(t)$.

We have found the system's free-vibration response to general initial conditions in Section 2.5. For underdamped systems, the response is given in eqn. (2.56) and is repeated here for convenience:

$$x_{IC}(t) = e^{-\zeta\omega_n t}\left[x_0\cos\omega_d t + \frac{v_0 + \zeta\omega_n x_0}{\omega_d}\sin\omega_d t\right] \tag{2.121}$$

This response would need to be added into every segment of the response.

■ Example 2.15: Transient Response to Saw-Tooth Loading

Assume that a mass–spring–dashpot system starts at rest with an initial displacement Δ. It is subjected to a ramp loading of the form $F_0 t/t_0$ up to $t = t_0$ and then abruptly unloads entirely, that is,

$$F(t) = \begin{cases} F_0 t/t_0 & \text{when } 0 \le t \le t_0 \\ 0 & \text{when } t > t_0 \end{cases}$$

Determine the transient response of the system.

□ Solution:

Denote the normalized forcing function as $f(t) = F(t)/m$ and $f_0 = F_0/m$. The forcing function can be decomposed into three components as depicted in Fig. 2.44.

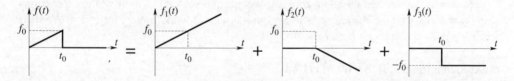

Figure 2.44   Decomposition of ramp-loading step-unloading into three loadings

It is also noted that the first two segments can be expressed by the same ramp loading, but in different slopes. The forcing is expressible as

$$f(t) = \begin{cases} f_{ramp}(t) & \text{when } 0 \le t \le t_0 \\ f_{ramp}(t) - f_{ramp}(t - t_0) - f_{step}(t - t_0) & \text{when } t > t_0 \end{cases} \tag{a}$$

where

$$f_{ramp}(t) = \frac{f_0}{t_0}t \qquad f_{step}(t) = f_0 H(t) \tag{b}$$

and $H(t)$ is the Heaviside step function.

Denote the responses due to forcing $f_{ramp}(t)$ and $f_{step}(t)$ and the trivial initial conditions as $x_{ramp}(t)$, $x_{step}(t)$, respectively. Additionally, denote the free-vibration response to the initial conditions as $x_{IC}(t)$. Then, the system's transient response can be constructed as

$$x(t) = \begin{cases} x_{IC}(t) + x_{ramp}(t) & \text{when } 0 \le t \le t_0 \\ x_{IC}(t) + x_{ramp}(t) - x_{ramp}(t - t_0) - x_{step}(t - t_0) & \text{when } t > t_0 \end{cases} \tag{c}$$

Loading $f_{\text{ramp}}(t)$ is a ramp loading to which the system's transient response under trivial initial conditions has been found in Example 2.11, which is adapted and written as

$$x_{\text{ramp}}(t) = \frac{f_0}{t_0 \omega_n^3} \left[\omega_n t - 2\zeta - \frac{e^{-\zeta \omega_n t}}{\omega_d} \sin(\omega_d t - \phi_{\text{ramp}}) \right] \tag{d}$$

where

$$\phi_{\text{ramp}} = \tan^{-1} \frac{2\zeta \sqrt{1 - \zeta^2}}{1 - 2\zeta^2} \tag{e}$$

Loading $f_{\text{step}}(t)$ is a step loading to which the system's transient response under trivial initial conditions has been found in Example 2.10 as

$$x_{\text{step}}(t) = -\frac{f_0}{\omega_n^2} \left[1 - \frac{e^{-\zeta \omega_n t}}{\omega_d} \cos(\omega_d t - \phi_{\text{step}}) \right] \tag{f}$$

where

$$\phi_{\text{step}} = \tan^{-1} \frac{\zeta}{\sqrt{1 - \zeta^2}} \tag{g}$$

The initial conditions are $x(0) = \Delta$ and $\dot{x}(0) = 0$. According to eqn. (2.121),

$$x_{\text{IC}}(t) = \Delta \frac{e^{-\zeta \omega_n t}}{\omega_d} \cos(\omega_d t - \phi_{\text{step}}) \tag{h}$$

Once the solutions to individual components have been found, the transient response of the system can be constructed based on eqn. (c). We might be able to "simplify" the expressions slightly if we expand the above expressions. But it is a lot of work for very little gain. We shall leave MATLAB to deal with it numerically.

□ **Exploring the Solution with MATLAB**

When nondimensionalizing the solution, we have two choices for nondimensionalizing the displacement: $f_0/\omega_n^2 = F_0/k$, the static deflection, or Δ, the initial displacement. Since this example is more about dynamic loading, we choose the former. Then, the latter also needs to be nondimensionalized. The nondimensionalized parameters are

$$\bar{x} = \frac{x(t)}{F_0/k} \qquad \bar{x}_0 = \frac{\Delta}{F_0/k} \qquad \bar{t} = \omega_n t \qquad r = \frac{t_0}{T_n} \qquad \bar{\omega}_d = \frac{\omega_d}{\omega_n} \tag{i}$$

Then, the solution in eqn. (c) can be nondimensionalized as

$$\bar{x}(\bar{t}) = \begin{cases} \bar{x}_{\text{IC}}(\bar{t}) + \bar{x}_{\text{ramp}}(\bar{t}) & \text{when} \quad 0 \le \bar{t} \le 2\pi r \\ \bar{x}_{\text{IC}}(\bar{t}) + \bar{x}_{\text{ramp}}(\bar{t}) - \bar{x}_{\text{ramp}}(\bar{t} - 2\pi r) - \bar{x}_{\text{step}}(\bar{t} - 2\pi r) & \text{when} \quad \bar{t} > 2\pi r \end{cases} \tag{j}$$

where

$$\bar{x}_{\text{IC}}(\bar{t}) = \bar{x}_0 \frac{e^{-\zeta \bar{t}}}{\bar{\omega}_d} \cos(\bar{\omega}_d \bar{t} - \phi_{\text{step}}) \tag{k}$$

$$\bar{x}_{\text{ramp}}(\bar{t}) = \frac{1}{2\pi r} \left[\bar{t} - 2\zeta - \frac{e^{-\zeta \bar{t}}}{\bar{\omega}_d} \sin(\bar{\omega}_d \bar{t} - \phi_{\text{ramp}}) \right] \tag{l}$$

$$\bar{x}_{\text{step}}(\bar{t}) = 1 - \frac{e^{-\zeta \bar{t}}}{\bar{\omega}_d} \cos(\bar{\omega}_d \bar{t} - \phi_{\text{step}}) \tag{m}$$

For comparison, the forcing is also nondimensionalized by $\bar{f} = f/f_0 = F/F_0$ and

$$\bar{f}_{ramp}(\bar{t}) = \frac{\bar{t}}{2\pi r} \qquad \bar{f}_{step}(\bar{t}) = 1 \tag{n}$$

and

$$\bar{f}(\bar{t}) = \begin{cases} \bar{f}_{ramp}(\bar{t}) & \text{when } 0 \leq \bar{t} \leq 2\pi r \\ \bar{f}_{ramp}(\bar{t}) - \bar{f}_{ramp}(\bar{t} - 2\pi r) - \bar{f}_{step}(\bar{t} - 2\pi r) & \text{when } \bar{t} > 2\pi r \end{cases} \tag{o}$$

The following MATLAB code implements the solution and plots the curves:

```
zeta=0.1;                                   % Damping ratio
r=.2;                                       % Duration of loading
x0 = 1;                                     % Initial disipacelemt
tmax = 40;                                  % Maximum time to plot
dt = .1;                                    % Time step size


omega_d = sqrt(1-zeta^2);                   % Damped natural frequency
phi_step = atan2( zeta, omega_d );          % Phase angle for step response
phi_ramp = atan2( 2*zeta*omega_d, 1-2*zeta^2); % Ph.angle for ramp response

% Anonymous functions for free-vibration, response, and forcing
xIC   = @(t) x0 * exp(-zeta*t)/omega_d .* cos(omega_d*t-phi_step);
xramp = @(t) (t-2*zeta - exp(-zeta*t)/omega_d .* sin(omega_d*t ...
              -phi_ramp))/2/pi/r;
xstep = @(t) 1 - exp(-zeta*t)/omega_d .* cos(omega_d*t-phi_step);
framp = @(t) t/2/pi/r;
fstep = @(t) ones(1,length(t));


t = [0:dt:tmax];                            % Time array
x1 = xIC(t) + xramp(t);                     % Response, first segment
x2 = x1 - xramp(t-2*pi*r) - xstep(t-2*pi*r); % Response, second segment
f1 = framp(t);                              % Forcing, first segment
f2 = f1 - framp(t-2*pi*r) - fstep(2-2*pi*r); % Forcing, second segment


% Packing to produce final system response and forcing
x = [x1(1:floor(2*pi*r/dt)) x2(floor(2*pi*r/dt)+1:length(t))];
f = [f1(1:floor(2*pi*r/dt)) f2(floor(2*pi*r/dt)+1:length(t))];

plot(t,x,'r', t,f,':m');
xlabel('\omega_n t');
ylabel('x(t)/(F_0/k)');
```

The structure of the code is similar to the previous example. Note that in the code, x1 and x2 are not the individual solutions. Instead, they are the response in the first and the second lines, respectively, of the right-hand side of eqn. (j) computed for all time instances in t.

Figures 2.45 through 2.47 show the system's responses at three different loading durations of $r = t_0/T_n = 0.2, 1$, and 2.5 for an initial displacement of Δ, the static deflection. Each figure contains the responses at three damping ratios of $\zeta = 0.1, 0.25$, and 0.75.

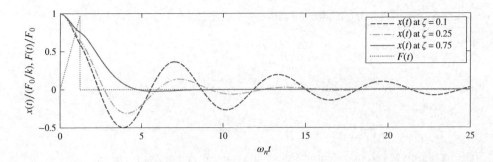

Figure 2.45 Transient responses to triangular loading duration of $t_0 = 0.2T_n$ and initial displacement of Δ at different damping ratios

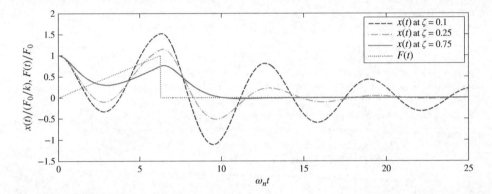

Figure 2.46 Transient responses to triangular loading duration of $t_0 = T_n$ and initial displacement of Δ at different damping ratios

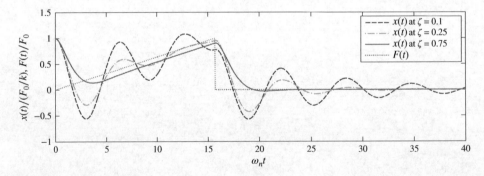

Figure 2.47 Transient responses to triangular loading duration of $t_0 = 2.5T_n$ and initial displacement of Δ at different damping ratios

We verify that all curves start from $\bar{x} = \bar{x}_0 = 1$ with a zero slope and that the response curves are smooth at the end of loading. We observe that when the loading duration is up to $r = 1$, the system's response shows no hint of the triangular shape of the forcing, which only becomes noticeable when the loading duration is long.

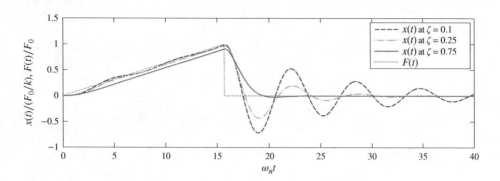

Figure 2.48 Transient responses to triangular loading duration of $t_0 = 2.5T_n$ and trivial initial displacement of Δ at different damping ratios

For comparison, the case for $r = 2.5$ is repeated without the initial displacement, by setting x0=0 in the code. The resulting curves are shown in Fig. 2.48. Comparing with the corresponding case of $\Delta = F_0/k$ in Fig. 2.47, it can be observed that the initial displacement makes a significant contribution to the system's response. Without the initial displacement, the system follows the forcing function rather closely until the moment of unloading. Curves after the unloading are similar for both cases of with and without the initial displacement.

2.7.5   Convolution Integral Method

The central idea of this method is also the principle of superposition. Suppose that we can divide the forcing function into many small pieces, as depicted in Fig. 2.49, then the response would be the sum of responses to all individual pieces. When each segment is small enough, the forcing during each step can be assumed as a constant. In the limit, the summation becomes an integration. Let us develop this idea in full.

First, we examine the response of the system due to a single piece of forcing acting during the time from $t = \tau$ to $t = \tau + \Delta\tau$ for a small duration $\Delta\tau$, as shown in Fig. 2.50. We are only interested in the system's response after the loading of this piece has ended, that is, when $t > \tau + \Delta\tau$.

We can directly utilize the solution for a square loading in Example 2.14 while noting the following characteristics of the loading: it has a constant forcing of $f(\tau)$, is delayed by τ, and has a duration of $\Delta\tau$. Specifically, we make the following changes to the solution in eqn. (g) in Example 2.14:

$$f_0 \rightarrow f(\tau) \qquad t \rightarrow t - \tau \qquad t_0 \rightarrow \Delta\tau$$

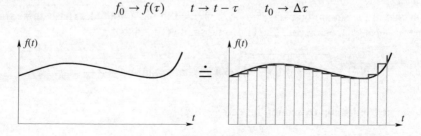

Figure 2.49 General loading $f(t)$ approximated by series of square loading

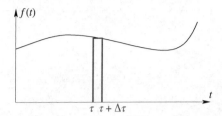

Figure 2.50   One single piece of square loading at time τ and of duration $\Delta\tau$

The response due to this piece of loading is an increment to the system's total response due to a small piece of loading that starts at time τ. Denote this incremental response as $\Delta x(t, \tau)$. Then, the adaption of eqn. (g) in Example 2.14 gives

$$\Delta x(t, \tau) = \frac{f(\tau)}{\omega_n^2 \sqrt{1 - \zeta^2}} \{ e^{-\zeta\omega_n(t-\tau-\Delta\tau)} \cos\left[\omega_d(t - \tau - \Delta\tau) - \phi\right]$$

$$- e^{-\zeta\omega_n(t-\tau)} \cos\left[\omega_d(t - \tau) - \phi\right] \} \tag{2.122}$$

where

$$\phi = \tan^{-1} \frac{\zeta}{\sqrt{1 - \zeta^2}} \tag{2.123}$$

Next, we linearize the above expression for $\Delta x(t, \tau)$ by keeping only up to the first-order small terms of Δt, since, in the end, we will set $\Delta\tau \to 0$. Using Taylor expansion for the two factors in eqn. (2.122) involving Δt and keeping only up to the first-order small terms of Δt,

$$e^{-\zeta\omega_n(t-\tau-\Delta\tau)} = e^{-\zeta\omega_n(t-\tau)} + \zeta\omega_n e^{-\zeta\omega_n(t-\tau)} \Delta\tau$$

$$\cos\left[\omega_d(t - \tau - \Delta\tau) - \phi\right] = \cos\left[\omega_d(t - \tau) - \phi\right] + \omega_d \sin\left[\omega_d(t - \tau) - \phi\right]\Delta\tau$$

Substituting them into eqn. (2.122) while keeping only up to the first-order terms of Δt gives

$$\Delta x(t, \tau) \approx f(\tau)\frac{e^{-\zeta\omega_n(t-\tau)}}{\omega_d} \left[\zeta \cos\omega_d(t - \tau - \phi)\right.$$

$$\left. + \sqrt{1 - \zeta^2} \sin\omega_d(t - \tau - \phi)\right] \Delta\tau \tag{2.124}$$

The contents in the square brackets can be expressed in the amplitude–phase angle form of $X \sin(t - \tau - \phi + \phi_1)$, using eqn. (2.21). Doing so, it is found that

$$X = \sqrt{\zeta^2 + \left(\sqrt{1 - \zeta^2}\right)^2} = 1 \qquad \phi_1 = \tan^{-1} \frac{\zeta}{\sqrt{1 - \zeta^2}} = \phi$$

Hence, the first-order approximation for eqn. (2.122) becomes

$$\Delta x(t, \tau) \approx f(\tau)\frac{e^{-\zeta\omega_n(t-\tau)}}{\omega_d} \sin\omega_d(t - \tau)\Delta\tau \tag{2.125}$$

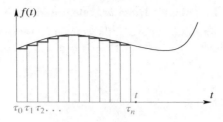

Figure 2.51 First n-prices of loadings that have been applied prior to time t

We now construct the total response of the system at any time t, denoted as $x(t)$, from the overall scheme of the things. The response $x(t)$ is the sum of responses to all pieces of loadings that have been applied by time t. That is, the first n pieces of the loading where n is such that $\tau_n \leq t < \tau_{n+1}$, as sketched in Fig. 2.51.

The total response can be written as

$$x(t) = \sum_{i=0}^{n} \Delta x(t, \tau_i) = \sum_{i=0}^{n} f(\tau_i) \frac{e^{-\zeta \omega_n (t - \tau_i)}}{\omega_d} \sin \omega_d (t - \tau_i) \Delta \tau \qquad (2.126)$$

In the limit as $\Delta \tau \to 0$, the summation becomes an integral; correspondingly, we write $\Delta \tau$ as $d\tau$, the discrete time τ_i as continuous τ, and the upper limit of the integral as time t. That is, the total response of the system at time t is

$$x(t) = \int_0^t f(\tau) \frac{e^{-\zeta \omega_n (t - \tau)}}{\omega_d} \sin \omega_d (t - \tau) d\tau \qquad (2.127)$$

More often, this integral is written as

$$x(t) = \int_0^t f(\tau) h(t - \tau) d\tau \qquad (2.128)$$

with

$$h(t) = \frac{e^{-\zeta \omega_n t}}{\omega_d} \sin \omega_d t \qquad (2.129)$$

Equation (2.128) is called the *Duhamel integral*. Mathematically, it is a special type of integral called the *convolution integral*, which has the following property:

$$x(t) = \int_0^t f(\tau) h(t - \tau) d\tau = \int_0^t f(t - \tau) h(\tau) d\tau \qquad (2.130)$$

Function $h(t)$ is called the system's *unit impulse response*. If we compare $h(t)$ with $\Delta x(t, \tau)$ in eqn. (2.125), we notice two changes. The first is that $f(\tau)$ and $\Delta \tau$ are dropped. Since $f(\tau) \Delta \tau$ constitutes an impulse, dropping $f(\tau) \Delta \tau$ means the response is normalized by the impulse. Second, $t - \tau$ becomes t, which implies that the impulse's application time is moved from $t = \tau$ to $t = 0$. Therefore, $h(t)$ is the system's response to a unit impulse applied at time $t = 0$.

Furthermore, according to the principle of linear momentum, applying an impulse to a mass at rest would impart a linear momentum to the mass. Since the time duration is zero, this is equivalent to applying an initial velocity to the mass. We shall leave this idea as an exercise for the readers to determine the appropriate initial velocity to obtain the system's unit impulse response function $h(t)$.

■ **Example 2.16: Square Loading via Duhamel Integral Method**

In this example, we demonstrate the use of the convolution integral method for the same square loading as in Example 2.14.

□ **Solution:**

The loading is

$$F(t) = \begin{cases} F_0 & \text{when } 0 \le t \le t_0 \\ 0 & \text{when } t > t_0 \end{cases} \quad \text{or} \quad f(t) = \begin{cases} f_0 & \text{when } 0 \le t \le t_0 \\ 0 & \text{when } t > t_0 \end{cases} \tag{a}$$

When $0 \le t \le t_0$, the loading is a constant f_0, Duhamel integral in eqn. (2.128) gives

$$x(t) = \frac{f_0}{\omega_d} \int_0^t e^{-\zeta\omega_n(t-\tau)} \sin\omega_d(t-\tau)d\tau \tag{b}$$

To evaluate this integral, we perform a change of variable $\xi = t - \tau$,

$$\int_0^t e^{-\zeta\omega_n(t-\tau)} \sin\omega_d(t-\tau)d\tau = \int_t^0 e^{-\zeta\omega_n\xi} \sin\omega_d\xi(-d\xi) = \int_0^t e^{-\zeta\omega_n\xi} \sin\omega_d\xi d\xi$$

which actually is the same as using eqn. (2.130). Recall the following indefinite integral

$$\int e^{ax} \sin bx dx = \frac{e^{ax}}{a^2 + b^2}(a \sin bx - b \cos bx) \tag{c}$$

Matching a with $-\zeta\omega_n$ and b with ω_d gives

$$\int_0^t e^{-\zeta\omega_n\xi} \sin\omega_d\xi d\xi = \frac{1}{(-\zeta\omega_n)^2 + \omega_d^2} \left[e^{-\zeta\omega_n\xi}(-\zeta\omega_n \sin\omega_d\xi - \omega_d \cos\omega_d\xi) \right]\Big|_0^t$$

$$= \frac{1}{\omega_n^2} \left[e^{-\zeta\omega_n t}(-\zeta\omega_n \sin\omega_d t - \omega_d \cos\omega_d t) + \omega_d \right] \tag{d}$$

Thus,

$$x(t) = \frac{f_0}{\omega_n^2} \left[1 - e^{-\zeta\omega_n t} \left(\cos\omega_d t + \frac{\zeta}{\sqrt{1-\zeta^2}} \sin\omega_d t \right) \right] \tag{e}$$

When $t > t_0$, the upper limit of the integration is limited to t_0, beyond which the loading vanishes

$$x(t) = \frac{f_0}{\omega_d} \int_0^{t_0} e^{-\zeta\omega_n(t-\tau)} \sin\omega_d(t-\tau)d\tau \tag{f}$$

We make the similar change of variables, by defining $\xi = t - \tau$,

$$x(t) = \frac{f_0}{\omega_d} \int_{t-t_0}^t e^{-\zeta\omega_n\xi} \sin\omega_d\xi d\xi \tag{g}$$

This is a familiar integral, which we had just integrated earlier, based on the indefinite integral of eqn. (c) with a different set of limits. Thus,

$$x(t) = \frac{f_0}{\omega_d[(-\zeta\omega_n)^2 + \omega_d^2]} \left[e^{-\zeta\omega_n\xi} \left(-\zeta\omega_n \sin\omega_d\xi - \omega_d \cos\omega_d\xi \right) \right]\Big|_{t-t_0}^{t}$$

$$= \frac{f_0}{\omega_n^2\omega_d} \{ e^{-\zeta\omega_n t} \left(-\zeta\omega_n \sin\omega_d t - \omega_d \cos\omega_d t \right)$$

$$- e^{-\zeta\omega_n(t-t_0)} \left[\zeta\omega_n \sin\omega_d(t-t_0) - \omega_d \cos\omega_d(t-t_0) \right] \}$$

$$= \frac{f_0}{\omega_n^2} \left\{ -e^{-\zeta\omega_n t} \left(\cos\omega_d t + \frac{\zeta}{\sqrt{1-\zeta^2}} \sin\omega_d t \right) \right.$$

$$\left. + e^{-\zeta\omega_n(t-t_0)} \left[\cos\omega_d(t-t_0) + \frac{\zeta}{\sqrt{1-\zeta^2}} \sin\omega_d(t-t_0) \right] \right\} \tag{h}$$

We verify that this is the same result we had obtained in Example 2.14.

□ **Exploring the Solution with MATLAB**

Here, we would not simply code the solution and plot the curves in MATLAB. This has been done in Example 2.14. Instead, we use MATLAB to perform the Duhamel integral numerically and to compare with the results in Example 2.14.

We introduce the following nondimensionalization parameters as usual:

$$\bar{x} = \frac{x}{f_0/\omega_n^2} \qquad \bar{t} = \omega_n t \qquad \bar{\omega}_d = \frac{\omega_d}{\omega_n} = \sqrt{1-\zeta^2} \tag{i}$$

$$\bar{f} = \frac{f}{f_0} = \frac{F}{F_0} \qquad r = \frac{t_0}{T_n} = \frac{\omega_n t_0}{2\pi} \quad \text{or} \quad \omega_n t_0 = 2\pi r \tag{j}$$

where T_n is the system's natural period, $T_n = 2\pi/\omega_n$.

Since we will perform the Duhamel integral numerically, we need to nondimensionalize the Duhamel integral first. The Duhamel integral can be written as, in view of the above-mentioned nondimensionalization parameters,

$$x(t) = \frac{f_0}{\omega_n^2} \int_0^t \frac{f(t)}{f_0} [\omega_n h(t-\tau)][\omega_n d\tau] \tag{k}$$

This suggests a change of variable for the integration: $\bar{\tau} = \omega_n\tau$, which changes the upper limit of integral to $\bar{t}$, and the following way of nondimensionalizing $h(t)$:

$$\bar{h}(\bar{t}) = \omega_n h(t) = \frac{\omega_n}{\omega_d} e^{-\zeta\bar{t}} \sin\bar{\omega}_d\bar{t} = \frac{e^{-\zeta\bar{t}}}{\bar{\omega}_d} \sin\bar{\omega}_d\bar{t} \tag{l}$$

Thus, eqn. (k) becomes the following nondimensionalized Duhamel integral:

$$\bar{x}(t) = \int_0^{\bar{t}} \bar{f}(\bar{t})\bar{h}(\bar{t}-\bar{\tau})d\bar{\tau} \tag{m}$$

Performing the integration numerically means we reverse the integral back into a summation, just like eqn. (2.126). Assume we discretize the time within the interested range from 0 to $\bar{t}_{\text{end}}$ into N_t equal steps. Denote the time at the beginning of the kth step as $\bar{t}_k$. Then, the numerical integration of the Duhamel integral in eqn. (m) can be written as

$$\bar{x}(\bar{t}_k) \approx \sum_{j=1}^{k} \bar{f}(\bar{t}_j)\bar{h}(\bar{t}_{k-j+1})\Delta\bar{t} \tag{n}$$

Note that this equation is in fact the nondimensionalized eqn. (2.126).

MATLAB has a build-in function conv that computes the following summation

$$\sum_{j=1}^{k} \bar{f}(\bar{t}_j)\bar{h}(\bar{t}_{k-j+1}) \tag{o}$$

and returns an array of the summed results for a series of k values. Such a summation is called a *convolution sum*. Furthermore, MATLAB simply runs this summation over all the possible values of j and k as long as the indices for $\bar{t}_j$ and $\bar{t}_{k-j+1}$ are within the specified range. This means that k will run up to $k = 2N_t - 1$. But, we are only interested in having k up to N_t. This means, when using the MATLAB function conv, we need to take care of the following two extra details: (1) the convolution sum must be multiplied by $\Delta\bar{t}$; and (2) the resulting array should be truncated to the first N_t elements.

The square loading can be written in Heaviside step function defined in eqn. (2.114) as

$$f(t) = f_0 \left[H(t) - H(t - t_0) \right] \tag{p}$$

and its nondimensionalized form is

$$\bar{f}(\bar{t}) = H(\bar{t}) - H(\bar{t} - 2\pi r) \tag{q}$$

MATLAB has a build-in function heaviside for the Heaviside step function.

With the above preparatory information, the following MATLAB code performs the numerical computation of the Duhamel integral.

```
zeta = .2;                                    % Damping ratio
r = .5;                                       % Loading duration
omega_d = sqrt(1-zeta^2);                     % Damped natural frequency
dt = 0.1;                                     % Time step
tmax = 40;                                    % End time

t = [dt:dt:tmax];                             % Time array
f = heaviside(t) - heaviside(t-2*pi*r);       % Forcing function
h = exp(-zeta*t) .* sin(omega_d*t) / omega_d; % Unit impulse response
x = conv( h, f ) * dt;                        % Convolution sum

plot(t,x(1:length(t)));                       % Truncating while plotting
xlabel('\omega_n t');
ylabel('x(t)/(F_0/k)');
```

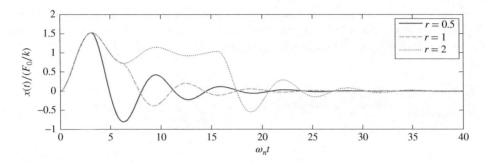

Figure 2.52 Transient responses to square loading for different loading durations represented by $r = t_0/T_n$ at damping ratio of $\zeta = 0.2$

In the above code, the convolution sum x is truncated to the same length of array t, by calling MATLAB function length, which returns the number of entries in the argument array. This ensures that the two arrays in the plot command have the same size.

Figure 2.52 compares the system's responses at different loading durations of $r = t_0/T_n = 0.5$, 1, and 2.5, at the same damping ratio of $\zeta = 0.2$. We verify that all curves start from $\bar{x} = 0$, with a slope of zero. We also verify that the curves are smooth at the end of loading. We observe that the curves for $r = 1$ and 2.5 are the same as the respective cases in Figs. 2.42 and 2.43. This is a very strong verification as the solutions in Figs. 2.42 and 2.43 are obtained in different methods.

We observe that, during a short period of time immediately after the application of the load, three curves coincide. This period ends at $\omega_n t = \pi$ or $t/T_n = 0.5$. This is because the forcing functions for the three cases are identical during this time.

———— · ————

In the above example, the analytical integration process seems extremely intimidating, especially the integration identity in eqn. (c) is one of those we probably have never seen previously; even that is just for a seemingly simple loading. We probably would thus conclude that this is the most difficult method among all. In fact, it is the most versatile method among all we have learned so far; and it can be used to handle the most general loading situations. The main advantage of this method is that, when the integration becomes too complicated for the analytical approach, a numerical scheme can be used. To a computer, performing a numerical integration is rather trivial. The MATLAB portion of the above example shows just that.

Let us spend a moment to savor this remarkable process, which demonstrates the beauty of the fundamental idea of the calculus. At the beginning, we are armed with the only "prior knowledge" of how the system responds to a step loading. From this humble beginning, we worked diligently, by wading through all the messy details, and smartly, by taking the limit to turn a summation into an integral, in the end, we become empowered with the ability of finding the system's response to *any* loading. This is quite an inspiration!

■ Example 2.17: Impulsive Loading

As the final example in this chapter, we numerically investigate the effects of different shapes of an impulsive loading on the responses of a system. The mass–spring–dashpot system is subjected to a time-varying force for a short duration of t_0. The force varies with time in the following four functional forms: a square shape, a triangular shape, a half-period of sine function, and a full period of sine-squared function, as shown in Fig. 2.53.

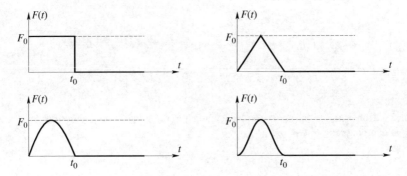

Figure 2.53   Four impulsive loadings of different shapes

□ Exploring the Solution with MATLAB

We define the following nondimensionalization parameters:

$$\bar{x} = \frac{x}{f_0/\omega_n^2} \qquad \bar{t} = \omega_n t \qquad r = \frac{t_0}{T_n} = \frac{\omega_n t_0}{2\pi} \qquad \bar{f} = \frac{f}{f_0} \tag{a}$$

All nondimensionalized forcing functions have a peak value of 1. Utilizing the Heaviside step function, the nondimensionalized forcing functions are expressible as

$$\text{Square:} \quad \bar{f}(\bar{t}) = H(\bar{t}) - H(\bar{t} - 2\pi r) \tag{b}$$

$$\text{Triangular:} \quad \bar{f}(\bar{t}) = \frac{1}{\pi r} \left[\bar{t} H(\bar{t}) - 2(\bar{t} - \pi r)H(\bar{t} - \pi r) + (\bar{t} - 2\pi r)H(\bar{t} - 2\pi r) \right] \tag{c}$$

$$\text{Sine:} \quad \bar{f}(\bar{t}) = \sin\frac{\bar{t}}{2r} \left[H(\bar{t}) - H(\bar{t} - 2\pi r) \right] \tag{d}$$

$$\text{Sine square:} \quad \bar{f}(\bar{t}) = \sin^2\frac{\bar{t}}{2r} \left[H(\bar{t}) - H(\bar{t} - 2\pi r) \right] \tag{e}$$

The following MATLAB code performs the numerical computations for the four forcing functions and plots the responses in the same figure using different colors.

```
zeta = 0.1;                                    % Damping ratio
r = 0.2;                                       % Loading duration
tmax = 40;                                     % Max time to be computed
dt = .01;                                      % Time step size
omega_d = sqrt(1-zeta^2);                      % Damped natural frequency


t = [0:dt:tmax];                               % Time series
h = exp(-zeta*t) .* sin(omega_d*t) / omega_d;  % Unit impulse response


% Forcing functions
f_rect = 1 - heaviside(t-2*pi*r);
```

```
f_triang = 1/pi/r *( t - 2*(t-pi*r) .* heaviside(t-pi*r) ...
    + (t-2*pi*r) .*heaviside(t-2*pi*r));
f_sin = sin(t/2/r) .* (1- heaviside(t-2*pi*r) );
f_sinsq = sin(t/2/r) .* sin(t/2/r) .* (1 - heaviside(t-2*pi*r));

% Convolution for the responses
x1 = conv( f_rect, h ) * dt;
x2 = conv( f_triang, h ) * dt;
x3 = conv( f_sin, h ) * dt;
x4 = conv( f_sinsq, h ) * dt;

plot(t,x1(1:length(t)),'r',t,x2(1:length(t)),'--g',t,x3(1:length(t)), ...
    '-.b',t,x4(1:length(t)),':m');
legend('Square','Triangle','Sine','Sine square');
xlabel('\omega_n t');
ylabel('x(t)/(F_0/k)');
```

In the above code, $H(\bar{t})$ is coded as 1 to avoid a bug in some versions of MATLAB that produces a undefined value NaN for heaviside(0), which ruins the computation.

Figures 2.54 through 2.59 show the responses of the system of damping ratio $\zeta = 0.2$ when it is subjected to these impulsive loadings for loading durations $r = t_0/T_n = 0.05, 0.1, 0.25, 0.5, 1$, and 2. The different loadings have been separately plotted and verified to be correct. The responses can be verified by checking the initial conditions.

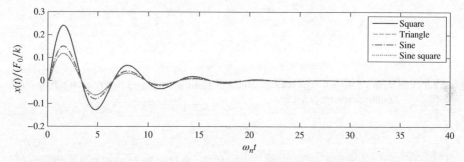

Figure 2.54 System's transient responses to impulsive loadings of four different shapes for a duration of $t_0 = 0.05T_n$ at damping ratio of $\zeta = 0.2$

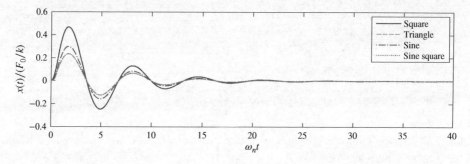

Figure 2.55 System's transient responses to impulsive loadings of four different shapes for a duration of $t_0 = 0.1T_n$ at damping ratio of $\zeta = 0.2$

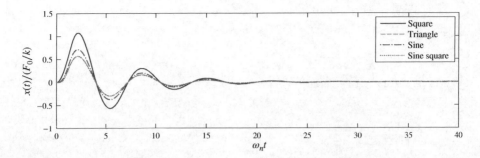

Figure 2.56 System's transient responses to impulsive loading of four different shapes for a duration of $t_0 = 0.25T_n$ at damping ratio of $\zeta = 0.2$

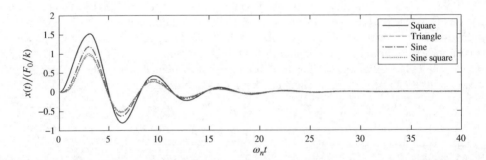

Figure 2.57 System's transient responses to impulsive loadings of four different shapes for a duration of $t_0 = 0.5T_n$ at damping ratio of $\zeta = 0.2$

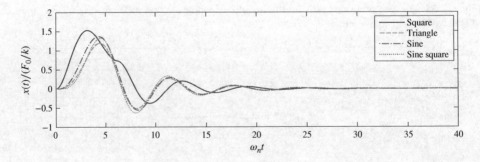

Figure 2.58 System's transient responses to impulsive loadings of four different shapes for a duration of $t_0 = T_n$ at damping ratio of $\zeta = 0.2$

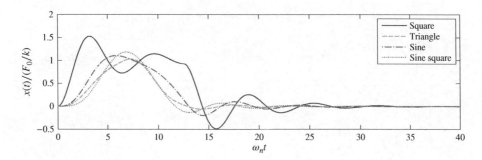

Figure 2.59 System's transient responses to impulsive loadings of four different shapes for a duration of $t_0 = 2T_n$ at damping ratio of $\zeta = 0.2$

From these responses, we make the following observations: (1) When the loading duration is very short (cases of $r = 0.05$ and 0.1), the shape of the loading is not important. The system's responses have almost identical shapes, but different amplitudes. The amplitude differences are due to the differences in the amount of energy (or linear impulse) imparted into the system during the loading. (2) When the loading duration is $r = 0.25$, the rise times show slight differences among the four cases. (3) When $r = 0.5$, the differences in the shapes of the response curves start to become noticeable. (4) The differences become pronounced when $r \geq 1$ and become very different when $r = 2$.

For Short Durations, Only Linear Impulse Matters

We want to investigate further our first observation, that is, for short durations, the shape of the loading is unimportant. This reminds us the principle of linear impulse and linear momentum in dynamics. In impact problems, what matters is the linear impulse. For the four loadings, if we integrate the loading over its entire duration, the resulting linear impulses are different:

$$I_{\text{rectangular}} = F_0 t_0 \tag{f}$$

$$I_{\text{triangular}} = \frac{1}{2} F_0 t_0 \tag{g}$$

$$I_{\text{sine}} = \int_0^{t_0} F_0 \sin \frac{\pi t}{t_0} dt = \frac{2}{\pi} F_0 t_0 \tag{h}$$

$$I_{\text{sine square}} = \int_0^{t_0} F_0 \sin^2 \frac{\pi t}{t_0} dt = \frac{1}{2} F_0 t_0 \tag{i}$$

If we normalized the responses by the linear impulse for each case, the relevant portion of the MATLAB code becomes

```
x1 = conv( f_rect, h ) * dt;
x2 = 2* conv( f_triang, h ) * dt;
x3 = pi/2 * conv( f_sin, h ) * dt;
x4 = 2* conv( f_sinsq, h ) * dt;
```

The normalized results for the cases of $r = 0.05, 0.1$, and 0.25 (and still at $\zeta = 0.2$) are shown in Figs. 2.60 through 2.62. Since we focus on the early stage of the response, only the range $\omega_n t \leq 10$ is shown.

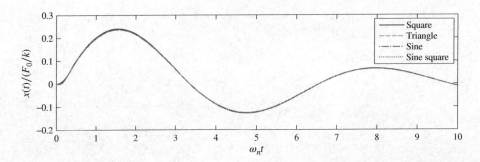

Figure 2.60 Normalized transient responses to impulsive loadings of four different shapes for a duration of $t_0 = 0.05T_n$ at damping ratio of $\zeta = 0.2$

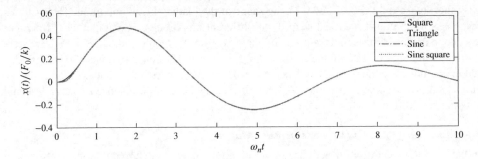

Figure 2.61 Normalized transient responses to impulsive loadings of four different shapes for a duration of $t_0 = 0.1T_n$ at damping ratio of $\zeta = 0.2$

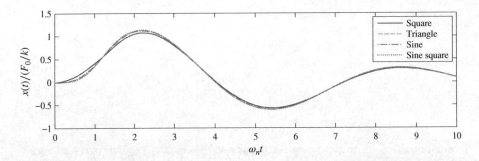

Figure 2.62 Normalized transient responses to impulsive loadings of four different shapes for a duration of $t_0 = 0.25T_n$ at damping ratio of $\zeta = 0.2$

From this set of plots, it can be observed that, for the case $r = 0.05$, the four curves are nondistinguishable. Slight distinction is observable for the case $r = 0.1$, and the distinction becomes more noticeable when $t = 0.25$. This confirms our general believe that for shock and impact problems, the exact form of loading history is not important, up to the duration of approximately $r = 0.1$.

Shock Response Spectrum

In many engineering applications, one way to characterize a system for its responses to different shocks is to produce the so-called *shock response spectrum* or the *shock spectrum*. This is a plot of the system's response peak versus the shock's characteristics. Typically, the choices for the system's response include the displacement and the acceleration, and the choices for the shock's characteristics include the duration and the frequency.

Since we are doing the analysis numerically, it is convenient to explore this concept. Here, we choose the peak displacement as the characteristics for the system's response and $T_n/t_0 = 1/r$ as the shock's characteristics. In addition, we use the half-period sine function form as the shock. These are common choices in engineering practice.

Our computational strategy is to simply compute the responses for a series of $1/r$ values and pick out the maximum value for each case. Finally, plot these values against $1/r$. The following MATLAB code implements this strategy. It computes the shock spectrum for T_n/t_0 from 0.1 to 10, with a damping ratio of $\zeta = 0.2$.

```
zeta = .2;                       % Damping ratio
omega_d = sqrt(1-zeta^2);        % Damped natural frequency
tmax = 5;                        % Max time to be computed
dt = 0.001;                      % Time step size

t = [dt:dt:tmax];                % Time array
h = exp(-zeta*t) .* sin(omega_d*t) / omega_d;   % Unit impulse response

rinv = [.1:.1:10];               % 1/r
for i=1:length(rinv)             % Circling through all rinv array
    r = 1/rinv(i);               % Reverse back to r: loading duration
    f = sin(t/2/r) .*(1 - heaviside(t-2*pi*r));   % Forcing
    x = conv( h, f ) * dt;       % Convolution for response
    peak(i) = max(abs(x));       % Find the max, pack into peak array
end;                             % End of "for" loop

plot(rinv,peak);
xlabel('T_n/t_0');
ylabel('x_max/(F_0/k)');
```

In the above code, a `for` loop is used to run the computation over each r value. Within the loop, the maximum value of $|x|$ is picked out by the `max` function and packed into the `peak` array. The resulting shock response spectrum for the system is shown in Fig. 2.63.

☐ Cautionary Note on Numerical Computations

A cautionary note is in order here. In this example and Example 2.16, we used a numerical scheme to perform the integration for us. Integration is a relatively simple process to be performed numerically. However, as it is true with any numerical computations, we must bear one more issue in mind apart from making sure things are working correctly: the numerical errors. There are many factors that could cause errors. The foremost in this example is the time step size Δt. As we are approximating the continuously varying functions by a series of steps, such as illustrated in Fig. 2.49, we are introducing errors: the difference between an exact value and an approximate value. A larger time step will cause a larger error. In general, we should

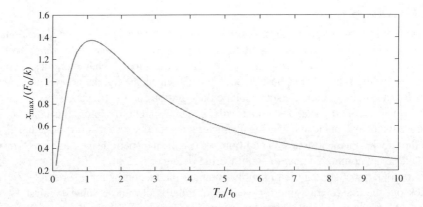

Figure 2.63   Shock response spectrum for single-DOF system at $\zeta = 0.2$

use a small time step to reduce this type of error. In the above example, we used 0.01 in the nondimensionalized time. We also need to bear in mind that "small" is a relative term. If we are looking at an impulsive loading that lasts only a small fraction of system's natural period, our time step may still be too large. For example, for the case $r = 0.05$, we only divide the entire loading into five steps. We should expect the error to be rather large. It is possible to keep track and estimate these errors. Interested readers are encouraged to modify the present code (such as using a time step of 0.001) to redo this exercise and compare the difference in the results. They are further suggested to look into a subject of study called *numerical analysis*. In this subject, different types of errors in numerical computations are explored, as well as ways to perform computational tasks more efficiently and more accurately, such as numerical integration, differentiation, finding eigenvalues for matrices, and even solving the differential equations numerically.

—————— · ——————

The above discussion on the Duhamel integral method, just like the decomposition method, works only for problems with the trivial initial conditions. For problems with nontrivial initial conditions, we can similarly add the system's free-vibration response to the system's response. Therefore, recall the free-vibration response for an underdamped system subjected to general initial conditions $x(0) = x_0$ and $\dot{x}(0) = v_0$, in eqn. (2.56), we can write

$$x(t) = \int_0^t h(\tau)h(t - \tau)d\tau + e^{-\zeta \omega_n t}\left[x_0 \cos \omega_d t + \frac{v_0 + \zeta \omega_n x_0}{\omega_d} \sin \omega_d t\right] \tag{2.131}$$

We shall leave it to the interested readers to prove this equation as an exercise.

2.8   Chapter Summary

2.8.1   Free Vibrations of Single-DOF Systems

Equation of Motion:

$$\ddot{x} + 2\zeta \omega_n \dot{x} + \omega_n^2 x = 0 \tag{2.132}$$

where x is the generalized coordinate used to describe the system, ζ is the damping ratio, and ω_n is the natural frequency.

General Solutions:

Case 1: Underdamped ($0 \leq \zeta < 1$):

$$x(t) = e^{-\zeta \omega_n t} \left(A \cos \omega_d t + B \sin \omega_d t \right) \tag{2.133}$$

where

$$\omega_d = \omega_n \sqrt{1 - \zeta^2} \tag{2.134}$$

is the damped natural frequency. If $\zeta = 0$, which can be considered as a special case of the underdamped case, we have an undamped system, and the solution becomes

$$x(t) = A \cos \omega_n t + B \sin \omega_n t \tag{2.135}$$

Case 2: Critically damped ($\zeta = 1$):

$$x(t) = (A + Bt)e^{-\omega_n t} \tag{2.136}$$

Case 3: Overdamped ($\zeta > 1$):

$$x(t) = e^{-\zeta \omega_n t} \left(Ae^{\omega_n \sqrt{\zeta^2 - 1}t} + Be^{-\omega_n \sqrt{\zeta^2 - 1}t} \right) \tag{2.137}$$

Initial Conditions: The initial conditions are given as $x(0) = x_0$ and $\dot{x}(0) = v_0$. They determine A and B.

Case 1 ($0 \leq \zeta < 1$) : $A = x_0$ $B = \dfrac{v_0 + \zeta \omega_n x_0}{\omega_d}$

Case 2 ($\zeta = 1$) : $A = x_0$ $B = v_0 + \omega_n x_0$

Case 3 ($\zeta > 1$) : $A = \dfrac{x_0}{2} + \dfrac{v_0 + \zeta \omega_n x_0}{2\omega_n \sqrt{\zeta^2 - 1}}$ $B = \dfrac{x_0}{2} - \dfrac{v_0 + \zeta \omega_n x_0}{2\omega_n \sqrt{\zeta^2 - 1}}$

Solution Procedure:
- Write the equation of motion in the normalized form as in eqn. (2.132).
- Determine the value of ζ and use the corresponding case of the general solution.
- Use the initial conditions to determine unknown constants A and B.

2.8.2   Steady-State Responses of Single-DOF Systems

Equation of Motion:
$$\ddot{x} + 2\zeta \omega_n \dot{x} + \omega_n^2 x = f(t) \tag{2.138}$$

where $f(t)$ is the forcing function, which is not necessarily a force. Steady-state response is the system's response when t is large and the forcing persists.

Harmonic Excitation: When $f(t) = f_0 \sin \Omega t$ or $f(t) = f_0 \cos \Omega t$, or their line combinations, the solution procedure is as follows:
- Assume the steady-state solution as $x(t) = A \cos \Omega t + B \sin \Omega t$.

- Substitute the assumed solution into the equation of motion:

$$(\omega_n^2 - \Omega^2)(A \cos \Omega t + B \sin \Omega t)$$

$$+2\zeta\omega_n\Omega(-A \sin \Omega t + B \cos \Omega t) = f(t) \tag{2.139}$$

- Equation (2.139) must hold at all times. Matching the coefficients for $\cos \Omega t$ and $\sin \Omega t$, respectively, gives two equations to determine A and B.
- Solve the resulting equations using Cramer's rule.
- If necessary, repack the solution into the forms $X \sin(\Omega t - \phi_s)$ or $X \cos(\Omega t - \phi_c)$ where

$$X = \sqrt{A^2 + B^2} \qquad \phi_s = -\tan^{-1}\frac{A}{B} \qquad \phi_c = \tan^{-1}\frac{B}{A} \tag{2.140}$$

- Often, the amplitude of steady-state response is of primary interest.

Non-sinusoidal Periodic Forcing: Any periodic forcing $f(t)$, of period T, can be expanded into Fourier series as

$$f(t) = \frac{a_0}{2} + \sum_{n=1}^{\infty} \left(a_n \cos \frac{2n\pi t}{T} + b_n \sin \frac{2n\pi t}{T} \right) \tag{2.141}$$

where

$$a_n = \frac{2}{T} \int_0^T f(t) \cos \frac{2n\pi t}{T} dt \qquad b_n = \frac{2}{T} \int_0^T f(t) \sin \frac{2n\pi t}{T} dt \tag{2.142}$$

The steady-state response due to $f(t)$ equals to the sum of the steady-state responses due to individual harmonics terms in eqn. (2.141) as the forcing function.

2.8.3   Transient Responses of Single-DOF Systems

Equation of Motion:

$$\ddot{x} + 2\zeta\omega_n\dot{x} + \omega_n^2 x = f(t) \tag{2.143}$$

Total Solution: The total solution consists of a particular solution $x_p(t)$ and the homogeneous solution $x_h(t)$. Physically they are the steady-state solution and the free-vibration solution, respectively (see previous two summaries).

$$x(t) = x_p(t) + x_h(t) \tag{2.144}$$

For an *underdamped system*,

$$x(t) = x_p(t) + e^{-\zeta\omega_n t}(A \sin \omega_d t + B \cos \omega_d t) \tag{2.145}$$

Analytical Method (for forcing functions having simple analytical expressions): In this method, assume the particular solution to be a linear combination of all functional forms involved in $f(t)$ and its first and second derivatives. Find the particular solution first; then use the initial conditions to determine A and B in eqn. (2.145).

Laplace Transform Method: Applying Laplace transform to the differential equation, along with the initial conditions, converts the differential equation to an algebraic equation, and the system's response in the transformed space is expressible as

$$\tilde{x}(s) = \frac{\tilde{f}(s) + (s + 2\zeta\omega_n)x(0) + \dot{x}(0)}{s^2 + 2\zeta\omega_n s + \omega_n^2} \tag{2.146}$$

where $\tilde{f}(s)$ is the Laplace transform of the forcing. To find the inverse of $\tilde{x}(s)$, one approach is to express $\tilde{x}(s)$ as a sum of partial fractions and then inverse. Special care is needed when a root in the denominator is a repeated root.

Decomposition Method (for loads that can be decomposed into simpler forms):

- Decompose the loading into components, each starting from a certain time but lasts forever.
- Find the total solution for each loading component subjected to trivial initial conditions. If a component's starting time is shifted by t_0, the solution is the same as loading starting at time $t = 0$ except that t is replaced by $t - t_0$.
- For each time segment, the solution is the sum of solutions to all loading components that are effective during the time segment.
- *Trivial initial conditions*: $x(0) = 0$ and $\dot{x}(0) = 0$. Vibrations of most mechanical systems start from such conditions. If the initial conditions of a problem are not trivial, the solution is the sum of (1) the transient solution for trivial initial conditions and (2) the free-vibration solution due to the given initial conditions.

Convolution Integral Method (for any time-varying loads): Duhamel integral is

$$x(t) = \int_0^t f(\tau)h(t - \tau)d\tau + e^{-\zeta\omega_n t}\left[x_0 \cos\omega_d t + \frac{v_0 + \zeta\omega_n x_0}{\omega_d}\sin\omega_d t\right] \tag{2.147}$$

where

$$h(t) = \frac{e^{-\zeta\omega_n t}}{\omega_d}\sin\omega_d t \tag{2.148}$$

is the system's response due to a unit impulse. During the integration, t is treated as a constant. The convolution integral can be performed numerically.

Problems

Problem 2.1: As discussed in the text, in general, a linear combination of $\sin\Omega t$ and $\cos\Omega t$ can be expressed in an amplitude–phase angle form such as

$$A\sin\Omega t + B\cos\Omega t = X\sin(\Omega t + \phi_s) = X\cos(\Omega t - \phi_c)$$

where X is positive and ϕ's have a period of 2π. Furthermore, ϕ's are often expressed as $\tan^{-1}(\cdot)$. However, the mathematical definition for $\tan^{-1}(\cdot)$ has its value ranging from $-\pi/2$ to $\pi/2$. A potential problem of this conflict has been demonstrated in Example 2.1. Discuss the options to resolve this conflict, that is, modifications that can be made to the two amplitude–phase angle expressions if ϕ_c and ϕ_s in the above expression are based on the mathematical definition of $\tan^{-1}(\cdot)$.

Problem 2.2: Find the natural frequency for the U-shaped tube in Problem 1.23.

Problem 2.3: Find the natural frequency for the semicircular cylinder in Problem 1.31.

Problem 2.4: Find the natural frequency for the rigid slender rod resting on a circular cylinder in Problem 1.33.

Problem 2.5: Find the natural frequency for the linked circular disks in Problem 1.41.

Problem 2.6: Find the natural frequency for the rigid rocker in Problem 1.43.

Problem 2.7: A pendulum consists of a mass m and a rigid massless link of length l, as depicted in Fig. 1.4. Find the subsequent motion when the bob of the pendulum is given a horizontal initial velocity of v_0 at its vertical hanging position.

Problem 2.8: The composite rigid-body pendulum in Problem 1.20 is hit by a bullet of mass m_0 traveling at a speed of v_0, which is aimed horizontally at the center of the circular disk. After the impact, the bullet becomes embedded in the disk. Find the subsequent motion of the pendulum. (*Hint: During the impact, the angular momentum of the system comprising the pendulum and the bullet about the hinge is conserved.*)

Problem 2.9: In Problem 1.4, the particle is initially located on the y-axis (cannot be exactly on the y-axis, but as close as possible) and is suddenly released at $t = 0$. Find the subsequent motions of the particle. Assume the rotational speed $\Omega^2 > g/R$. Note that the linearized equation of motion has been obtained in Problem 1.49.

Problem 2.10: For a mass–spring–dashpot system, assume $m = 1\,\text{kg}$, $k = 100\,\text{N/m}$. Determine the dashpot constant for the system to be critically damped (and denote this dashpot constant as c_c). What is the unit for c?

Problem 2.11: Continue on Problem 2.10, determine the mass's displacement as a function of time $x(t)$ for the initial conditions $x(0) = 0.1\,\text{m}$ and $v(0) = 0$ for the following three cases: (1) $c = 0.1c_c$, (2) $c = c_c$, and (3) $c = 10c_c$. Use MATLAB to plot two versions of the system's response. In one version, the numerical values as computed are simply plotted. In the second version, plot the nondimensionalized response. In both versions, plot all three cases in a single figure for up to at least 5 s.

Problem 2.12: Continue on Problem 2.10, determine the mass's displacement as a function of time $x(t)$ for the initial conditions $x(0) = 0$ and $v(0) = 1\,\text{m/s}$ for the following three cases: (1) $c = 0.1c_c$, (2) $c = c_c$, and (3) $c = 10c_c$. Use MATLAB to plot two versions of the system's response. In one version, the numerical values as computed are simply plotted. In the second version, plot the nondimensionalized response. In both versions, plot all three cases in a single figure, for up to at least 5 s.

Problem 2.13: The mass–spring–dashpot system in Fig. P2.13 is subjected to initial conditions $x(0) = \Delta$ and $\dot{x}(0) = v_0$. Find the kinetic energy, the potential energy, and the total mechanical energy in the system as functions of time. Assume that the system is underdamped. Use MATLAB to plot the variations of these energies with time, for the following cases: $\zeta = 0.1, 0.5$, and 0.99; and $(v_0/\omega_n)/\Delta = 0.1, 1$, and 10. For each combination of the two parameters, plot all three energies in the same figure.

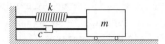

Figure P2.13   Horizontally oriented mass–spring–dashpot system

Problem 2.14: Continue on Problem 2.13: assume the system is critically damped. Use MATLAB to plot the variations of these energies with time for the cases of $(v_0/\omega_n)/\Delta = 0.1$, 1, and 10. Plot all curves in the same figure.

Problem 2.15: Continue on Problem 2.13: assume the system is overdamped. Use MATLAB to plot the variations of these energies with time for the following cases: $\zeta = 1.01$, 2, and 5; and $(v_0/\omega_n)/\Delta = 0.1$, 1, and 10. For each combination of the two parameters, plot all three energies in the same figure.

Problem 2.16: In a bungee jumping practice, a jumper of mass m is harnessed to a cable and jumps off a ledge. The cable has an effective spring constant k and an unstretched length L. Assume that there is no initial velocity when he jumps off and that there is no obstacle in the course of the fall. Find the position of the jumper as a function of time and use MATLAB to plot the time history of his position for L/Δ ratio of 10 and 100, where Δ is the static deflection of the cable due to the jumper's weight. (*Hint: There are two stages of the jumper's motions.*)

Problem 2.17: For the system described in Example 1.20, if $x(t) = X_0 \sin \Omega t$, find the steady-state response of the system. Use MATLAB to plot the amplitude of the steady-state response as a function of frequency ratio $r = \Omega/\omega_n$.

Problem 2.18: For the pendulum-in-frame system described in Problem 1.11, the frame undergoes a specified motion $x(t) = X_0 \sin \Omega t$, where x is measured from a reference configuration. Find the steady-state response of the system. Use MATLAB to plot the amplitude and the phase lag of the steady-state response as functions of frequency ratio $r = \Omega/\omega_n$.

Problem 2.19: For the rotating rod-on-collar system described in Problem 1.17, if the rod rotates at a constant angular velocity of Ω, determine the steady-state response of the collar. Use MATLAB to plot the amplitude and the phase lag of the steady-state response as functions of frequency ratio $r = \Omega/\omega_n$.

Problem 2.20: For the pivoted-rod system described in Problem 1.27, determine the steady-state response of the system when the right end of the rod is subjected to a harmonic excitation of the form $y(t) = Y_0 \sin \Omega t$, where y is measured upward from the hinge pin on the left end. Use an absolute displacement of the mass as the generalized coordinate. Use MATLAB to plot the amplitude and the phase lag of the steady-state response as functions of frequency ratio $r = \Omega/\omega_n$.

Problem 2.21: Solve Problem 2.20 by using a relative displacement of the mass with respect to the right end of the beam as the generalized coordinate.

Problem 2.22: For the mass-on-cart system described in Problem 1.10, the cart undergoes a specified motion $x(t) = X_0 \sin \Omega t$, which is measured from a reference configuration. Find the steady-state response of the system. Use MATLAB to plot the amplitude and the phase lag of the steady-state response as functions of frequency ratio $r = \Omega/\omega_n$. Include cases of $\zeta = 0, 0.1, 0.25, 0.5$, and 0.75 in the same figure.

Problem 2.23: For the rod-pulley system described in Problem 1.26, under the small motion assumption, the right end of the rod moves according to $y(t) = Y_0 \cos \Omega t$, where y is measured upward from the position of the hinge pin on the left end. Find the steady-state response of the system. Use MATLAB to plot the amplitude and the phase lag of the steady-state response as functions of frequency ratio $r = \Omega/\omega_n$. Include cases of $\zeta = 0$, $0.1, 0.25, 0.5$, and 0.75 in the same figure.

Problem 2.24: Find the steady-state motion of the cylinder-on-cart system described in Part (b) of Problem 1.29. Use MATLAB to plot the amplitude and the phase lag of the steady-state response as functions of frequency ratio $r = \Omega/\omega_n$. Include cases of $\zeta = 0$, $0.1, 0.25, 0.5$, and 0.75 in the same figure.

Problem 2.25: In modeling a vehicle's ride quality, the vehicle is modeled as a mass–spring–dashpot system and is being tested on a road. The road surface can be approximated as a sinusoidal curve, with an amplitude H, and a period L that is much larger than the vehicle size, as shown in Fig. P2.25. The vehicle travels at a constant speed of v_0. Find the steady-state vertical displacement of the vehicle's body (mass m) relative to the road surface. Use MATLAB to plot the amplitude and the phase lag of the steady-state response as functions of frequency ratio $r = \Omega/\omega_n$. Include cases of $\zeta = 0.1, 0.5, 1$, and 2 in the same figure. (*Hint: The system is critically damped and overdamped, respectively, for the last two damping ratios.*)

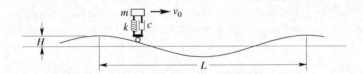

Figure P2.25   Vehicle model traveling on sinusoidal road surface

Problem 2.26: Solve Problem 2.25 but measuring the vehicle's vertical displacement relative to the horizontal reference line instead of the road surface.

Problem 2.27: An underdamped mass–spring–dashpot system is subjected to an excitation of $F(t) = F_0 \sin(\Omega t + \pi/3)$. Determine the steady-state response of the system. Use MATLAB to plot the amplitude and the phase lag of the steady-state response as functions of frequency ratio $r = \Omega/\omega_n$. Include cases of $\zeta = 0, 0.1, 0.25, 0.5$, and 0.75 in the same figure.

Problem 2.28: An underdamped mass–spring–dashpot system is subjected to an excitation of $F(t) = F_0(1 + \cos \Omega t)$. Determine the steady-state response of the system. Use MATLAB to plot the amplitude and the phase lag of the steady-state response as functions of frequency ratio $r = \Omega/\omega_n$. Include cases of $\zeta = 0, 0.1, 0.25, 0.5$, and 0.75 in the same figure.

Problem 2.29: An underdamped mass–spring–dashpot system is subjected to an excitation of $F(t) = F_0[\cos \Omega t + \cos(1.05\Omega t)]$. Determine the steady-state response of the system. Assume $\zeta = 0.1$. Use MATLAB to plot the steady-state response, as well as the forcing, for frequency ratios $\Omega/\omega_n = 0.5, 1, 2,$ and 5. For each frequency ratio, plot the response and the forcing in the same figure.

Problem 2.30: An underdamped mass–spring–dashpot system is subjected to an excitation of $F(t) = F_0 \cos \Omega t \cos(1.05\Omega t)$. Determine the steady-state response of the system. Assume $\zeta = 0.1$. Use MATLAB to plot the steady-state response, as well as the forcing, for frequency ratios $\Omega/\omega_n = 0.5, 1, 2,$ and 5. For each frequency ratio, plot the response and the forcing in the same figure.

Problem 2.31: An underdamped mass–spring–dashpot system is subject to a periodic force $F(t)$ of a period T and a saw-tooth form, as shown in Fig. P2.31. Assume $\zeta = 0.1$.

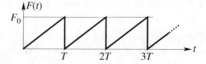

Figure P2.31   Periodic loading of saw-tooth shape

(a) Obtain the Fourier series expansion for the force.
(b) Find the Fourier series expansion of the system's steady-state response.
(c) For $T/T_n = 0.5$, where T_n is the natural period of the system, use MATLAB to plot 1-term, 2-term, 5-term, and 25-term approximations for the forcing and the system's steady-state response. Plot all curves in the same figure for comparison.
(d) Repeat Part (c) for $T/T_n = 1$.
(e) Repeat Part (c) for $T/T_n = 2$.
(f) Repeat Part (c) for $T/T_n = 5$.

In the solution process, the following integrals might be useful:

$$\int x \sin ax\, dx = \frac{1}{a^2} \sin ax - \frac{x}{a} \cos as$$

$$\int x \cos ax\, dx = \frac{1}{a^2} \cos ax + \frac{x}{a} \sin ax$$

Problem 2.32: Repeat the same tasks listed in Problem 2.31 if the system is subjected to a periodic force $F(t)$ of a triangular form, as shown in Fig. P2.32, and $T = 2t_0$.

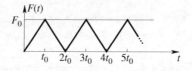

Figure P2.32   Periodic loading of triangular shape

Problem 2.33: Repeat the same tasks listed in Problem 2.31 if the system is subjected to a periodic force $F(t)$ of a half-sine-function form, as shown in Fig. P2.33. In the solution process, the following integrals might be useful:

$$\int \sin ax \sin bx dx = \frac{1}{2}\left[\frac{\sin(a-b)x}{a-b} - \frac{\sin(a+b)x}{a+b}\right]$$

$$\int \sin ax \cos bx dx = -\frac{1}{2}\left[\frac{\cos(a-b)x}{a-b} + \frac{\cos(a+b)x}{a+b}\right]$$

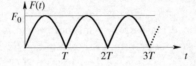

Figure P2.33   Periodic loading of half–sine–function shape

Problem 2.34: Repeat the same tasks listed in Problem 2.31 if the system is subjected to a periodic force $F(t)$ as shown in Fig. P2.34. Note that $F(t)$ has a period of T, but within each period, a constant force F_0 is applied only during an initial duration of t_0. Within each task from Part (c) to Part (f), produce a separate figure for each of the following cases: $t_0/T = 0.1, 0.5$, and 0.9.

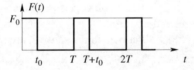

Figure P2.34   Periodic loading of a rectangular shape

Problem 2.35: Repeat Problem 2.34 if the system is subjected to a periodic force $F(t)$ as shown in Fig. P2.35. Note that $F(t)$ has a period of T, but within each period, only during the initial duration of t_0 the force varies in accordance with a half-period of the sine function of amplitude F_0 and the force vanishes afterward.

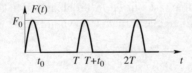

Figure P2.35   Periodic loading of half–sine–function shape

Problem 2.36: Repeat Problem 2.34 if the system is subjected to a periodic force $F(t)$ as shown in Fig. P2.36. Note that $F(t)$ has a period of T; but within each period, only during an initial duration of t_0 the force varies in accordance with a full period of a sine-squared function of amplitude F_0, and the force vanishes afterward. In the solution process, the following integrals might be useful:

$$\int \sin^2 ax \cos bx \, dx = \frac{\sin bx}{2b} - \frac{1}{4}\left[\frac{\sin(2a-b)x}{2a-b} + \frac{\sin(2a+b)x}{2a+b}\right]$$

$$\int \sin^2 ax \sin bx \, dx = -\frac{\cos bx}{2b} + \frac{1}{4}\left[\frac{\cos(2a-b)x}{2a-b} - \frac{\cos(2a+b)x}{2a+b}\right]$$

$$\int \sin^2 ax \, dx = \frac{x}{2} - \frac{\sin 2ax}{4a}$$

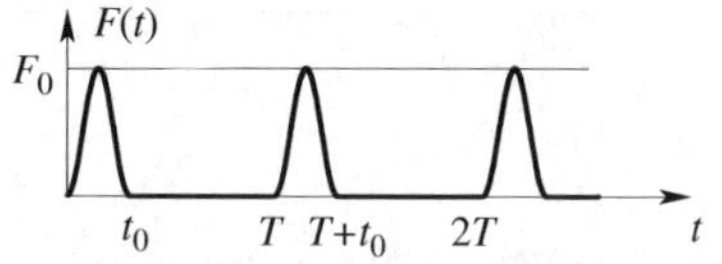

Figure P2.36 Periodic loading of sine-squared function shape

Problem 2.37: For an underdamped mass–spring–dashpot system, the mass is subjected to a force $F(t) = F_0 e^{-t/\tau}$. The system is initially at rest. Find the transient response of the system. Use MATLAB to plot the response for the ratios $\tau/T_n = 0.2, 0.5, 1, 2$, and 5, where T_n is the system's natural period: produce one figure for each τ/T_n ratio and include cases $\zeta = 0.05, 0.1, 0.25, 0.5$, and 0.75 in the same figure.

Problem 2.38: For an underdamped mass–spring–dashpot system, the mass is subjected to a force $F(t) = F_0(1 - e^{-t/\tau})$. The system is initially at rest. Find the transient response of the system. Use MATLAB to plot the response for the ratios $\tau/T_n = 0.2, 0.5, 1, 2$, and 5, where T_n is the system's natural period: produce one figure for each τ/T_n ratio and include cases $\zeta = 0.05, 0.1, 0.25, 0.5$, and 0.75 in the same figure.

Problem 2.39: For an underdamped mass–spring–dashpot system, the mass is subjected to a force $F(t) = a + b\sin\Omega t$. The system is initially at rest. Find the transient response of the system. Use MATLAB to plot the response for the ratios $\Omega/\omega_n = 0.5, 1$, and 2, and the ratios $a/b = -1.5, 0$, and 1.5: produce one figure for each combination of the two ratios, and include cases $\zeta = 0.05, 0.1, 0.25, 0.5$, and 0.75 in the same figure.

Problem 2.40: Use the *analytical method* to find the transient response of an underdamped mass–spring–dashpot system subjected to a so-called *ramped step loading* as shown in Fig. P2.40. The system is initially at rest. Use MATLAB to plot the response for the ratios $t_0/T_n = 0.5, 1$, and 2, where T_n is the system's natural period: produce one figure for each t_0/T_n ratio and include cases $\zeta = 0.05, 0.1, 0.25, 0.5$, and 0.75 in the same figure.

Figure P2.40 Transient ramped step loading

Problem 2.41: Use the *analytical method* to find the transient response of an underdamped mass–spring–dashpot system subjected to one period of sine loading, of period t_0 and amplitude F_0, as shown in Fig. P2.41. The system is initially at rest. Use MATLAB to plot the response for the ratios $t_0/T_n = 0.5$, 1, and 2, where T_n is the system's natural period: produce one figure for each t_0/T_n ratio and include cases $\zeta = 0.05, 0.1, 0.25, 0.5,$ and 0.75 in the same figure.

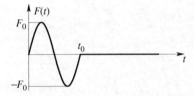

Figure P2.41   Transient loading of one period of sine function

Problem 2.42: A car is running over a speed bump at a constant speed of v_0. The bump can be described by the curve $y = h[1 - (x/a)^2]$ within the range $x \in [-a, a]$, where a is the half-span of the bump and h is the height, as shown in Fig. P2.42. Use the *analytical method* to find the car's vertical motion as it runs across the speed bump. Assume that the car runs smoothly over the bump without causing impact. Use MATLAB to plot the car's vertical motion for the speed–frequency ratio $(v_0/a)/\omega_n$ of values 0.25, 0.5, 1, and 2: produce one figure for each speed–frequency ratio and include cases $\zeta = 0.1, 0.5, 1,$ and 5 in the same figure. Note that the system is critically damped and overdamped, respectively, for the last two damping ratios.

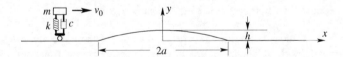

Figure P2.42   Speed bump with parabolic profile

Problem 2.43: An underdamped mass–spring–dashpot system is subjected to a step loading of F_0 starting at $t = 0$. The mass has an initial velocity of v_0 and a zero initial displacement. Use the *Laplace transform method* to find the transient response of the system. Use MATLAB to plot the system's response and include cases $\zeta = 0.05, 0.1, 0.25, 0.5,$ and 0.75 in the same figure.

Problem 2.44: Use the *Laplace transform method* to solve Problem 2.37. The system is subjected to an initial displacement of Δ and an initial velocity of v_0.

Problem 2.45: Use the *Laplace transform method* to solve Problem 2.38. The system is subjected to an initial displacement of Δ and an initial velocity of v_0.

Problem 2.46: Use the *Laplace transform method* to solve Problem 2.40. The system is subjected to an initial displacement of Δ and an initial velocity of v_0.

Problem 2.47: Use the *Laplace transform method* to find the system's transient response to *unit impulse* applied at time $t = 0$. The system is initially at rest. The unit impulse is an instantaneous force whose time duration is infinitesimally small, and yet the integral over time is unity. Mathematically, the unit impulse is described by the *Dirac delta function* $\delta(t)$, defined as the following:

$$\delta(t) = \begin{cases} \infty & \text{when } t = 0 \\ 0 & \text{when } t \neq 0 \end{cases} \quad \text{and} \quad \int_{-\epsilon}^{\epsilon} \delta(t)dt = 1$$

where ϵ is an arbitrarily small positive number.

Problem 2.48: Use the *decomposition method* to solve Problem 2.40.

Problem 2.49: Use the *decomposition method* to solve Problem 2.41.

Problem 2.50: For an underdamped mass–spring–dashpot system, use the *decomposition method* to find the system's transient response due to a loading of a triangular shape, as shown in Fig. P2.50. The system is initially at rest. Use MATLAB to plot the response for the cases $t_0/T_n = 0.2, 0.5, 1, 2$, and 5, where T_n is the system's natural period: produce one figure for each t_0/T_n ratio and include cases $\zeta = 0.05, 0.1, 0.25, 0.5$, and 0.75 in the same figure.

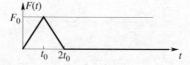

Figure P2.50   Transient triangle loading

Problem 2.51: For an underdamped mass–spring–dashpot system, use the *decomposition method* to find the system's transient response due to a loading of a reverse triangle shape, as shown in Fig. P2.51. The system is initially at rest. Use MATLAB to plot the response for the cases $t_0/T_n = 0.2, 0.5, 1, 2$, and 5, where T_n is the system's natural period: produce one figure for each t_0/T_n ratio and include cases $\zeta = 0.05, 0.1, 0.25, 0.5$, and 0.75 in the same figure.

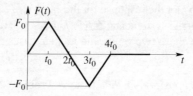

Figure P2.51   Transient triangle reversal loading

Problem 2.52: For an underdamped mass–spring–dashpot system, use the *decomposition method* to find the system's transient response due to a loading of the shape of a half-period of the sine function, as shown in Fig. P2.52. Use MATLAB to plot the response for the cases $t_0/T_n = 0.2, 0.5, 1, 2$, and 5, where T_n is the system's natural period:

produce one figure for each t_0/T_n ratio and include cases $\zeta = 0.05, 0.1, 0.25, 0.5$, and 0.75 in the same figure.

Figure P2.52 Transient loading of a half-period of sine function

Problem 2.53: For an underdamped mass–spring–dashpot system, use the *decomposition method* to find the system's transient response due to a loading of the shape of a period of the sine-squared function, as shown in Fig. P2.53. The system is initially at rest. Use MATLAB to plot the response for the cases $t_0/T_n = 0.2, 0.5, 1, 2$, and 5, where T_n is the system's natural period: produce one figure for each t_0/T_n ratio, include cases $\zeta = 0.05$, 0.1, 0.25, 0.5, and 0.75 in the same figure.

Figure P2.53 Transient loading of a period of sine-squared function

Problem 2.54: Use the *decomposition method* to find the transient response of an underdamped mass–spring–dashpot system due to an square loading of force F_0 for a duration of t_0. The system has an initial velocity of v_0. Use MATLAB to plot the response for $t_0/T_n = 0.2, 0.5, 1, 2$, and 5 where T_n is the natural period of the system; and $(v_0/\omega_n)/(F_0/k) = -1.5, 0$, and 1.5. Produce one figure for each combination of these two ratios, and include the cases for $\zeta = 0.05, 0.1, 0.25, 0.5$, and 0.75 in the same figure.

Problem 2.55: Solve Problem 2.40 with an initial displacement of Δ using the *decomposition method*. Use MATLAB to plot the response for the cases $t_0/T_n = 0.2, 0.5, 1, 2$, and 5 where T_n is the natural period of the system; and $\Delta/(F_0/k) = -1.5, 0$, and 1.5. Produce one figure for each combination of these two ratios and include cases $\zeta = 0.05, 0.1, 0.25, 0.5$, and 0.75 in the same figure.

Problem 2.56: Use the *decomposition method* to solve Problem 2.53 when the system has an initial velocity of v_0. Use MATLAB to plot the response for the cases $t_0/T_n = 0.2, 0.5, 1, 2$, and 5 where T_n is the natural period of the system, and $(v_0/\omega_n)/(F_0/k) = -1.5, 0$, and 1.5. Produce one figure for each combination of these two ratios and include cases $\zeta = 0.05$, 0.1, 0.25, 0.5, and 0.75 in the same figure.

Problem 2.57: Use the *Laplace transform method* to verify that, for the case of nontrivial initial conditions, the Duhamel integral is given in eqn. (2.131).

Problem 2.58: Determine the initial velocity such that the free vibration of a underdamped mass–spring–dashpot system produces the so-called *unit impulse response function*, the $h(t)$ in the Duhamel integral. Use this initial velocity to determine the unit impulse response functions for critically damped and overdamped systems.

Problem 2.59: Determine the unit impulse response function, the $h(t)$ in the Duhamel integral, for a rigid link pendulum in Example 1.23.

Problem 2.60: Use the *convolution integral method* (numerical) to solve Problem 2.42.

Problem 2.61: Use the *convolution integral method* (numerical) to solve Problem 2.48.

Problem 2.62: Use the *convolution integral method* (numerical) to solve Problem 2.49.

Problem 2.63: Use the *convolution integral method* (numerical) to solve Problem 2.50.

Problem 2.64: Use the *convolution integral method* (numerical) to solve Problem 2.51.

3

Lumped-Parameter Modeling

3.1 Objectives

This chapter presents methods to construct lumped-parameter models for simple engineering structures. The main focus is modeling simple structures into single-degree-of-freedom (single-DOF) systems having one equivalent spring and one equivalent mass. The procedures for obtaining these equivalent system characteristic parameters for simple structures are illustrated by examples. Several methods are explored, including the conventional method based on the *Mechanics of Materials*, the Castigliano method, and the Rayleigh–Ritz method. More complicated structures could be modeled by combining established lumped-parameter modeling results. Finally, vibrations of several lumped-parameter models are analyzed, including examples exploring an interesting phenomenon called the *whirling of shafts*.

3.2 Modeling

Modeling is the process of building a model to represent a physical system. A model is a simplified, and typically idealized, approximation to the original system. Modeling is done for the following two purposes: (1) to simplify a physical system such that it can be analyzed mathematically and (2) to capture the essence of the physics of the system.

Modeling can be done at different levels. Very often, modeling is done at the level of mathematics that we are capable of handling, and yet preserve the purpose of capturing the essence of the physics. In this chapter, we will be modeling at a level that allows us to perform the analysis with a pen and paper. In a later chapter (Chapter 5), we will model systems at a level that is suitable for numerical analyses using computers.

Models for mechanical systems can be classified into lumped-parameter models and distributed-parameter models. Many mechanical systems are constructed from materials that have continuously distributed parameters, such as masses, stiffnesses, and frictions. Such idealized models without changing the nature of the distributed parameters are called the distributed-parameter models. In a lumped-parameter model, the physical parameters are lumped into a certain locations selected *a priori*. Typically, a distributed-parameter model requires a continuous function to describe the spatial distribution of the parameters; whereas

Fundamentals of Mechanical Vibrations, First Edition. Liang-Wu Cai.
© 2016 John Wiley & Sons, Ltd. Published 2016 by John Wiley & Sons, Ltd.
Companion Website: www.wiley.com/go/cai/fundamentals_mechanical_vibrations

in a lumped-parameter model, distributed system characteristics are lumped into and observed from a few discrete locations. All the examples we have seen so far in this book, from the systems we have found the equations of motion in Chapter 1 to single-DOF systems we studied in Chapter 2, are lumped-parameter models.

3.3   Idealized Elements

We first review the main characteristics of idealized elements that have been used in our lumped-parameter models. As mentioned earlier, vibration is the process of converting the energy from one form to another. This requires a system to have devices that can store both kinetic energy and potential energy. Mass elements are the device for storing kinetic energy. They can also store gravitational potential energy if the geometry permits. Spring elements are another device for storing potential energy. We will review these elements from the energy storage perspective.

3.3.1   Mass Elements

3.3.1.1   Particle

A *particle* is an idealized object that has a mass but has no measurable size. It is also called a *point mass*. The kinetic energy stored in a particle is

$$T_{\text{particle}} = \frac{1}{2}mv^2 \tag{3.1}$$

where m is the mass of the particle, and v is its velocity in a fixed reference frame or the *absolute velocity*. The gravitational potential energy stored in a particle is

$$V_{\text{particle}} = mgh \tag{3.2}$$

where h is the vertical height of the particle's location measured from a fixed datum point. Obviously different choices of the datum point result in different values in the potential energy. As long as the datum point is fixed, the difference in the form of a constant is immaterial.

3.3.1.2   Rigid Body

A *rigid body* is an idealized object that has both a mass and a measurable size, but is undeformable. Because of the size, it also has an orientation.

A rigid body is said to be undergoing a *planar motion* if trajectories of all points in the rigid body remain in parallel planes at all times. For a rigid body undergoing a planar motion, the stored kinetic energy can be written as

$$T_{\text{rigid body}} = \frac{1}{2}mv_C^2 + \frac{1}{2}I_C\omega^2 \tag{3.3}$$

where m is the mass of the rigid body; I_C is the *mass moment of inertia* of the rigid body about its center of mass, called the *centroid*; v_C is the velocity of its centroid relative to a fixed reference frame; and ω is its angular velocity.

Figure 3.1   Linear spring element

The gravitational potential energy stored in a rigid body is

$$V_{\text{rigid body}} = mgh_C \tag{3.4}$$

where h_C is the vertical height of the centroid measured from a fixed datum point.

3.3.2   Spring Elements

3.3.2.1   Linear Spring

A *linear spring* is an idealized massless deformable object. Its typical graphical representation is depicted in Fig. 3.1. The meaning of the word linear is twofold: (1) its deformation is along a straight line to which it is aligned, measurable as a linear displacement; (2) the relation between the force and the deformation, generally called the *constitutive relation*, is linear, which can be written as

$$F_{\text{spring}} = k\Delta \tag{3.5}$$

where k is the *spring constant*, also called the *spring rate*, and Δ is the amount of deformation in the spring. The potential energy stored in a linear spring is

$$V_{\text{spring}} = \frac{1}{2}k\Delta^2 \tag{3.6}$$

Multiple springs can be connected in parallel or in series to form a compound linear spring, to be represented by an equivalent spring constant. When springs are connected in series, the force acting on each spring is the same and the deflections are cumulative; this gives the equivalent spring constant as

$$k_{\text{equiv}} = \frac{1}{\dfrac{1}{k_1} + \dfrac{1}{k_2} + \cdots} \tag{3.7}$$

where k_i's are the spring constants of individual springs. When springs are connected in parallel, the deformation in all the springs is the same while the forces are cumulative; this gives the equivalent spring constant as

$$k_{\text{equiv}} = k_1 + k_2 + \cdots \tag{3.8}$$

The total potential energy stored in connected springs is the sum of all energies stored in individual springs, regardless of the way they are connected.

3.3.2.2   Torsional Spring

A *torsional spring* is an idealized massless deformable object whose deformation is an angular displacement of one end with respect to the other. Its typical graphical representation

Figure 3.2   A torsional spring

is depicted in Fig. 3.2. A linear torsional spring has a linear constitutive relation between the moment (torque) and the angular deformation as

$$M_{\text{torsional spring}} = k_t \Delta\theta \tag{3.9}$$

where k_t is the *torsional spring constant* or the *torsional spring rate*, and $\Delta\theta$ is the angular displacement of one end of the torsional spring with respect to the other end. The potential energy stored in a deformed linear torsional spring is

$$V_{\text{torsional spring}} = \frac{1}{2} k_t (\Delta\theta)^2 \tag{3.10}$$

Torsional springs can also be connected in parallel or in series, and the resulting equivalent spring constant is similar to that of linear springs.

3.3.3   Damping Elements

3.3.3.1   Dashpot

A *dashpot* is an idealized massless deformable object that can deform along a straight line to which it is aligned. Its typical graphical representation is depicted in Fig. 3.3. It produces a force along the dashpot, with a magnitude proportional to the velocity difference between its two ends, and in the direction resisting the relative motion. This linear constitutive relationship can be written as

$$F_{\text{dashpot}} = c\Delta v \tag{3.11}$$

where c is the *dashpot constant* and Δv is the velocity difference between its two ends.

A dashpot dissipates energy, which is often measured by the energy dissipation during one complete period of sinusoidal motion. For a displacement x, within one period of motion, the energy dissipated by the dashpot is

$$E_{\text{dashpot}} = \int_{t_0}^{T+t_0} -c\dot{x}^2 \, dt \tag{3.12}$$

where t_0 is an arbitrary starting time and T is the period of the motion. If x is a sinusoidal motion expressible as $x(t) = X_0 \sin(\Omega t + \phi)$, the energy dissipated by the dashpot during one period of motion is

$$E_{\text{dashpot}} = -c\pi\Omega X_0^2 \tag{3.13}$$

Figure 3.3   A typical linear dashpot

Figure 3.4   A torsional dashpot

3.3.3.2   Torsional Dashpot

A *torsional dashpot* is an idealized massless deformable object whose deformation is an angular displacement of one end with respect to the other end. Its typical graphical representation is depicted in Fig. 3.4. It produces a torque (or moment) that is linearly proportional to the difference of the angular velocity between its two ends. This linear constitutive relationship can be written as

$$T_{\text{dashpot}} = c_t \Delta\omega \tag{3.14}$$

where c_t is called the *torsional dashpot constant* and $\Delta\omega$ is the angular velocity difference between its two ends.

For a sinusoidal angular motion of amplitude Θ and frequency Ω, in a process similar to the case for the dashpot, it can be found that the energy dissipated by the torsional dashpot during one period of sinusoidal motion is

$$E_{\text{torsional dashpot}} = -c_t \pi \Omega \Theta^2 \tag{3.15}$$

3.4   Lumped-Parameter Modeling of Simple Components and Structures

In engineering practice, we often encounter distributed-parameter systems. Lumped-parameter models for such systems, although very crude, often capture the essence of the physics of the systems.

For vibration analysis, from Chapter 2, we have come to a good understanding of the importance of a system's natural frequency. A distributed-parameter system, which is also called a *continuous system*, has an infinite number of natural frequencies. But the most important one is the lowest natural frequency, and its importance has earned itself a special name called the *fundamental natural frequency*. Simple lumped-parameter models, which can be analyzed with the back-of-envelope calculations, can give a very good estimate of the fundamental natural frequency of the system. For this simple reason, the lumped-parameter modeling holds its special place in engineering practices in the days of the computer proliferation.

In this section, we introduce a procedure based on the *Mechanics of Materials* for obtaining characteristic parameters for a lumped-parameter model. We examine the equivalent spring constant and the equivalent mass separately. In this procedure, we first choose a point in a system where the parameters to be lumped at. We shall refer to this point as the *point of interest*. We illustrate the modeling process by using extensive examples.

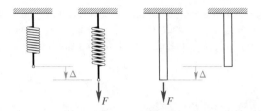

Figure 3.5   Equivalence between spring and rod under uniaxial loading

3.4.1   Equivalent Spring Constants

We start our lump-parameter modeling with spring elements, simply because it is likely a more familiar concept for us. Since taking the *Mechanics of Materials* course, we have been familiar with the equivalence between a spring and a rod under a uniaxial extension, as illustrated in Fig. 3.5. For a rod of Young's modulus E, length L, and cross-sectional area A, its equivalent spring constant is well known as

$$k_{\text{equiv}} = \frac{EA}{L} \tag{3.16}$$

We shall derive this well-known result as an example to demonstrate the procedure for obtaining the equivalent spring constant. The procedure includes the following steps:

Step 1: *Identify the point of interest and apply a force or a moment at the point of interest.*
    In this example, we apply a tensile force F at the end of the rod.
Step 2: *Obtain the expression for the resulting deformation in the structure.*
    This step often employs the analyses and approximations of the *Mechanics of Materials*. We can directly use known results if they are available (often tabulated at the end of many textbooks).
    The rod under a uniaxial loading produces a uniform uniaxial stress

$$\sigma = \frac{F}{A}$$

Hooke's law relates the stress and the strain in uniaxial loading as

$$\epsilon = \frac{\sigma}{E}$$

and the strain is related to the displacement as

$$\epsilon = \frac{du}{dx}$$

where $u(x)$ is the displacement in the rod. Combining these pieces of information gives

$$\frac{du}{dx} = \frac{\sigma}{E} = \frac{F}{EA} \tag{3.17}$$

Integrating the above equation gives

$$u(x) = \frac{F}{EA}x + C \tag{3.18}$$

where C is the constant of integration, which can be determined by the displacement at the fixed end of the rod, that is, $u = 0$ when $x = 0$. This is called the *boundary condition*, which, in this particular case, gives $C = 0$. Thus, the deformation in the rod is

$$u(x) = \frac{F}{EA}x \tag{3.19}$$

Step 3: *Obtain the expression for the displacement at the point of interest.*

Denote the displacement at the end of the rod as Δ. From eqn. (3.19),

$$\Delta \equiv u(L) = \frac{FL}{EA} \tag{3.20}$$

Step 4: *Compare the displacement at the point of interest to the constitutive relation for a linear spring in eqn. (3.5) or a torsional spring in eqn. (3.9) and obtain the expression for the equivalent spring constant.*

Doing so, we arrive at the well-known result of eqn. (3.16).

In the following, we will follow this procedure to obtain the equivalent spring constants for a few simple engineering structures.

■ Example 3.1: Equivalent Spring Constant for a Shaft Under Torsion

A uniform circular shaft of radius R and length L is undergoing a torsional motion. One end of the shaft is fixed and the other end is free. Assume that the shaft is made of an elastic material of shear modulus G. The free end of the shaft is the point of interest. Find the equivalent spring constant for the shaft.

□ **Solution:**

• We apply a torque T at the free end of the shaft, as shown in Fig. 3.6, in which we also introduced a coordinate system to facilitate the mathematical description of the problem.

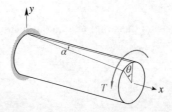

Figure 3.6   Shaft with one fixed end and one free end subjected to torque

• According to *Mechanics of Materials*, the torque produces a twist angle α as illustrated. By definition, the twist angle equals the engineering shear strain on the surface of the shaft. Furthermore, the shear strain varies linearly in the radial direction. Combining these two pieces of information gives the engineering shear strain in the shaft as

$$\gamma(r) = \alpha \frac{r}{R} \tag{a}$$

where r is the radial distance from the center of the shaft. Hooke's law relates the shear stress and the engineering shear strain as the following:

$$\tau = G\gamma \tag{b}$$

The balance of the moment requires that, at the cross section at x, the torque due to the shear stress equals the externally applied torque, that is,

$$T = \int_A \tau r\, dA = \int_A G\alpha \frac{r}{R} r\, dA = \frac{\alpha G}{R} \int_A r^2\, dA \tag{c}$$

The integral in the last expression is the *polar moment of the area*, denoted as J, that is,

$$J = \int_A r^2\, dA \tag{d}$$

Then, the twist angle α can be expressed in terms of applied torque as

$$\alpha = \frac{TR}{GJ} \tag{e}$$

For a twist angle α, it causes a point on the surface of the shaft at x to move an arc of length αx. This means the angular displacement of the cross section is expressible as

$$\theta(x) = \frac{\alpha x}{R} = \frac{Tx}{GJ} \tag{f}$$

- Denote the angular displacement at the point of interest as Θ, then

$$\Theta \equiv \theta(L) = \frac{TL}{GJ} \tag{g}$$

- Comparing eqn. (g) with the moment–angular displacement relation for a torsional spring in eqn. (3.9), the equivalent spring constant for the shaft is

$$k_{t\text{ equiv}} = \frac{GJ}{L} \tag{h}$$

■ **Example 3.2: Equivalent Spring Constant for Cantilever Beam**

A uniform cantilever beam of length L is made of an elastic material of Young's modulus E and has a moment of area I. The free end of the beam is the point of interest. Find the equivalent spring constant for the beam.

□ **Solution:**

- We apply a force F at the free end of the beam, as shown in Fig. 3.7, in which we have also introduced a Cartesian coordinate system to facilitate the analysis.

Figure 3.7 Cantilever beam with force F applied to its free end

- Recall the *beam equation* in Mechanics of Materials:

$$EI \frac{d^2u(x)}{dx^2} = M(x) \tag{a}$$

where $u(x)$ is the deflection of the beam, $M(x)$ is the moment distribution, E is the Young's modulus, and I is called the *area moment of inertia* or the *moment of area* for the cross section. EI is also called the *bending stiffness* of the beam.

To find the moment distribution, we draw two free-body diagrams: one for the entire beam isolated from its support, and the other for a segment of the isolated beam of length x starting from the clamped end, as shown in Fig. 3.8. In these free-body diagrams, R and M_0 are the resultant force and moment, respectively, at the cantilevered end; and $Q(x)$ and $M(x)$ are the shear force and bending moment, respectively, at x.

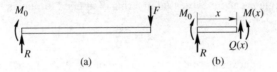

Figure 3.8   Free-body diagrams for the cantilever beam: (a) for the entire beam and (b) for a segment of beam of length x

From the free-body diagram for the entire beam, the static equilibrium requires

$$R = F \qquad \text{and} \qquad M_0 = -FL \tag{b}$$

From the free-body diagram for the segment, the balance of moments about the cut gives

$$M(x) = M_0 + Rx = F(x - L) \tag{c}$$

where eqns. (b) have been used. We can use the balance of forces to obtain the expression for $Q(x)$. But we do not need it. Substituting eqn. (c) into the beam equation gives

$$EIu'' = F(x - L) \tag{d}$$

Integrating once gives

$$EIu' = F\left(\frac{1}{2}x^2 - Lx\right) + C_1 \tag{e}$$

Integrating again gives

$$EIu = F\left(\frac{1}{6}x^3 - \frac{1}{2}Lx^2\right) + C_1 x + C_2 \tag{f}$$

where C_1 and C_2 are the constants of integration, which will be determined by the boundary conditions. The beam is *cantilevered*, also called *clamped*, at its left end, meaning that both its deflection and slope vanish. Mathematically, the boundary conditions can be written as $u(0) = 0$ and $u'(0) = 0$, which give $C_1 = C_2 = 0$. Hence, the beam's deflection can be written as

$$u(x) = \frac{Fx^2}{6EI}(x - 3L) \tag{g}$$

- Denote the deflection at the point of interest as Δ, that is,

$$\Delta \equiv -u(L) = \frac{FL^3}{3EI} \tag{h}$$

In beam theory, the downward deflection is negative; yet for a spring we consider the deflection along the direction of the force as positive, whence the negative sign is added.
- Comparing eqn. (h) with the constitutive relation for a linear spring in eqn. (3.5) gives

$$k_{\text{equiv}} = \frac{3EI}{L^3} \tag{i}$$

■ Example 3.3: Equivalent Spring Constant for Simply-Supported Beam

Assume that a uniform beam, of length L, Young's modulus E, and area moment of inertia I, is simply supported at both ends. The center of the beam is the point of interest. Find the equivalent spring constant for the beam.

□ Solution:

- We apply a force F at the center of the beam, as shown in Fig. 3.9, in which we have also introduced a Cartesian coordinate system to facilitate the analysis.

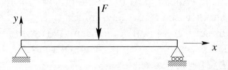

Figure 3.9   A simply-supported beam with force F applied to its center

- We draw two free-body diagrams: one for the entire beam isolated from its supports, and the other for a segment of the isolated beam of length x ($0 \leq x \leq L/2$) starting from its left end, as shown in Fig. 3.10. The second free-body diagram is only valid for $x \leq L/2$ because of a force acting at $x = L/2$. But since we only need deflection at $x = L/2$, this suffices. In these free-body diagrams, R is the resultant force at both supports and $Q(x)$ and $M(x)$ are the shear force and bending moment, respectively. By denoting the resultant forces at both ends by the same R, we have utilized the symmetry of the problem.

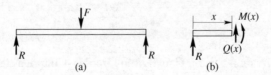

Figure 3.10   Free-body diagrams for simply-supported beam: (a) for the entire beam and (b) for a segment of beam of length x ($0 \leq x \leq L/2$)

From the free-body diagram for the entire beam, the static equilibrium requires

$$R = \frac{1}{2}F \tag{a}$$

From the free-body diagram for the segment, the balance of moment about the cut gives

$$M(x) = Rx = \frac{1}{2}Fx \tag{b}$$

where eqn. (a) has been used. Then, the beam equation becomes

$$EIu'' = \frac{1}{2}Fx \tag{c}$$

Integrating twice gives

$$EIu' = \frac{1}{4}Fx^2 + C_1 \tag{d}$$

$$EIu = \frac{1}{12}Fx^3 + C_1x + C_2 \tag{e}$$

where C_1 and C_2 are the constants of integration to be determined by the boundary conditions. The boundary condition at $x = 0$ is $u(0) = 0$, which gives $C_2 = 0$. The boundary condition at $x = L$ cannot be used because the above expression for $u(x)$ is only valid for $0 \le x \le L/2$.

To determine C_1, we can either repeat the same process for the right half of the beam, starting from drawing another free-body diagram for $L/2 \le x \le L$ to using that boundary condition, or use the symmetry condition of the beam. We use the symmetry condition here. In order for the deformed beam to be symmetric about $x = L/2$, the slope at $x = L/2$ must vanish. This means $u'(L/2) = 0$, that is,

$$0 = \frac{1}{4}F\left(\frac{L}{2}\right)^2 + C_1 \quad \text{or} \quad C_1 = -\frac{FL^2}{16} \tag{f}$$

Thus, the deflection for the left half of the beam can be written as

$$EIu = \frac{1}{12}Fx^3 - \frac{FL^2}{16}x = \frac{Fx}{48}(4x^2 - 3L^2) \tag{g}$$

- Denote the deflection at the point of interest as Δ, that is,

$$\Delta \equiv -u\left(\frac{L}{2}\right) = \frac{FL^3}{48EI} \tag{h}$$

- Comparing eqn. (h) with the constitutive relation for a linear spring in eqn. (3.5) gives

$$k_{\text{equiv}} = \frac{48EI}{L^3} \tag{i}$$

■ **Example 3.4: Equivalent Spring Constant for Doubly-Clamped Beam**

Assume both ends of a uniform beam of length L are clamped. The bending stiffness of the beam is EI. The center of the beam is the point of interest. Find the equivalent spring constant for the beam.

□ **Solution:**

- We apply a force F at the center of the beam, as shown in Fig. 3.11, in which we have also introduced a Cartesian coordinate system to facilitate the analysis.

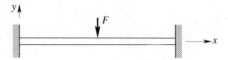

Figure 3.11   Doubly-clamped beam with force F applied to its center

- We draw two free-body diagrams: one for the entire beam isolated from its supports, and the other for a segment of the isolated beam of length x ($0 \leq x \leq L/2$) starting from its left end, as shown in Fig. 3.12. In these free-body diagrams, R and M_0 are the resultant forces and moments, respectively, at both ends, and $Q(x)$ and $M(x)$ are the shear force and bending moment, respectively. By denoting the reaction forces and moments at both ends by identical symbols, we have utilized the symmetry of the problem.

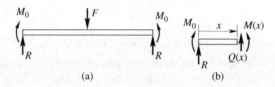

(a)   (b)

Figure 3.12   Free-body diagrams for doubly-clamped beam: (a) for entire beam and (b) for segment of beam of length x ($0 \leq x \leq L/2$)

From the free-body diagram for the entire beam, because of the symmetry, the static equilibrium gives $R = F/2$. However, we do not have enough equations from static equilibrium to determine the moment M_0. Such a structure is said to be *statically indeterminate,*, which typically has more supports (boundary conditions) than required for achieving the static equilibrium. M_0 will be determined later using boundary conditions. For now, we carry it through the analysis as an undetermined constant.

From the free-body diagram for the segment, the moment balance about the cut gives

$$M(x) = M_0 + Rx = M_0 + \frac{1}{2}Fx \tag{a}$$

Thus, the beam equation becomes

$$EIu'' = M_0 + \frac{1}{2}Fx \tag{b}$$

Integrating twice gives

$$EIu' = M_0 x + \frac{1}{4}Fx^2 + C_1 \tag{c}$$

$$EIu = \frac{1}{2}M_0 x^2 + \frac{1}{12}Fx^3 + C_1 x + C_2 \tag{d}$$

where C_1 and C_2 are the constants of integration to be determined by the boundary conditions.

The boundary conditions at $x = 0$ are $u(0) = 0$ and $u'(0) = 0$, which give $C_1 = C_2 = 0$. Additionally, as the beam is symmetric about $x = L/2$, its slope must vanish at the symmetry line, that is, $u'(L/2) = 0$. According to eqn. (c),

$$0 = M_0 \frac{L}{2} + \frac{1}{4}F\frac{L^2}{4} \qquad \text{or} \qquad M_0 = -\frac{FL}{8} \tag{e}$$

Finally, the deflection of the beam can be written as

$$EIu = -\frac{1}{16}FLx^2 + \frac{1}{12}Fx^3 = \frac{Fx^2}{48EI}(4x - 3L) \tag{f}$$

- Denote the deflection at the point of interest as Δ, that is,

$$\Delta \equiv -u\left(\frac{L}{2}\right) = -\frac{F}{48EI}\frac{L^2}{4}(2L - 3L) = \frac{FL^3}{192EI} \tag{g}$$

- Comparing eqn. (g) with the constitutive relation for a linear spring in eqn. (3.5) gives

$$k_{\text{equiv}} = \frac{192EI}{L^3} \tag{h}$$

■ Example 3.5: Equivalent Spring Constant for Clamped–Simply-Supported Beam

A uniform beam of length L has its left end clamped and its right end simply supported on a roller. The beam's bending stiffness is EI. The center of the beam is the point of interest. Find the equivalent spring constant for the beam.

□ Solution:

- We apply a force F at the center of the beam, as shown in Fig. 3.13, in which we have also introduced a Cartesian coordinate system to facilitate the analysis.

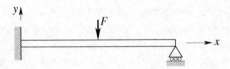

Figure 3.13   Clamped–simply-supported beam

- We draw the free-body diagram for the entire beam isolated from its supports, as shown in Fig. 3.14, where R_1 and R_2 are the resultant forces and M_0 is the resultant moment. From this free-body diagram, the static equilibrium requires

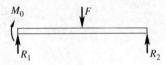

Figure 3.14   Free-body diagram for clamped–simply-supported beam

$$R_1 + R_2 - F = 0 \tag{a}$$

$$M_0 + F\frac{L}{2} - R_2 L = 0 \tag{b}$$

Note that this is also a statically indeterminate structure, meaning that the unknowns cannot be completely determined by static equilibrium alone.

We draw two more free-body diagrams for a segment of the isolated beam of length x: one for $0 \le x \le \frac{1}{2}L$, and the other for $\frac{1}{2}L \le x \le L$, as shown in Fig. 3.15. They will be used to find the moment distributions on the left and right halves of the beam, for which we denote their deflections as u_L and u_R, respectively.

From the free-body diagram for $0 \le x < \frac{1}{2}L$ in Fig. 3.15a, the balance of moment about the cut gives

(a)   (b)

Figure 3.15   Free-body diagrams for beam segment, of length x, of clamped–simply-supported beam: (a) $0 \le x \le \frac{1}{2}L$ and (b) $\frac{1}{2}L \le x \le L$

$$M(x) = M_0 + R_1 x$$

and consequently the beam equation becomes

$$EIu_L'' = M_0 + R_1 x \tag{c}$$

Integrating it twice gives

$$EIu_L' = M_0 x + \frac{1}{2}R_1 x^2 + C_1 \tag{d}$$

$$EIu_L = \frac{1}{2}M_0 x^2 + \frac{1}{6}R_1 x^3 + C_1 x + C_2 \tag{e}$$

where C_1 and C_2 are constants of integration to be determined by the boundary conditions. The boundary conditions at $x = 0$ are $u(0) = 0$ and $u'(0) = 0$, which give $C_1 = C_2 = 0$. Thus, for the left half of the beam,

$$EIu_L = \frac{1}{2}M_0 x^2 + \frac{1}{6}R_1 x^3 \tag{f}$$

For the right half of the beam, from the free-body diagram in Fig. 3.15b, the balance of moment about the cut gives

$$M(x) = M_0 + R_1 x - F\left(x - \frac{L}{2}\right) = R_2(L - x)$$

where eqn. (a) has been used. Consequently, the beam equation becomes

$$EIu_R'' = R_2(L - x) \tag{g}$$

Integrating twice gives

$$EIu_R' = R_2\left(Lx - \frac{1}{2}x^2 + C_3\right) \tag{h}$$

$$EIu_R = R_2\left(\frac{1}{2}Lx^2 - \frac{1}{6}x^3 + C_3x + C_4\right) \tag{i}$$

where C_3 and C_4 are constants of integration to be determined by the boundary conditions. The boundary condition at $x = L$ is $u_R(L) = 0$, which gives

$$\frac{L^3}{2} - \frac{L^3}{6} + C_3L + C_4 = 0 \tag{j}$$

Since we have split repressions for the left and right halves of the beam, yet physically it is a single piece, the beam must be continuous at $x = L/2$. This requires the continuities of both u and u': $u_L(L/2) = u_R(L/2)$ and $u_L'(L/2) = u_R'(L/2)$, that is,

$$\frac{1}{2}M_0\frac{L^2}{4} + \frac{1}{6}R_1\frac{L^3}{8} = R_2\left(\frac{1}{2}\frac{L^3}{4} - \frac{1}{6}\frac{L^3}{8} + C_3\frac{L}{2} + C_4\right) \tag{k}$$

$$M_0\frac{L}{2} + \frac{1}{2}R_1\frac{L^2}{4} = R_2\left(\frac{L^2}{2} - \frac{1}{2}\frac{L^2}{4} + C_3\right) \tag{l}$$

Equations (j) through (l), along with equilibrium conditions in eqns. (a) and (b), form a set of five equations for five unknowns. Solving these equations gives

$$C_3 = -\frac{2L^2}{5} \qquad C_4 = \frac{L^3}{15} \qquad M_0 = -\frac{3FL}{16} \qquad R_1 = \frac{11}{16}F \qquad R_2 = \frac{5}{16}F$$

Finally, the deflection of the beam can be written as

$$EIu_L(x) = -\frac{3FL}{32}x^2 + \frac{11F}{96}x^3 = \frac{FL^3}{96}\left[-9\left(\frac{x}{L}\right)^2 + 11\left(\frac{x}{L}\right)^3\right] \tag{m}$$

$$EIu_R(x) = \frac{FL^3}{96}\left[-5\left(\frac{x}{L}\right)^3 + 15\left(\frac{x}{L}\right)^2 - 12\left(\frac{x}{L}\right) + 2\right] \tag{n}$$

- Denote the deflection at the point of interest as Δ. At $x = L/2$, the deflection can be found by using either expression for $u(x)$ as

$$\Delta \equiv -u\left(\frac{L}{2}\right) = \frac{7}{768}\frac{FL^3}{EI} \tag{o}$$

- Comparing eqn. (o) with the constitutive relation for a linear spring in eqn. (3.5) gives

$$k_{equiv} = \frac{768EI}{7L^3} \tag{p}$$

■ **Example 3.6: Equivalent Spring Constant for L-Shaped Cantilever Beam**

Consider a uniform L-shaped cantilever beam whose two legs, made of the same material, form a right angle. Leg AB, of length a, is cantilevered to a wall, leg BC, of length b, is parallel to the wall. The beam has a circular cross section of radius r. The material has Young's modulus E and shear modulus G. The point of interest is the free end of the beam. Find the equivalent spring constant for the beam.

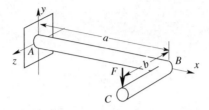

Figure 3.16   L-shaped cantilever beam

□ Solution 1: Using Conventional Procedure

- We apply a force F at the free end of the L-shaped beam.
- We draw two free-body diagrams for the two legs isolated from their supports, as shown in Fig. 3.17, in which double arrows represent twisting torques.

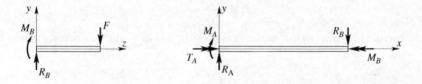

Figure 3.17   Free-body diagrams for two legs of L-shaped beam

The static equilibrium requires

$$R_B = R_A = F \qquad M_B = -Fb \qquad M_A = -Fa$$

We draw two more free-body diagrams, as shown in Fig. 3.18: one for a segment of leg AB, of length x; and the other for a segment of leg BC, of length z.

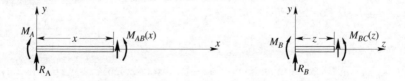

Figure 3.18   Free-body diagrams for two leg segments of L-shaped beam

- From the free-body diagram for the segment of AB, the balance of moments about the cut gives

$$M_{AB}(x) = M_A + R_A x = -Fa + Fx$$

Consequently, the beam equation becomes

$$EIu''_{AB} = F(x - a) \tag{a}$$

and integrating twice gives

$$EIu'_{AB} = -Fax + \frac{1}{2}Fx^2 + C_1 \tag{b}$$

$$EIu_{AB} = -\frac{1}{2}Fax^2 + \frac{1}{6}Fx^3 + C_1 x + C_2 \tag{c}$$

where C_1 and C_2 are constants of integration to be determined by the boundary conditions. The boundary conditions at $x = 0$ is $u_{AB}(0) = 0$ and $u'_{AB}(0) = 0$, which give $C_1 = C_2 = 0$. Thus,

$$u_{AB}(x) = \frac{Fx^2}{6EI}(x - 3a) \tag{d}$$

Leg AB is further twisted by torque $M_B = -Fb$. This torque causes a twist angle of $M_B/(GJ)$, which is the shear strain on the surface of the shaft. This is similar to the shaft in Example 3.1. Along that shaft, the angular displacement is

$$\theta(x) = \frac{M_B x}{GJ} = -\frac{Fbx}{GJ} \tag{e}$$

At the end of the shaft, twist causes an angular displacement for the cross section at B

$$\Theta_B = \frac{M_B a}{GJ} = -\frac{Fab}{GJ} \tag{f}$$

From the free-body diagram for a segment of leg BC, the balance of moments about the cut gives

$$M_{BC}(z) = M_B + R_B z = -Fb + Fz$$

which is substituted into the beam equation and subsequently integrated twice to give

$$EIu'_{BC} = -Fbz + \frac{1}{2}Fz^2 + C_3 \tag{g}$$

$$EIu_{BC} = -\frac{1}{2}Fbz^2 + \frac{1}{6}Fz^3 + C_3 z + C_4 \tag{h}$$

The constants of integration C_3 and C_4 are to be determined by the continuity conditions at junction B ($z = 0$). For the deflection,

$$u_{BC}\big|_{z=0} = u_{AB}\big|_{x=a} = -\frac{1}{3}Fa^3 \tag{i}$$

Leg BC's slope at B is determined by the angular displacement of leg AB, due to twisting, at this point. Note that a positive angular displacement (with respect to the x-axis) produces an upward tilt for leg BC and hence a positive slope. Thus,

$$u'_{BC}\big|_{z=0} = \Theta = -\frac{Fab}{GJ} \tag{j}$$

Evaluating eqns. (g) and (h) for the boundary conditions in eqns. (i) and (j) gives

$$C_3 = -\frac{EIFab}{GJ} \qquad C_4 = -\frac{1}{3}Fa^3 \tag{k}$$

Hence, eqn. (h) becomes

$$u_{BC} = \frac{F}{6EI}\left(-3bz^2 + z^3 - 2a^3 + 6\frac{ab}{GJ}z\right) \tag{l}$$

- Denote the deflection at the point of interest as Δ, that is,

$$\Delta \equiv -u_{BC}|_{z=b} = F\left(\frac{a^3 + b^3}{3EI} + \frac{Fab^2}{GJ}\right) \tag{m}$$

- Comparing eqn. (m) with the constitutive relation for a linear spring in eqn. (3.5) gives

$$k_{equiv} = \frac{1}{\frac{a^3 + b^3}{3EI} + \frac{ab^2}{GJ}} = \frac{3EGIJ}{(a^3 + b^3)GJ + 3ab^2 EI} \tag{n}$$

For a beam with a circular cross section of radius r, $I = \frac{1}{4}\pi r^4$ and $J = \frac{1}{2}\pi r^4$, then

$$k_{equiv} = \frac{3EG\pi r^4}{2[2(a^3 + b^3)G + 3ab^2 E]} \tag{o}$$

□ **Solution 2: Using Known Lumped-Parameter Equivalences**

As an alternative approach, we may use the known lumped-parameter equivalences to obtain the equivalent spring constant of the L-shaped beam. In this process, it is helpful to refer to the free-body diagrams for the two legs in Fig. 3.17.

Leg AB is a cantilever beam, which has an equivalent spring constant of $3EI/a^3$, as shown in Example 3.2. Leg BC is also a cantilever beam since junction B is capable of supporting a moment, and hence has an equivalent spring constant of $3EI/b^3$. The displacements due to these two springs are cumulative.

There is one complication at junction B: leg AB is also being twisted. Thus, AB has an equivalent torsional spring constant of GJ/a based on Example 3.1. The twisting torque Fb produces an angular displacement of Fba/GJ at junction B, according to eqn. (g) in Example 3.1. This angular displacement causes a linear displacement at point C by an arm length b, giving a displacement of Fab^2/GJ at Point C. In other words, the twist adds a linear displacement as if there is one more linear spring of spring constant $GJ/(ab^2)$.

The total deflection at the point of interest (Point C) comprises three components, as illustrated in Fig. 3.19. This suggests that the three springs are connected in series. Therefore, the effective spring constant k_{equiv} is, according to eqn. (3.7),

$$\frac{1}{k_{equiv}} = \frac{a^3}{3EI} + \frac{b^3}{3EI} + \frac{ab^2}{GJ} \tag{p}$$

which will give the same result as in eqn. (n).

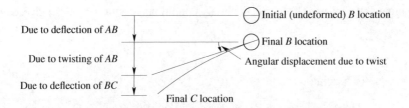

Figure 3.19   Composition of deflection at free end of L-shaped beam

3.4.2   Equivalent Masses

The equivalent mass is typically determined through estimating the kinetic energy stored in the system. Once we have found the system's deformation, a key step in the process is to assume that, when the system is in motion, the velocity distribution in the structure follows the same distribution as the deformation. This effectively sets the velocity for every point in the structure. Then, the kinetic energy of the system is calculated and compared with the formal expression of the kinetic energy for a particle or a rigid body.

Again, we use the uniform rod under uniaxial tension as an example to illustrate the procedure.

Step 1: *Reexpress the deformation of the structure in terms of the deformation at the point of interest, Δ, as $u(x) = \Delta f(x)$.*

The displacement throughout the rod has been found in eqn. (3.19), and the deflection at the point of interest has been found in eqn. (3.20). Thus, $u(x)$ can be reexpressed as

$$u(x) = \Delta \frac{x}{L} \tag{3.21}$$

By going through this step, the applied force F is eliminated from the expression for the deformation, and $f(x)$ is called its *form function*.

Step 2: *Assume that, when velocity of the point of interest is $\dot{\Delta}$, the velocity in the structure is expressible by the same form function as $v(x) = \dot{\Delta} f(x)$.*

We assume that the velocity throughout the rod is expressible as

$$v(x) = \dot{\Delta} \frac{x}{L} \tag{3.22}$$

Step 3: *Calculate the kinetic energy for the entire structure.*

Assuming that the rod has a mass density ρ and a cross-sectional area A, the total kinetic energy stored in the rod can be calculated as

$$T = \int_{V_{\text{rod}}} \frac{1}{2} \rho v^2 dV_{\text{rod}} = \frac{1}{2} \rho A \dot{\Delta}^2 \int_0^L \left(\frac{x}{L}\right)^2 dx = \frac{1}{6} \rho A \dot{\Delta}^2 L \tag{3.23}$$

where V_{rod} is the volume of the rod. Note that the total mass of the rod is $m = \rho AL$. Thus,

$$T = \frac{1}{6} m \dot{\Delta}^2 \tag{3.24}$$

Step 4: *Obtain the expression for the equivalent mass by comparing the resulting expression with the kinetic energy expression for a particle or a rigid body.*

Comparing the above expression with the kinetic energy for a particle in eqn. (3.1) gives the equivalent mass for the rod as

$$m_{\text{equiv}} = \frac{m}{3} \tag{3.25}$$

In the following, we follow this process to determine the equivalent mass for all the examples for which we have found the equivalent spring constants.

■ Example 3.7: Equivalent Mass for Shaft under Torsion

Assume that the shaft in Example 3.1 is uniform and has a total mass of m. The free end is the point of interest. Find the equivalent mass moment of inertia for the shaft.

□ Solution:

- In Example 3.1, we have found the angular displacement of the entire shaft $\theta(x)$ and the angular displacement at the point of interest Θ as

$$\theta(x) = \frac{Mx}{GJ} \quad \text{and} \quad \Theta \equiv \theta(L) = \frac{ML}{GJ} \tag{a}$$

Thus, the angular displacement of the shaft can be reexpressed in terms of Θ as

$$\theta(x) = \Theta\frac{x}{L} \tag{b}$$

- Assume that, when the point of interest rotates at an angular velocity of Ω, the angular velocity throughout the shaft is

$$\omega(x) = \Omega\frac{x}{L} \tag{c}$$

- Assume that the shaft has a mass density ρ and a cross-sectional area A. For an infinitesimally thin slice of the shaft located at x, of thickness dx, the differential element is undergoing a rotation of angular velocity $\omega(x)$ about its center. The moment of inertia for the differential element is $\frac{1}{2}(dm)R^2$ where $dm = \rho A dx$. Thus, the kinetic energy in the shaft is

$$T = \int_0^L \frac{1}{2}\frac{1}{2}\rho A R^2 \omega^2 dx = \frac{1}{4}\rho A R^2 \Omega^2 \int_0^L \left(\frac{x}{L}\right)^2 dx = \frac{1}{4}\frac{1}{3}mR^2\Omega^2 = \frac{1}{6}I\Omega^2 \tag{d}$$

where $m = \rho A L$ and the moment of inertia of the shaft $I = \frac{1}{2}mR^2$ have been used.
- Comparing eqn. (d) with the kinetic energy for a rigid body rotation about a fixed axis in eqn. (3.3), we have

$$I_{\text{equiv}} = \frac{I}{3} = \frac{1}{6}mR^2 \tag{e}$$

■ Example 3.8: Equivalent Mass for Cantilever Beam

Assume that the uniform cantilever beam in Example 3.2 has mass m. The end of the beam is the point of interest. Find the equivalent mass for the beam.

□ Solution:

- The deflection of the cantilever beam has been found in Example 3.2 as

$$u(x) = \frac{Fx^2}{6EI}(x - 3L) \tag{a}$$

from which the deflection at the point of interest is

$$\Delta \equiv -y(L) = \frac{FL^3}{3EI} \tag{b}$$

Thus, the deflection of the left half of the beam can be expressed in terms of Δ as

$$u(x) = \frac{\Delta}{2}\left(\frac{x}{L}\right)^2 \left(\frac{x}{L} - 3\right) \tag{c}$$

- Assume that, when the point of interest has a velocity of $\dot{\Delta}$, the velocity distribution throughout the beam is

$$v(x) = \frac{\dot{\Delta}}{2}\left(\frac{x}{L}\right)^2 \left(\frac{x}{L} - 3\right) \tag{d}$$

- The total kinetic energy stored in the beam can be calculated as

$$T = \frac{1}{2}\int_{V_{beam}} \rho v^2 dV_{beam} = \frac{\rho A}{2}\left(\frac{\dot{\Delta}}{2}\right)^2 \int_0^L \left(\frac{x}{L}\right)^4 \left(\frac{x}{L} - 3\right)^2 dx \tag{e}$$

where ρ and A are the mass density and the cross-sectional area, respectively, of the beam, and $m = \rho A L$. Making a change of variable $\xi = x/L$, then

$$T = \frac{m}{8}\dot{\Delta}^2 \int_0^1 \left(\xi^6 - 6\xi^5 + 9\xi^4\right) d\xi = \frac{33m}{280}\dot{\Delta}^2 \tag{f}$$

- Comparing the final expression in eqn. (f) with eqn. (3.1) gives

$$m_{equiv} = \frac{33m}{140} \approx 0.2357m \tag{g}$$

■ Example 3.9: Equivalent Mass for Simply-Supported Beam

Assume that the uniform simply-supported beam in Example 3.3 has mass m. The center of the beam is the point of interest. Find the equivalent mass for the beam.

□ **Solution:**

- It has been found in Example 3.3 that, for the left half of the beam,

$$u(x) = \frac{Fx}{48EI}(4x^2 - 3L^2) \tag{a}$$

from which the displacement at the point of interest is

$$\Delta \equiv -u\left(\frac{L}{2}\right) = \frac{FL^3}{48EI} \tag{b}$$

Thus, the deflection of the left half of the beam can be expressed in terms of Δ as

$$u(x) = \Delta\left[4\left(\frac{x}{L}\right)^3 - 3\frac{x}{L}\right] \tag{c}$$

- Assume that, when the point of interest has a velocity of $\dot{\Delta}$, the velocity distribution throughout the beam is

$$v(x) = \dot{\Delta}\left[4\left(\frac{x}{L}\right)^3 - 3\frac{x}{L}\right] \tag{d}$$

- The above expressions for the displacement and the velocity are only for the left half of the beam. We can use the symmetry to account for the kinetic energy stored in the right half of the beam. Thus, the kinetic energy of the beam can be found as

$$T = 2 \int_{V_{\text{left-half beam}}} \frac{1}{2} \rho v^2 dV_{\text{beam}} = \rho A \dot{\Delta}^2 \int_0^{\frac{L}{2}} \left[4 \left(\frac{x}{L} \right)^3 - 3 \frac{x}{L} \right]^2 dx \qquad (e)$$

where ρ and A are the mass density and the cross-sectional area, respectively, of the beam. Making a change of variable $\xi = x/L$ and noting $m = \rho A L$, then

$$T = m \dot{\Delta}^2 \int_0^{\frac{1}{2}} \left(16 \xi^6 - 24 \xi^4 + 9 \xi^2 \right) d\xi = \frac{17}{70} m \dot{\Delta}^2 \qquad (f)$$

- Comparing the final expression in eqn. (f) with eqn. (3.1) gives

$$m_{\text{equiv}} = \frac{17}{35} m \approx 0.4857 m \qquad (g)$$

■ Example 3.10: Equivalent Mass for Doubly-Clamped Beam

Assume that the uniform doubly-clamped beam in Example 3.4 has mass m. The center of the beam is the point of interest. Find the equivalent mass for the beam.

□ Solution:

- In Example 3.4, the deflection for the left half of the beam has been found as

$$u(x) = \frac{Fx^2}{48EI} (4x - 3L) \qquad (a)$$

from which the deflection at the point of interest is

$$\Delta \equiv u \left(\frac{L}{2} \right) = \frac{FL^3}{192EI} \qquad (b)$$

Thus, the deflection of the left half of the beam can be reexpressed in terms of Δ as

$$\frac{u(x)}{\Delta} = 4 \left(\frac{x}{L} \right)^2 \left(4 \frac{x}{L} - 3 \right) \qquad (c)$$

- Assume that, when the point of interest has a velocity of $\dot{\Delta}$, the velocity distribution throughout the beam is

$$v(x) = 4 \dot{\Delta} \left(\frac{x}{L} \right)^2 \left(4 \frac{x}{L} - 3 \right) \qquad (d)$$

- The above expressions for the displacement and the velocity are only for the left half of the beam. We can use the symmetry to account for the other half. Thus, the total kinetic energy for the beam is

$$T = 2 \int_{V_{\text{left-half beam}}} \frac{1}{2} \rho v^2 dV_{\text{beam}} = 16 \rho A \dot{\Delta}^2 \int_0^{\frac{L}{2}} \left(\frac{x}{L} \right)^4 \left(4 \frac{x}{L} - 3 \right)^2 dx \qquad (e)$$

where ρ and A are the mass density and the cross-sectional area, respectively, of the beam. Making a change of variable $\xi = x/L$ and noting $m = \rho AL$,

$$T = 16m\dot{\Delta}^2 \int_0^{\frac{1}{2}} \left(16\xi^6 - 24\xi^5 + 9\xi^4\right) d\xi = \frac{13}{70}m\dot{\Delta}^2 \tag{f}$$

- Comparing the final expression in eqn. (f) with eqn. (3.1) gives

$$m_{\text{equiv}} = \frac{13}{35}m \approx 0.3714m \tag{g}$$

■ **Example 3.11: Equivalent Mass for Clamped–Simply-Supported Beam**

Assume the uniform clamped–simply-supported beam in Example 3.5 has mass m. The center of the beam is the point of interest. Find the equivalent mass for the beam.

□ **Solution:**

- In Example 3.5, we have found deflection of the beam as

$$u_L(x) = -\frac{FL^3}{96EI}\left[-9\left(\frac{x}{L}\right)^2 + 11\left(\frac{x}{L}\right)^3\right] \tag{a}$$

$$u_R(x) = \frac{FL^3}{96EI}\left[-5\left(\frac{x}{L}\right)^3 + 15\left(\frac{x}{L}\right)^2 - 12\left(\frac{x}{L}\right) + 2\right] \tag{b}$$

where subscripts L and R signify the left half ($0 \le x \le \frac{L}{2}$) and right half ($\frac{L}{2} \le x \le L$), respectively, of the beam. Furthermore, the deflection at the point of interest is

$$\Delta \equiv -u\left(\frac{L}{2}\right) = \frac{7FL^3}{768EI} \tag{c}$$

Thus, the deflection of the beam can be reexpressed in terms of Δ as

$$u_L(x) = \frac{8\Delta}{7}\left[-9\left(\frac{x}{L}\right)^2 + 11\left(\frac{x}{L}\right)^3\right] \tag{d}$$

$$u_R(x) = \frac{8\Delta}{7}\left[-5\left(\frac{x}{L}\right)^3 + 15\left(\frac{x}{L}\right)^2 - 12\left(\frac{x}{L}\right) + 2\right] \tag{e}$$

- Assume that, when the point of interest has a velocity of $\dot{\Delta}$, the velocity distribution throughout the beam is

$$v_L(x) = \frac{8\dot{\Delta}}{7}\left[-9\left(\frac{x}{L}\right)^2 + 11\left(\frac{x}{L}\right)^3\right] \tag{f}$$

$$v_R(x) = \frac{8\dot{\Delta}}{7}\left[-5\left(\frac{x}{L}\right)^3 + 15\left(\frac{x}{L}\right)^2 - 12\left(\frac{x}{L}\right) + 2\right] \tag{g}$$

- The total kinetic energy stored in the beam can be calculated as

$$T = \frac{1}{2}\rho A \left(\frac{8\dot{\Delta}}{7}\right)^2 \int_0^{L/2} \left[-9\left(\frac{x}{L}\right)^2 + 11\left(\frac{x}{L}\right)^3\right]^2 dx$$

$$+ \frac{1}{2}\rho A \left(\frac{8\dot{\Delta}}{7}\right)^2 \int_{L/2}^{L} \left[-5\left(\frac{x}{L}\right)^3 + 15\left(\frac{x}{L}\right)^2 - 12\left(\frac{x}{L}\right) + 2\right]^2 dx \qquad \text{(h)}$$

where ρ and A are the mass density and the cross-sectional area, respectively, of the beam and $m = \rho A L$. Making a change of variable $\xi = x/L$ gives

$$T = \frac{1}{2}m\frac{64\dot{\Delta}^2}{49}\left[\int_0^{\frac{1}{2}} \left(81\xi^4 - 198\xi^5 + 121\xi^6\right) d\xi\right.$$

$$\left. + \int_{\frac{1}{2}}^{1} \left(25\xi^6 - 150\xi^5 + 345\xi^4 - 380\xi^3 + 204\xi^2 - 48\xi + 4\right) d\xi\right]$$

$$= \frac{1}{2}\frac{764}{1715}m\dot{\Delta}^2 \qquad \text{(i)}$$

- Comparing the final expression in eqn. (i) with eqn. (3.1) gives

$$m_{\text{equiv}} = \frac{764}{1715}m \approx 0.4455m \qquad \text{(j)}$$

■ **Example 3.12: Equivalent Mass for L-Shaped Cantilever Beam**

Assume that the L-shaped beam in Example 3.6 is uniform and of mass m. The free end of the beam is the point of interest. Find the equivalent mass for the beam.

□ **Solution:**

- In Example 3.6, the deflections of legs AB and BC have been found as

$$u_{AB}(x) = \frac{Fx^2}{6EI}(x - 3a) \qquad \text{(a)}$$

$$u_{BC}(z) = \frac{F}{6EI}\left(z^3 - 3bz^2 - 2a^3\right) - F\frac{ab}{GJ}z \qquad \text{(b)}$$

and the angular displacement of leg AB along the x-axis as

$$\theta_{AB} = -\frac{Fbx}{GJ} \qquad \text{(c)}$$

For this structure, the expression for the deflection at the point of interest, Δ, is rather complicated. Since the main purpose is to normalize the deflection such that it is expressed in terms of Δ instead of the applied force F, we can use the force–deflection relation $F = k_{\text{equiv}}\Delta$ to replace all F's in the above expressions, and k_{equiv} can be referred to Example 3.6 when needed.

- Assume that, when the point of interest has a velocity of $\dot{\Delta}$ throughout the beam, the velocity and angular velocity are expressible as

$$v_{AB}(x) = \frac{k_{\text{equiv}}\dot{\Delta}}{6EI}x^2(x - 3a) \tag{d}$$

$$v_{BC}(z) = k_{\text{equiv}}\dot{\Delta}\left[\frac{1}{6EI}(z^3 - 3bz^2 - 2a^3) - \frac{ab}{GJ}z\right] \tag{e}$$

$$\omega_{AB}(x) = -k_{\text{equiv}}\dot{\Delta}\frac{b}{GJ}x \tag{f}$$

- The kinetic energy can be calculated as follows: for leg AB, we look at an infinitesimally thin cross-sectional slice of the beam located at x of thickness dx. We treat this differential element as a rigid body, whose angular velocity is $\omega_{AB}(x)$ and centroidal velocity is $v_{AB}(x)$. It has mass $dm = \rho A dx$ and moment of inertia $dI = \frac{1}{2}r^2 dm$. Thus,

$$
\begin{aligned}
T_{AB} &= \int_0^a \left(\frac{1}{2}\rho A v_{AB}^2 + \frac{1}{2}\frac{1}{2}\rho A r^2 \omega_{AB}^2\right) dx \\
&= \frac{\rho A}{2}\left(k_{\text{equiv}}\dot{\Delta}\right)^2 \int_0^a \left[\frac{1}{(6EI)^2}(x^6 - 6ax^5 + 9a^2x^4) + \frac{1}{2(GJ)^2}b^2r^2x^2\right] dx \\
&= \frac{\rho A}{2}\left(k_{\text{equiv}}\dot{\Delta}\right)^2 \left[\frac{11}{420(EI)^2}a^7 + \frac{1}{6(GJ)^2}a^3b^2r^2\right]
\end{aligned} \tag{g}
$$

For leg BC, we look at an infinitesimally thin cross-sectional slice of the beam of thickness dz located at z. We treat this differential element as a rigid body, whose angular velocity is $\omega_{AB}|_{x=a}$ and centroidal velocity is $v_{BC}(z)$. Note that the rotation is about a diameter, and the corresponding moment of inertia $dI = \frac{1}{4}r^2 dm$, where $dm = \rho A dz$. Thus,

$$
\begin{aligned}
T_{BC} &= \int_0^b \left(\frac{1}{2}\rho A v_{BC}^2 + \frac{1}{2}\frac{1}{4}\rho A r^2 \omega_{AB}^2(a)\right) dz \\
&= \frac{\rho A}{2}\left(k_{\text{equiv}}\dot{\Delta}\right)^2 \int_0^b \left\{\left[\frac{z^3 - 3bz^2 - 2a^3}{6EI} - \frac{abz}{GJ}\right]^2 + \frac{a^2b^2r^2}{4(GJ)^2}\right\} dz \\
&= \frac{\rho A}{2}\left(k_{\text{equiv}}\dot{\Delta}\right)^2 \left\{\int_0^b \left[\left(\frac{abz}{GJ}\right)^2 - \frac{abz(z^3 - 3bz^2 - 2a^3)}{3EIGJ}\right.\right. \\
&\quad + \left.\frac{z^6 + 9b^2z^4 + 4a^6 - 6bz^5 - 4a^3z^3 + 12a^3bz^2}{(6EI)^2}\right] dz + \frac{a^2b^3r^2}{4(GJ)^2}\right\} \\
&= \frac{\rho A}{2}\left(k_{\text{equiv}}\dot{\Delta}\right)^2 \left[\frac{ab^3}{3EIGJ}\left(\frac{11}{20}b^3 + a^3\right) + \frac{1}{(GJ)^2}a^2b^3\left(\frac{1}{4}r^2 + \frac{1}{3}b^2\right)\right. \\
&\quad + \left.\frac{1}{(EI)^2}\left(\frac{11}{420}b^7 + \frac{1}{9}a^6b + \frac{1}{12}a^3b^4\right)\right]
\end{aligned} \tag{h}
$$

Summing up, the total kinetic energy stored in the L-beam is

$$T = \frac{\rho A}{2}\left(k_{\text{equiv}}\dot{\Delta}\right)^2 \left\{ \frac{ab^3}{3EIGJ}\left(\frac{11}{20}b^3 + a^3\right) + \frac{1}{(EI)^2}\left[\frac{11}{420}\left(a^7 + b^7\right)\right.\right.$$
$$\left. + \frac{1}{9}a^6b + \frac{1}{12}a^3b^4\right] + \frac{1}{(GJ)^2}a^2b^2\left(\frac{1}{6}ar^2 + \frac{1}{4}br^2 + \frac{1}{3}b^3\right)\right\} \tag{i}$$

Noting that $J = \frac{1}{2}\pi r^4$ and $I = \frac{1}{4}\pi r^2 = \frac{1}{2}J$, and, from eqn. (o) in Example 3.6,

$$k_{\text{equiv}} = \frac{3EG\pi r^4}{2[2(a^3 + b^3)G + 3ab^2E]} \tag{j}$$

The total kinetic energy is

$$T = \frac{\rho A}{2}\left(\frac{\dot{\Delta}}{2(a^3 + b^3)G + 3ab^3E}\right)^2 \left\{G^2\left[\frac{33}{35}\left(a^7 + b^7\right) + 4a^6b + 3a^3b^4\right]\right.$$
$$\left. + E^2a^2b^2\left(\frac{3}{2}ar^2 + \frac{9}{4}br^2 + 3b^4\right) + EGab^3\left(\frac{33}{10}b^3 + 6a^3\right)\right\}$$
$$= \frac{1}{2}\dot{\Delta}^2\frac{m}{140}\left(\frac{[132(a^7 + b^7) + 140a^3b(4a^3 + 3b^3)]G^2}{+105a^2b^2(2ar^2 + 3br^2 + 4b^3)E^2 + 42ab^3(20a^3 + 11b^3)EG}\right)$$

where $m = \rho A(a + b)$ has been used.

- Compared to the expression for a point mass, the equivalent mass can be written as

$$m_{\text{equiv}} = \frac{m}{140}\frac{\left([132(a^7 + b^7) + 140a^3b(4a^3 + 3b^3)]G^2\right.}{\left.+105a^2b^2(2ar^2 + 3br^3 + 4b^3)E^2 + 42ab^3(20a^3 + 11b^3)EG\right)}{(a + b)[2(a^3 + b^3)G + 3ab^2E]^2} \tag{k}$$

□ **Exploring the Solution with MATLAB**

We are curious about what effects the different parameters have in the above expression. We introduce the following nondimensionalization parameters

$$\alpha = \frac{r}{a} \qquad \beta = \frac{b}{a} \qquad \overline{m} = \frac{m_{\text{equiv}}}{m}$$

For elastic materials, E and G are related by the following relation:

$$E = 2G(1 + v) \tag{l}$$

where v is Poisson's ratio. For most materials, $v \approx 1/3$, which gives $E \approx 8G/3$. Using this approximation allows us to reduce one parameter. The expression for the equivalent mass can be simplified to the following:

$$\overline{m} = \frac{\left(3[33(1 + \beta^7) + 35\beta(4 + 3\beta^3)]\right.}{\left.+560\beta^2(2\alpha^2 + 3\beta\alpha^3 + 4\beta^3) + 28\beta^3(20 + 11\beta^3)\right)}{420(1 + \beta)(1 + 4\beta^2 + \beta^3)^2} \tag{m}$$

The $\overline{m} \sim \beta$ relation is plotted Fig. 3.20 for the cases $\alpha = 0.01$, 0.1, and 0.2. Note that the α values represent a range that spans the extreme ends of the possible aspect ratios for a beam, from extremely thin to where the "long beam" assumption may not be valid. Yet, the three curves are very close to each other. This indicates that the parameter α plays only a negligible role. In other words, the contribution by the twist in leg AB is negligible.

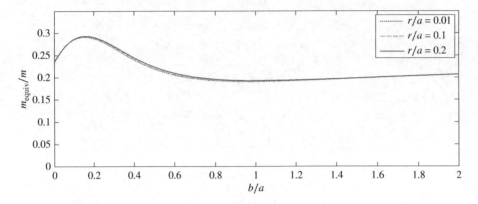

Figure 3.20   Equivalent mass of L-shaped beam at different $\alpha = r/a$ ratios

□ **Discussion:**

In Example 3.6, we were able to directly use previous lumped-parameter modeling results to determine the equivalent spring constant by constructing a new lumped-parameter model of connected springs. However, the accounting for contributions to the kinetic energy is more complicated. In this example, the complication comes from estimating the kinetic energy for leg BC since the entire leg BC is moving. It is possible, through careful considerations, to construct such a model. However, such "careful considerations" often lead to complicated calculations that are no simpler, if not more complicated, than following through the above process. Thus, for finding the equivalent mass for compound structures, we would not attempt to construct a model based on previous lump-parameter models.

3.4.3   Damping Models

The main function of damping elements is to dissipate energy, as a way to capture the physics behind the energy losses in vibration systems.

The lumped-parameter models for energy-dissipation elements include linear and torsional dashpots, which are an idealization of a class physical devices consisting of a plunger confined to move in a viscous fluid, such as the shock absorber used in cars and light trucks. Mathematically, they fill the niche of the first-order derivative term in a second-order ordinary differential equation.

Structural materials, from which distributed-parameter systems are constructed, generally dissipate energy to a certain degree, but its mechanism is different from the dashpot. One peculiar behavior of the dashpot makes it unsuitable for modeling damping in structural materials:

the energy dissipated by a dashpot is proportional to the frequency, as shown in eqn. (3.13), which does not corroborate well with experimental observations of structural materials.

3.4.3.1 Structural Damping Model

Damping in real structural materials is often frequency dependent, but this frequency dependence is rather weak, as a secondary effect. The first-order approximation for such damping would be to assume the energy dissipation being frequency independent. A model based on this approximation is to assume $c\Omega$ in eqn. (3.13) as a constant. That is, the dashpot coefficient c is no longer a constant but is proportional to the reciprocal of the driving frequency, such as

$$c = h/\Omega \tag{3.26}$$

where h is the new constant.

One reason that makes this a plausible damping model for structural materials is that it also exhibits an interesting *hysteresis loop* that has been observed in force–deflection curves in testing for material properties during a cyclic loading of many structural materials, as illustrated in Fig. 3.21: the loading and unloading portions of the curve seem to follow slightly different paths, forming a small loop.

Let us examine the force exerted to the mass in this model in the steady state. Assume the steady-state displacement is expressible as $x(t) = X_0 \sin(\Omega t + \phi)$. The damping force can be written as $F = -h\dot{x}/\Omega$. The total force on the mass includes the forces exerted by the spring and by the dashpot and is expressible as

$$F(t) = kX_0 \sin(\Omega t + \phi) - hX_0 \cos(\Omega t + \phi)$$

$$= kX_0 \sin(\Omega t + \phi) \mp hX_0\sqrt{1 - \sin^2(\Omega t + \phi)}$$

$$= kx \mp h\sqrt{X_0^2 - x^2} \tag{3.27}$$

This force–displacement relation is shown in Fig. 3.22 for $h/k = 0, 0.1, 0.2$, and 0.5. For small values of h/k ratio, the curves in the first quadrant is similar to the curve in Fig. 3.21. For this reason, this model has been called the *hysteretic damping* model, and consequently, h is called the *hysteretic constant*.

One difficulty with this model is that if the forcing does not explicitly contain the frequency information, we would not know how to write the damping coefficient h. To resolve this

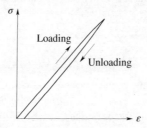

Figure 3.21   Stress–strain curve showing hysteresis loop in loading and unloading

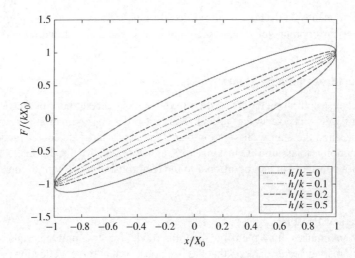

Figure 3.22   Hysteresis loop of hysteretic damping model at different h/k values

difficulty, another model has been introduced: Borrowing from the complex notation, if the excitation force is written as $F_0 e^{i\Omega t}$, then the steady-state response would be $x(t) = X_0 e^{i(\Omega t + \phi)}$, and, in turn, $\dot{x} = i\Omega x(t)$. This suggests that the $c\dot{x}$ term in the equation of motion can be written as $ci\Omega x(t) = ihx(t)$, that is, the equation of motion can be written as

$$m\ddot{x} + (k + ih)x = F_0 e^{i\Omega t} \tag{3.28}$$

This way, the left-hand side of the equation does not explicitly involve the frequency of the driving force. Taking this idea one step further, if the right-hand side is allowed to be any forcing function, the equation of motion will become

$$m\ddot{x} + k(1 + i\eta)x = F(t) \tag{3.29}$$

In this equation, the first-order derivative term disappears from the left-hand side, while the stiffness becomes complex. This model is called the *structural damping model* or the *complex stiffness model*. In this model, $(k + ih) = k(1 + i\eta)$ is the *complex stiffness*, and η is the *structural damping factor*.

So far this is the most appealing approach to incorporate material damping effects into vibration analyses. However, in this course, we will focus on the real material parameters and thus will not explore this model further.

3.4.3.2   Coulomb Damping

Another common type of damping is the friction that occurs when two surfaces come into contact. This is called the *dry friction* or the *Coulomb damping*. There are static and kinetic frictions characterized by coefficients μ_s and μ_k, respectively. Typically, $\mu_s > \mu_k$. The static friction is the force that needs to be overcome in order for the object to move on the surface; whereas the kinetic friction is the frictional force that drags down the moving object.

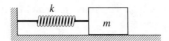

Figure 3.23   Mass–spring system with dry friction

From an energy perspective, the energy dissipated by the dry friction during one period of sinusoidal motion of amplitude X_0 can be found as

$$E_{\text{Dry Friction}} = -2\mu_k N(2X_0) = -4\mu_k N X_0 \tag{3.30}$$

where N is the normal force, the first 2 accounts for both back and forth directions, and the second 2 accounts for the total distance between stops.

For some simple scenarios, the damping level is extremely low and the normal force N is a constant; before the motion is overcome by the static friction, eqn. (3.30) can be compared with eqn. (3.13) to obtain an "equivalent" dashpot constant. Such an equivalency needs to be approached with extreme care, as the underlying physical mechanisms are completely different.

The mathematical challenge in dealing with this type of damping is that the friction force is in the opposite direction of the motion. Using the horizontally oriented mass–spring system, as shown in Fig. 3.23, as an example, to account for the dry friction, the equation of motion would become

$$m\ddot{x} + \text{sign}(\dot{x})\mu_k mg + kx = F(t) \tag{3.31}$$

where

$$\text{sign}(\dot{x}) = \begin{cases} 1 & \text{when} \quad \dot{x} > 0 \\ -1 & \text{when} \quad \dot{x} < 0 \end{cases} \quad \text{or} \quad \text{sign}(\dot{x}) = \frac{\dot{x}}{|\dot{x}|} \tag{3.32}$$

The $\text{sign}(\dot{x})$ function makes eqn. (3.31) a nonlinear equation. This type of nonlinearity is called piece-wise linear: it can be split into two linear equations:

$$m\ddot{x} + kx = F(t) - \mu_k mg \qquad \text{when } \dot{x} > 0 \tag{3.33}$$

$$m\ddot{x} + kx = F(t) + \mu_k mg \qquad \text{when } \dot{x} < 0 \tag{3.34}$$

In analyzing the motion, we can alternately solve the two equations of motion while tracking the value of $\dot{x}$: whenever $\dot{x}$ reaches 0, signaling a change of its sign, we need to evaluate x and $kx - F(t) - \text{sign}(\dot{x})\mu_s mg$. The latter is used to judge whether the mass would be able to overcome the static friction to continue to move. If it does, the alternate equation should be used while the x value, along with $\dot{x} = 0$, is used as the initial condition for the subsequent motion.

This process is not too difficult to track for such a simple (or the simplest) case of dry friction, as we shall explore with MATLAB in the following example. But the situation could become much more involved if the normal force varies.

■ Example 3.13: Free Vibration with Dry Friction

Assume that dry friction exists between the mass and the horizontal floor in the horizontal spring–mass system, as shown in Fig. 3.23. Denote the static and kinetic coefficients of friction as μ_s and μ_k, respectively, with $\mu_s > \mu_k$. The mass is subjected to initial conditions $x(0) = \Delta$ and $\dot{x}(0) = 0$. Find the subsequent motion.

□ **Solution:**

The equation of motion for the system has been given in eqns. (3.33) and (3.34). Given $F(t) = 0$, the normalized equation of motion can be written as

$$\ddot{x} + \omega_n^2 x = \mp \mu_k g \tag{a}$$

The equation of motion is not homogeneous. It is similar to a transient vibration problem. The particular solution can be readily found as $\mp \Delta_\mu$ where

$$\Delta_\mu = \frac{\mu_k g}{\omega_n^2} \tag{b}$$

and hence the total solution can thus be assumed to be

$$x(t) = A \cos \omega_n t + B \sin \omega_n t \mp \Delta_\mu \tag{c}$$

The solution process is to solve for constants A and B every time $\dot{x}$ switches its sign.

At the beginning, as soon as the mass is released, the spring forces the mass to return to its unstretched position and the friction will be in the opposite direction. Thus, for the initial stage of motion, the $+$ sign in eqn. (a) should be used. The initial conditions lead to $B = 0$ and $A = \Delta - \Delta_\mu$, and the solution can be written as

$$x(t) = (\Delta - \Delta_\mu) \cos \omega_n t + \Delta_\mu \tag{d}$$

Switching occurs when the speed reaches 0 for the first time, denoted as t_1, which can be obtained by and setting the derivative to eqn. (d) to zero. This gives $\omega_n t_1 = \pi$, or $t_1 = \pi/\omega_n$ and

$$x(t_1) = -\Delta + 2\Delta_\mu \qquad \dot{x}(t_1) = 0 \tag{e}$$

After this switch, the equation takes the negative sign in eqn. (a). Using the conditions at t_1 above, A and B can be found and, in turn,

$$x_1(t) = (\Delta - 3\Delta_\mu) \cos \omega_n t - \Delta_\mu \tag{f}$$

The process can be continued. For the ith time, the sign flips at $t_i = i\pi/\omega_n$, and the subsequent motion can be found as

$$x_i(t) = [\Delta - (2i + 1)\Delta_\mu] \cos \omega_n t + (-1)^i \Delta_\mu \tag{g}$$

The last remaining question is when will the motion stop. Obviously, the motion stops only when the velocity reaches 0, that is, at one of the times when the frictional force flips the sign. Denote that time as t_e, corresponding to impending i_e's switch. The motion stops because the spring force is insufficient to overcome the static friction, that is, $k|x(t_e)| < \mu_s mg$. The expression $x(t_e)$ is to be found from the one after the previous switch, that is, $x_{i_e-1}(t)$. Thus,

$$k\left|[\Delta - 2(i_e - 1)\Delta_\mu](-1)^{i_e} + (-1)^{i_e-1}\Delta_\mu\right| = k\left(\Delta - 2i_e\Delta_\mu\right) < \mu_s mg \tag{h}$$

Since i_e is an integer, it must be the smallest integer to satisfy the above relation, which can be written as, with slight rearrangement,

$$i_e = \left\lceil \frac{1}{2} \left(\frac{\omega_n^2 \Delta}{\mu_k g} - \frac{\mu_s}{\mu_k} \right) \right\rceil \tag{i}$$

where $\lceil \cdot \rceil$ is the *ceiling function*: the smallest integer larger than the enclosed argument.

□ **Exploring the Solution with MATLAB**

Introduce the following nondimensionalized parameters

$$\bar{x} = \frac{x}{\Delta} \qquad \bar{t} = \omega_n t \qquad r_\mu = \frac{\Delta_\mu}{\Delta} = \frac{\mu_k g}{\omega_n^2 \Delta} \qquad r_s = \frac{\mu_s}{\mu_k}$$

Then, after the ith switch

$$\bar{x}(\bar{t}) = [1 - (2i + 1)r_\mu] \cos \bar{t} + (-1)^i r_\mu \tag{j}$$

The number of times of switching i can be found as

$$i = \left\lfloor \frac{t}{\pi} \right\rfloor \tag{k}$$

where $\lfloor \cdot \rfloor$ is the *floor function* (MATLAB's `floor`): the largest integer smaller than the argument; and the motion stops as $i = i_e$, where

$$i_e = \left\lceil \frac{1}{2} \left(\frac{1}{r_\mu} - r_s \right) \right\rceil \tag{l}$$

where the ceiling function in MATLAB is `ceil`.

The following MATLAB code implements the above solution. Since we have to track the motion at almost every step, the matrix computation mechanism of MATLAB is not used.

```
rs = 1.2;        % Ratio of static coef of frction to kinetic CoF
rmu = .05;       % Parameter for mu_k
dt = .1;         % Time step
tmax = 60;       % max time

ie = ceil((1/rmu(iN)-rs)/2); % last switch, stuck!
t = 0:dt:tmax;              % Time array
for step=1:length(t)       % Loop through every entry in the time array
    i = floor(t(step)/pi);    % Number of switches has occured
    if i<ie                % If it still moves
        x(step) = ( 1-(2*i+1)*rmu )*cos (t(step)) + (-1)^i * rmu;
    else                    % If it stops, stuck in last switch's potion
        x(step) = ( 1-(2*ie+1)*rmu )*cos (ie*pi) + (-1)^ie * rmu;
    end
end
plot(t, x);
xlabel('\omega_n t');
ylabel('x(t)/\Delta');
```

Results for the case $r_s = \mu_s/\mu_k = 1.5$ at a series of values of r_μ, from 0.02 to 0.05, are shown in Fig. 3.24. The most distinctive feature, compared to the viscous damping of dashpots in Chapter 2, is that the amplitude of the vibration decreases linearly: it decreases by a constant amount after each switch of direction, until it becomes "stuck" in a dead zone, when the spring force is insufficient to overcome the static friction. The motion does not stop at the same location. Rather, it stops within a region defined by $|x| < \mu_s mg/k$, which is independent of the initial conditions.

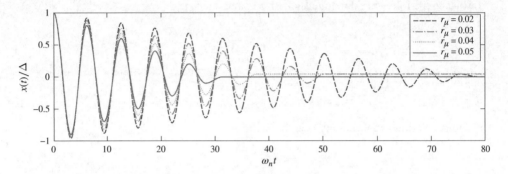

Figure 3.24 Free-vibration responses for mass–spring system with dry friction with $\mu_s/\mu_k = 1.5$ for a series of r_μ values

3.5 Alternative Methods

In the examples in the previous section, the mathematics can sometimes be rather involved, although the physics on which they are based are simple and intuitive. Yet, we should be reminded that, despite these length calculations, the end results are just approximations. This is especially true when evaluating the equivalent masses.

There are other methods that might be mathematically simpler but less intuitive. Here, we explore a few such methods, including experimental approaches.

3.5.1 Castigliano Method for Equivalent Spring Constants

In the theory of linear elasticity, *Castigliano's Theorem* states that the deflection at a point where a load is applied equals to the derivative of the system's total strain energy with respect to that load. This can be written mathematically as

$$\Delta = \frac{\partial U}{\partial F} \tag{3.35}$$

where U is the strain energy, F is the amplitude of the applied force F, and Δ is the displacement at the point where F acts upon and in the direction of F.

A few points need to be noted here. First, the strain energy is the potential energy by another name. It is the mechanical energy stored in a deformable body through its deformation. Second, F is not necessary the sole force acting on the body. It can be one among many.

Third, the theorem is applicable to not only the deflection–force pair, it can be the angular displacement–moment or torque pair, as long as the product of the pair is work or energy.

For a uniform rod under a uniform axial loading by force F, the strain energy is

$$U_{\text{rod}} = \int_0^L \frac{1}{2} \frac{F^2}{EA} dx = \frac{F^2 L}{2EA} \tag{3.36}$$

where E is the Young's modulus, A is the cross-sectional area, and L is the length. If we take the partial derivative about F, the theorem gives

$$\Delta_{\text{rod}} = \frac{FL}{EA} \tag{3.37}$$

which is the same as in eqn. (3.20), with which we have been familiar.

For a shaft under a torque T, the strain energy can be similarly written as

$$U_{\text{shaft}} = \frac{T^2 L}{2GJ} \tag{3.38}$$

where G is the shear modulus and J is the polar moment of area. It can be verified that the theorem would give the same equivalent torsional spring constant as in Example 3.1.

For beams, if it is not too short, the strain energy can be assumed to be entirely contributed by the bending moment M, as

$$U_{\text{beam}} = \frac{1}{2} \int_0^L \frac{M^2}{EI} dx \tag{3.39}$$

The theorem gives, after exchanging the orders of partial derivative and the integral,

$$\Delta_{\text{beam}} = \int_0^L \frac{M}{EI} \frac{\partial M}{\partial F} dx \tag{3.40}$$

Equation (3.40) can be directly used to find the deflection at the point of interest. This would replace the steps of solving the beam equations to find the deflection for the entire beam and then evaluating the expression at the point of interest. We will use three examples to illustrate this process.

■ **Example 3.14: Equivalent Spring Constant for Cantilever Beam, Revisited**

Use the Castigliano method to find the equivalent spring constant for the cantilever beam in Example 3.2. The free end is the point of interest.

□ **Solution:**

We can follow the same process as in Example 3.2, up to the point of finding the expression for the bending moment, as given eqn. (c) in Example 3.2,

$$M(x) = F(x - L) \tag{a}$$

Then, since

$$\frac{\partial M}{\partial F} = x - L \tag{b}$$

the deflection at the point of interest (where F is applied) can be found by using eqn. (3.40),

$$\Delta = \int_0^L \frac{F(x-L)}{EI}(x-L)dx = \frac{F}{EI}\int_0^L (x-L)^2 dx = \frac{FL^3}{3EI} \tag{c}$$

which is the same as eqn. (h) in Example 3.2.

_____ · _____

The Castigliano method can also be used for statically indeterminate structures. In such structures, the static equilibrium does not provide enough conditions to solve for all resultant forces and moments, and we need to enlist boundary conditions to provide the extra conditions. The boundary conditions can be evaluated by Castigliano's theorem. In such cases, Castigliano's theorem is invoked multiple times at different locations.

■ Example 3.15: Spring Constant for Clamped–Simply-Supported Beam, Revisited

Use the Castigliano method to find the equivalent spring constant for the clamped–simply-supported beam in Example 3.5. The center of the beam is the point of interest.

□ Solution:

We follow the same process as in Example 3.5, up to the point of finding the expression for the bending moment. In the process, the static equilibrium gives

$$\begin{cases} R_1 + R_2 - F = 0 \\ M_0 + F\frac{L}{2} - R_2 L = 0 \end{cases} \tag{a}$$

where the resultant forces R_1 and R_2 and moment M_0 are shown in the free-body diagram Fig. 3.25, which is a duplicate of Fig. 3.14, and the bending moments have been found as

$$M(x) = \begin{cases} M_0 + R_1 x & \text{for } 0 \le x \le L/2 \\ R_2(L-x) & \text{for } L/2 \le x \le L \end{cases} \tag{b}$$

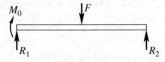

Figure 3.25   Free-body diagram for clamped–simply-supported beam

This is a statically indeterminate structure. We cannot resolve the three resultant forces and moment (R_1, R_2, and M_0) by equilibrium conditions in eqns. (a) alone. We need one more condition to help, and this would come from the boundary conditions.

There are three boundary conditions: $u(0) = 0$, $u'(0) = 0$, and $u(L) = 0$. Not coincidentally, there is a force corresponding to each of these conditions. We can pick any of these boundary conditions to be evaluated by Castigliano's theorem. Here, we choose $u(L) = 0$. Consequently, R_2 needs to be treated as an applied force, and all other resultant force and moment are to be expressed in terms of R_2 and F. Then, according to eqns. (a),

$$R_1 = F - R_2 \qquad\qquad M_0 = \left(R_2 - \frac{1}{2}F\right)L \tag{c}$$

Furthermore, the bending moment can be written as

$$M(x) = \begin{cases} R_2(L-x) + F\left(x - \frac{1}{2}L\right) & \text{for} \quad 0 \le x \le L/2 \\ R_2(L-x) & \text{for} \quad L/2 \le x \le L \end{cases} \tag{d}$$

and, in both halves of the beam,

$$\frac{\partial M}{\partial R_2} = L - x \tag{e}$$

Then, evaluating the boundary condition $u(L) = 0$ using Castigliano's theorem gives

$$0 = \int_0^{L/2} \frac{R_2(L-x) + F\left(x - \frac{1}{2}L\right)}{EI}(L-x)dx + \int_{L/2}^{L} \frac{R_2(L-x)}{EI}(L-x)dx$$

$$= \frac{R_2}{EI}\int_0^L (L-x)^2 dx + \frac{F}{EI}\int_0^{L/2}\left(x - \frac{1}{2}L\right)(L-x)dx = \frac{R_2 L^3}{3EI} - \frac{5FL^3}{48EI}$$

which gives

$$R_2 = \frac{5}{16}F \tag{f}$$

Then, R_1 and M_0 can be found by combining this result with the equilibrium conditions in eqns. (a). More importantly, the bending moment can be reexpressed in terms of the applied force F only. Doing so, eqn. (d) becomes

$$M(x) = \begin{cases} \dfrac{1}{16}F(11x - 3L) & \text{for} \quad 0 \le x \le L/2 \\ \dfrac{5}{16}F(L-x) & \text{for} \quad L/2 \le x \le L \end{cases} \tag{g}$$

We also need $\partial M/\partial F$, as

$$\frac{\partial M}{\partial F} = \begin{cases} \dfrac{1}{16}(11x - 3L) & \text{for} \quad 0 \le x \le L/2 \\ \dfrac{5}{16}(L-x) & \text{for} \quad L/2 \le x \le L \end{cases} \tag{h}$$

Then, using Castigliano's theorem in eqn. (3.40) again, the deflection at the point of interest can be found as

$$\Delta = \frac{F}{16^2 EI}\int_0^{L/2}(11x - 3L)^2 dx + \frac{5^2 F}{16^2 EI}\int_{L/2}^L (L-x)^2 dx$$

$$= \frac{F}{16^2 EI}\left[\int_0^{L/2}(121x^2 - 66xL + 9L^2)dx + 5^2\int_{L/2}^L (L^2 - 2xL + x^2)dx\right]$$

$$= \frac{7FL^3}{768EI} \tag{i}$$

which is the same as eqn. (o) in Example 3.5.

■ **Example 3.16: Equivalent Spring Constant for Doubly-Clamped Beam, Revisited**

Use the Castigliano method to find the equivalent spring constant for the doubly-clamped beam in Example 3.4. The center of the beam is the point of interest.

□ **Solution:**

We follow through the same process as in Example 3.4, up to the point of finding the expression for the bending moment. In the process, the equilibrium conditions have been used to conclude that the resultant forces at both ends are the same and equal to $F/2$, and that the moments at both ends are the same, denoted as M_0. But M_0 cannot be determined because the system is statically indeterminate. The bending moment for $0 \le x \le \frac{L}{2}$ is

$$M(x) = M_0 + \frac{1}{2}Fx \tag{a}$$

At $x = 0$ where M_0 acts, the boundary condition is $u'(0) = 0$. Castigliano theorem can be used for this slope–moment pair. Noting $\partial M/\partial M_0 = 1$ and the symmetry of the beam, Castigliano's theorem gives

$$u'(0) = 0 = 2\int_0^{L/2} \frac{M}{EI}\frac{\partial M}{\partial M_0}dx = 2\frac{1}{EI}\int_0^{L/2}\left(M_0 + \frac{1}{2}Fx\right)dx$$

$$= \frac{L}{EL}\left(\frac{M_0 L}{2} + \frac{FL^2}{16}\right) \tag{b}$$

This gives

$$M_0 = -\frac{FL}{8} \tag{c}$$

which is the same as in Example 3.4. By this point, all the resultant forces and moments have been determined and expressed in terms of F.

Next, the bending moment is also reexpressed in terms of F as

$$M = -\frac{FL}{8} + \frac{1}{2}Fx = \frac{F}{2}\left(x - \frac{1}{4}L\right) \tag{d}$$

and

$$\frac{\partial M}{\partial F} = \frac{1}{2}\left(x - \frac{1}{4}L\right) \tag{e}$$

Then, applying Castigliano's theorem again at the point of interest gives

$$\Delta = 2\frac{F}{4EI}\int_0^{L/2}\left(x - \frac{1}{4}L\right)^2 dx$$

$$= \frac{F}{2EI}\int_0^{L/2}\left(x^2 - \frac{1}{2}xL + \frac{1}{16}L^2\right)dx = \frac{FL^3}{192EI} \tag{f}$$

Again, this is the same as eqn. (g) in Example 3.4.

————————— · —————————

As seen from the above examples, this method does not require finding the deflection for the entire beam. It directly finds the deflection at the point of interest. It can also handle statically indeterminate structures in a systematic manner. The results are identical to those following the conventional Mechanics of Materials approach. The reason is that Castigliano's theorem is an exact theory.

The limitation of this method is, due to the lack of (or more precisely, the avoidance of obtaining) the expression of the deflection for the entire beam, it would not be possible to go further toward finding the equivalent mass using this method.

3.5.2   Rayleigh–Ritz Method for Equivalent Masses

In this method, instead of working through the Mechanical of Materials modeling details involving the beam equation, we make an educated guess for the beam's deflection. As long as the boundary conditions are satisfied, the essential physics are captured, and we are assured of reasonably good results. This may sound too good to be true, but it is. It is one of those rarities! In the following, we use examples to illustrate how to make the *educated* guesses.

■ **Example 3.17: Equivalent Mass for Cantilever Beam, Revisit**

Use the Rayleigh–Ritz method to find the equivalent mass for the cantilever beam in Example 3.2.

□ **Solution 1: Using a Sinusoidal Function**

The boundary conditions are that both the deflection and the slope vanish at the clamped end. As the first attempt, we look at a cosine function, which starts from a zero slope. We can shift it to force a zero starting value, as $\cos \theta - 1$, as shown in Fig. 3.26.

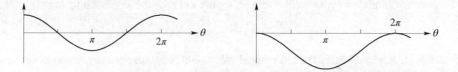

Figure 3.26   Shapes of $\cos \theta$ function and its shifted form $\cos \theta - 1$

On the right end of the beam, it is unconstrained; by physical intuition we know that it has both nonzero deflection and nonzero slope. We should be careful not to over constrain it, such as locating the right end at $\theta = \pi$, where the slope vanishes. We choose to locate the right end of the beam at $\theta = \pi/2$ so that it has nonzero deflection and nonzero slope. That is, let $x = L$ correspond to $\theta = \pi/2$. Thus, we assume

$$u(x) = C \left(\cos \frac{\pi x}{2L} - 1 \right) \tag{a}$$

where C is a constant to be determined. Then, we denote the deflection at the point of interest as Δ as earlier, that is, $\Delta = u(L)$. This gives $C = -\Delta$. Then, we have arrived at our *educated guess* for the form function for the beam's deflection:

$$u(x) = \Delta \left(1 - \cos \frac{\pi x}{2L} \right) \tag{b}$$

Assume that, when the point of interest has a velocity of $\dot{\Delta}$, the velocity in the beam is expressible as

$$v(x) = \dot{\Delta} \left(1 - \cos \frac{\pi x}{2L} \right) \tag{c}$$

We can calculate the kinetic energy as

$$T = \int_0^L \frac{1}{2}\rho A \dot{\Delta}^2 \left(1 - \cos\frac{\pi x}{2L}\right)^2 dx$$

$$= \frac{1}{2}\rho A \dot{\Delta}^2 \int_0^L \left(1 + \cos^2\frac{\pi x}{2L} - 2\cos\frac{\pi x}{2L}\right) dx$$

$$= \frac{1}{2}\rho A L \dot{\Delta}^2 \left(1 + \frac{1}{2} - 2\frac{2}{\pi}\right) = \frac{1}{2}\left(\frac{3}{2} - \frac{4}{\pi}\right) m\dot{\Delta}^2 \qquad (d)$$

This gives

$$m_{\text{equiv}} = \left(\frac{3}{2} - \frac{4}{\pi}\right) m = 0.22676m \qquad (e)$$

□ **Discussions:**

In our educated guess in eqn. (a), we have one undetermined constant C. This constant is determined using the condition $\Delta = u(L)$. This means that when we choose a form function, we can always leave one undetermined constant.

A curious reader might ask: what happens if we locate the right end at $\theta = \pi/3$? The curious reader can follow the same process to find that, in this case, $m_{\text{equiv}} = 0.2110m$.

□ **Solution 2: Using a Polynomial**

By this point, we might wonder what would happen if we chose a different form function. So here we try another way: we will construct a polynomial, whose general form is

$$u(x) = a + bx + cx^2 + dx^3 + ex^4 + \cdots \qquad (f)$$

where a, b, etc. are constants to be determined. We would like to use as many terms as possible. Being a form function, we can leave one constant undetermined.

We shall look into the boundary conditions much harder than earlier. We already know that the clamped end requires $u(0) = 0$, $u'(0) = 0$. Previously we noted that the right end is "unconstrained," but a force F is applied there. Recall the following relations for a beam:

$$M(x) = EIu''(x) \qquad (g)$$

$$Q(x) = M'(x) = EIu'''(x) \qquad (h)$$

These equations allow us to look into loadings as boundary conditions. The first is actually the beam equation. For the free end of the beam, we have a vertical force applied, but there is no moment. Thus, one more boundary condition can be written as $u''(L) = 0$.

Thus, we have come up with three boundary conditions. Adding the allowance of one extra coefficient, we would be able to include four terms in the polynomial. Thus, we assume

$$u(x) = a + bx + cx^2 + dx^3 \qquad (i)$$

and then

$$u'(x) = b + 2cx + 3dx^2 \qquad (j)$$

$$u''(x) = 2c + 6dx \qquad (k)$$

The three boundary conditions are used to determine the constants: $u(0) = 0$ gives $a = 0$; $u'(0) = 0$ gives $b = 0$; and $u''(L) = 0$ gives $d = -c/(3L)$. Thus,

$$u(x) = c\left(x^2 - \frac{x^3}{3L}\right) \tag{l}$$

Finally, denoting the deflection at the point of interest as Δ gives

$$\Delta \equiv u(L) = c\left(L^2 - \frac{L^3}{3L}\right) = \frac{2}{3}cL^2 \quad \text{or} \quad c = \frac{3\Delta}{2L^2} \tag{m}$$

Thus, the deflection is assumed to be

$$u(x) = \frac{3\Delta}{2}\left[\left(\frac{x}{L}\right)^2 - \frac{1}{3}\left(\frac{x}{L}\right)^3\right] \tag{n}$$

Now, assume that, when the point of interest has a velocity of $\dot{\Delta}$, the velocity in the beam is expressible as

$$v(x) = \frac{3\dot{\Delta}}{2}\left[\left(\frac{x}{L}\right)^2 - \frac{1}{3}\left(\frac{x}{L}\right)^3\right] \tag{o}$$

The kinetic energy can be calculated as

$$T = \int_0^L \frac{1}{2}\rho A\left(\frac{3}{2}\dot{\Delta}\right)^2\left[\left(\frac{x}{L}\right)^2 - \frac{1}{3}\left(\frac{x}{L}\right)^3\right]^2 dx$$

$$= \frac{9}{8}\rho A\dot{\Delta}^2 \int_0^L \left[\left(\frac{x}{L}\right)^4 + \frac{1}{9}\left(\frac{x}{L}\right)^6 - \frac{2}{3}\left(\frac{x}{L}\right)^5\right] dx = \frac{33}{280}m\dot{\Delta}^2 \tag{p}$$

where $m = \rho AL$ has been used. This gives

$$m_{equiv} = \frac{33}{140}m = 0.2357m \tag{q}$$

which is exactly the same as what we obtained in Example 3.8.

□ **Discussion:**

We should not be surprised to see that the result is exactly the same as in Example 3.8, since the form function in eqn. (n) is exactly the same as in eqn. (c) in Example 3.8. We started from a generic polynomial, and we are able to argue our way through, while incorporating known boundary conditions, to pin down the same expression for the deflection as in Mechanics of Materials process. Is this not a remarkable process?!

Since this is probably the first time we encounter with the idea of treating the beam's loading conditions as boundary conditions, it is a good time to look at such boundary conditions for other types of constraints. Table 3.1 summarizes all four types of boundary conditions (from zeroth to third derivatives of the deflection) for common end supports of the beams.

■ **Example 3.18: Equivalent Mass for Doubly-Clamped Beam, Revisit**

Use the Rayleigh–Ritz method to find the equivalent mass for the doubly-clamped beam in Example 3.10.

Table 3.1   Common end supports (boundary conditions) for beams

End support	Deflection	Slope	Moment	Force
Pinned or simple support	$u = 0$	—	$u'' = 0$	—
Clamped or fixed or cantilevered	$u = 0$	$u' = 0$	—	—
Guided	—	$u' = 0$	—	$u''' = 0$
Free	—	—	$u'' = 0$	$u''' = 0$

□ **Solution: Using a Sinusoidal Function**

For the doubly-clamped beam, the boundary conditions are that both the deflection and the slope at both ends vanish. Looking at the function $\cos\theta - 1$ sketched in Fig. 3.26 in Example 3.17, we can locate the right end of the beam at $\theta = 2\pi$. Thus, we assume

$$u(x) = C\left(\cos\frac{2\pi x}{L} - 1\right) \tag{a}$$

At the point of interest, $\Delta = u(L/2) = -2C$, which gives $C = -\Delta/2$. Thus,

$$u(x) = \frac{\Delta}{2}\left(1 - \cos\frac{2\pi x}{L}\right) \tag{b}$$

Now, assume that, when the point of interest has a velocity of $\dot\Delta$, the velocity in the beam is expressible as

$$v(x) = \frac{\dot\Delta}{2}\left(1 - \cos\frac{2\pi x}{L}\right) \tag{c}$$

Then, the kinetic energy of the beam can be calculated as

$$T = \int_0^L \frac{1}{2}\rho A \frac{\dot\Delta^2}{4}\left(1 - \cos\frac{2\pi x}{L}\right)^2 dx$$

$$= \frac{1}{2}\rho A \frac{\dot\Delta^2}{4}\int_0^L \left(1 + \cos^2\frac{2\pi x}{L} - 2\cos\frac{2\pi x}{L}\right) dx = \frac{1}{2}\frac{3}{8}m\dot\Delta^2 \tag{d}$$

which gives

$$m_{\text{equiv}} = \frac{3}{8}m = 0.375m \tag{e}$$

□ **Solution 2: Using a Polynomial**

The beam has four boundary conditions. Adding the allowance for an extra, we can use a polynomial up to five terms, that is,

$$u = a + bx + cx^2 + dx^3 + ex^4 \tag{f}$$

and

$$u' = b + 2cx + 3dx^2 + 4ex^3 \tag{g}$$

The boundary conditions $u(0) = 0$ and $u'(0) = 0$ give $a = b = 0$. The other two give

$$u(L) = 0 = cL^2 + dL^3 + eL^4 \tag{h}$$

$$u'(L) = 0 = 2cL + 3dL^2 + 4eL^3 \tag{i}$$

Expressing d and e in terms of c gives

$$d = -2c/L \quad \text{and} \quad e = c/L^2 \tag{j}$$

Hence,

$$u(x) = c\left(x^2 - \frac{2x^3}{L} + \frac{x^4}{L^2}\right) \tag{k}$$

Denoting the deflection at the point of interest $(x = L/2)$ as Δ gives

$$u(L/2) \equiv \Delta = c\left[\left(\frac{L}{2}\right)^2 - \frac{2}{L}\left(\frac{L}{2}\right)^3 + \frac{1}{L^2}\left(\frac{L}{2}\right)^4\right] = \frac{cL^2}{16} \tag{l}$$

or, $c = 16\Delta/L^2$. Finally the deflection throughout the beam can be written as

$$u(x) = 16\Delta\left[\left(\frac{x}{L}\right)^2 - 2\left(\frac{x}{L}\right)^3 + \left(\frac{x}{L}\right)^4\right] \tag{m}$$

Assume that, when the point of interest has a velocity of $\dot\Delta$, the velocity in the beam is expressible as

$$v(x) = 16\dot\Delta\left[\left(\frac{x}{L}\right)^2 - 2\left(\frac{x}{L}\right)^3 + \left(\frac{x}{L}\right)^4\right] \tag{n}$$

Then, the kinetic energy of the beam can be calculated as

$$T = \frac{1}{2}\rho A (16\dot\Delta)^2 \int_0^L \left[\left(\frac{x}{L}\right)^2 - 2\left(\frac{x}{L}\right)^3 + \left(\frac{x}{L}\right)^4\right]^2 dx = \frac{1}{2}\frac{128}{315}m\dot\Delta^2 \tag{o}$$

which gives

$$m_{\text{equiv}} = \frac{128}{315}m = 0.4063m \tag{p}$$

3.5.3 Rayleigh–Ritz Method for Equivalent Spring Constants

When using the Rayleigh–Ritz method, we make educated guesses for the form function for the deflection without applying any forces. Hence, we cannot use the force–deflection relation to find the equivalent spring constant. But, since we have the deformed shape, we can evaluate the strain energy stored in the beam and retrieve the spring constant from the potential energy.

The beam equation relates the deflection to the moment as

$$M(x) = EIu'' \tag{3.41}$$

Substituting eqn. (3.41) into eqn. (3.39) gives the strain energy in a beam in terms of the deflection as

$$U = \int_0^L \frac{[EIu''(x)]^2}{2EI}dx = \frac{1}{2}\int_0^L EI[u''(x)]^2 dx \tag{3.42}$$

As mentioned earlier, the strain energy is merely another name for the potential energy. If $u(x)$ in eqn. (3.42) is expressed in terms of the deflection at the point of interest Δ, the resulting expression for U above can be compared with the expression for a potential energy for a spring in eqn. (3.6) to obtain the equivalent spring constant for the beam. Let us revisit the examples in the previous section.

■ **Example 3.19: Equivalent Spring Constant for Cantilever Beam, Re-Revisit**

Continuing with Example 3.17, use the Rayleigh–Ritz method to find the equivalent spring constant for the cantilever beam in Example 3.2.

☐ **Solution 1: Using a Sinusoidal Function**

From Example 3.17, using a sinusoidal function, we have assumed the deflection of the beam being expressible as

$$u(x) = \Delta \left(1 - \cos \frac{\pi x}{2L}\right) \tag{a}$$

Then, taking the derivatives with respect to x twice gives

$$u''(x) = \Delta \left(\frac{\pi}{2L}\right)^2 \cos \frac{\pi x}{2L} \tag{b}$$

The potential energy stored in the beam can be calculated, according to eqn. (3.42), as

$$V = \frac{EI}{2} \int_0^L \Delta^2 \left(\frac{\pi}{2L}\right)^4 \cos^2 \frac{\pi x}{2L} dx = \frac{1}{2} \frac{EI}{L^3} \frac{\pi^4}{2^5} \Delta^2 \tag{c}$$

Thus, comparing with eqn. (3.6), the equivalent spring constant for the beam is

$$k_{\text{equiv}} = \frac{\pi^4}{2^5} \frac{EI}{L^3} = 3.044 \frac{EI}{L^3} \tag{d}$$

☐ **Solution 2: Using a Polynomial**

Also from Example 3.17, using a polynomial function, we have assumed the deflection of the beam being expressible as

$$u(x) = \frac{3\Delta}{2} \left[\left(\frac{x}{L}\right)^2 - \frac{1}{3}\left(\frac{x}{L}\right)^3\right] \tag{e}$$

Consequently,

$$u''(x) = \frac{3\Delta}{L^2} \left[1 - \left(\frac{x}{L}\right)\right] \tag{f}$$

The potential energy stored in the beam can be calculated, according to eqn. (3.42), as

$$V = \frac{1}{2} EI \frac{9\Delta^2}{L^4} \int_0^L \left[1 - \left(\frac{x}{L}\right)\right]^2 dx$$

$$= \frac{9}{2} \frac{EI\Delta^2}{L^4} \int_0^L \left[1 - 2\left(\frac{x}{L}\right) + \left(\frac{x}{L}\right)^2\right] dx = \frac{3}{2} \frac{EI\Delta^2}{L^3} \tag{g}$$

Thus, according to eqn. (3.6), the equivalent spring constant for the beam is

$$k_{\text{equiv}} = \frac{3EI}{L^3} \tag{h}$$

which is the same as what we obtained in Example 3.2.

■ **Example 3.20: Equivalent Spring Constant for Doubly-Clamped Beam, Re-Revisit**

Continuing with Example 3.18, use the Rayleigh–Ritz method to find the equivalent spring constant for the doubly-clamped beam in Example 3.4.

□ **Solution 1: Using a Sinusoidal Function**

From Example 3.18, using a sinusoidal function, the deflection in the beam has been assumed as

$$u = \frac{\Delta}{2}\left(1 - \cos\frac{2\pi x}{L}\right) \tag{a}$$

Taking the derivative with respect to x twice gives

$$u''(x) = \frac{2\pi^2\Delta}{L^2}\cos\frac{2\pi x}{L} \tag{b}$$

The potential energy stored in the beam can be calculated, according to eqn. (3.42), as

$$V = \frac{EI}{2}\left(\frac{2\pi^2\Delta}{L^2}\right)^2\int_0^L\cos^2\frac{2\pi x}{L}dx = \frac{\pi^4 EI}{L^3}\Delta^2 \tag{c}$$

Thus, according to eqn. (3.6), the equivalent spring constant for the beam is

$$k_{equiv} = \frac{2\pi^4 EI}{L^3} = 194.82\frac{EI}{L^3} \tag{d}$$

□ **Solution 2: Using a Polynomial**

From Example 3.18, using a polynomial function, the deflection in the beam has been assumed to be expressible as

$$u(x) = 16\Delta\left[\left(\frac{x}{L}\right)^2 - 2\left(\frac{x}{L}\right)^3 + \left(\frac{x}{L}\right)^4\right] \tag{e}$$

Taking the derivatives with respect to x twice gives

$$u''(x) = \frac{32\Delta}{L^2}\left[1 - 6\frac{x}{L} + 6\left(\frac{x}{L}\right)^2\right] \tag{f}$$

The potential energy stored in the beam can be calculated, according to eqn. (3.42), as

$$V = \frac{EI}{2}\left(\frac{32\Delta}{L^2}\right)^2\frac{L}{5} \tag{g}$$

Thus, according to eqn. (3.6), the equivalent spring constant for the beam is

$$k_{equiv} = \frac{32^2}{5}\frac{EI}{L^3} = 204.8\frac{EI}{L^3} \tag{h}$$

3.5.4   Rayleigh–Ritz Method for Natural Frequencies

We have seen that the Rayleigh–Ritz method can be used to determine both the equivalent spring constant and the equivalent mass. In fact, the Rayleigh–Ritz method was originally conceived for estimating the fundamental natural frequencies for engineering structures, based on reasonably good approximations of deformed shapes. Having the equivalent spring constant and mass, the natural frequency can be estimated without much work.

■ **Example 3.21: Fundamental Natural Frequency of Cantilever Beam**

In this example, we combine the modeling results for the cantilever beam so far, estimate its fundamental natural frequency, and compare these estimates.

□ **Solution:**

Using the Mechanics of Materials method in Examples 3.2 and 3.8, we have determined equivalent spring constant and mass, denoted as $k_{\text{equiv}}^{\text{MM}}$ and $m_{\text{equiv}}^{\text{MM}}$, respectively, for the cantilever beam. From these results, the beam's natural frequency can be estimated as

$$\omega_n^{\text{MM}} = \sqrt{\frac{k_{\text{equiv}}^{\text{MM}}}{m_{\text{equiv}}^{\text{MM}}}} = \sqrt{\frac{\frac{3EI}{L^3}}{\frac{33m}{140}}} = 3.5675\sqrt{\frac{EI}{mL^3}} \tag{a}$$

Using the Rayleigh–Ritz method in Examples 3.17 and 3.19, we determined the equivalent spring constant and mass, denoted as $k_{\text{equiv}}^{\text{RR}}$ and $m_{\text{equiv}}^{\text{RR}}$, respectively, twice: once using a sinusoidal function and once using a polynomial. The latter gave the same results as Mechanics of Materials method. Using the results from the former gives

$$\omega_n^{\text{RR}} = \sqrt{\frac{k_{\text{equiv}}^{\text{RR}}}{m_{\text{equiv}}^{\text{RR}}}} = \sqrt{\frac{\frac{3.044EI}{L^3}}{0.22676m}} = 3.6639\sqrt{\frac{EI}{mL^3}} \tag{b}$$

In comparison,

$$\frac{\omega_n^{\text{RR}}}{\omega_n^{\text{MM}}} = 1.0270 \tag{c}$$

The difference in the two estimates of the natural frequency is within 3%.

■ **Example 3.22: Fundamental Natural Frequency of Doubly-Clamped Beam**

In this example, we combine the modeling results for the doubly-clamped beam so far, estimate its fundamental natural frequency, and compare these estimates.

□ **Solution:**

Using the Mechanics of Materials method in Examples 3.4 and 3.10, we determined the equivalent spring constant and mass, denoted as $k_{\text{equiv}}^{\text{MM}}$ and $m_{\text{equiv}}^{\text{MM}}$, respectively, for the cantilever beam. From these results, the beam's natural frequency can be estimated as

$$\omega_n^{\text{MM}} = \sqrt{\frac{k_{\text{equiv}}^{\text{MM}}}{m_{\text{equiv}}^{\text{MM}}}} = \sqrt{\frac{\frac{192EI}{L^3}}{\frac{13m}{35}}} = 22.73\sqrt{\frac{EI}{mL^3}} \tag{a}$$

Using the Rayleigh–Ritz method in Examples 3.20 and 3.18, we determined the equivalent spring constant and mass twice. From the results using a sinusoidal function, denote the equivalent spring constant and mass as k_{equiv}^{RR1} and m_{equiv}^{RR1}, respectively; the fundamental natural frequency can be estimated as

$$\omega_n^{RR1} = \sqrt{\frac{k_{equiv}^{RR1}}{m_{equiv}^{RR1}}} = \sqrt{\frac{\frac{194.82EI}{L^3}}{\frac{3m}{8}}} = 22.793\sqrt{\frac{EI}{mL^3}} \qquad (b)$$

From the results using a polynomial, denote the equivalent spring constant and mass as k_{equiv}^{RR2} and m_{equiv}^{RR2}, respectively; the fundamental natural frequency can be estimated as

$$\omega_n^{RR2} = \sqrt{\frac{k_{equiv}^{RR2}}{m_{equiv}^{RR2}}} = \sqrt{\frac{\frac{204.8EI}{L^3}}{0.4063m}} = 22.45\sqrt{\frac{EI}{mL^3}} \qquad (c)$$

In comparison,

$$\frac{\omega_n^{RR1}}{\omega_n^{MM}} = 1.003 \quad \frac{\omega_n^{RR2}}{\omega_n^{MM}} = 0.9877 \qquad (d)$$

The differences in the estimates of the natural frequency for the doubly-clamped beam are within 0.3% and 1.2%, respectively.

These examples show the amazing power of the Rayleigh–Ritz method. The beauty is that the exact expression for the deformed shape is not needed at all. One might wonder, as an inquisitive engineer, why this method works so well? The key to its success is that it only looks at the overall deformed shape, by examining the total kinetic and potential energies stored in the deformed structure.

As these systems are distributed-parameter systems, they have an infinite number of natural frequencies. What we have done so far is to estimate the lowest natural frequency called the *fundamental natural frequency* for such systems. The Rayleigh–Ritz method can also be used to estimate other natural frequencies, as long as the chosen form function is reasonably close to the *mode shape* of a particular *mode*. We shall learn the meanings of those two terms in next chapter.

3.5.5   Determining Lumped Parameters Through Experimental Measurements

3.5.5.1   Static Deflection

A direct and convenient way to measure the equivalent stiffness of a structure or a system component is to take the measurement of the so-called *static deflection*. For a spring, at the static equilibrium, the following relation holds:

$$k\Delta = mg \qquad (3.43)$$

where Δ is the amount of deformation in the spring at equilibrium, which is called the *static deflection*. For a structure or a system component, it is feasible to load the system at the point of interest with a known dead weight W, which can be its own weight or an added static load, and measure the resulting static deflection Δ. Then, the effective spring constant can be estimated according to

$$k_{\text{equiv}} = \frac{W}{\Delta} \tag{3.44}$$

3.5.5.2 Logarithmic Decrement

As mentioned earlier, the dashpot remains a useful approximation for structural damping for low levels of damping. The effective damping ratio usually has to be obtained through experimental measurements. One approach is to induce a free vibration in the system and measure the decay of the amplitude.

Assume that two amplitudes X_1 and X_2 have been measured at two different times t_1 and t_2: at both times, the displacement reaches the peak. According to the system's response for a single-DOF system, the displacements can be written as

$$X_1 = x\left(t_1\right) = e^{-\zeta \omega_n t_1} \left(A \cos \omega_d t_1 + B \sin \omega_d t_1\right) \tag{3.45}$$

$$X_2 = x\left(t_2\right) = e^{-\zeta \omega_n t_2} \left(A \cos \omega_d t_2 + B \sin \omega_d t_2\right) \tag{3.46}$$

Since t_1 and t_2 are the times when the displacement reaches the peak, they are an integer number n periods apart, that is, $t_2 = t_1 + nT_d$ where $T_d = 2\pi/\omega_d$ is the *damped natural period* of the system. The terms in the parentheses in eqns. (3.45) and (3.46) are identical, and hence

$$\frac{X_1}{X_2} = \frac{e^{-\zeta \omega_n t_1}}{e^{-\zeta \omega_n t_2}} = \frac{e^{-\zeta \omega_n t_1}}{e^{-\zeta \omega_n (t_1 + nT_d)}} = e^{n\zeta \omega_n T_d} \tag{3.47}$$

Taking the natural logarithm (logarithm with base e) gives

$$\ln \frac{X_1}{X_2} = n\zeta \omega_n T_d = \frac{2n\pi \zeta \omega_n}{\omega_d} = \frac{2n\pi \zeta}{\sqrt{1-\zeta^2}} \tag{3.48}$$

When the damping ratio ζ is small, the above expression can be approximated as

$$\ln \frac{X_1}{X_2} \approx 2n\pi \zeta \tag{3.49}$$

which is called the *logarithmic decrement*, and can be used to determine the damping ratio as

$$\zeta \approx \frac{1}{2n\pi} \ln \frac{X_1}{X_2} \tag{3.50}$$

3.5.5.3 Quality Factor at Resonance

When the resonance behavior of a system is important, the damping ratio is often measured by observing the system's resonance behavior. Recall that, in the steady-state response spectrum,

$$\text{Peak height} \approx Q = \frac{1}{2\zeta} \tag{3.51}$$

$$\text{Peak width} \approx \frac{1}{Q} = 2\zeta \tag{3.52}$$

If either one or both can be measured, the damping ratio ζ can be obtained using the above relations. However, obtaining a response spectrum could be a rather complicated process, as it requires the system to reach the steady state with a wide range of forcing frequencies. Furthermore, when the system's characteristics are unknown, its resonance frequency must be approached with great care.

Measurement around the resonance also means that the natural frequency can be measured. Or, more precisely, the damped natural frequency ω_d can be measured, which can then be converted to the natural frequency ω_n with the information on the damping ratio ζ. If the natural frequency can be determined experimentally, and the equivalent spring constant can be estimated either experimentally or through modeling, the equivalent mass can be determined by

$$m_{equiv} \approx \frac{k_{equiv}}{\omega_{peak}^2} \tag{3.53}$$

3.5.5.4   Structural Damping and Complex Stiffness

In the structural hysteretic damping model, the hysteresis loop can be found to have the following intersections with the axes: setting $F = 0$ in eqn. (3.27) gives the intersection with the abscissa as

$$x|_{F=0} = \pm \frac{\eta X_0}{\sqrt{1 - \eta^2}} \tag{3.54}$$

and setting $x = 0$ gives

$$F|_{x=0} = \pm h X_0 \tag{3.55}$$

Furthermore, for most structural materials, $\eta = h/k$ is very small, in which case, $F_{max} \approx k X_0$, and the above two equations give

$$\frac{x|_{F=0}}{x_{max}} \approx \eta \qquad\qquad \frac{F|_{x=0}}{F_{max}} \approx \eta \tag{3.56}$$

This suggests a way of determining the structural damping ratio: matching these intersection points with those in the hysteresis loops in the force–displacement curves obtained from cyclic loading of a given material.

3.6   Examples with Lumped-Parameter Models

In this section, we analyze a few systems that are constructed based on lumped-parameter models of simple structures.

■ Example 3.23: Unbalanced Washing Machine Mounted on Floor

We continue to deal with the unbalanced washing machine we have investigated earlier in Example 2.7, in which we let the machine stand on its own feet. To reduce the vibration, we decide to bolt the machine to the floor. This effectively makes the machine rigidly attached to the floor. However, the floor itself can flex. We can model the floor as a doubly-clamped beam on which the washing machine is mounted at the center, as shown in Fig. 3.27. The floor is modeled as a uniform beam of length L, mass m_f, and bending stiffness EI. The machine itself has a width d, a mass M for the housing, and m for the clothes and drum together. Estimate the natural frequency of the setup.

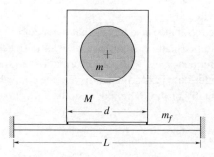

Figure 3.27   Washing machine of width d bolted to floor of length L

□ **Solution:**

Assume that the washing machine has two feet that are a distance d apart, symmetric about the center. We use the Rayleigh–Ritz method for its simplicity, and we continue to use the shifted cosine function in Example 3.18 as the form function:

$$u(x) = C\left(\cos\frac{2\pi x}{L} - 1\right) \tag{a}$$

There are two symmetrically located points of interest, at $x = (L \pm d)/2$. But they are undergoing the same motion. Equating the deflection at the points of interest to Δ gives

$$\Delta = C\left(\cos\frac{2\pi\frac{L\pm d}{2}}{L} - 1\right) = C\left[\cos\left(\pi \pm \frac{\pi d}{L}\right) - 1\right] = -C\left(\cos\frac{\pi d}{L} + 1\right) \tag{b}$$

Or

$$C = -\frac{\Delta}{1 + \cos\frac{\pi d}{L}} \tag{c}$$

Then, the form function and its double derivative are

$$u(x) = \frac{\Delta}{1 + \cos\frac{\pi d}{L}}\left(1 - \cos\frac{2\pi x}{L}\right) \tag{d}$$

$$u''(x) = -\frac{\Delta}{1 + \cos\frac{\pi d}{L}}\left(\frac{2\pi}{L}\right)^2 \cos\frac{2\pi x}{L} \tag{e}$$

The potential energy stored in the floor can be calculated as

$$V = \frac{EI}{2}\int_0^L \left[\frac{\Delta}{1 + \cos\frac{\pi d}{L}}\left(\frac{2\pi}{L}\right)^2\right]^2 \cos^2\frac{2\pi x}{L}\,dx = \frac{4\pi^4 EI}{L^3}\frac{\Delta^2}{\left(1 + \cos\frac{\pi d}{L}\right)^2} \tag{f}$$

Thus, the equivalent spring constant for the floor is

$$k_{equiv} = \frac{8\pi^4 EI}{L^3 \left(1 + \cos\frac{\pi d}{L}\right)^2} \tag{g}$$

Assuming that, when the point of interest moves at the velocity of $\dot{\Delta}$, the velocity distribution in the beam is

$$v(x) = \frac{\dot{\Delta}}{1 + \cos\frac{\pi d}{L}} \left(1 - \cos\frac{2\pi x}{L}\right) \tag{h}$$

Then, the kinetic energy stored in the beam can be calculated as

$$T = \frac{1}{2}\rho A \left(\frac{\dot{\Delta}}{1 + \cos\frac{\pi d}{L}}\right)^2 \int_0^L \left(1 - \cos\frac{2\pi x}{L}\right)^2 dx = \frac{3m_f}{4} \frac{\dot{\Delta}^2}{\left(1 + \cos\frac{\pi d}{L}\right)^2} \tag{i}$$

where ρ and A are the mass density and cross-sectional area, respectively, of the floor, and $m_f = \rho AL$. Hence, the equivalent mass of the floor is

$$m_{equiv} = \frac{3m_f}{2} \frac{1}{\left(1 + \cos\frac{\pi d}{L}\right)^2} \tag{j}$$

The floor and the washing machine's feet can be viewed as two springs connected in series. But, with the washing machine's feet bolted to the floor, the stiffness of the feet is essentially infinity. Thus, the floor is the only component contributing to the spring constant. For the mass, the washing machine's masses can be viewed as lumped masses directly added into the equivalent mass of the floor. Thus, the natural frequency of the setup is

$$\omega_n = \sqrt{\frac{k_{equiv}}{M + m + m_{equiv}}} = \sqrt{\frac{\dfrac{8\pi^4 EI}{L^3 \left(1 + \cos\frac{\pi d}{L}\right)^2}}{M + m + \dfrac{3m_f}{2}\dfrac{1}{\left(1 + \cos\frac{\pi d}{L}\right)^2}}}$$

$$= \frac{4\pi^2}{L} \sqrt{\frac{EI}{L\left[2(M + m)\left(1 + \cos\frac{\pi d}{L}\right)^2 + 3m_f\right]}} \tag{k}$$

Since the equivalent spring from the washing machine's feet does not appear in the above expression, it is impossible to compare the resulting natural frequency with the machine's stand-lone natural frequency.

■ **Example 3.24: Cantilever Beam in MEMS Devices**

MEMS stands for micro-electro-mechanical systems. They are microscopic mechanical systems built onto silicon chips, along with integrated circuits, to perform various tasks such as sensors and actuators. Cantilever beams are the most common type of mechanical springs used in MEMS devices because of their structural simplicity. Typically, a beam is etched out from a silicon layer deposited on the top of a support layer or the substrate, and the remainder of the layer serves as the support, modeled as a cantilevered end. Figure 3.28 depicts a cantilever beam designed for MEMS applications, where a plate is built-in at the end of the two thin legs. The plate could serve as a receptor as well as an added mass. This particular beam can be used in two ways: in-plane applications in which the beam swings sideways, and out-of-plane applications in which the beam flexes in the thickness direction. In this example, we shall investigate the relations between the natural frequencies in both directions and the various geometrical parameters.

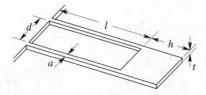

Figure 3.28   Cantilever beam for MEMS applications

□ Solution: In-Plane Applications

For in-plane applications, the plate constrains the two legs such that the legs have zero slope at the junctions. The two legs, modeled as springs, are connected in parallel to the plate, which will be modeled as a lumped mass since it only undergoes translational motion.

We use Rayleigh–Ritz method to model the legs. For each leg, there are three boundary conditions; hence, we use a polynomial up to the third order, as

$$u(x) = a + bx + cx^2 + dx^3 \tag{a}$$

where x starts from the root of the beam. The boundary conditions are $u(0) = 0$, $u'(0) = 0$, and $u'(l) = 0$. Additionally, $u(l) = \Delta$. Using these conditions leads to

$$u(x) = \Delta\left(\frac{x}{l}\right)^2 \left(3 - 2\frac{x}{l}\right) \tag{b}$$

Then,

$$u''(x) = \frac{6\Delta}{l^2}\left(1 - 2\frac{x}{l}\right) \tag{c}$$

We also assume that, when the plate moves at a velocity of $\dot{\Delta}$, the velocity in the beam is expressible as

$$v(x) = \dot{\Delta}\left(\frac{x}{l}\right)^2 \left(3 - 2\frac{x}{l}\right) \tag{d}$$

The potential energy stored in a leg is, according to eqn. (3.42),

$$V = \frac{EI}{2}\left(\frac{6\Delta}{l^2}\right)^2 \int_0^l \left(1 - 2\frac{x}{l}\right)^2 dx = \frac{6EI}{l^3} \tag{e}$$

and the kinetic energy is

$$T = \frac{1}{2}\rho A \dot{\Delta}^2 \int_0^l \left(\frac{x}{l}\right)^4 \left(3 - 2\frac{x}{l}\right)^2 dx = \frac{1}{2}\frac{13}{35}\rho A l \dot{\Delta}^2 \tag{f}$$

Thus, the equivalent spring constant and the equivalent mass for each leg are

$$k_{equiv} = \frac{12EI}{l^3} = \frac{Ea^3t}{l^3} \qquad\qquad m_{equiv} = \frac{13}{35}\rho Al = \frac{13}{35}\rho atl \qquad\qquad (g)$$

where, based on the geometry, $A = at$ and $I = \frac{1}{12}a^3t$ have been used.

Therefore, the natural frequency of the beam used in in-plane applications is

$$\omega_{in\text{-}plane} = \sqrt{\frac{\frac{2Ea^3t}{l^3}}{2\frac{13}{35}\rho atl + \rho(d + 2a)ht}} = \sqrt{\frac{70Ea^3}{\rho l^3[26al + 35(d + 2a)h]}} \qquad (h)$$

Note that the thickness t does not appear in the final expression for the natural frequency.

□ **Solution: Out-of-Plane Applications**

For out-of-plane applications, it is possible to directly use previous lump-parameter modeling results: each leg can be modeled as a cantilever beam; two legs are connected in parallel; and the plate is a lumped mass attached to the two springs. This way, each leg has, according to Examples 3.2 and 3.8,

$$k_{equiv} = \frac{3EI}{l^3} = \frac{Eat^3}{4l^3} \qquad\qquad m_{equiv} = \frac{33}{140}\rho Al = \frac{33}{140}\rho atl \qquad (i)$$

where $A = at$ and $I = \frac{1}{12}at^3$ have been used. Then, the natural frequency is

$$\omega_{out\text{-}of\text{-}plane} = \sqrt{\frac{2\frac{Eat^3}{4l^3}}{2\frac{33}{140}\rho atl + \rho(d + 2a)ht}} = \sqrt{\frac{35Eat^2}{\rho l^3[33al + 70(d + 2a)h]}} \qquad (j)$$

This approach may be slightly too simplistic: if the plate is large, the deflection in the plate will need to be taken into account. A proper modeling would be to treat the entire structure, including the plate, as a cantilever beam of variable cross sections. One portion contains the two legs and the other portion contains just the plate; and the two portions are continuous in both deflection and the slope. We shall leave this approach as an exercise for the readers.

□ **Exploring the Solution with MATLAB**

There are a few parameters in the expressions for the natural frequencies. We limit our observations to the case where $t = a$. We introduce the following nondimensionalization parameters:

$$\bar{d} = \frac{d}{a} \qquad \bar{h} = \frac{h}{a} \qquad \bar{l} = \frac{l}{a} \qquad \bar{\omega} = \frac{\omega_n}{\frac{1}{a}\sqrt{E/\rho}} \qquad (k)$$

Then, the nondimensionalized natural frequencies are

$$\bar{\omega}_{in\text{-}plane} = \sqrt{\frac{70}{\bar{l}^3[26\bar{l} + 35(\bar{d} + 2)\bar{h}]}} \qquad \bar{\omega}_{out\text{-}of\text{-}plane} = \sqrt{\frac{35}{\bar{l}^3[33\bar{l} + 70(\bar{d} + 2)\bar{h}]}} \qquad (l)$$

The following MATLAB code plots the two nondimensionalized natural frequencies. The variations of two $\bar{\omega}$'s with respect to $\bar{l}$ are shown in Fig. 3.29 for a series of values of $\bar{h} = h/a$

ranging from 10 to 70, at an increment of 10, and a fixed value of $\overline{d} = d/a = 10$. In both cases, the top curve corresponds to $\overline{h} = h/a = 10$.

```
d = 10;                          % d/a ratio
h = 70;                          % h/a ratio
l = .2:.01:5;                    % l/a ratio as x-axis
omega_in  = sqrt( 70 ./ l .^3 ./(26*l+35*(d+2)*h));
omega_out = sqrt( 35 ./ l .^3 ./(33*l+70*(d+2)*h));
subplot(1,2,1)
plot( l,omega_in,'b');
ylabel('\omega_{in}/\omega_0');
xlabel('l/a');
subplot(1,2,2)
plot( l,omega_out,'r');
ylabel('\omega_{out}/\omega_0');
xlabel('l/a');
```

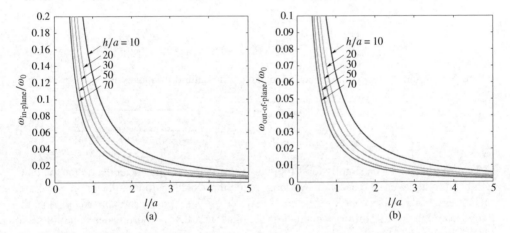

Figure 3.29 Variations of (a) in-plane and (b) out-of-plane natural frequencies with leg length l for a series of h/a values

All curves in Fig. 3.29 show the strong dominance of the factor $l^{-3/2}$, making it difficult to observe the effects of other parameters. But one interesting feature is that, although the curves in two figures are similar, the scales on the vertical axes differ by a factor of 2. Thus, we would like to compare the two frequencies. Dividing the two frequencies gives

$$\frac{\omega_{\text{in-plane}}}{\omega_{\text{out-of-plane}}} = \sqrt{\frac{2[33\overline{l} + 70(\overline{d}+2)\overline{h}]}{26\overline{l} + 35(\overline{d}+2)\overline{h}}} \tag{m}$$

There are two limiting cases when the mass of the plate is dominating or negligible:

$$\lim_{h\to\infty} \frac{\omega_{\text{in-plane}}}{\omega_{\text{out-of-plane}}} = 2 \qquad \lim_{h\to 0} \frac{\omega_{\text{in-plane}}}{\omega_{\text{out-of-plane}}} = \sqrt{\frac{33}{13}} = 1.5935 \tag{n}$$

Between these two extremes, Fig. 3.30 shows the variation of the frequency ratio with h/a for a series of d/a values, ranging from 1 to 50, for a fixed value of $l/a = 50$.

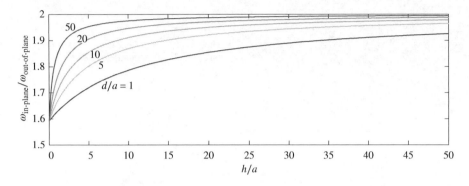

Figure 3.30   Frequency ratio at various d/a values, for the case $l/a = 50$

■ Example 3.25: Rotating Shaft

Consider a uniform circular shaft supported by two bearings, as shown in Fig. 3.31. Its left end is driven by a motor at a constant angular velocity Ω; its right end rotates freely in the bearing. A wheel is mounted at the shaft's mid-span. Assume that the shaft has a Young's modulus E, shear modulus G, mass m, length L, and radius r; and the wheel has a mass M and a moment of inertia about its centroid I. Assume that there is perfect alignment and no relative motion between the wheel and the shaft. Find the natural frequencies of the system in bending and twisting modes.

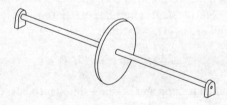

Figure 3.31   Wheel on rotating shaft supported by two bearings

□ Solution:

Equations of Motion

- *Preparatory Setup*: For reference purpose, we paint a radius on the wheel such that this radius initially lies in the horizontal direction. We also paint a radius on the shaft at the driving end, initially aligned with the one on the wheel. This way, the radius painted on the shaft follows the driving motion, while the radius painted on the wheel represents the sum of the rotation and the twist in the shaft. The geometry is shown in Fig. 3.32. Additionally, we set up a Cartesian coordinate system such that the origin is located at the left bearing;

the z-axis runs through the center of the undeformed shaft; and the x-axis aligned with the painted radii at $t = 0$.

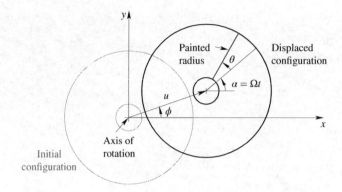

Figure 3.32   Geometry of wheel mounted on rotating shaft

- *Generalized Coordinates*: We define $\{\, x, y, \theta \,\}$ as a set of complete and independent generalized coordinates, where x and y are Cartesian coordinates of the center of the wheel and θ is the shaft's twisting counterclockwise angular displacement (angle between the painted radii) at its mid-span.
- *Admissible Variations*: We verify that $\{\, \delta x, \delta y, \delta\theta \,\}$ is a complete and independent set of admissible variations of this set of generalized coordinates.
- *Holonomicity*: We conclude that the system is holonomic and has three DOFs.
- *Generalized Forces*: There is no nonconservative force or element in the system: $\delta W^{\text{n.c.}} = 0$. Thus, $\Xi_x = 0$, $\Xi_y = 0$, and $\Xi_\theta = 0$.
- *Lagrangian*: In this example, we ignore the kinetic energy stored in the shaft. The wheel stores the kinetic energy, and the shaft stores the potential energy. For the wheel, the velocity at its center is $\dot{x}\boldsymbol{i} + \dot{y}\boldsymbol{j}$ and its angular velocity is $\Omega + \dot{\theta}$. Thus,

$$T = \frac{1}{2}M(\dot{x}^2 + \dot{y}^2) + \frac{1}{2}I(\Omega + \dot{\theta})^2 \tag{a}$$

The shaft undergoes both deflection and twisting simultaneously. Hence,

$$V = \frac{1}{2}k_{\text{equiv}}(x^2 + y^2) + \frac{1}{2}k_{t\,\text{equiv}}\theta^2 \tag{b}$$

where we can use the lumped-parameter modeling results from Example 3.3 for k_{equiv} and from Example 3.1 for $k_{t\,\text{equiv}}$. Combining eqns. (a) and (b) gives

$$\mathcal{L} = \frac{1}{2}M(\dot{x}^2 + \dot{y}^2) + \frac{1}{2}I(\Omega + \dot{\theta})^2 - \frac{1}{2}k_{\text{equiv}}(x^2 + y^2) - \frac{1}{2}k_{t\,\text{equiv}}\theta^2 \tag{c}$$

- *Lagrange's Equation*: The system has three equations of motion: one for each generalized coordinate. For a generalized coordinate denoted as ξ, the Lagrange's equation is

$$\frac{d}{dt}\left(\frac{\partial\mathcal{L}}{\partial\dot{\xi}}\right) - \frac{\partial\mathcal{L}}{\partial\xi} = \Xi_\xi \tag{d}$$

Without going into details of the process, here we simply write the resulting equations of motion as the following:

$$x\text{-Equation:} \qquad M\ddot{x} + k_{equiv}x = 0 \qquad\qquad (e)$$

$$y\text{-Equation:} \qquad M\ddot{y} + k_{equiv}y = 0 \qquad\qquad (f)$$

$$\theta\text{-Equation:} \qquad I\ddot{\theta} + k_{t\,equiv}\theta = 0 \qquad\qquad (g)$$

Natural Frequencies

Although the system has three DOFs, each equation of motion involves only one generalized coordinate. They can be analyzed separately as three unrelated equations. Such equations are said to be *decoupled*. The x- and y-equations are identical, making the x- and y-coordinates exchangeable, representing the deflection. They have the same natural frequency. The θ-equation has a different natural frequency. The natural frequencies are given as the following:

$$\omega_{dn} = \sqrt{\frac{k_{equiv}}{M}} \qquad \omega_{rn} = \sqrt{\frac{k_{t\,equiv}}{I}} \qquad\qquad (h)$$

where ω_{dn} is called the *deflection natural frequency* and ω_{rn} is called the *rotational natural frequency*. The natural frequencies are also called the *critical speeds* of the shaft.

□ Discussion: What If the Mass of the Shaft Is Considered?

If the mass of the shaft is considered, it adds two terms in the kinetic energy in the same way as the wheel. To the end, it simply modifies the M and I of the wheel slightly and does not alter the physics to the slightest bit.

■ Example 3.26: Whirling of Shaft—Elementary Phenomenon

The *whirling* of a shaft describes a phenomenon observed on a rotating shaft: the shaft seems to be pronouncedly bended outward while rotating, and sometime it produces a distinctive whirling sound. It is often attributed to mass imbalance: the mass center of either the wheel or the shaft does not align with the axis of rotation. In this example, we explore the effects of mass imbalance of the wheel. Consider the shaft in Example 3.25. The wheel has the same mass but its mass center is located at a distance e, called the *eccentricity*, from its geometric center. Assume that the centroidal mass moment of inertia of the wheel is I. Such a model is called the *Jeffcott rotor model*. We investigate the transient and steady-state vibration of the shaft when it starts up from rest, up to the constant angular velocity Ω.

□ Solution:

Equations of Motion

- *Preparatory Setup*: For reference, we paint a line on the wheel that connects its mass center with its geometric center. We also paint a radius on the shaft at the driving end so that the two painted radii are aligned at $t = 0$, just as in Example 3.25. We also set a Cartesian coordinate system in the same way. This geometry is shown in Fig. 3.33.

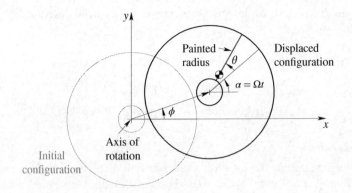

Figure 3.33   Geometry of unbalanced wheel mounted on rotating shaft

- *Generalized Coordinate*: We define $\{\, x, y, \theta \,\}$ as a set of complete and independent generalized coordinates, where x and y are the Cartesian coordinates of the geometric center, and θ is the angle between the two painted radii, as shown in Fig. 3.33.
- *Admissible Variations*: We verify that $\{\, \delta x, \delta y, \delta\theta \,\}$ is a complete and independent set of admissible variations of this set of generalized coordinates.
- *Holonomicity*: We conclude that the system is holonomic and has three DOFs.
- *Generalized Forces*: There is no nonconservative force or element in the system: $\delta W^{\text{n.c.}} = 0$. Thus, $\Xi_x = 0$, $\Xi_y = 0$, and $\Xi_\theta = 0$.
- *Lagrangian*: The wheel undergoes two rotations combined: a rigid-body rotation at angular velocity Ω driven by the motor and the twist in the shaft at angular velocity $\dot\theta$. The two angular velocities are in the same direction. The geometric center is located at Cartesian coordinates (x, y). Denote $\alpha = \Omega t$. The mass center is located at

$$x_C = x + e \, \cos\,(\alpha + \theta) \tag{a}$$

$$y_C = y + e \, \sin\,(\alpha + \theta) \tag{b}$$

Taking a time derivative of eqns. (a) and (b) gives the velocity of the mass center

$$\dot{x}_C = \dot{x} - e(\Omega + \dot\theta)\,\sin\,(\alpha + \theta)$$

$$\dot{y}_C = \dot{y} + e(\Omega + \dot\theta)\,\cos\,(\alpha + \theta)$$

Then, the kinetic energy of the wheel can be written as

$$T = \frac{1}{2} M v_C^2 + \frac{1}{2} I (\Omega + \dot\theta)^2$$

$$= \frac{1}{2} M \left[\dot{x}^2 + \dot{y}^2 + e^2 (\Omega + \dot\theta)^2 - 2e\dot{x}(\Omega + \dot\theta)\,\sin\,(\alpha + \theta) \right.$$

$$\left. + 2e\dot{y}(\Omega + \dot\theta)\,\cos\,(\alpha + \theta) \right] + \frac{1}{2} I (\Omega + \dot\theta)^2 \tag{c}$$

The potential energy stored in the shaft is the same as in Example eqn. (3.25). Thus,

$$\mathcal{L} = \frac{1}{2}I(\Omega + \dot{\theta})^2 + \frac{1}{2}M\left[\dot{x}^2 + \dot{y}^2 + e^2(\Omega + \dot{\theta})^2 - 2e\dot{x}(\Omega + \dot{\theta})\sin(\alpha + \theta)\right.$$

$$\left. +2e\dot{y}(\Omega + \dot{\theta})\cos(\alpha + \theta)\right] - \frac{1}{2}k_{equiv}(x^2 + y^2) - \frac{1}{2}k_{t\,equiv}\theta^2 \qquad (d)$$

- *Lagrange's Equations*: The system has three equations of motion: one for each generalized coordinate. For a generalized coordinate denoted as ξ, the Lagrange's equation is

$$\frac{d}{dt}\left(\frac{\partial \mathcal{L}}{\partial \dot{\xi}}\right) - \frac{\partial \mathcal{L}}{\partial \xi} = \Xi_\xi \qquad (e)$$

Without going into details of the process, here we simply write the resulting equations of motion as the following:

$$x\text{-Equation:} \quad M\ddot{x} - Me\ddot{\theta}\sin(\alpha + \theta) - Me(\Omega + \dot{\theta})^2\cos(\alpha + \theta) + k_{equiv}u = 0 \qquad (f)$$

$$y\text{-Equation:} \quad M\ddot{y} + Me\ddot{\theta}\cos(\alpha + \theta) - Me(\Omega + \dot{\theta})^2\sin(\alpha + \theta) + k_{equiv}y = 0 \qquad (g)$$

$$\theta\text{-Equation:} \quad (Me^2 + I)\ddot{\theta} - Me\ddot{x}\sin(\alpha + \theta) + Me\ddot{y}\cos(\alpha + \theta) + k_{t\,equiv}\theta = 0 \qquad (h)$$

Linearization of Equations of Motion

This is a set of rather complicated nonlinear differential equations. In order to observe the elementary whirling phenomenon, we assume small motions and a small eccentricity. That is, x, y, and e are small compared to other dimensions in the problem such as the length of the shaft and θ is also small. Then, the equations of motion can be linearized into

$$M\ddot{x} + k_{equiv}x = Me\Omega^2 \cos \Omega t$$

$$M\ddot{y} + k_{equiv}y = Me\Omega^2 \sin \Omega t$$

$$I\ddot{\theta} + k_{t\,equiv}\theta = 0$$

which is a set of completely uncoupled equations. There is no forcing in the third equation. We do not anticipate any notable unusual behaviors, and hence will leave it out of the following discussions. The other two can be normalized to

$$\ddot{x} + \omega_{dn}^2 x = e\Omega^2 \cos \Omega t \qquad (i)$$

$$\ddot{y} + \omega_{dn}^2 y = e\Omega^2 \sin \Omega t \qquad (j)$$

where ω_{dn} is the deflection natural frequency defined in Example 3.25.

Steady-State Whirling

Equations (i) and (j) represent forced vibrations of undamped systems. The steady-state solutions can be readily found as

$$x(t) = \frac{er^2}{1 - r^2} \cos \Omega t \qquad y(t) = \frac{er^2}{1 - r^2} \sin \Omega t \qquad (k)$$

where $r = \Omega/\omega_{dn}$ is the frequency ratio. Individually, the response is the same as the unbalanced washing machine in Example 2.7, with damping ratio set to 0.

The deflection, denoted as u, can be calculated as

$$u(t) = \sqrt{x^2(t) + y^2(t)} = \frac{er^2}{|1 - r^2|} \tag{1}$$

which, interestingly, is time independent, although both x and y components are time dependent. This suggests that, in the steady state, the shaft bows out and maintains the shape while rotating. This is the essential phenomenon of shaft whirling.

Furthermore, at low frequencies, as $r \to 0$, $u \to 0$. The wheel rotates about the geometric center. But at high frequencies, as $r \to \infty$, $u \to e$,

$$x(t) \to -e \cos \Omega t \qquad \text{and} \qquad y(t) \to -e \cos \Omega t$$

Substituting them into expressions for x_C and y_C in eqns. (a) and (b),

$$x_C(t) \to e\,[\cos(\Omega t + \theta) - \cos \Omega t] \approx 0 \tag{m}$$

$$y_C(t) \to e\,[\sin(\Omega t + \theta) - \sin \Omega t] \approx 0$$

That is, the wheel is self-aligning at high frequencies by moving its mass center toward the axis of rotation. In either of these extremes, the deflection is small, and hence the whirling is not very noticeable. But in the resonance range, the deflection can be large and noticeable.

Transient Whirling

We assume that initially the wheel is at rest. The total solution for the linearized x-equation, eqn. (i), can be assumed as

$$x(t) = A_x \cos \omega_{dn} t + B_x \sin \omega_{dn} t + \frac{er^2}{1 - r^2} \cos \Omega t$$

Using the trivial initial conditions gives

$$A_x = -\frac{er^2}{1 - r^2} \qquad B_x = 0$$

Thus,

$$x(t) = \frac{er^2}{1 - r^2} \left[\cos \Omega t - \cos \omega_{dn} t \right] \tag{n}$$

By a similar process, the solution for the linearized y-equation, eqn. (j), can be found as

$$y(t) = \frac{er^2}{1 - r^2} \left[\sin \Omega t - r \sin \omega_{dn} t \right] \tag{o}$$

□ Exploring the Solution with MATLAB

Introducing the following nondimensionalization parameters:

$$\bar{x} = \frac{x}{e} \qquad \bar{y} = \frac{y}{e} \qquad \bar{t} = \omega_{dn} t$$

Then, the nondimensionalized total solutions for $x(t)$ and $y(t)$ in eqns. (n) and (o) can be written as

$$\bar{x}(\bar{t}) = \frac{r^2}{1 - r^2} \left(\cos r\bar{t} - \cos \bar{t} \right) \tag{p}$$

$$\bar{y}(\bar{t}) = \frac{r^2}{1 - r^2} \left(\sin r\bar{t} - r \sin \bar{t} \right) \tag{q}$$

We are also interested in the motion of the mass center, whose coordinates are given in eqns. (a) and (b). Since we no longer consider θ, we can set $\theta = 0$. Thus,

$$\bar{x}_C(\bar{t}) = \frac{r^2}{1 - r^2} \left(\cos r\bar{t} - \cos \bar{t} \right) + \cos r\bar{t} = \frac{1}{1 - r^2} \left(\cos r\bar{t} - r^2 \cos \bar{t} \right) \tag{r}$$

$$\bar{y}_C(\bar{t}) = \frac{r^2}{1 - r^2} \left(\sin r\bar{t} - r \sin \bar{t} \right) + \sin r\bar{t} = \frac{1}{1 - r^2} \left(\sin r\bar{t} - r^3 \sin \bar{t} \right) \tag{s}$$

Steady-State Whirling

We first observe the motions in the steady state, in which the coordinates for the geometric are given in eqns. (p) and (q) without the transient term,

$$\bar{x}(\bar{t}) = \frac{r^2}{1 - r^2} \cos r\bar{t} \qquad \bar{y}(\bar{t}) = \frac{r^2}{1 - r^2} \sin r\bar{t} \tag{t}$$

and the mass center, given in eqns. (r) and (s) without the transient term,

$$\bar{x}_C(\bar{t}) = \frac{1}{1 - r^2} \cos r\bar{t} \qquad \bar{y}_C(\bar{t}) = \frac{1}{1 - r^2} \sin r\bar{t} \tag{u}$$

From these expressions, it is obvious that both centers move in circles. The following MATLAB code plots the trajectories of both centers of the wheel in the steady state.

```
r = .75;                             % Frequency ratio
t = [0 : 0.1/r : 2.1*pi/r];          % Time array
x = r^2/(1-r^2) * cos(r*t);          % x- & y-coord. for geometric center
y = r^2/(1-r^2) * sin(r*t);
xC = 1/(1-r^2) * cos(r*t);           % x- & y-coord. for mass center
yC = 1/(1-r^2) * sin(r*t);
plot(x,y,':b', xC,yC, 'r');
xlabel('x/e, x_C/e');
ylabel('y/e, y_C/e');
```

The trajectories of both centers at a series of frequency ratios, $r = 0.2, 0.75, 1.5,$ and 7.5, are shown in Fig. 3.34. In this figure, the dotted curve is the trajectory of the geometric center

(x, y) and the solid curve is that of the mass center (x_C, y_C). We can validate these curves by noting that both curves appear to be circular, as suggested by the expressions in eqns. (t) and (u), and that the difference in the radii of the two circles is 1.

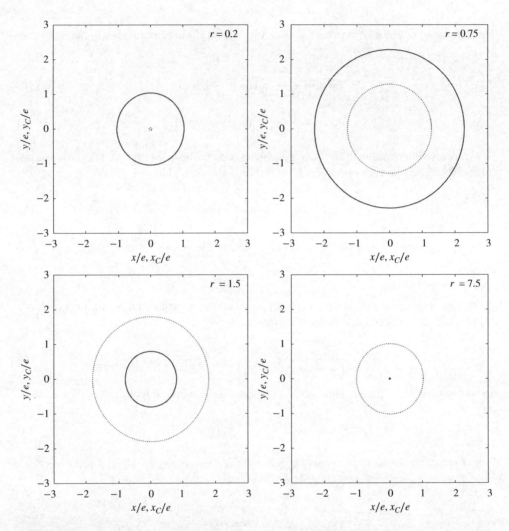

Figure 3.34   Trajectories of geometric center (dotted) and mass center (solid) of steady-state whirling wheel at different frequency ratios r

What is interesting in this figure is that the two circles swap their positions when the frequency ratio crosses the critical value of $r = 1$. When $r < 1$, the solid circle is in the outside, enclosing the dotted circle; when $r > 1$, the dotted circle is in the outside, enclosing the solid circle. We also observe that, at the lowest frequency ratio shown ($r = 0.2$), the dotted circle shrinks to almost a dot; while at the highest frequency shown ($r = 7.5$), the solid curve shrinks to almost a dot.

A more intuitive way to observe this "swapping" of positions between the dotted and solid circles is to see how the line connecting the two centers moves with time. We plot a series of snap shots of this line segment in the *xy*-plane over a full period of motion at a fixed time interval, like a motion picture. This is done by the following MATLAB code:

```
r = 1.5;                              % Frequency ratio
N = 200;                              % Number of instances in one period
t = [0 : 2*pi/r/N : 2*pi/r];          % Time array
x = r^2/(1-r^2) * cos(r*t);           % x- & y- coord. of geometric center
y = r^2/(1-r^2) * sin(r*t);
xC = 1/(1-r^2) * cos(r*t);            % x- & y-coord. of mass center
yC = 1/(1-r^2) * sin(r*t);
OldColor=[1 .6 1];                    % Color for oldest snapshot
for i=1:N                             % Circle through all time instance
    color = (1-.8*i/N)*OldColor;      % Determine the color for current
    plot([x(i) xC(i)],[y(i) yC(i)],'color',color,); % Plot CC-Line
    hold on;                          % Turn on "hold" for next plot cmd
    plot([xC(i)],[yC(i)],'o','color',color);   % Circle at mass center,
    plot([x(i)],[y(i)],'.','color',color);     % Dot at geometric center
end
axis([-3 3 -3 3]);
xlabel('x/e, x_C/e');
ylabel('y/e, y_C/e');
```

In the above code, the computation for the coordinates is the same as earlier, but the plotting portion is changed. It runs through every time instant, and within each instant, it plots a line segment that connects the mass center and the geometric center; a circle is placed at the mass center; and a small dot is placed at the geometric center. Furthermore, the color is varied such that the earlier the time instant, the color has faded more (becomes lighter). This is done by setting the faintest color, stored in `OldColor`, and then multiplying this base color by a timed intensifying factor and placing it in the `color` specification of `plot` command.

The resulting plots are shown in Fig. 3.35 for frequency ratios $r = 0.75$ and 1.5. It confirms that, when $r < 1$, the mass center lies on the outside, the rotation is said to be *centrifugal*; whereas when $r > 1$, the mass center lies on the inside, and the rotation is said to be *centripetal*. Figure 3.35 also shows that lighter colors are shown on the upper half of the circle for $r = 0.75$ but in the lower half for $r = 1.5$. This is consistent with the location of the mass center relative to the geometric center, as the motion is driven in the counterclockwise direction.

Moreover, at all instants, the line segment is always in the radial direction. When the shaft has completed one rotation, the line connecting the two centers has also completed one rotation. That is, the whirling is at the same frequency of the driving frequency. Such a whirling is called the *synchronous whirling*.

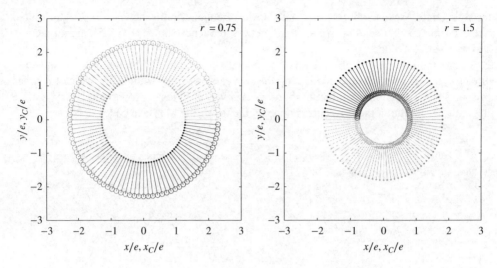

Figure 3.35 Snapshots of line connecting mass center (circles) and geometric center (dots) of wheel during one period of steady-state whirling. Darker colors indicate more recent times

Transient Whirling

For the transient phase, the following MATLAB code plots the trajectories of both centers:

```
r = .75;                                  % Frequency ratio
t = [0 : 0.1/r : 20.01*pi/r];             % Time array
x = r^2/(1-r^2) * (cos(r*t)-cos(t));      % Coord. for geometric center
y = r^2/(1-r^2) * (sin(r*t)-r*sin(t));
xC = x + cos(r*t);                        % Coord. for mass center
yC = y + sin(r*t);
plot(x,y,':b', xC,yC, 'r');
xlabel('x/e, x_C/e');
ylabel('y/e, y_C/e');
```

The trajectories of the two centers are shown in Fig. 3.36 for the first 10 periods from the beginning, at the same set of driving frequencies: $r = 0.2, 0.75, 1.5$, and 7.5.

In these figures, the shape of the curves are much more complicated; only the curves in the case $r = 0.2$ show trajectories that resemble circles. But they all essentially form closed curves, except the case of $r = 7.5$. A closed curve represents a periodic motion. The reason we see closed curves is that all the frequency ratios are *rational numbers*: a number can be represented as the division of two integers, namely, $0.2 = 1/5$, $0.75 = 3/4$, $1.5 = 3/2$, and $7.5 = 15/2$. For example, for the case $r = 3/4$, the transient motion will complete a period in a duration of three periods of the driving motion, which equals four periods of the natural frequency. On the other hand, the reason we do not see a closed curve for the case of $r = 7.5$ is that it would take 15 periods of the driving motion to complete one period; but we have only plotted the first 10 periods.

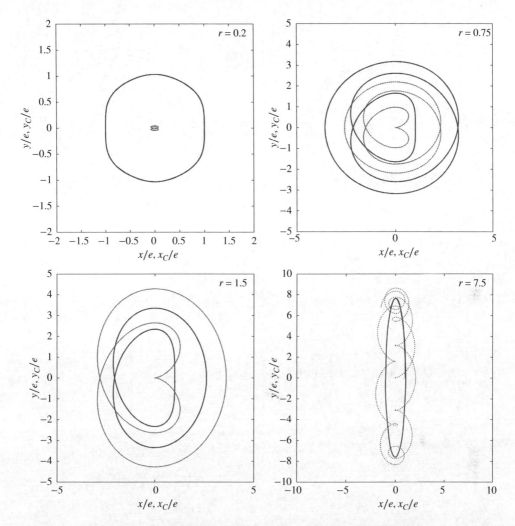

Figure 3.36   Trajectories of the geometric center (dotted) and mass center (solid) of the transient whirling wheel during the first 10 periods of driving motion at different frequency ratios

With this in mind, we can observe these trajectories at a different set of frequency ratios, all nonrational (called *irrational*) numbers. The trajectories for the two centers at frequency ratios of $r = \pi/10 \approx 0.314$, $\pi/4 \approx 0.785$, $\pi/2 \approx 1.57$, and $2\pi \approx 6.26$ are shown in Fig. 3.37. In this case, we see that almost all curves are not closed, as we can identify the beginning and end points of each curve. For the case $r = 0.2$, the curves are too close to each other to tell them apart. But, looking into the numerical values reveals that the curve is not closed either. In such cases, the motion is aperiodic.

It would be helpful and illustrious to see how the orientation of the wheel changes as time progresses. For this, we use the same plotting strategy as in Fig. 3.35 to illustrate the line that connects the two centers. We plot this line for the first 10 periods of loading, at 10 evenly

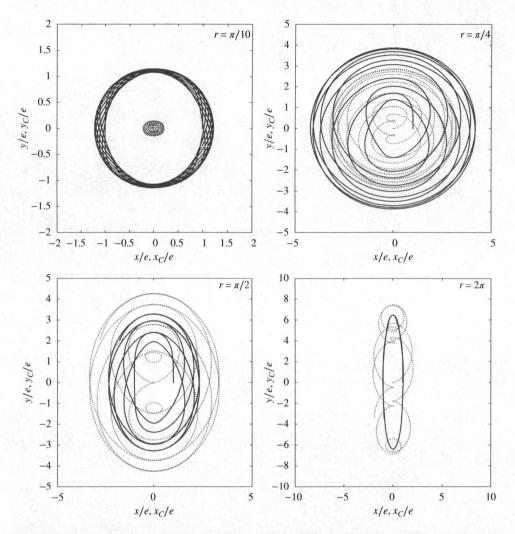

Figure 3.37   Trajectories of geometric center (dotted) and mass center (solid) of transient whirling wheel during first 10 periods of driving motion at different frequency ratios

spaced instants in each period. The plotting portion of MATLAB code is the same as the one used to produce Fig. 3.35. The results are shown in Fig. 3.38.

The resulting plots are graphically pretty but hard to summarize. If we have to use only one word to describe it, it would be "wondering": the line segment wonders all over the place in a small vicinity of the rotating axis. It can be in any orientation, at any location! But, there is one exception: the case of $r = \pi/10 \approx 0.314$. In this case, the driving frequency is low, and movement of the line is rather synchronous, similar to synchronous steady-state whirling in Fig. 3.35.

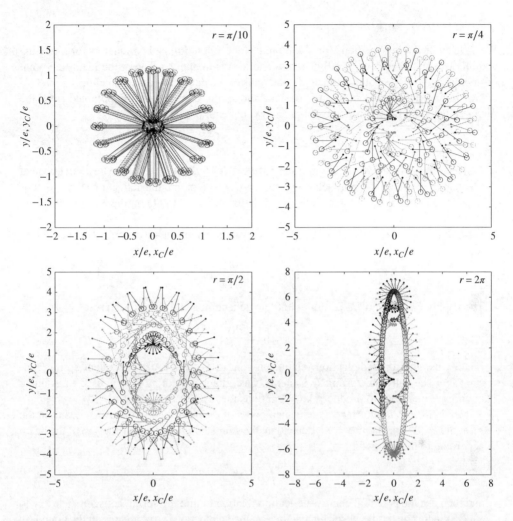

Figure 3.38 Snapshots of line connecting mass center (circles) and geometric center (dots) of transient whirling wheel for first 10 periods of driving motion at different frequency ratios. Darker colors indicate more recent times

The wondering of the wheel is the loud whirling we often hear when an unbalanced rotary machinery starts up. For actual machinery, the transient period will die down soon and be replaced by the synchronous steady-state whirling. In this example, we did not consider any damping in the system. In such a system, the transient motion will last forever, and we do not have the opportunity to observe how the transient phase subsides.

In actual systems, damping always exists in one way or another. In fact, the damping will cause other curious whirling phenomena. Interested readers are encouraged to explore further into the vast literature on this topic.

3.7   Chapter Summary

We have studied a few methods for obtaining the equivalent lumped parameters for modeling continuous components, especially beams. The idea is to apply a load at the identified point of interest and subsequently estimate the energies or the deflection due to loading.

For finding the equivalent spring constant, the spring constant can be obtained from the force–deflection relation or the potential energy–deflection relation:

$$F = k\Delta \qquad V = \frac{1}{2}k\Delta^2 \tag{3.57}$$

- *Conventional Method* or the *Mechanics of Materials Method*: Analyze the beam in classical Mechanics of Materials approach using the beam equation, find the deflection at the point of interest, and then use the force–deflection relation of spring to obtain the equivalent spring constant.
- *Castigliano Method*: Calculate the strain energy stored in the beam according to the following:

$$U = \int_0^L \frac{M}{EI} \frac{\partial M}{\partial x} dx \tag{3.58}$$

Then apply Castigliano's theorem to find the deflection at the point of interest

$$\Delta|_F = \frac{\partial U}{\partial F} \tag{3.59}$$

Finally, use the force–deflection relation in eqn. (3.5) to find the equivalent spring constant. For statically indeterminate systems, Castigliano's theorem can be used to evaluate boundary conditions to produce extra equations to solve for resultants.

- *Rayleigh–Ritz Method*: Make an educated guess of the deformed shape of the beam, in the form of $u = \Delta f(x)$ where $\Delta = u|_F$, calculate the strain energy (potential energy) of the beam according to the following:

$$V = \frac{1}{2} \int_0^L EI[u''(x)]^2 dx \tag{3.60}$$

and then use the potential energy–deflection relation to find the spring constant. The key to success in this method is ensuring that the assumed deflection form satisfies all the boundary conditions. Apart from typical deflection and slope boundary conditions, bending moment and lateral force boundary conditions as summarized in Table 3.1 should also be looked into.

For finding equivalent mass, all methods boil down to the same key assumption: when the system is in motion, the velocity distribution follows the same form function of the deflection distribution. Then the following kinetic energy–velocity relation is used to find the equivalent mass or equivalent moment of inertia:

$$T = \frac{1}{2}m\dot{\Delta}^2 \qquad \text{or} \qquad T = \frac{1}{2}I\Omega^2 \tag{3.61}$$

This can be done in continuation of finding the equivalent spring constant. The exception is that the Castigliano method cannot be used.

The equivalent spring constant and the equivalent mass depend on the location of the point of interest. As a reference, Figs. 3.39 and 3.40 summarize the lumped-parameter modeling

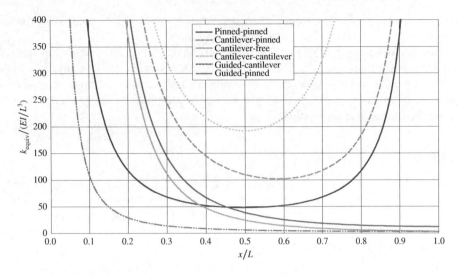

Figure 3.39 Equivalent spring constant for uniform beams in various end supports when location of point of interest x varies throughout the length of beam L

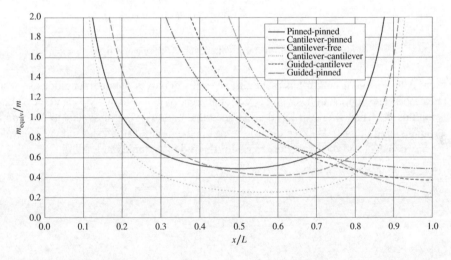

Figure 3.40 Equivalent mass for uniform beams in various end supports when location of point of interest x varies throughout the length of beam L

results for a uniform beam with common end supports when the point of interest, located at x, varies throughout the beam, of length L, mass m, and bending rigidity EI.

For structures made of connected components, the equivalent spring constant can also be obtained by converting each component into a lumped-parameter model so that the structure becomes a set of connected springs. However, for the equivalent mass, only when the boundary conditions for all components in the connected model remain the same as the stand-alone models, we can utilize the lumped-parameter models for each component. If a nonmoving point in a stand-alone lumped-parameter model becomes moving in the composite structure, we need to reconstruct the velocity distribution form function.

For modeling damping in continuous structures, we explored two potential models: the complex stiffness model and dry friction model. The former leads to the analysis involving complex variables and the latter leads to nonlinear, but piece-wise linear, equation(s) of motion. In many cases, especially at very low level of damping, the dashpot model can still be used, as least as an approximation.

Problems

Problem 3.1:   A thin-walled circular shaft has its left end fixed to the wall and its right end free, as shown in Fig. P3.1. The shaft has a mean radius R, a thickness t, and a length L. The material properties include the mass density ρ and the shear modulus G. Find the equivalent torsional spring constant and the equivalent moment of inertia if the free end of the shaft is the point of interest.

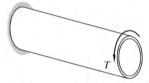

Figure P3.1   Hallow thin-walled shaft with one end fixed

Problem 3.2:   A rod of length L is made of two uniform halves of equal length, as shown in Fig. P3.2. The upper half has a cross-sectional area A_1, a mass density ρ_1, and the Young's modulus E_1. The lower half has a cross-sectional area of A_2, a mass density ρ_2, and the Young's modulus E_2. The free end of the rod is the point of interest. Find the equivalent spring constant and the equivalent mass for the rod under uniaxial loading.

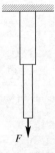

Figure P3.2   Rod of difference cross-sectional areas under uniaxial loading

Problem 3.3:   A circular rod, of mass m and length L, is made of a uniform material with a linearly varying diameter, from D at the top to d at the bottom, as shown in Fig. P3.3. The Young's modulus is E. The free end of the rod is the point of interest. Find the equivalent spring constant and the equivalent mass for the rod under uniaxial loading.

Figure P3.3   Rod of nonuniform cross-sectional area under uniaxial loading

Problem 3.4:   A uniform beam, of mass m and length L, has two pin-supports located $\frac{1}{4}L$ from the ends, as shown in Fig. P3.4. Assume that the Young's modulus is E, and the moment of area is I. Find the equivalent spring constant and the equivalent mass of the beam if the center of the beam is the point of interest.

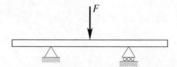

Figure P3.4   Simply-supported beam with supports located $\frac{1}{4}L$ from both ends

Problem 3.5:   The *four-point bending* test is one of standard methods for testing mechanical properties of materials. In this setup, the test piece, treated as a uniform beam, is simply supported on both ends. A loading fixture is placed on top of the test piece at its center and the load is applied at the center of the loading fixture, as shown in Fig. P3.5. Assume that the loading fixture of length $\frac{1}{2}L$ and mass m_0 is rigid. The test piece has the mass m, the length L, Young's modulus E, and moment of area I. Find the equivalent spring constant and the equivalent mass for the setup if the loading point (the center of the loading fixture) is the point of interest.

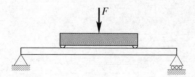

Figure P3.5   Four-point bending test setup

Problem 3.6:   A uniform beam of mass m and length L is pin-supported at its center, as shown in Fig. P3.6. A loading mechanism (not shown) distributes the applied load equally to its both ends while preventing any rotation. The beam has a bending stiffness EI. Find the equivalent spring constant and the equivalent mass of the beam if the loading points are the points of interest.

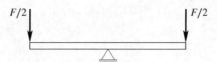

Figure P3.6   Beam pin-supported at center and equally loaded at both ends

Problem 3.7:   *Leaf springs* are commonly used in suspension systems for trucks and freight trains. It is typically comprised of slender strips of spring steel, of rectangular cross section but different lengths, stacked, and bound together. The thickest part is at the center, which is tied to the axle; and the thinnest parts are at its two ends, tied to the body. Due to the symmetry, the load is evenly distributed to the two ends. A simplified model is shown in Fig. P3.7. In this simplified model, the leaf spring comprises only two strips: the center portion, of a half of the total length, has two strips stack together. A single strip has a cross-sectional area of A and a moment of area I; and the center portion has a cross-sectional area $2A$ and a moment of area $8I$. The material has a mass density ρ and the Young's modulus E. Find the equivalent spring constant and the equivalent mass if the ends of the beam are the point of interest.

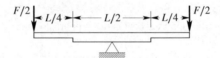

Figure P3.7   Simplified model for leaf spring equally loaded at both ends

Problem 3.8:   Figure P3.8 represents an alternative way of using a leaf spring: its two ends are simply supported while the center is the loading point. Assuming all parameters remain the same as in Problem 3.7, find the equivalent spring constant and the equivalent mass if the center of the leaf spring is the point of interest.

Figure P3.8   Simplified model for leaf spring in alternate loading arrangement

Problem 3.9:   An overhung simply-supported uniform beam, of mass m and length L, has its right support located at a distance $\frac{1}{3}L$ from its right end, as shown in Fig. P3.9. The bending stiffness is EI. Find the equivalent spring constant and the equivalent mass for the beam if its right end is the point of interest.

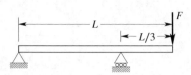

Figure P3.9   Overhung simply-supported beam

Problem 3.10: One of common forms of beam support is the cantilever end, which inhibits both the deflection and the slope. Such an end support is sometimes called *build-in*, which represents a physical realization of a cantilever support: a tight-fitting groove is preformed in the wall, and the end of the beam is inserted by some means. Over times, the fitting could become loose. A loose-fitting build-in end can be modeled as two pin-supports, as shown in Fig. P3.10. Assume that the beam is uniform, and the mass and the length of the portion of the beam outside the wall, for which a typical cantilever beam accounts, are m and L, respectively. The build-in portion has a length of $\frac{1}{8}L$. The beam has a bending stiffness EI. Find the equivalent spring constant and the equivalent mass if the free end of the beam is the point of interest.

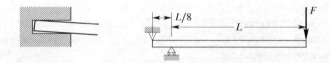

Figure P3.10   Loose-fitting build-in end modeled as two pin-supports

Problem 3.11: A *guided end* is a type of end supports for beams where the slope of the beam is maintained at a prescribed value, typically zero, while its lateral deflection is allowed. A *guided cantilever beam* has one end cantilevered and the other guided, as shown in Fig. P3.11. Assume the uniform beam, of mass m and length L, has a bending stiffness of EI. Find the equivalent spring constant and the equivalent mass for the beam if the guided end is the point of interest.

Figure P3.11   Guided cantilever beam

Problem 3.12: A uniform cantilever beam, of mass m and length L, has an extra roller support at its right end, as shown in Fig. P3.12. The beam's bending stiffness is EI. Find the equivalent spring constant and the equivalent mass if the point of interest is located at $\frac{2}{3}L$ from the wall.

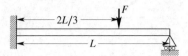

Figure P3.12   Clamped–simply-supported beam

Problem 3.13: A uniform cantilever beam, of mass m and length L, has an extra roller support at $\frac{2}{3}L$ from the wall, as shown in Fig. P3.13. The beam's bending stiffness is EI. Find the equivalent spring constant and the equivalent mass if the center of the beam is the point of interest.

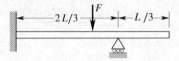

Figure P3.13   Overhung clamped–simply-supported beam

Problem 3.14: A uniform simply-supported beam, of mass m and length L, has an extra roller support at its center, as shown in Fig. P3.14. The bending stiffness of the beam is EI. The point of interest is located at $\frac{1}{4}L$ from the left end. Find the equivalent spring constant and the equivalent mass for the beam.

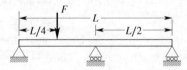

Figure P3.14   Multispan simply-supported beam

Problem 3.15: A structure is modeled as comprising of three beam segments pin-joined into a single piece. The two ends of the joined beam are cantilevered, as shown in Fig. P3.15. The two cantilevered segments have mass $\frac{1}{2}m$ and length of $\frac{1}{2}L$, and the center piece has mass m and length L. The bending stiffness is EI for all segments. The center of the structure is the point of interest. Find the equivalent spring constant and the equivalent mass for the structure.

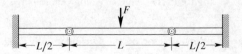

Figure P3.15   Three-segment double-cantilever beam with pin-joints

Problem 3.16: Two uniform beams, each of mass m, length L, and bending stiffness EI, are pin-joined into one piece. One end of the joined beam is cantilevered, and other end is roller-supported, as shown in Fig. P3.16. The point of interest is located at center of the beam member whose end is roller-supported. Find the equivalent spring constant and the equivalent mass for the joined beam.

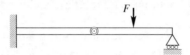

Figure P3.16   Cantilever beam pinned to simply-supported beam

Problem 3.17: A uniform cantilever beam, of mass m and length L, has a bending stiffness of EI. Determine the equivalent spring constant and the equivalent mass if the point of interest is located at a distance a from the wall, as shown in Fig. P3.17. Use MATLAB to plot the variations of both the equivalent spring constant and the equivalent mass as a varies from 0 to L.

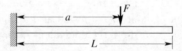

Figure P3.17   Cantilever beam loaded at distance a from wall

Problem 3.18: A uniform simply-supported beam, of mass m and length L, has a bending stiffness of EI. Determine the equivalent spring constant and the equivalent mass if the point of interest is located at a distance a from the left end, as shown in Fig. P3.18. Use MATLAB to plot the variations of both the equivalent spring constant and the equivalent mass as a varies from 0 to L.

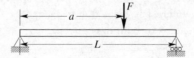

Figure P3.18   Simply-supported beam loaded at arbitrary location

Problem 3.19: A uniform simply-supported beam, of mass m and length L, has its right roller support located at a distance a from its left end, as shown in Fig. P3.19. The bending stiffness is EI. Find the equivalent spring constant and the equivalent mass for the beam if its right end is the point of interest. Use MATLAB to plot the variations of both the equivalent spring constant and the equivalent mass as a varies from 0 to L.

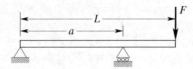

Figure P3.19   Overhung simply-supported beam loaded at arbitrary location

Problem 3.20: A uniform beam, of mass m, length L, and bending stiffness EI, has its left end clamped and right end on a roller support, as shown in Fig. P3.20. Determine the equivalent spring constant and the equivalent mass if the point of interest is located at a distance a from the wall. Use MATLAB to plot the variations of both the equivalent spring constant and the equivalent mass as a varies from 0 to L.

Figure P3.20   Clamped–simply-support beam loaded at arbitrary location

Problem 3.21: A uniform beam, of mass m, length L, and bending stiffness EI, has both ends clamped, as shown in Fig. P3.21. Determine the equivalent spring constant and the equivalent mass when the point of interest is located at a distance a from the left end. Use MATLAB to plot the variations of both the equivalent spring constant and the equivalent mass as a varies from 0 to L.

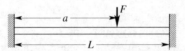

Figure P3.21   Doubly-clamped beam loaded at arbitrary location

Problem 3.22: A uniform beam, of mass m, length L, and bending stiffness EI, has one end clamped and the other end guided, as shown in Fig. P3.22. (See Problem 3.11 for an explanation for a *guided end.*) Determine the equivalent spring constant and the equivalent mass when the point of interest is located at a distance a from the clamped end. Use MATLAB to plot the variations of both the equivalent spring constant and the equivalent mass as a varies from 0 to L.

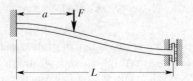

Figure P3.22   Guided cantilever beam loaded at arbitrary location

Problem 3.23: A uniform beam, of mass m, length L, and bending stiffness EI, has its left end simply supported and its right end guided, as shown in Fig. P3.23. (See Problem 3.11 for an explanation for a *guided end.*) Determine the equivalent spring constant and the equivalent mass when the point of interest is located at a distance a from the left end. Use MATLAB to plot the variations of both the equivalent spring constant and the equivalent mass as a varies from 0 to L.

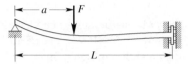

Figure P3.23   A guided simply-supported beam loaded at an arbitrary location

Problem 3.24: A uniform L-shaped beam, whose geometry is shown in Fig. P3.24, is clamped at the end of the longer arm. The total mass is m. All arms are made of the same material, of bending stiffness EI. The point of interest is the free end of the shorter arm. Find the equivalent spring constant and the equivalent mass of the L-beam if it is loaded in the vertical direction. Neglect the deformation due to axial compression, if any.

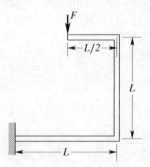

Figure P3.24   L-shaped beam with longer arm cantilevered

Problem 3.25: Repeat Problem 3.24 if the beam is loaded in the horizontal direction.

Problem 3.26: A uniform U-shaped beam, whose geometry is shown in Fig. P3.26, has one of its ends cantilevered and the other end free. All three arms are made of the same material, of bending stiffness EI. The total mass is m. The point of interest is the free end of the beam. Find the equivalent spring constant and the equivalent mass if it is loaded in the vertical direction as shown. Neglect the deformation due to axial compression, if any. Use MATLAB to plot the variations of both the equivalent spring constant and the equivalent mass as b/a ratio varies from 0 to 5.

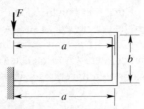

Figure P3.26   Cantilevered U-shaped beam

Problem 3.27: A uniform cantilever beam, of mass m and length L, is additionally supported by a spring k at its center, as shown in Fig. P3.27. The beam's bending stiffness is EI. Find the equivalent spring constant and the equivalent mass for the beam if its free end is the point of interest.

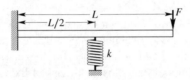

Figure P3.27   Cantilever beam with spring support at the mid-span

Problem 3.28: A uniform cantilever beam, of mass m and length L, is harnessed at its mid-span by a cable, whose other end is tied to the wall, as shown in Fig. P3.28. The beam's bending stiffness is EI. At equilibrium, the cable forms a $45°$ angle with the beam, which is horizontal. The cable has a Young's modulus E_2 and a cross-sectional area A_2 but of negligible mass. Find the equivalent spring constant and the equivalent mass for the beam if its free end is the point of interest.

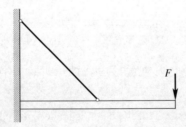

Figure P3.28   Cantilever beam harnessed by elastic cable at the mid-span

Problem 3.29: Use the Castigliano method to find the equivalent spring constant for the nonuniform rod described in Problem 3.3.

Problem 3.30: Use the Castigliano method to find the equivalent spring constant for the overhung simply-supported beam described in Problem 3.4.

Problem 3.31: Use the Castigliano method to find the equivalent spring constant for the beam under four-point bending setup described in Problem 3.5.

Problem 3.32: Use the Castigliano method to find the equivalent spring constant for the pin-supported beam described in Problem 3.6.

Problem 3.33: Use the Castigliano method to find the equivalent spring constant for the leaf-spring model under end loading described in Problem 3.7.

Problem 3.34: Use the Castigliano method to find the equivalent spring constant for the leaf-spring model under central loading described in Problem 3.8.

Problem 3.35: Use the Castigliano method to find the equivalent spring constant for the cantilever beam described in Problem 3.17.

Problem 3.36: Use the Castigliano method to find the equivalent spring constant for the L-shaped beam described in Problem 3.24.

Problem 3.37: Use the Castigliano method to find the equivalent spring constant for the U-shaped beam described in Problem 3.26.

Problem 3.38: Use the Castigliano method to find the equivalent spring constant for the guided cantilever beam described in Problem 3.11.

Problem 3.39: Use the Castigliano method to find the equivalent spring constant for the pin-supported cantilever beam described in Problem 3.12.

Problem 3.40: Use the Castigliano method to find the equivalent spring constant for the pin-supported cantilever beam described in Problem 3.20.

Problem 3.41: Use the Castigliano method to find the equivalent spring constant for the doubly-clamped beam described in Problem 3.21.

Problem 3.42: Use the Rayleigh–Ritz method to find the equivalent spring constant and the equivalent mass for the overhung simply-supported beam described in Problem 3.4. Note the boundary conditions for an unconstrained unloaded end are that both moment and shear force vanish. That is, $u'' = 0$ and $u''' = 0$.

Problem 3.43: Use the Rayleigh–Ritz method to find the equivalent spring constant and the equivalent mass for the beam under four-point bending described in Problem 3.5. Note that for a simply-supported end, the bending moment vanishes, that is, $u'' = 0$.

Problem 3.44: Use the Rayleigh–Ritz method to find the equivalent spring constant and the equivalent mass for the centrally pin-supported beam described in Problem 3.6. Note that at both ends of the beam, the bending moment vanishes, that is, $u'' = 0$.

Problem 3.45: Use the Rayleigh–Ritz method to find the equivalent spring constant and the equivalent mass for the overhung simply-supported beam described in Problem 3.9. Note that for the loading point where the beam is otherwise unconstrained, the bending moment vanishes, that is, $u'' = 0$.

Problem 3.46: Use the Rayleigh–Ritz method to find the equivalent spring constant and the equivalent mass for the guided cantilever beam described in Problem 3.11.

Problem 3.47: Use the Rayleigh–Ritz method to find the equivalent spring constant and the equivalent mass for the pin-supported cantilever beam described in Problem 3.12. Note that for the pin-supported end, the bending moment vanishes, that is, $u'' = 0$.

Problem 3.48: A uniform cantilever beam, of mass m and length L, has a mass m_0 affixed at a distance a from the wall, as shown in Fig. P3.48. The beam's bending stiffness is EI. Find the equivalent spring constant and the equivalent mass for the beam if its free end is the point of interest.

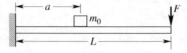

Figure P3.48   Cantilever beam with attached mass

Problem 3.49: A uniform cantilever beam, of mass m and length L, is subjected to a uniformly distributed pressure w_0 over a length a starting from its left end, as shown in Fig. P3.49. The beam's bending stiffness is EI. Find the equivalent spring constant and the equivalent mass for the beam if its free end is the point of interest.

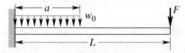

Figure P3.49   Cantilever beam loaded by distributed force

Problem 3.50: Two identical uniform cantilever beams, of mass m and length L, have their noncantilever ends stacked one on the top of the other, as shown in Fig. P3.50. The point of interest is the overlapping point, where the two beams are assumed to remain in contact at all times. The bending stiffness for each beam is EI. Find the equivalent spring constant and the equivalent mass for the stacked beams.

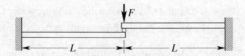

Figure P3.50   Cantilever beams on top of each other

Problem 3.51: A uniform beam, of mass m and length L, is suspended at its two ends by two identical massless springs, of spring constant k, as shown in Fig. P3.51. The bending stiffness of the beam is EI. Find the equivalent spring constant and the equivalent mass if the center of the beam is the point of interest.

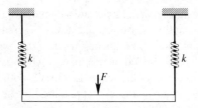

Figure P3.51   Beam suspended from two identical springs

Problem 3.52: A uniform beam, of mass m and length L, is suspended by two massless extensible cables, of lengths l_1 and l_2, respectively, at its two ends, as shown in Fig. P3.52. Both cables are of Young's modulus E_0 and cross-sectional area A_0. The beam's bending stiffness is EI. Find the equivalent spring constant and the equivalent mass if the center of the beam is the point of interest.

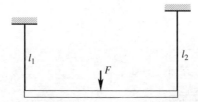

Figure P3.52   Beam suspended from elastic cables of uneven lengths

Problem 3.53: A massless rope is tightly wound around a uniform circular shaft, whose one end is fixed to the wall. After countless rounds, it dangles down a length b from the side of the shaft at a distance a from the wall, as shown in Fig. P3.53. The shaft has mass m, length L, and a radius R. The material properties for the shaft are Young's modulus E_1 and shear modulus G_1. The rope has Young's modulus E_2 and cross-sectional area A_2. Assume that, due to the tight winding, only the dangled portion of the rope is extensible. The free end of the rope is the point of interest. Find the equivalent spring constant and the equivalent mass of the shaft.

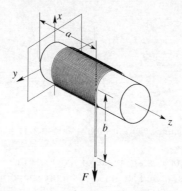

Figure P3.53   Rope wraps around and dangles from circular shaft

Problem 3.54: In MEMS designs, *folded beams* are a common spring–mass vibration subsystem that has found many motion-sensing applications. Figure P3.54 illustrates a design for micro-accelerometer applications. In this design, the *proof mass* at the center is attached to four symmetrically located folded beams, one of which is demarcated by the dashed line. The *comb fingers* are a part of the sensor that moves with the proof mass in the direction perpendicular to the fingers, with respect to a meshing set of fixed comb fingers (not shown) such that the capacitance between the fingers is changed, which is subsequently amplified and read out by on-board circuitry. From mechanics perspective, the comb fingers only act as an added mass to the proof mass. The entire structure is etched out of a silicone layer and is designed to float in space except the four anchor points. Assume that the mass density and the Young's modulus of the silicon layer are ρ and E, respectively. The thickness of the silicon layer is h. The total mass of the proof mass and comb fingers is m_0. Find the equivalent spring constant and the equivalent mass of the subsystem, with the proof mass being the point of interest. Use MATLAB to explore the relations between the equivalent spring constant and the equivalent mass and the geometric parameters.

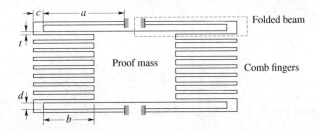

Figure P3.54   Folded beam design for MEMS accelerometer applications

Problem 3.55: Three uniform beams are made of the same material, of bending stiffness EI. Two of the beams, both of mass m and length L, are simply supported and laid in parallel on the ground at a distance $\frac{1}{2}L$ apart. The third beam, of mass $\frac{1}{2}m$ and length $\frac{1}{2}L$, rests on the top of the two beams on the ground, at their mid-spans and perpendicular to them, as shown in Fig. P3.55. The point of interest is the center of the shorter beam on the top. Find the equivalent spring constant and the equivalent mass for the three-beam assembly.

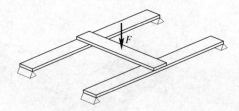

Figure P3.55   Three beams in simply-supported H-shape arrangement

Problem 3.56: Two uniform beams are made of the same material, of bending stiffness EI. The shorter beam, of mass m and length L, is simply supported on the ground. The longer beam, of mass $2m$ and length $2L$, sits on the top of the shorter beam at the mid-spans of both beams. One end of the longer beam is simply supported while the other end is free, which is the point of interest. The two beams are perpendicular to each other, as shown in Fig. P3.56. Find the equivalent spring constant and the equivalent mass for the two-beam assembly.

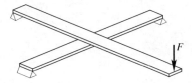

Figure P3.56   Two beams form cross arrangement

Problem 3.57: Two uniform beams are made of the same material, of bending stiffness EI. The longer beam, of mass m and length L, is simply supported on the ground. The shorter beam, of mass $\frac{1}{2}m$ and length $\frac{1}{2}L$, has one end cantilevered to a wall while the other end rests on top of the simply-supported beam at its mid-span, as shown in Fig. P3.57. Find the equivalent spring constant and the equivalent mass for the two-beam assembly if the point of interest is the free end of the shorter beam.

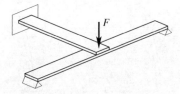

Figure P3.57   Cantilever beam with its free end resting on simply-supported beam

Problem 3.58: A frame mounted horizontally on a ledge forms two sides of an equilateral triangle, of side length L, as shown in Fig. P3.58. The mounting can be assumed as clamped. Each arm of the frame has mass m and bending stiffness EI. Find the equivalent spring constant and the equivalent mass for the frame if the point of interest is the free end of the frame, loaded in the direction perpendicular to the frame.

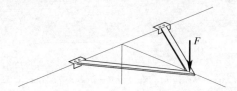

Figure P3.58   Triangular-shaped cantilever beam clamped to ledge

Problem 3.59: Repeat Problem 3.58 if the frame further propped by a slender rod, also of length L, but of Young's modulus E_0 and cross-sectional area A_0. The rod is pin-joined in both ends, which are located in the vertical symmetry plane of the frame, as shown in Fig. P3.59.

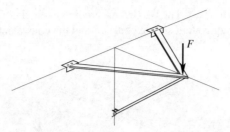

Figure P3.59   Triangular-shaped cantilever beam with support from underneath

Problem 3.60: A simple model for a one-story building is shown in Fig. P3.60. The roof of mass m_0 is supported by four identical uniform columns (only two are shown) of mass m, length L, and bending stiffness EI. Each column cantilevers its ends to the roof and the ground, while the roof end can slide horizontally. Find the fundamental natural frequency of the building model.

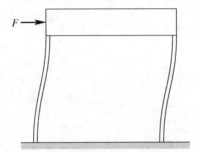

Figure P3.60   Model for small building with guided cantilever columns

Problem 3.61: An electric motor of mass M is mounted at the free end of a uniform cantilever beam of mass m, length L, and bending stiffness EI, as shown in Fig. P3.61. As the motor runs at a constant angular velocity of Ω, the imbalance in the motor imparts a force into the beam. The vertical component of the force is expressible as $F_0 \sin \Omega t$. Find the steady-state vibration of the beam due to the imbalance. Consider only small transverse deflections of the beam.

Figure P3.61   Motor mounted at the free end of cantilever beam

Problem 3.62: As a modeling exercise, artificially add a damping ratio $\zeta < 1$ into the whirling shaft in Example 3.26 and use MATLAB to explore the transient whirling morphing into steady whirling in the range $0.01 \leq \zeta \leq 0.1$.

4

Vibrations of Multi-DOF Systems

4.1 Objectives

In this chapter, basic concepts and theories for vibrations of multiple-degree-of-freedom (multi-DOF) systems are examined. Although most examples are focused on two-DOF systems, all analysis methods are presented in a matrix form so that they can be readily used for general multi-DOF systems. Apart from the same types of vibration analyses as in single-DOF systems, namely, free vibrations, steady-state, and transient responses of the systems, a few concepts unique to multi-DOF systems are also explored: the modal analysis, vibration analyses using principal coordinates. As with single-DOF systems, all examples of vibration analyses are presented with a dedicated *Exploring the Solution with MATLAB* section. Empowered by MATLAB, we also explore a few many-DOF examples whose analyses are only possible with the help of computational tools, including a fascinating *Slinky levitation* phenomenon, and examples of constructing a many-DOF system to approach the continuum limit.

4.2 Matrix Equation of Motion

For vibration analysis of multi-DOF systems, it is often more convenient and notationally advantageous to write the equations of motion in a matrix form. We shall use an example to illustrate this notation.

Consider a simple mass–spring two-DOF system shown in Figure 4.1. Both masses are identical, of mass m, and all springs are identical, of spring constant k. All contacts are frictionless. Initially, the system is in equilibrium.

We define $\{x_1, x_2\}$ as a complete and independent set of generalized coordinates for the system, where x_1 and x_2 and are the displacements of the masses on the left and on the right, respectively, both measured from their respective equilibrium positions with respect to ground. The derivation is rather straightforward. Without much ado, we simply write the following equations of motion:

$$m\ddot{x}_1 + 2kx_1 - kx_2 = 0 \tag{4.1}$$

$$m\ddot{x}_2 - kx_1 + 2kx_2 = 0 \tag{4.2}$$

Fundamentals of Mechanical Vibrations, First Edition. Liang-Wu Cai.
© 2016 John Wiley & Sons, Ltd. Published 2016 by John Wiley & Sons, Ltd.
Companion Website: www.wiley.com/go/cai/fundamentals_mechanical_vibrations

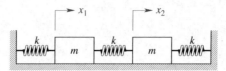

Figure 4.1   Two-mass mass–spring system

The above set of equations can be written in a matrix form such as

$$[M]\{\ddot{x}\} + [K]\{x\} = \{0\} \tag{4.3}$$

where, for the present two-DOF system,

$$\{x\} = \begin{Bmatrix} x_1 \\ x_2 \end{Bmatrix} \tag{4.4}$$

Using eqn. (4.3) as a template, the set of equations of motion in eqns. (4.1) and (4.2) can be written as

$$\begin{bmatrix} m & 0 \\ 0 & m \end{bmatrix} \begin{Bmatrix} \ddot{x}_1 \\ \ddot{x}_2 \end{Bmatrix} + \begin{bmatrix} 2k & -k \\ -k & 2k \end{bmatrix} \begin{Bmatrix} x_1 \\ x_2 \end{Bmatrix} = \begin{Bmatrix} 0 \\ 0 \end{Bmatrix} \tag{4.5}$$

or

$$[M] = \begin{bmatrix} m & 0 \\ 0 & m \end{bmatrix} \qquad [K] = \begin{bmatrix} 2k & -k \\ -k & 2k \end{bmatrix} \tag{4.6}$$

In matrix notation, $[M]$ and $[K]$ are called the *mass matrix* and the *stiffness matrix*, respectively.

If we assume a linear dashpot, of dashpot constant c, is connected in parallel to each spring, as shown in Fig. 4.2, the equations of motion for the system become

$$[M]\{\ddot{x}\} + [C]\{\dot{x}\} + [K]\{x\} = \{0\} \tag{4.7}$$

where

$$[C] = \begin{bmatrix} 2c & -c \\ -c & 2c \end{bmatrix} \tag{4.8}$$

is called the *damping matrix*.

Equation (4.7) is the most general form in matrix notation for a set of linear second-order ordinary differential equations. For many systems, the equations of motion may be nonlinear. They must be linearized before they can be written in the form of eqn. (4.7). This should not be an issue since we are studying small motions anyway. But it is a necessary step. For a system of N DOFs, the $[M]$, $[K]$, and $[C]$ matrices are square matrices of size $N \times N$.

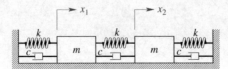

Figure 4.2   Two-mass mass–spring–dashpot system

In the following, we reexamine the equations of motion for a few systems that will be used as examples in this chapter.

■ Example 4.1: Double Pendulum

Write the equations of motion for the double-pendulum system in Example 1.22 in matrix form for small motions.

□ **Solution:**

The equations of motion have been derived in Example 1.22. They are repeated here:

$$2ml^2\ddot{\theta}_1 + ml^2\ddot{\theta}_2 \cos(\theta_2 - \theta_1) - ml^2\dot{\theta}_2^2 \sin(\theta_2 - \theta_1) + 2mgl \sin\theta_1 = 0 \tag{a}$$

$$ml^2\ddot{\theta}_2 + ml^2\ddot{\theta}_1 \cos(\theta_2 - \theta_1) - ml^2\dot{\theta}_1^2 \sin(\theta_2 - \theta_1) + mgl \sin\theta_2 = 0 \tag{b}$$

where θ_1 and θ_2 are the respective counterclockwise angular displacements of the pendulums with respect to the vertical. For small motions, we assume that not only both θ's are small but also their time derivatives are small. Thus, we drop $\dot{\theta}^2$ terms and replace sine functions by their arguments and cosine functions by 1. Then, eqns. (a) and (b) become

$$2l\ddot{\theta}_1 + l\ddot{\theta}_2 + 2g\theta_1 = 0 \tag{c}$$

$$l\ddot{\theta}_1 + l\ddot{\theta}_2 + g\theta_2 = 0 \tag{d}$$

where both sides have been divided by ml^2. In matrix form, the equations of motion can be written as

$$l \begin{bmatrix} 2 & 1 \\ 1 & 1 \end{bmatrix} \begin{Bmatrix} \ddot{\theta}_1 \\ \ddot{\theta}_2 \end{Bmatrix} + g \begin{bmatrix} 2 & 0 \\ 0 & 1 \end{bmatrix} \begin{Bmatrix} \theta_1 \\ \theta_2 \end{Bmatrix} = \begin{Bmatrix} 0 \\ 0 \end{Bmatrix} \tag{e}$$

■ Example 4.2: Connected Planar Pendulums

Consider two planar pendulums, both of mass m and length l, linked via a spring of spring constant k. Furthermore, each pendulum is connected to a side wall by an identical spring, as sketched in Fig. 4.3. All the springs are unstretched when both pendulums are in vertical position. Obtain the equation(s) of motion for the system.

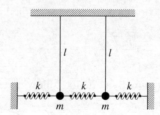

Figure 4.3   Planar pendulums linked by springs

□ **Solution:**

In this example, we need to derive the equation(s) of motion for the system first.

- *Generalized Coordinates*: We define $\{\,\theta_1, \theta_2\,\}$ as a set of complete and independent generalized coordinates, where θ_1 and θ_2 are the angular displacements of the two pendulums with respect to the vertical, measured in the counterclockwise direction.

- *Admissible Variations*: We verify that $\{\ \delta\theta_1, \delta\theta_2\ \}$ is a complete and independent set of admissible variations in this set of generalized coordinates.
- *Holonomicity*: We conclude that the system is holonomic and has two DOFs.
- *Generalized Forces*: There is no nonconservative force in the system: $\delta W^{\text{n.c.}} = 0$. Thus,

$$\Xi_{\theta_1} = 0 \qquad \Xi_{\theta_2} = 0 \tag{a}$$

- *Kinetic Energy*: The two masses have speeds of $l\dot{\theta}_1$ and $l\dot{\theta}_2$, respectively. Thus,

$$T = \frac{1}{2}ml^2\dot{\theta}_1^2 + \frac{1}{2}ml^2\dot{\theta}_2^2 \tag{b}$$

- *Potential Energy*: We choose the ceiling as the datum for gravitational potential energy. For the potential energy in the springs, for small motions, we need to include horizontal positions of two masses to determine the amounts of stretch in the springs. Thus,

$$V = -mgl\cos\theta_1 - mgl\cos\theta_2$$
$$+ \frac{1}{2}kl^2\sin^2\theta_1 + \frac{1}{2}kl^2(\sin\theta_1 - \sin\theta_2)^2 + \frac{1}{2}kl^2\sin^2\theta_2 \tag{c}$$

Using the small motion assumption, we keep terms up to the second order of angular displacements

$$V = mgl\left(-2 + \frac{1}{2}\theta_1^2 + \frac{1}{2}\theta_2^2\right) + \frac{1}{2}kl^2\left[\theta_1^2 + \theta_2^2 + (\theta_1 - \theta_2)^2\right] \tag{d}$$

- *Lagrangian*: Combining eqns. (b) and (d), according to eqn. (d),

$$\mathcal{L} = T - V = \frac{1}{2}ml^2\dot{\theta}_1^2 + \frac{1}{2}ml^2\dot{\theta}_2^2$$
$$- mgl\left(-2 + \frac{1}{2}\theta_1^2 + \frac{1}{2}\theta_2^2\right) - \frac{1}{2}kl^2\left[\theta_1^2 + \theta_2^2 + (\theta_1 - \theta_2)^2\right] \tag{e}$$

- *Lagrange's Equation*: The system has two equations of motion, one for each generalized coordinate. For a generalized coordinate denoted as ξ, the Lagrange's equation is

$$\frac{d}{dt}\left(\frac{\partial\mathcal{L}}{\partial\dot{\xi}}\right) - \frac{\partial\mathcal{L}}{\partial\xi} = \Xi_\xi \tag{f}$$

Without going into details, we directly write the following equations of motion for the system:

$$\theta_1\text{-Equation:} \qquad ml\ddot{\theta}_1 + mg\theta_1 + kl(2\theta_1 - \theta_2) = 0 \tag{g}$$
$$\theta_2\text{-Equation:} \qquad ml\ddot{\theta}_2 + mg\theta_2 + kl(2\theta_2 - \theta_1) = 0 \tag{h}$$

In matrix form, the equations of motion can be written as

$$ml\begin{bmatrix} 1 & 0 \\ 0 & 1 \end{bmatrix}\begin{Bmatrix} \ddot{\theta}_1 \\ \ddot{\theta}_2 \end{Bmatrix} + \begin{bmatrix} mg + 2kl & -kl \\ -kl & mg + 2kl \end{bmatrix}\begin{Bmatrix} \theta_1 \\ \theta_2 \end{Bmatrix} = \begin{Bmatrix} 0 \\ 0 \end{Bmatrix} \tag{i}$$

4.3 Modal Analysis: Natural Frequencies and Mode Shapes

Starting from this section, we shall proceed to analyze three types of vibrations similar to those for single-DOF systems: free vibrations, steady-state responses, and transient responses. But, there is a significant difference between single-DOF systems and multi-DOF systems. For a single-DOF system, if we normalized the equation of motion, the natural frequency of the system immediately becomes apparent. Multi-DOF systems do not possess this convenience. Instead, we have to go through a process called *modal analysis* to find the natural frequencies for a system.

Since the natural frequencies are inherent characteristics of a system, the modal analysis is performed to a system stripped off its damping and externally applied loads. In such cases, the matrix equation of motion of the system is

$$[M]\{\ddot{x}\} + [K]\{x\} = 0 \tag{4.9}$$

Similar to finding the free vibration in single-DOF systems, we assume the response of the system to be of the form

$$\{x\} = \{X\}\, e^{\lambda t} \tag{4.10}$$

where $\{X\}$ and λ are constants. Substituting eqns. (4.10) into eqn. (4.9) gives

$$[M]\lambda^2\{X\}e^{\lambda t} + [K]\{X\}\, e^{\lambda t} = 0 \tag{4.11}$$

Reorganizing and canceling out the factor $e^{\lambda t}$, since it is not always zero, give

$$\left(\lambda^2[M] + [K]\right)\{X\} = 0 \tag{4.12}$$

An obvious solution is $\{X\} = 0$. This is called the *trivial solution,* which is not what we are interested in, since we are looking for inherent oscillatory motions of the system. With motions, $\{X\}$ cannot vanish, at least not all its elements may vanish simultaneously. Such a solution, if exists, is called a *nontrivial solution.*

Equation (4.12) represents a set of linear equations. The unique thing is that its right-hand side is zero. According to matrix theories, eqn. (4.12) is called a *homogeneous linear equation system*. For such a system to have nontrivial solution, the determinant of its *system matrix*, which is the square matrix in front of $\{X\}$, must vanish, that is,

$$\det\left(\lambda^2[M] + [K]\right) = 0 \tag{4.13}$$

We shall continue with the example shown in Fig. 4.1, whose equations of motion is in eqn. (4.5), to see what this process entails. Following the process just outlined, we assume the response of the system to be of the form

$$\begin{Bmatrix} x_1 \\ x_2 \end{Bmatrix} = \begin{Bmatrix} X_1 \\ X_2 \end{Bmatrix} e^{\lambda t} \tag{4.14}$$

Substituting this assumed solution into eqn. (4.5) gives

$$\begin{bmatrix} m & 0 \\ 0 & m \end{bmatrix} \begin{Bmatrix} X_1 \\ X_2 \end{Bmatrix} \lambda^2 e^{\lambda t} + \begin{bmatrix} 2k & -k \\ -k & 2k \end{bmatrix} \begin{Bmatrix} X_1 \\ X_2 \end{Bmatrix} e^{\lambda t} = \begin{Bmatrix} 0 \\ 0 \end{Bmatrix} \tag{4.15}$$

Since $e^{\lambda t}$ is not always zero, the above equation can be simplified by eliminating $e^{\lambda t}$

$$\begin{bmatrix} 2k + m\lambda^2 & -k \\ -k & 2k + m\lambda^2 \end{bmatrix} \begin{Bmatrix} X_1 \\ X_2 \end{Bmatrix} = \begin{Bmatrix} 0 \\ 0 \end{Bmatrix} \tag{4.16}$$

Requiring the determinant of the system matrix to vanish gives

$$\left(2k + m\lambda^2\right)^2 + (-k)^2 = 0 \tag{4.17}$$

Similar to the single-DOF case, eqn. (4.17) is called the *characteristic equation*.

Equation (4.17) is a quadratic equation of λ^2. Its roots can be written as

$$\lambda^2 = \pm\frac{k}{m} - 2\frac{k}{m} \tag{4.18}$$

Equation (4.18) represents four possible purely imaginary roots for λ, namely,

$$\lambda_{1,2} = \pm i\sqrt{-\frac{k}{m} + 2\frac{k}{m}} = \pm i\sqrt{\frac{k}{m}} \tag{4.19}$$

$$\lambda_{3,4} = \pm i\sqrt{\frac{k}{m} + 2\frac{k}{m}} = \pm i\sqrt{\frac{3k}{m}} \tag{4.20}$$

where $i = \sqrt{-1}$. For notational simplicity, we denote

$$\lambda_{1,2} = \pm i\omega_1 \qquad \lambda_{3,4} = \pm i\omega_2 \tag{4.21}$$

where

$$\omega_1 = \sqrt{\frac{k}{m}} \quad \text{and} \quad \omega_2 = \sqrt{\frac{3k}{m}} \tag{4.22}$$

We can now proceed to find the nontrivial solutions for X_1 and X_2. To do this, we convert the matrix form of eqn. (4.16) back into the algebraic form as

$$(2k + m\lambda^2)X_1 - kX_2 = 0 \tag{4.23}$$

$$-kX_1 + (2k + m\lambda^2)X_2 = 0 \tag{4.24}$$

Now here comes the "Gocha!" moment: if we substitute any of the λ's into the above two equations, we found that the two equations become identical! According to the matrix theories, since we rejected the trivial solution, we now can have an infinite number of solutions, as long as a condition is met! The condition comes in the form of eqn. (4.17), which makes the two equations identical. Sometimes, it makes one of equations to become trivial, such as $0 = 0$.

In order to meaningfully write the solutions, we typically choose one of the unknowns, say X_1, as the basis and express all others in terms of this chosen one. Substituting $\lambda_{1,2}^2 = -\omega_1^2 = -k/m$ into either equation, say eqn. (4.23), gives

$$\left.\frac{X_2}{X_1}\right|_{\lambda=\lambda_{1,2}} = \frac{2k - m\frac{k}{m}}{k} = 1 \tag{4.25}$$

Similarly, substituting $\lambda_{3,4}^2 = -\omega_2^2 = -3k/m$ into eqn. (4.23) gives

$$\left.\frac{X_2}{X_1}\right|_{\lambda=\lambda_{3,4}} = \frac{2k - 3m\frac{k}{m}}{k} = -1 \tag{4.26}$$

Note that it is λ^2 that enters into these equations. Hence, we only make two substitutions while we have four roots. Then, the general solution can be written as

$$\left\{\begin{matrix} x_1 \\ x_2 \end{matrix}\right\} = C_1 \left\{\begin{matrix} 1 \\ \frac{X_2}{X_1}\Big|_{\lambda=\lambda_1} \end{matrix}\right\} e^{\lambda_1 t} + C_2 \left\{\begin{matrix} 1 \\ \frac{X_2}{X_1}\Big|_{\lambda=\lambda_2} \end{matrix}\right\} e^{\lambda_2 t}$$

$$+ C_3 \left\{\begin{matrix} 1 \\ \frac{X_2}{X_1}\Big|_{\lambda=\lambda_3} \end{matrix}\right\} e^{\lambda_3 t} + C_4 \left\{\begin{matrix} 1 \\ \frac{X_2}{X_1}\Big|_{\lambda=\lambda_4} \end{matrix}\right\} e^{\lambda_4 t}$$

$$= C_1 \left\{\begin{matrix} 1 \\ 1 \end{matrix}\right\} e^{i\omega_1 t} + C_2 \left\{\begin{matrix} 1 \\ 1 \end{matrix}\right\} e^{-i\omega_1 t} + C_3 \left\{\begin{matrix} 1 \\ -1 \end{matrix}\right\} e^{i\omega_2 t} + C_4 \left\{\begin{matrix} 1 \\ -1 \end{matrix}\right\} e^{-i\omega_2 t} \tag{4.27}$$

As we have seen in the single-DOF case, these pairs of exponential functions with purely imaginary but conjugate exponents represent harmonic motions and can be replaced by the corresponding pair of sine and cosine functions. Since we prefer real expressions, the general solution in eqn. (4.27) can be equivalently written as

$$\left\{\begin{matrix} x_1 \\ x_2 \end{matrix}\right\} = a_1 \left\{\begin{matrix} 1 \\ 1 \end{matrix}\right\} \cos\omega_1 t + b_1 \left\{\begin{matrix} 1 \\ 1 \end{matrix}\right\} \sin\omega_1 t$$

$$+ a_2 \left\{\begin{matrix} 1 \\ -1 \end{matrix}\right\} \cos\omega_2 t + b_2 \left\{\begin{matrix} 1 \\ -1 \end{matrix}\right\} \sin\omega_2 t \tag{4.28}$$

where a_1, a_2, b_1, and b_2 are a new set of real constants yet to be determined.

With this hindsight, we shall rewind. Let us restart the process, with real notions and sine and cosine functions in mind. From the very beginning, at the place of eqn. (4.14), we should start with the assumption that

$$\left\{\begin{matrix} x_1 \\ x_2 \end{matrix}\right\} = \left\{\begin{matrix} X_1 \\ X_2 \end{matrix}\right\} e^{i\omega t} \tag{4.29}$$

Then we should arrive at, instead of eqn. (4.16),

$$\begin{bmatrix} 2k - m\omega^2 & -k \\ -k & 2k - m\omega^2 \end{bmatrix} \left\{\begin{matrix} X_1 \\ X_2 \end{matrix}\right\} = \left\{\begin{matrix} 0 \\ 0 \end{matrix}\right\} \tag{4.30}$$

and subsequently, instead of eqn. (4.17),

$$\left(2k - m\omega^2\right)^2 + (-k)^2 = 0 \tag{4.31}$$

which shall now be called the *frequency equation*, whose roots would give the system's natural frequencies. In solving for the natural frequencies, we should note that the system's natural frequency is nonnegative, but we would later make up the two solutions (associated with $\pm\omega$) by a pair of $\sin \omega t$ and $\cos \omega t$. This would lead to the same solutions as eqn. (4.22). Corresponding to the two natural frequencies, we shall denote the solutions for X_1 and X_2 as

$$\left\{X^{(1)}\right\} = \left\{ \begin{array}{c} \frac{X_1}{X_1} \\ \frac{X_2}{X_1} \end{array} \right\}\Bigg|_{\omega=\omega_1} = \left\{ \begin{array}{c} 1 \\ 1 \end{array} \right\} \tag{4.32}$$

$$\left\{X^{(2)}\right\} = \left\{ \begin{array}{c} \frac{X_1}{X_1} \\ \frac{X_2}{X_1} \end{array} \right\}\Bigg|_{\omega=\omega_2} = \left\{ \begin{array}{c} 1 \\ -1 \end{array} \right\} \tag{4.33}$$

Then the solution should be assumed as

$$\left\{ \begin{array}{c} x_1 \\ x_2 \end{array} \right\} = a_1 \left\{X^{(1)}\right\} \cos \omega_1 t + b_1 \left\{X^{(1)}\right\} \sin \omega_1 t$$

$$+ a_2 \left\{X^{(2)}\right\} \cos \omega_2 t + b_2 \left\{X^{(2)}\right\} \sin \omega_2 t \tag{4.34}$$

which is the same as eqn. (4.28).

In this solution, $\left\{X^{(1)}\right\}$ and $\left\{X^{(2)}\right\}$ are called the *mode shapes* of the system corresponding to their respective natural frequencies. The mode shapes are often presented graphically. There is no "standard" way of drawing the mode shapes. The idea is to plot the shapes that resemble the configurations of the system in motion, such as the one shown in Fig. 4.4 for the two-mass system. Note that a mode shape is not unique, as it only represents the ratio. Very often, mode shapes are presented in some normalized way. Collectively, a natural frequency with its corresponding mode shape is called a *natural mode* of the system, or simply, a *mode*.

The meanings of the mode shapes are that, without an external forcing, the system can sustain motions of only these few modes. As in a single-DOF system, without externally imposing the frequency, the system can move only in one frequency. For two-DOF system, the system can only move in these two frequencies.

For the present two-DOF system, the frequency equation in eqn. (4.30) is a quadratic equation for ω^2. For an N-DOF system, the system matrix would be of a size $N \times N$, and the frequency equation would be an Nth power equation for ω^2. Correspondingly, the system has N natural modes. Conventionally, modes are sorted by their frequencies from low to high such that $0 \leq \omega_1 \leq \omega_2 \leq \omega_3 \leq \cdots \leq \omega_N$.

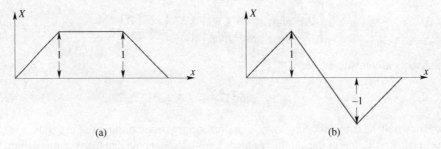

(a)   (b)

Figure 4.4   Mode shapes of two-mass mass–spring system: (a) Mode 1 and (b) Mode 2

The process of finding the natural modes of a system is generally referred to as the *modal analysis*. Its procedure can be summarized as the following:

- Rewrite the equations of motion in matrix form while discarding the damping.
- Assume a solution of the form $\{x\} = \{X\}e^{i\omega t}$, the equations of motion become

$$\left([K] - \omega^2[M]\right)\{X\} = 0 \tag{4.35}$$

- Solve the following frequency equation to obtain the natural frequencies:

$$\det\left([K] - \omega^2[M]\right) = 0 \tag{4.36}$$

and arrange the natural frequencies orderly from low to high.
- For each natural frequency ω_i, substitute ω_i into eqn. (4.35). Identify and discard the equation that becomes either identical to another or trivial. Solve the remaining $N - 1$ equations by treating one of X_i's as known. Packing the solutions into a column matrix gives the ith mode shape of the system, $\{X^{(i)}\}$.
- If desired, plot the mode shapes in a form that is illustrative of the system's displaced configuration.

In the following, we conduct modal analysis for a few example systems.

■ **Example 4.3: Natural Modes of Double Pendulum**

Conduct the modal analysis (finding the natural frequencies and mode shape) for the double pendulum in Example 4.1.

□ **Solution:**

The matrix form of the equations of motion has been found in Example 4.1 as

$$l\begin{bmatrix} 2 & 1 \\ 1 & 1 \end{bmatrix}\begin{Bmatrix} \ddot{\theta}_1 \\ \ddot{\theta}_2 \end{Bmatrix} + g\begin{bmatrix} 2 & 0 \\ 0 & 1 \end{bmatrix}\begin{Bmatrix} \theta_1 \\ \theta_2 \end{Bmatrix} = \begin{Bmatrix} 0 \\ 0 \end{Bmatrix} \tag{a}$$

Assume a solution of the form

$$\begin{Bmatrix} \theta_1 \\ \theta_2 \end{Bmatrix} = \begin{Bmatrix} \Theta_1 \\ \Theta_2 \end{Bmatrix}e^{i\omega t} \tag{b}$$

Then, the equations of motion become, according to eqn. (4.35),

$$\begin{bmatrix} 2g - 2l\omega^2 & -l\omega^2 \\ -l\omega^2 & g - l\omega^2 \end{bmatrix}\begin{Bmatrix} \Theta_1 \\ \Theta_2 \end{Bmatrix} = \begin{Bmatrix} 0 \\ 0 \end{Bmatrix} \tag{c}$$

The frequency equation, according to eqn. (4.36), is

$$\det\begin{bmatrix} 2g - 2l\omega^2 & -\omega^2 l \\ -l\omega^2 & g - \omega^2 l \end{bmatrix} = 0 \tag{d}$$

which is a quadratic equation for ω^2:

$$l^2\omega^4 - 4gl\omega^2 + 2g^2 = 0 \tag{e}$$

Solving the above equation gives

$$\omega_{1,2}^2 = \frac{4gl \pm \sqrt{16(gl)^2 - 8g^2l^2}}{2l^2} = \left(2 \pm \sqrt{2}\right)\frac{g}{l} \tag{f}$$

When taking the square root, we pick only the ones with the positive sign. Thus,

$$\omega_1 = \sqrt{2 - \sqrt{2}}\sqrt{\frac{g}{l}} \qquad \omega_2 = \sqrt{2 + \sqrt{2}}\sqrt{\frac{g}{l}} \tag{g}$$

Since eqn. (c) contains two equations, we can pick any one of the two to solve. Say, we pick the first equation, that is,

$$\left(2g - 2l\omega^2\right)\Theta_1 - l\omega^2\Theta_2 = 0 \tag{h}$$

Then, at $\omega = \omega_1$,

$$\left.\frac{\Theta_2}{\Theta_1}\right|_{\omega=\omega_1} = \frac{2g - 2l\omega_1^2}{l\omega_1^2} = \frac{2g - 2l\left(2 - \sqrt{2}\right)\frac{g}{l}}{l\left(2 - \sqrt{2}\right)\frac{g}{l}} = \sqrt{2} \tag{i}$$

and at $\omega = \omega_2$,

$$\left.\frac{\Theta_2}{\Theta_1}\right|_{\omega=\omega_2} = \frac{2g - 2l\omega_2^2}{l\omega_2^2} = \frac{2g - 2l\left(2 + \sqrt{2}\right)\frac{g}{l}}{l\left(2 + \sqrt{2}\right)\frac{g}{l}} = -\sqrt{2} \tag{j}$$

Thus, the natural modes for the system are

$$\text{Mode 1:} \qquad \omega_1 = \sqrt{2 - \sqrt{2}}\sqrt{\frac{g}{l}} \qquad \{\Theta^{(1)}\} = \left\{\begin{matrix} 1 \\ \sqrt{2} \end{matrix}\right\} \tag{k}$$

$$\text{Mode 2:} \qquad \omega_2 = \sqrt{2 + \sqrt{2}}\sqrt{\frac{g}{l}} \qquad \{\Theta^{(2)}\} = \left\{\begin{matrix} 1 \\ -\sqrt{2} \end{matrix}\right\} \tag{l}$$

The mode shapes are schematically sketched in Fig. 4.5, where we have attempted to draw the mode shapes to be illustrative of the configurations of the double pendulum in motion.

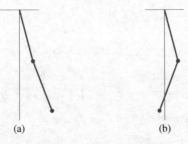

(a)   (b)

Figure 4.5   Mode shapes of double pendulum: (a) Mode 1 and (b) Mode 2

■ **Example 4.4: Natural Modes of Connected Pendulums**

Conduct the modal analysis (finding the natural frequencies and mode shapes) for the two connected pendulums in Example 4.2.

□ **Solution:**

The equations of motion for the two connected pendulum have been found in Example 4.2 as, in matrix form,

$$ml \begin{bmatrix} 1 & 0 \\ 0 & 1 \end{bmatrix} \begin{Bmatrix} \ddot{\theta}_1 \\ \ddot{\theta}_2 \end{Bmatrix} + \begin{bmatrix} mg + 2kl & -kl \\ -kl & mg + 2kl \end{bmatrix} \begin{Bmatrix} \theta_1 \\ \theta_2 \end{Bmatrix} = \begin{Bmatrix} 0 \\ 0 \end{Bmatrix} \tag{a}$$

Assume a solution of the form

$$\begin{Bmatrix} \theta_1 \\ \theta_2 \end{Bmatrix} = \begin{Bmatrix} \Theta_1 \\ \Theta_2 \end{Bmatrix} e^{i\omega t} \tag{b}$$

Then the equations of motion become, according to eqn. (4.35),

$$\begin{bmatrix} mg + 2kl - ml\omega^2 & -kl \\ -kl & mg + 2kl - ml\omega^2 \end{bmatrix} \begin{Bmatrix} \Theta_1 \\ \Theta_2 \end{Bmatrix} = \begin{Bmatrix} 0 \\ 0 \end{Bmatrix} \tag{c}$$

The frequency equation for the system is, according to eqn. (4.36),

$$\left(mg + 2kl - ml\omega^2 \right)^2 - (kl)^2 = 0 \tag{d}$$

which is a quadratic equation for ω^2:

$$(ml)^2\omega^4 - 2(mg + 2kl) ml\omega^2 + (mg)^2 + 4mgkl + 3(kl)^2 = 0 \tag{e}$$

Solving it gives

$$\omega_{1,2}^2 = \frac{2(mg + 2kl) ml \pm \sqrt{4(ml)^2(mg + 2kl)^2 - 4(ml)^2 \left[(mg)^2 + 4mgkl + 3(kl)^2 \right]}}{2(ml)^2}$$

$$= \frac{g}{l} + 2\frac{k}{m} \pm \frac{k}{m} \tag{f}$$

that is,

$$\omega_1^2 = \frac{g}{l} + \frac{k}{m} \qquad \omega_2^2 = \frac{g}{l} + 3\frac{k}{m} \tag{g}$$

To find the mode shape, we again use the first equation in eqn. (c). Hence,

$$\left. \frac{\Theta_2}{\Theta_1} \right|_{\omega=\omega_1} = \frac{mg + 2kl - ml\omega_1^2}{kl} = \frac{mg + 2kl - ml\left(\frac{g}{l} + \frac{k}{m} \right)}{kl} = 1 \tag{h}$$

$$\left. \frac{\Theta_2}{\Theta_1} \right|_{\omega=\omega_2} = \frac{mg + 2kl - ml\omega_2^2}{kl} = \frac{mg + 2kl - ml\left(\frac{g}{l} + 3\frac{k}{m} \right)}{kl} = -1 \tag{i}$$

Thus, the natural modes for the system are

$$\text{Mode 1:} \qquad \omega_1 = \sqrt{\frac{g}{l} + \frac{k}{m}} \qquad \{X^{(1)}\} = \begin{Bmatrix} 1 \\ 1 \end{Bmatrix} \tag{j}$$

$$\text{Mode 2:} \qquad \omega_2 = \sqrt{\frac{g}{l} + 3\frac{k}{m}} \qquad \{X^{(2)}\} = \begin{Bmatrix} 1 \\ -1 \end{Bmatrix} \tag{k}$$

The mode shapes are sketched in Fig. 4.6.

(a)   (b)

Figure 4.6   Mode shapes of planar pendulums linked by springs: (a) Mode 1 and (b) Mode 2

■ Example 4.5: Natural Modes of Unconstrained System

Consider the two masses m on a frictionless floor as shown in Fig. 4.7. The masses are connected to each other by a spring k but otherwise unconstrained. Determine the natural modes of the system.

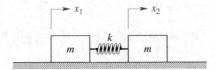

Figure 4.7   Unrestrained two-mass mass–spring system

□ Solution:

We define $\{x_1, x_2\}$ as a complete and independent set of generalized coordinates, where x_1 and x_2 are the displacements of the masses with respect to the ground, measured from the position when the springs are unstretched and the system is at rest. This system is similar to the one in Fig. 4.1, with two springs connecting the masses to the walls removed. Without too much of diversion, we leave the detailed derivation to the reader and directly write the equations of motion for the system as

$$m\ddot{x}_1 + k(x_1 - x_2) = 0 \tag{a}$$

$$m\ddot{x}_2 + k(x_2 - x_1) = 0 \tag{b}$$

or, in matrix form,

$$\begin{bmatrix} m & 0 \\ 0 & m \end{bmatrix} \begin{Bmatrix} \ddot{x}_1 \\ \ddot{x}_2 \end{Bmatrix} + \begin{bmatrix} k & -k \\ -k & k \end{bmatrix} \begin{Bmatrix} x_1 \\ x_2 \end{Bmatrix} = \begin{Bmatrix} 0 \\ 0 \end{Bmatrix} \tag{c}$$

Assume a solution of the form

$$\begin{Bmatrix} x_1 \\ x_2 \end{Bmatrix} = \begin{Bmatrix} X_1 \\ X_2 \end{Bmatrix} e^{i\omega t} \tag{d}$$

Then the equations of motion become, according to eqn. (4.35),

$$\begin{bmatrix} k - m\omega^2 & -k \\ -k & k - m\omega^2 \end{bmatrix} \begin{Bmatrix} X_1 \\ X_2 \end{Bmatrix} = \begin{Bmatrix} 0 \\ 0 \end{Bmatrix}$$ (e)

The frequency equation is, according to eqn. (4.36),

$$- 2km\omega^2 + m^2\omega^4 = 0$$ (f)

It can be solved for

$$\omega_1^2 = 0 \qquad \omega_2^2 = 2\frac{k}{m}$$ (g)

To find the mode shape, we chose to use the first equation in eqn. (e), then,

$$\left.\frac{X_2}{X_1}\right|_{\omega=\omega_1} = \frac{k - m\omega_1^2}{k} = \frac{k - 0}{k} = 1$$ (h)

$$\left.\frac{X_2}{X_1}\right|_{\omega=\omega_2} = \frac{k - m\omega_2^2}{k} = \frac{k - m2\frac{k}{m}}{k} = -1$$ (i)

The natural modes of the system can be summarized as

$$\text{Mode 1:} \qquad \omega_1 = 0 \qquad \{X^{(1)}\} = \begin{Bmatrix} 1 \\ 1 \end{Bmatrix}$$ (j)

$$\text{Mode 2:} \qquad \omega_2 = \sqrt{\frac{2k}{m}} \qquad \{X^{(2)}\} = \begin{Bmatrix} 1 \\ -1 \end{Bmatrix}$$ (k)

□ **Discussion:**

From this example, we can see that the natural frequency of a mode is zero. What is the meaning of a natural frequency of 0? At this "frequency," there is no oscillatory motion. It is no surprise that the corresponding mode shape is $\{1, 1\}^T$, which means the two masses move in unison as if they are one piece. In other words, there is no relative motion between parts of the system. For this reason, this mode is often called the *rigid-body mode*, and such a system is called a *semidefinite system*.

■ **Example 4.6: Natural Modes of Three-DOF System**

Consider three identical masses m moving on a frictionless floor as shown in Fig. 4.8. The masses are connected to each other by a spring of spring constant k, and two masses at both ends are also connected to the wall. Determine the natural modes of the system.

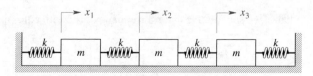

Figure 4.8   Three-mass mass–spring system

□ **Solution:**

We define $\{x_1, x_2, x_3\}$ as a complete and independent set of the generalized coordinates, where x_1, x_2, and x_3 are the displacements of the three masses with respect to the ground, measured from their respective equilibrium positions. We directly write the equations of motion for the system without detailing the derivation process as

$$m\ddot{x}_1 + k(2x_1 - x_2) = 0 \tag{a}$$

$$m\ddot{x}_2 + k(2x_2 - x_1 - x_3) = 0 \tag{b}$$

$$m\ddot{x}_3 + k(2x_3 - x_2) = 0 \tag{c}$$

or in matrix form,

$$\begin{bmatrix} m & 0 & 0 \\ 0 & m & 0 \\ 0 & 0 & m \end{bmatrix} \begin{Bmatrix} \ddot{x}_1 \\ \ddot{x}_2 \\ \ddot{x}_3 \end{Bmatrix} + \begin{bmatrix} 2k & -k & 0 \\ -k & 2k & -k \\ 0 & -k & 2k \end{bmatrix} \begin{Bmatrix} x_1 \\ x_2 \\ x_3 \end{Bmatrix} = \begin{Bmatrix} 0 \\ 0 \\ 0 \end{Bmatrix} \tag{d}$$

Assume a solution of the form

$$\begin{Bmatrix} x_1 \\ x_2 \\ x_3 \end{Bmatrix} = \begin{Bmatrix} X_1 \\ X_2 \\ X_3 \end{Bmatrix} e^{i\omega t} \tag{e}$$

The equations of motion become, according to eqn. (4.35),

$$\begin{bmatrix} 2k - m\omega^2 & -k & 0 \\ -k & 2k - m\omega^2 & -k \\ 0 & -k & 2k - m\omega^2 \end{bmatrix} \begin{Bmatrix} X_1 \\ X_2 \\ X_3 \end{Bmatrix} = \begin{Bmatrix} 0 \\ 0 \\ 0 \end{Bmatrix} \tag{f}$$

The frequency equation is, according to eqn. (4.36),

$$(2k - m\omega^2)^3 - 2k^2(2k - m\omega^2) = 0 \tag{g}$$

which is a cubic equation about ω^3. Fortunately, it can be factored as

$$(2k - m\omega^2)(2k^2 - 4km\omega^2 + m^2\omega^4) = 0 \tag{h}$$

Thus, the solutions to the frequency equation are

$$\omega_2^2 = \frac{2k}{m} \qquad \omega_{1,3}^2 = \frac{4km \pm \sqrt{16k^2m^2 - 8k^2m^2}}{2m^2} = \left(2 \pm \sqrt{2}\right)\frac{k}{m} \tag{i}$$

To find the mode shape, we substitute a natural frequency back into eqn. (f). At $\omega = \omega_1$, the eqn. (f) becomes

$$\begin{bmatrix} \sqrt{2}k & -k & 0 \\ -k & \sqrt{2}k & -k \\ 0 & -k & \sqrt{2}k \end{bmatrix} \begin{Bmatrix} X_1 \\ X_2 \\ X_3 \end{Bmatrix} = \begin{Bmatrix} 0 \\ 0 \\ 0 \end{Bmatrix} \tag{j}$$

There is no obvious trivial equation or identical equations. But they are not independent since we set the determinant of the system matrix to 0. In such cases, we can drop any one to form a two-equation system. Choosing the first two equations, canceling the common factor k, and moving X_1's to the right-hand side give

$$\begin{bmatrix} -1 & 0 \\ \sqrt{2} & -1 \end{bmatrix} \begin{Bmatrix} X_2 \\ X_3 \end{Bmatrix} = \begin{Bmatrix} -\sqrt{2}X_1 \\ X_1 \end{Bmatrix}$$ (k)

which can be solved for

$$\left. \frac{X_2}{X_1} \right|_{\omega=\omega_1} = \sqrt{2} \qquad \left. \frac{X_3}{X_1} \right|_{\omega=\omega_1} = 1$$ (l)

Similarly, at $\omega = \omega_2$ and $\omega = \omega_3$,

$$\left. \frac{X_2}{X_1} \right|_{\omega=\omega_2} = 0 \qquad \left. \frac{X_3}{X_1} \right|_{\omega=\omega_2} = -1$$ (m)

and

$$\left. \frac{X_2}{X_1} \right|_{\omega=\omega_3} = -\sqrt{2} \qquad \left. \frac{X_3}{X_1} \right|_{\omega=\omega_3} = 1$$ (n)

Finally, the natural modes of the system can be summarized as

$$\text{Mode 1:} \qquad \omega_1 = \sqrt{2 - \sqrt{2}} \sqrt{\frac{k}{m}} \qquad \{X^{(1)}\} = \begin{Bmatrix} 1 \\ \sqrt{2} \\ 1 \end{Bmatrix}$$ (o)

$$\text{Mode 2:} \qquad \omega_1 = \sqrt{\frac{2k}{m}} \qquad \{X^{(2)}\} = \begin{Bmatrix} 1 \\ 0 \\ -1 \end{Bmatrix}$$ (p)

$$\text{Mode 3:} \qquad \omega_1 = \sqrt{2 + \sqrt{2}} \sqrt{\frac{k}{m}} \qquad \{X^{(3)}\} = \begin{Bmatrix} 1 \\ -\sqrt{2} \\ 1 \end{Bmatrix}$$ (q)

The mode shapes are schematically shown in Fig. 4.9.

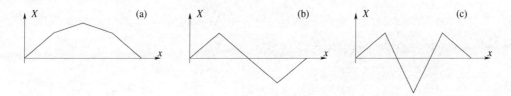

Figure 4.9   Mode shapes of three-mass mass–spring system: (a) Mode 1, (b) Mode 2, and (c) Mode 3

□ **Discussion**

Among the mode shapes, we see that, in the second mode, $X_2/X_1 = 0$. This means that the center mass does not move in this mode. Such a point is called a *node*. What happens if we choose X_2 as the basis to express the mode shapes? We would encounter a problem with Mode 2: X_2 would not appear in the equations that are used to determine the mode shape. In such cases, we would have to choose another one, such as X_3, as the basis.

4.4   Free Vibrations

4.4.1   Free Vibrations of Undamped Systems

We now return to the free-vibration analyses. We have obtained the natural modes as well as the general solutions for the undamped and unforced systems. The general solution is precisely the one for the free-vibration problems of the system.

For the two-DOF system we studied at the beginning of this chapter, the general solution has been found as given in eqn. (4.34), which is repeated here with a slight rearrangement:

$$\left\{ \begin{array}{c} x_1(t) \\ x_2(t) \end{array} \right\} = \{X^{(1)}\} (a_1 \cos \omega_1 t + b_1 \sin \omega_1 t) + \{X^{(2)}\} (a_2 \cos \omega_2 t + b_2 \sin \omega_2 t) \qquad (4.37)$$

It has the following compositions: for each mode i, a mode shape $\{X^{(i)}\}$ is followed by a linear combination of $\sin \omega_i t$ and $\cos \omega_i t$ pair, which is the real version of the complex solution pair $e^{\pm i\omega_i t}$. In general, for an N-DOF system, the general solution for free vibrations can be written as, for the nth generalized coordinate x_n,

$$x_n(t) = \sum_{m=1}^{N} X_n^{(m)} \left(a_m \cos \omega_m t + b_m \sin \omega_m t \right) \qquad (4.38)$$

where $X_n^{(m)}$ denotes the nth row of the mode m's mode shape, which has a natural frequency of ω_m, and a_m and b_m are constants to be determined.

Furthermore, we want to express the general solution in eqn. (4.38) in matrix form. Equation (4.38) strongly resembles the following matrix summation:

$$\{x_n(t)\} = \sum_{m=1}^{N} [V_{nm}]\{s_m(t)\} \qquad (4.39)$$

which, in symbolic form, is the following:

$$\{x(t)\} = [V]\{s(t)\} \qquad (4.40)$$

Indeed, comparing eqns. (4.38) to (4.39) reveals that

$$[V] = \begin{bmatrix} X_1^{(1)} & X_1^{(2)} & X_1^{(3)} & \cdots & X_1^{(N)} \\ X_2^{(1)} & X_2^{(2)} & X_2^{(3)} & \cdots & X_2^{(N)} \\ \vdots & \vdots & \vdots & \ddots & \vdots \\ X_N^{(1)} & X_N^{(2)} & X_N^{(3)} & \cdots & X_N^{(N)} \end{bmatrix} = \left[\{X^{(1)}\} \; \{X^{(2)}\} \cdots \{X^{(N)}\} \right] \qquad (4.41)$$

and

$$\{s(t)\} = \begin{Bmatrix} a_1 \cos \omega_1 t + b_1 \sin \omega_1 t \\ a_2 \cos \omega_2 t + b_2 \sin \omega_2 t \\ \vdots \\ a_N \cos \omega_N t + b_N \sin \omega_N t \end{Bmatrix} \tag{4.42}$$

Equation (4.41) shows that $[V]$ comprises mode shapes that are orderly packed column by column into a square matrix, which is hence called the *modal matrix*, and $\{s(t)\}$ is called the *modal temporal factor*.

The constants a_i and b_i are to be determined by the initial conditions, which generally include initial positions and initial velocities of all system components:

$$\{x(0)\} = \{x_0\} \qquad \{\dot{x}(0)\} = \{v_0\} \tag{4.43}$$

In order to evaluate the velocities, we take a time derivative of eqn. (4.40), giving

$$\{\dot{x}(t)\} = [V]\{\dot{s}(t)\} \tag{4.44}$$

where

$$\{\dot{s}(t)\} = \begin{Bmatrix} \omega_1(-a_1 \sin \omega_1 t + b_1 \cos \omega_1 t) \\ \omega_2(-a_2 \sin \omega_2 t + b_2 \cos \omega_2 t) \\ \vdots \\ \omega_N(-a_N \sin \omega_N t + b_N \cos \omega_N t) \end{Bmatrix} \tag{4.45}$$

Evaluating $\{s(t)\}$ and $\{\dot{s}(t)\}$ at time $t = 0$ gives

$$\{s(0)\} = \begin{Bmatrix} a_1 \\ a_2 \\ \vdots \\ a_N \end{Bmatrix} \equiv \{a\} \qquad \{\dot{s}(0)\} = \begin{Bmatrix} \omega_1 b_1 \\ \omega_2 b_2 \\ \vdots \\ \omega_N b_N \end{Bmatrix} \equiv \{\beta\} \tag{4.46}$$

where

$$\beta_i = \omega_i b_i \qquad \text{or} \qquad b_i = \frac{\beta_i}{\omega_i} \tag{4.47}$$

Then, the initial conditions in eqn. (4.43) gives the following relations:

$$\{x_0\} = [V]\{a\} \qquad \{v_0\} = [V]\{\beta\} \tag{4.48}$$

They can be solved to give

$$\{a\} = [V]^{-1}\{x_0\} \qquad \{\beta\} = [V]^{-1}\{v_0\} \tag{4.49}$$

from which a_i's and b_i's can be found.

■ Example 4.7: Free Vibration of Two-DOF Mass–Spring System

Consider the system we have considered in Fig. 4.1. Assume initially the mass on the left is moved to the right by an amount Δ and then released from rest, find the subsequent motion of the system.

□ **Solution:**

The natural modes of this system have been analyzed in the previous section: specifically, natural frequencies in eqn. (4.22), and mode shapes in eqns. (4.32) and (4.33). They are summarized as the following:

$$\text{Mode 1} \quad \omega_1 = \sqrt{\frac{k}{m}} \quad \{X^{(1)}\} = \begin{Bmatrix} 1 \\ 1 \end{Bmatrix} \tag{a}$$

$$\text{Mode 2:} \quad \omega_2 = \sqrt{\frac{3k}{m}} \quad \{X^{(2)}\} = \begin{Bmatrix} 1 \\ -1 \end{Bmatrix} \tag{b}$$

Hence, the modal matrix and its inverse are

$$[V] = \begin{bmatrix} 1 & 1 \\ 1 & -1 \end{bmatrix} \quad [V]^{-1} = \frac{1}{2}\begin{bmatrix} 1 & 1 \\ 1 & -1 \end{bmatrix} \tag{c}$$

The initial conditions can be written as

$$\{x(0)\} = \begin{Bmatrix} \Delta \\ 0 \end{Bmatrix} \quad \{\dot{x}(0)\} = \{0\} \tag{d}$$

Using eqns. (4.49), we find that $\{\beta\} = 0$ and

$$\{a\} = \frac{1}{2}\begin{bmatrix} 1 & 1 \\ 1 & -1 \end{bmatrix}\begin{Bmatrix} \Delta \\ 0 \end{Bmatrix} = \frac{\Delta}{2}\begin{Bmatrix} 1 \\ 1 \end{Bmatrix} \tag{e}$$

Hence, according to eqn. (4.40),

$$\begin{Bmatrix} x_1(t) \\ x_2(t) \end{Bmatrix} = \frac{\Delta}{2}\begin{bmatrix} 1 & 1 \\ 1 & -1 \end{bmatrix}\begin{Bmatrix} \cos\omega_1 t \\ \cos\omega_2 t \end{Bmatrix} = \frac{\Delta}{2}\begin{Bmatrix} \cos\omega_1 t + \cos\omega_2 t \\ \cos\omega_1 t - \cos\omega_2 t \end{Bmatrix} \tag{f}$$

or

$$x_1(t) = \frac{\Delta}{2}\left[\cos\left(\sqrt{\frac{k}{m}}t\right) + \cos\left(\sqrt{\frac{3k}{m}}t\right)\right] \tag{g}$$

$$x_2(t) = \frac{\Delta}{2}\left[\cos\left(\sqrt{\frac{k}{m}}t\right) - \cos\left(\sqrt{\frac{3k}{m}}t\right)\right] \tag{h}$$

□ **Exploring the Solution with MATLAB**

We introduce the following nondimensionalized parameters, for $i = 1, 2$,

$$\bar{x}_i = \frac{x_i}{\Delta} \quad \bar{t} = \omega_1 t = \sqrt{\frac{k}{m}}t \quad \text{and} \quad \omega_2 t = \sqrt{3}\bar{t} \tag{i}$$

The nondimensionalized solution can be written as

$$\bar{x}_1(t) = \frac{1}{2}\left(\cos\bar{t} + \cos\sqrt{3}\bar{t}\right) \tag{j}$$

$$\bar{x}_2(t) = \frac{1}{2}\left(\cos\bar{t} - \cos\sqrt{3}\bar{t}\right) \tag{k}$$

The following MATLAB code plots the nondimensionalized responses of the two masses, which are shown in Fig. 4.10.

```
t = [0:.01:25];
x1 = .5*(cos(t) + cos(sqrt(3)*t));
x2 = .5*(cos(t) - cos(sqrt(3)*t));
plot(t,x1,'r',t,x2,'--b');
xlabel('\omega_1 t');
ylabel('x_{1,2}(t)/\Delta');
```

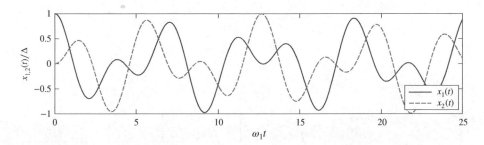

Figure 4.10   Displacements of two-mass mass–spring system subjected to an initial displacement Δ on the left mass

The preliminary verification for the correctness of the code and the curves can be made by observing the initial conditions: the two curves started from different displacements that match the initial displacement conditions and their slopes are both zero, which match the initial velocity conditions. The responses exhibit strong oscillatory motions in both masses, and yet there is no clear periodicity. This is typical of motions of mixed frequencies in which the difference in the frequencies is not too great and yet not too small.

■ Example 4.8: Double Pendulum with Initial Velocity

Assume the double pendulum in Example 4.1 is initially at rest, and both pendulums are in the vertical orientation. Then, the first mass is given an initial velocity v_0 in the horizontal direction. Find the subsequent response of the system.

□ Solution:

The natural modes of the double pendulum have been found in Example 4.3 as

$$\text{Mode 1:} \qquad \omega_1 = \sqrt{2 - \sqrt{2}}\sqrt{\frac{g}{l}} \qquad \{\Theta^{(1)}\} = \left\{\begin{matrix} 1 \\ \sqrt{2} \end{matrix}\right\} \qquad \text{(a)}$$

$$\text{Mode 2:} \qquad \omega_2 = \sqrt{2 + \sqrt{2}}\sqrt{\frac{g}{l}} \qquad \{\Theta^{(2)}\} = \left\{\begin{matrix} 1 \\ -\sqrt{2} \end{matrix}\right\} \qquad \text{(b)}$$

Hence, the modal matrix and its inverse are

$$[V] = \begin{bmatrix} 1 & 1 \\ \sqrt{2} & -\sqrt{2} \end{bmatrix} \qquad [V]^{-1} = \frac{1}{4}\begin{bmatrix} 2 & \sqrt{2} \\ 2 & -\sqrt{2} \end{bmatrix} \qquad \text{(c)}$$

Note that the generalized coordinates are angular displacements and their derivatives are angular velocities. Hence, the given initial velocity needs to be converted into the angular velocity of the first pendulum. Doing so, the initial conditions can be written as

$$\{\theta(0)\} = 0 \qquad \{\dot{\theta}(0)\} = \frac{v_0}{l}\begin{Bmatrix} 1 \\ 0 \end{Bmatrix} \tag{d}$$

Using eqn. (4.49), we find that $\{a\} = 0$ and

$$\{\beta\} = \frac{v_0}{l}\frac{1}{4}\begin{bmatrix} 2 & \sqrt{2} \\ 2 & -\sqrt{2} \end{bmatrix}\begin{Bmatrix} 1 \\ 0 \end{Bmatrix} = \frac{v_0}{2l}\begin{Bmatrix} 1 \\ 1 \end{Bmatrix} \tag{e}$$

and consequently

$$b_1 = \frac{v_0}{2l\omega_1} \qquad b_2 = \frac{v_0}{2l\omega_2} \tag{f}$$

Then, according to eqn. (4.40),

$$\begin{Bmatrix} \theta_1(t) \\ \theta_2(t) \end{Bmatrix} = \frac{v_0}{2l}\begin{bmatrix} 1 & 1 \\ \sqrt{2} & -\sqrt{2} \end{bmatrix}\begin{Bmatrix} \frac{1}{\omega_1}\sin\omega_1 t \\ \frac{1}{\omega_2}\sin\omega_2 t \end{Bmatrix} \tag{g}$$

$$= \frac{v_0}{2\sqrt{gl}}\begin{Bmatrix} \frac{1}{\sqrt{2-\sqrt{2}}}\sin\omega_1 t + \frac{1}{\sqrt{2+\sqrt{2}}}\sin\omega_2 t \\ \frac{\sqrt{2}}{\sqrt{2-\sqrt{2}}}\sin\omega_1 t - \frac{\sqrt{2}}{\sqrt{2+\sqrt{2}}}\sin\omega_2 t \end{Bmatrix} \tag{h}$$

or

$$\theta_1(t) = \frac{v_0}{2\sqrt{2-\sqrt{2}}\sqrt{gl}}\left[\sin\omega_1 t + \frac{\sqrt{2-\sqrt{2}}}{\sqrt{2+\sqrt{2}}}\sin\omega_2 t\right] \tag{i}$$

$$\theta_2(t) = \frac{v_0}{\sqrt{2}\sqrt{2-\sqrt{2}}\sqrt{gl}}\left[\sin\omega_1 t - \frac{\sqrt{2-\sqrt{2}}}{\sqrt{2+\sqrt{2}}}\sin\omega_2 t\right] \tag{j}$$

□ **Exploring the Solution with MATLAB**

We define the following nondimensionalized parameters: for $i = 1, 2$,

$$\bar{\theta}_i = \frac{\theta_i}{v_0/(l\omega_1)} \qquad \bar{t} = \omega_1 t \qquad \omega_2 t = \frac{\omega_2}{\omega_1}\bar{t} = \left(1 + \sqrt{2}\right)\bar{t} \tag{k}$$

Then, the nondimensionalized responses of the masses are

$$\bar{\theta}_1\left(\bar{t}\right) = \frac{1}{2\sqrt{2}}\left[\sin\bar{t} + \left(\sqrt{2}-1\right)\sin\left(1 + \sqrt{2}\right)\bar{t}\right] \tag{l}$$

$$\bar{\theta}_2\left(\bar{t}\right) = \frac{1}{2}\left[\sin\bar{t} - \left(\sqrt{2}-1\right)\sin\left(1 + \sqrt{2}\right)\bar{t}\right] \tag{m}$$

The following MATLAB code produces the plot of the responses of the two masses, which are shown in Fig. 4.11.

```
t = [0:.01:25];
x1 = .5*(sin(t) + (sqrt(2)-1)*sin((1+sqrt(2))*t));
x2 = .5*sqrt(2)*(sin(t) - (sqrt(2)-1)*sin((1+sqrt(2))*t));
plot(t,x1,'r',t,x2,'--b');
xlabel('\omega_1 t');
ylabel('\theta_{1,2}(t)/(v_0/\omega_1l)');
```

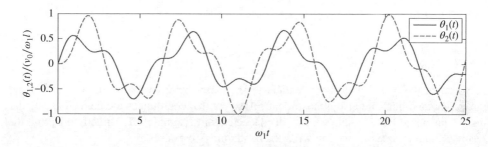

Figure 4.11   Angular displacements of double pendulum subjected to initial velocity v_0 at upper mass

The curves can be checked for the initial conditions: both pendulums start from zero angular displacement; one curve has a nonzero slope, as being subjected to an initial velocity; and the other curve has zero slope. The system's response is again strongly oscillatory and yet without clearly defined periods.

■ **Example 4.9: Beat Phenomenon in Linked Pendulum**

We have encountered a beat phenomenon in a single-DOF system, when it is subjected to a sinusoidal forcing that is close to the system's natural frequency. Here, we demonstrate the beat phenomenon of a different mechanism. Consider the connected pendulum in Example 4.4. Assume the springs in the system are very weak such that $\sqrt{k/m}$ is only a small fraction, denoted as γ, of $\sqrt{g/l}$. If the pendulums on the left is given an initial angular displacement Θ, find the subsequent responses of the two pendulums.

□ **Solution:**

In Example 4.4, the natural modes for the connected pendulums have been found as

$$\text{Mode 1:}\qquad \omega_1 = \sqrt{\frac{g}{l}+\frac{k}{m}}\qquad \{X^{(1)}\} = \begin{Bmatrix}1\\1\end{Bmatrix}\qquad\text{(a)}$$

$$\text{Mode 2:}\qquad \omega_2 = \sqrt{\frac{g}{l}+3\frac{k}{m}}\qquad \{X^{(2)}\} = \begin{Bmatrix}1\\-1\end{Bmatrix}\qquad\text{(b)}$$

Given that $\sqrt{k/m} = \gamma\sqrt{g/l}$, the natural frequencies can be written as

$$\omega_1 = \sqrt{\frac{(1+\gamma)g}{l}} \qquad \omega_2 = \sqrt{\frac{(1+3\gamma)g}{l}} \qquad \text{(c)}$$

We note that the mode shapes do not contain any parameters such as k, m, l, or g. This means that the mode shapes remain unchanged. The modal matrix and its inverse are

$$[V] = \begin{bmatrix} 1 & 1 \\ 1 & -1 \end{bmatrix} \qquad [V]^{-1} = \frac{1}{2}\begin{bmatrix} 1 & 1 \\ 1 & -1 \end{bmatrix} \qquad \text{(d)}$$

The initial conditions can be written as

$$\{\theta(0)\} = \begin{Bmatrix} \Theta \\ 0 \end{Bmatrix} \qquad \{\dot{\theta}(0)\} = \mathbf{0} \qquad \text{(e)}$$

Note that the modal matrix is identical to the mass–spring system in Example 4.7; and the initial conditions are also almost identical except that coordinates are defined as angular displacements in the example, and as displacements in Example 4.7. Directly adapting the results from Example 4.7, we can write the pendulums' responses as

$$\theta_1(t) = \frac{\Theta}{2}\left[\cos\left(\sqrt{\frac{(1+\gamma)g}{l}}t\right) + \cos\left(\sqrt{\frac{(1+3\gamma)g}{l}}t\right)\right] \qquad \text{(f)}$$

$$\theta_2(t) = \frac{\Theta}{2}\left[\cos\left(\sqrt{\frac{(1+\gamma)g}{l}}t\right) - \cos\left(\sqrt{\frac{(1+3\gamma)g}{l}}t\right)\right] \qquad \text{(g)}$$

□ **Exploring the Solution with MATLAB**

We define the following nondimensionalization parameters: for $i = 1, 2$,

$$\bar{\theta}_i = \frac{\theta_i}{\Theta} \qquad \bar{t} = \omega_1 t \qquad \omega_2 t = \frac{\omega_2}{\omega_1}\bar{t} = \sqrt{\frac{1+3\gamma}{1+\gamma}}\bar{t} \qquad \text{(h)}$$

Then, the system's nondimensionalized responses are

$$\bar{\theta}_1(\bar{t}) = \frac{1}{2}\left[\cos\bar{t} + \cos\left(\sqrt{\frac{1+3\gamma}{1+\gamma}}\bar{t}\right)\right] \qquad \text{(i)}$$

$$\bar{\theta}_2(\bar{t}) = \frac{1}{2}\left[\cos\bar{t} - \cos\left(\sqrt{\frac{1+3\gamma}{1+\gamma}}\bar{t}\right)\right] \qquad \text{(j)}$$

The following MATLAB code plots the responses of the two pendulums. Figures 4.12 through 4.14 show the responses for three cases: $\gamma = 0.05, 0.1$, and 0.2.

```
gamma = 0.1;                        % Spring/pendulum ratio
t = [0:.05:200];                    % Time array
q1 = .5*(cos(t)+cos(sqrt((1+3*gamma)/(1+gamma))*t));
q2 = .5*(cos(t)-cos(sqrt((1+3*gamma)/(1+gamma))*t));
subplot('Position', [.1 .6 .85, .35]);
plot(t,q1,'r');
xlabel('\omega_1 t');
ylabel('\theta_1(t)/\Theta');
subplot('Position', [.1 .1 .85, .35]);
plot(t,q2,'b');
xlabel('\omega_1 t');
ylabel('\theta_2(t)/\Theta');
```

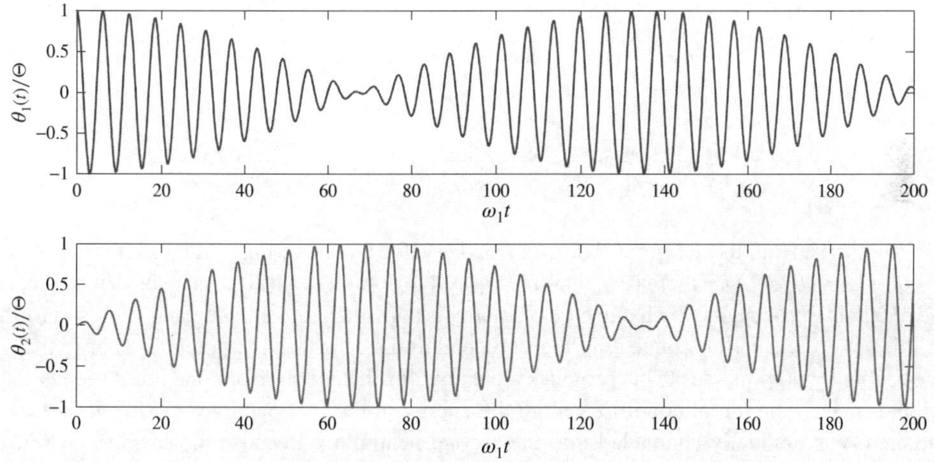

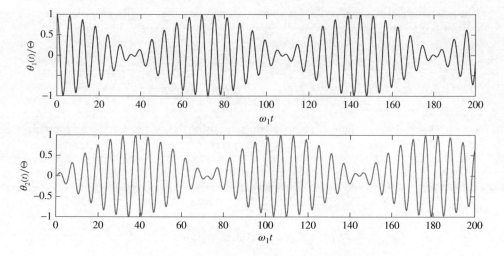

Figure 4.12 Responses of weakly linked pendulums at $\sqrt{k/m} = 0.05\sqrt{g/l}$

Figure 4.13 Responses of weakly linked pendulum at $\sqrt{k/m} = 0.1\sqrt{g/l}$

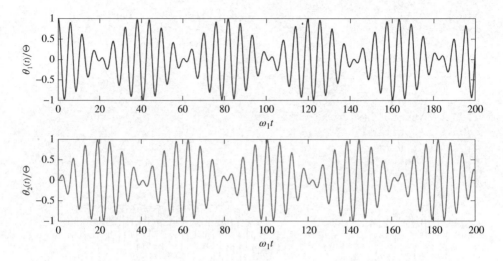

Figure 4.14   Responses of weakly linked pendulum at $\sqrt{k/m} = 0.2\sqrt{g/l}$

We observe from these figures that, at a small γ value (weak springs), it takes many periods to complete a *beat*. As γ increases, the time needed to complete a beat decreases. Vibration, as we recall, is a process in which one type of energy is converted into another, and back and forth. In this set up, the spring connecting the two pendulums can store the potential energy that is accessible by both masses. This provides a channel for the energy from one mass to leak into another mass. The initial conditions store the energy in the first pendulum. After its release, the energy is gradually channeled into the second pendulum. Because the spring is weak, it takes many periods to complete the transfer of the energy from one mass into another. The weaker the spring, the slower the energy transfer rate, and it takes more periods to complete a beat.

The mathematical explanation for this phenomenon is the same as for the case of single-DOF systems subjected to near-resonance excitation. Using the trigonometry identities:

$$\cos\alpha + \cos\beta = 2\cos\frac{\alpha+\beta}{2}\cos\frac{\alpha-\beta}{2}$$

$$\cos\alpha - \cos\beta = 2\sin\frac{\alpha+\beta}{2}\sin\frac{\alpha-\beta}{2}$$

The responses in eqns. (i) and (j) can be rewritten as, using the case $\gamma = 0.1$ as an example,

$$\sqrt{\frac{1+3\gamma}{1+\gamma}} = 1.0871$$

and

$$\bar{\theta}_1(\bar{t}) = \cos 1.0436\bar{t}\cos 0.0436\bar{t} \tag{k}$$

$$\bar{\theta}_2(\bar{t}) = -\sin 1.0436\bar{t}\sin 0.0436\bar{t} \tag{l}$$

The second harmonic factor in each of the above expressions is a much slower varying function, which forms a modulation (graphically an envelope) to the much faster harmonic response dominated by the first factor. The two factors in eqn. (k) is illustrated in Fig. 4.15, which is generated by the following MATLAB code:

```
t = [0:.02:100];
p1 = cos(1.0436*t);
p2 = cos(0.0436*t);
plot(t,p1,':b', t,p2,'g--', t,p1.*p2,'r');
xlabel('\omega_1 t');
ylabel('\theta(t)/\Theta');
```

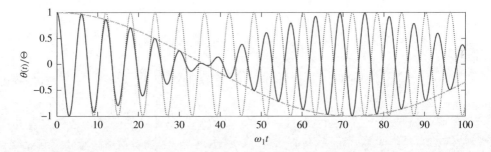

Figure 4.15   Formation of beat (solid) due to two cosinusoidal functions of significantly different frequencies: lower frequency component (dashed) forms modulation over higher frequency component (dotted)

In Fig. 4.15, the dotted and the dashed curves are those two individual factors. When they are multiplied, in the solid blue curve, it looks like a faster oscillation riding on the top of a slower oscillation. A complete beat is formed by a half-period of the slower oscillation.

4.4.2   Free Vibrations of Undamped Unconstrained Systems

In the next, we observe the motion of an unconstrained system, which is also called a *semidefinite system*. For an unconstrained system, its first natural frequency ω_1 is 0, such as in Example 4.5. A subscript 1 is used since 0 is the lowest possible natural frequency, and thus it is always the first mode. This natural frequency corresponds to the so-called rigid-body mode in which there is no relative motion among components of the system; and correspondingly, the mode shape is all 1's in the column matrix.

With $\omega_1 = 0$, the corresponding solutions are $\cos \omega_1 t = 1$ and $\sin \omega_1 t = 0$. In complex notation, $\lambda_{1,2} = 0$ is a repeated root. In such cases, according to the theory of differential equations, the second independent solution is t times the first solution. That is, the temporal factor corresponding to this mode is

$$s_1(t) = a_1 + b_1 t \tag{4.50}$$

which makes a lot of physical sense: $s_1(t)$ represents a rectilinear motion at a constant speed of b_1. Then, the corresponding initial conditions are

$$\{x^{(1)}(0)\} = a_1\{X^{(1)}\} \qquad \{\dot{x}^{(1)}(0)\} = b_1\{X^{(1)}\} \tag{4.51}$$

The procedure we have outlined earlier remains essentially unchanged, except that, after having $\{a\}$ and $\{\beta\}$ determined via eqn. (4.49), $b_1 = \beta_1$ and $x_1(t)$ in eqn. (4.50) is used in $\{s(t)\}$ in eqn. (4.40). We illustrate the solution process for an unconstrained system through the following example.

■ **Example 4.10: Unconstrained System Subjected to Initial Velocity**

Consider the unconstrained mass–spring system in Example 4.5. Assume that the mass on the left is given an initial velocity of v_0. Determine the system's subsequent motion.

□ **Solution:**

In Example 4.5, the natural modes of this system have been found as

$$\text{Mode 1:} \qquad \omega_1 = 0 \qquad \{X^{(1)}\} = \begin{Bmatrix} 1 \\ 1 \end{Bmatrix} \tag{a}$$

$$\text{Mode 2:} \qquad \omega_2 = \sqrt{\frac{2k}{m}} \qquad \{X^{(2)}\} = \begin{Bmatrix} 1 \\ -1 \end{Bmatrix} \tag{b}$$

and the modal matrix and its inversion are

$$[V] = \begin{bmatrix} 1 & 1 \\ 1 & -1 \end{bmatrix} \qquad [V]^{-1} = \frac{1}{2}\begin{bmatrix} 1 & 1 \\ 1 & -1 \end{bmatrix} \tag{c}$$

The initial conditions are

$$\{x(0)\} = \mathbf{0} \qquad \{\dot{x}(0)\} = \begin{Bmatrix} v_0 \\ 0 \end{Bmatrix} \tag{d}$$

Using eqn. (4.49), we found that $\{a\} = 0$, and

$$\{\beta\} = \frac{1}{2}\begin{bmatrix} 1 & 1 \\ 1 & -1 \end{bmatrix}\begin{Bmatrix} v_0 \\ 0 \end{Bmatrix} = \frac{v_0}{2}\begin{Bmatrix} 1 \\ 1 \end{Bmatrix} \tag{e}$$

Then, we have

$$b_1 = \beta_1 = \frac{v_0}{2} \qquad b_2 = \frac{\beta_2}{\omega_2} = \frac{v_0}{2}\sqrt{\frac{m}{2k}} \tag{f}$$

and the responses can be written as, according to eqn. (4.40) and replacing the first entry in $\{s(t)\}$ by eqn. (4.50),

$$\begin{Bmatrix} x_1(t) \\ x_2(t) \end{Bmatrix} = \frac{v_0}{2}\begin{bmatrix} 1 & 1 \\ 1 & -1 \end{bmatrix}\begin{Bmatrix} t \\ \sqrt{\frac{m}{2k}}\sin\sqrt{\frac{2k}{m}}t \end{Bmatrix} \tag{g}$$

or

$$x_1(t) = \frac{v_0}{2}\left(t + \sqrt{\frac{m}{2k}}\sin\sqrt{\frac{2k}{m}}t\right) \tag{h}$$

$$x_2(t) = \frac{v_0}{2}\left(t - \sqrt{\frac{m}{2k}}\sin\sqrt{\frac{2k}{m}}t\right) \tag{i}$$

☐ Exploring the Solution with MATLAB

We define the following nondimensionalization parameters: for $i = 1, 2$,

$$\bar{x}_i = \frac{x_i(t)}{v_0/\omega_2} \qquad \bar{t} = \omega_2 t \tag{j}$$

Note that, since $\omega_1 = 0$, we have to use ω_2 to nondimensionalize the time t. Then, the nondimensionalized responses can be written as

$$\bar{x}_1(\bar{t}) = \frac{1}{2}\left(\bar{t} + \sin\bar{t}\right) \tag{k}$$

$$\bar{x}_2(\bar{t}) = \frac{1}{2}\left(\bar{t} - \sin\bar{t}\right) \tag{l}$$

The following MATLAB code plots the nondimensionalized responses of the two masses, which are shown in Fig. 4.16.

```
t = [0:.05:20];
x1 = .5*(t+sin(t));
x2 = .5*(t-sin(t));
plot(t,x1,'r',t,x2,'--b');
xlabel('\omega_2 t');
ylabel('x_{1,2}(t)/(v_0/\omega_2)');
```

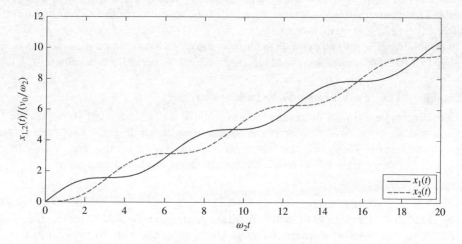

Figure 4.16 Responses of unconstrained two-mass mass–spring system subjected to initial velocity v_0 on the left mass

Recall that for a system that is comprised of particles, the center of the mass of the system can be obtained via averaging the coordinates of the two particles. Thus,

$$x_C(t) = \frac{1}{2}\left[x_1(t) + x_2(t)\right] = \frac{v_0}{2}t \qquad\qquad (m)$$

where x_C is the displacement of the center of the mass of the two-mass system. This means that the center of the masses is moving at a constant velocity that is half of the initial velocity given to the first mass, which, for the system of twice the mass, is the average initial velocity.

4.4.3   Free Vibrations of Systems of Many DOFs

So far we have focused on systems having two or three DOFs. However, the procedures described here are valid for any number of DOFs. The only difference between two-DOF systems and many-DOF systems is that, for the latter, analytical calculation often becomes impractical.

For such systems of more than two DOFs, using matrix notation becomes a requirement. In the matrix theory, the mode shapes we obtained are called the *eigenvectors*, and the corresponding λ's, as we initially introduced in eqn. (4.14), are called the *eigenvalues*. The problem of finding the natural frequencies, specifically, finding the nontrivial solution for eqn. (4.35), is called an *eigenvalue problem*. Using eigenvalue terminology, the frequency equation is called the *characteristic equation* and can be written as

$$\det\left([K] - \lambda[M]\right) = 0 \qquad\qquad (4.52)$$

Note that the natural frequencies are the nonnegative square roots of eigenvalues. This would become obvious by comparing the above equation with eqn. (4.36).

In MATLAB, a built-in function `eig(K,M)` is specifically designed for dealing with eigenvalue problems. This function determines the eigenvalues and eigenvectors and returns the eigenvectors in a square modal matrix and the corresponding eigenvalues λ in a diagonal matrix. The eigenvectors are normalized such that the modulus (or the *second norm* in the matrix theory) of each eigenvector equals 1.

In the following, we use an example to illustrate the procedure outlined in this section being applied to a 10-DOF system and use MATLAB to perform numerical computations.

■ Example 4.11: Free Vibration of 10-DOF System

Consider a system similar to Example 4.6, except that it now consists of 10 masses and 11 springs. It is simply a self-repetition of the configuration shown in Fig. 4.8. All masses and springs are identical. We assume that initially the left-most mass is moved to the right by an amount Δ and then released from rest. Find the subsequent motion of the system.

□ Solution:

We can extend the equations of motion for the system from what we had for the three-DOF system rather than deriving them from scratch. The equations of motion can be written as

$$[M]\{\ddot{x}\} + [K]\{x\} = 0 \qquad\qquad (a)$$

where

$$[M] = \begin{bmatrix} m & 0 & 0 & \cdots & 0 \\ 0 & m & 0 & \cdots & 0 \\ 0 & 0 & m & \cdots & 0 \\ \vdots & \vdots & \vdots & \ddots & \vdots \\ 0 & 0 & 0 & \cdots & m \end{bmatrix} \qquad [K] = \begin{bmatrix} 2k & -k & 0 & \cdots & 0 \\ -k & 2k & -k & \cdots & 0 \\ 0 & -k & 2k & \cdots & 0 \\ \vdots & \vdots & \vdots & \ddots & \vdots \\ 0 & 0 & 0 & \cdots & 2k \end{bmatrix} \qquad \text{(b)}$$

are both 10×10 matrices and

$$\{x\} = \{x_1(t) \quad x_2(t) \quad x_3(t) \quad \cdots \quad x_{10}(t)\}^T \tag{c}$$

The initial conditions can be written as

$$\{x(0)\} = \{\Delta \quad 0 \quad 0 \quad \cdots \quad 0\}^T \qquad \{\dot{x}(0)\} = \mathbf{0} \tag{d}$$

The eigenvalue problem can be written as, according to eqn. (4.36),

$$\left([K] - \omega^2[M]\right)\{X\} = \mathbf{0} \tag{e}$$

We shall use MATLAB to explore the problem and its solutions.

□ **Exploring the Solution with MATLAB**

We need to first define a set of nondimensionalization parameters. In the past, we used a natural frequency to nondimensionalize the time. However, in this problem, we have no idea about the system's natural frequencies, which are often used for the nondimensionalization.

A prominent feature in the system is masses m connected to springs k. From past experiences, we know that $\sqrt{k/m}$ has the unit of a frequency. We can denote it as a nominal frequency $\omega_0 = \sqrt{k/m}$ and use it to nondimensionalize the time t. Thus, we define the following normalization parameters:

$$\bar{x}_i = \frac{x_i}{\Delta} \qquad \bar{t} = \omega_0 t = \sqrt{\frac{k}{m}} t \tag{f}$$

The latter also means that

$$\frac{d\square}{dt} = \frac{d\square}{d\bar{t}} \frac{d\bar{t}}{dt} = \omega_0 \frac{d\square}{d\bar{t}}$$

That is, if we let the overdot to represent the derivative with respect to the nondimensionalized time $\bar{t}$ in the nondimensionalized equations, a factor of ω_0 must be added to every differentiation with respect to time. This way, the equation of motion becomes

$$\omega_0^2[M]\{\ddot{\bar{x}}\} + [K]\{\bar{x}\} = \mathbf{0} \tag{g}$$

Note that every entry in $[M]$ contains m and every entry in $[K]$ contains k. The above equation can be written as

$$\omega_0^2 m[\bar{M}]\{\ddot{\bar{x}}\} + k[\bar{K}]\{\bar{x}\} = \mathbf{0} \tag{h}$$

where

$$[\bar{K}] = \begin{bmatrix} 2 & -1 & 0 & \cdots & 0 \\ -1 & 2 & -1 & \cdots & 0 \\ 0 & -1 & 2 & \cdots & 0 \\ \vdots & \vdots & \vdots & \ddots & \vdots \\ 0 & 0 & 0 & \cdots & 2 \end{bmatrix} \qquad [\bar{M}] = \begin{bmatrix} 1 & 0 & 0 & \cdots & 0 \\ 0 & 1 & 0 & \cdots & 0 \\ 0 & 0 & 1 & \cdots & 0 \\ \vdots & \vdots & \vdots & \ddots & \vdots \\ 0 & 0 & 0 & \cdots & 1 \end{bmatrix} \qquad \text{(i)}$$

Dividing both sides by $m\omega_0^2$ nondimensionalizes the equations of motion to

$$[\overline{M}]\{\ddot{\overline{x}}\} + [\overline{K}]\{\overline{x}\} = 0 \tag{j}$$

which looks almost identical to eqn. (a).

The initial conditions can be written as

$$\{\overline{x}(0)\} = \{1 \quad 0 \quad 0 \quad \cdots \quad 0\}^T \qquad \{\dot{\overline{x}}(0)\} = 0 \tag{k}$$

The following MATLAB code performs all the computations from finding the natural frequencies and mode shapes to finding the responses of the system.

```
% Part 1: System Setup
m=eye(10);
k=2*eye(10);
for i=1:9
    k(i,i+1)=-1;
    k(i+1,i)=-1;
end;

% Part 2: Calculate Eigenvalues and Eigenvectors
[V,lambda]=eig(k,m);
omega=zeros(10,1);
for i=1:10
    omega(i)=sqrt(lambda(i,i));
end;

% Part 3: Set Initial Conditions
x0=zeros(10,1);
v0=zeros(10,1);
x0(1,1) = 1;
a = V\x0;
beta = V\v0;
b = beta ./ omega;

% Part 4: Calculate Responses
t = [0:.1:50];
for i=1:10
    s(i,:) = a(i) .* cos( omega(i) * t) + b(i) .* sin(omega(i)*t);
end;
x = V * s;

% Part 5: Plotting
for i=1:10
    subplot('Position',[.1 1.0-.092*i .85 .08])
    plot(t,x(i,:),'r');
    if i<10 set(gca,'XTickLabel',''); end;
    yname = strcat( 'x_{', int2str(i), '}/\Delta');
    ylabel(yname);
end;
xlabel('\omega_0 t');
```

The code is divided into five parts. The ideas behind each part are explained as follows.

In Part 1, matrices $[\overline{M}]$ and $[\overline{K}]$ are set up. MATLAB function `eye(N)` is used to create an $N \times N$ identity matrices. Then, the nonzero elements of $[K]$ are filled in explicitly.

In Part 2, eigenvalues and eigenvectors are computed. The function `eig(k,m)` returns two matrices: `V` is the square modal matrix and `lambda` is a diagonal matrix whose diagonal elements contain the eigenvalues. The eigenvalues are then converted into natural frequencies and stored in a row matrix named `omega`. The eigenvalues are found to be

$$\omega_n^2 = 0.0810, 0.3175, 0.6903, 1.1692, 1.7154, 2.2846, 2.8308, 3.3097, 3.6825, 3.9190$$

and the corresponding eigenvectors are, in the form of $[V]$ matrix,

$$[V] = \begin{bmatrix}
0.1201 & -0.2305 & -0.3223 & 0.3879 & -0.4221 & 0.4221 \\
0.2305 & -0.3879 & -0.4221 & 0.3223 & -0.1201 & -0.1201 \\
0.3223 & -0.4221 & -0.2305 & -0.1201 & 0.3879 & -0.3879 \\
0.3879 & -0.3223 & 0.1201 & -0.4221 & 0.2305 & 0.2305 \\
0.4221 & -0.1201 & 0.3879 & -0.2305 & -0.3223 & 0.3223 \\
0.4221 & 0.1201 & 0.3879 & 0.2305 & -0.3223 & -0.3223 \\
0.3879 & 0.3223 & 0.1201 & 0.4221 & 0.2305 & -0.2305 \\
0.3223 & 0.4221 & -0.2305 & 0.1201 & 0.3879 & 0.3879 \\
0.2305 & 0.3879 & -0.4221 & -0.3223 & -0.1201 & 0.1201 \\
0.1201 & 0.2305 & -0.3223 & -0.3879 & -0.4221 & -0.4221
\end{bmatrix}$$

$$\begin{bmatrix}
0.3879 & 0.3223 & -0.2305 & 0.1201 \\
-0.3223 & -0.42210 & 0.3879 & -0.2305 \\
-0.1201 & 0.2305 & -0.4221 & 0.3223 \\
0.4221 & 0.1201 & 0.3223 & -0.3879 \\
-0.2305 & -0.3879 & -0.1201 & 0.4221 \\
-0.2305 & 0.3879 & -0.1201 & -0.4221 \\
0.4221 & -0.1201 & 0.3223 & 0.3879 \\
-0.1201 & -0.2305 & -0.4221 & -0.3223 \\
-0.3223 & 0.4221 & 0.3879 & 0.2305 \\
0.3879 & -0.3223 & -0.2305 & -0.1201
\end{bmatrix} \qquad (1)$$

In Part 3, the initial conditions are set up, followed by finding the coefficients a_i and b_i according to eqn. (4.49). The code does not take the possible scenario of zero-natural frequency into account. Accommodating this possibility is rather simple and will be shown in a later example. In this example, we find that $\{\beta\} = 0$ and

$$\{a\} = \begin{Bmatrix}
0.1201 \\
-0.2305 \\
-0.3223 \\
0.3879 \\
-0.4221 \\
0.4221 \\
0.3879 \\
0.3223 \\
-0.2305 \\
0.1201
\end{Bmatrix} \qquad (m)$$

In Part 4, responses of the system are calculated according to eqn. (4.40). It first creates a time array t. It then loops over the 10 modes and calculates $\{\bar{s}(t)\}$ (s in the code) for each mode. At the end, s is a $10 \times N_t$ matrix, where N_t is the number of time instants. This way, each row represents one mass and each column represents a time instant. It is then right-multiplied with the modal matrix v according to eqn. (4.40).

Finally, in Part 5, the responses are plotted. The response for each mass (one row in $\{x(t)\}$ and x in the code) is plotted individually in a subplot for clarity. To reduce the clutter, the $\bar{t}$-axis's tick mark labels as well as axis labels are omitted except for the last subplot. The responses of all the 10 masses for time up to $\bar{t} = 50$ are illustrated in Fig. 4.17.

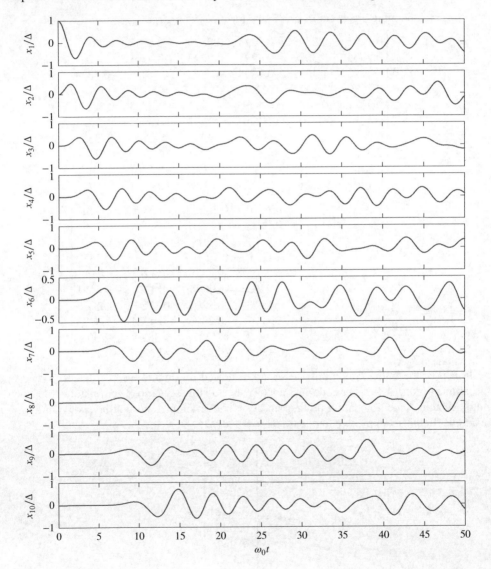

Figure 4.17   Responses of 10-mass mass–spring system to initial displacement Δ at the left-most mass for time up to $\bar{t} = \omega_0 t = 50$

The correctness of the results can be preliminarily verified by the initial conditions: all curves start from a 0 displacement except the first mass, which starts from $\bar{x}_1 = 1$; all curves have zero slope at the beginning, indicating zero initial velocity.

One of the interesting phenomena can be observed from these curves is that there are obvious delays in time when each mass starts to move or feels the disturbance coming from the first mass. Looking at the first peak in mass 10 and backward toward the first mass, we can see that these peaks fall on a straight line. Based on the first-peak times, we can estimate that it takes about a time interval of $\omega_0 \Delta t = 1$ or $\Delta t = 1/\omega_0$ for the disturbance to propagate from one mass to its neighbor. Another interesting observation is that, again, the curves do not exhibit clear periods although the motion is highly oscillatory.

4.4.3.1 To Infinity! And Beyond!!

One of the curious observations from Example 4.11 is that the first natural frequency is $\bar{\omega}_1 = \sqrt{0.0810} = 0.2846$. Recall that we have analyzed two similar but simpler systems. In an earlier case with two masses, the first natural frequency is $\bar{\omega}_1 = 1$, as given in eqn. (4.22). In Example 4.6 with three masses, the first natural frequency is $\bar{\omega}_1 = \sqrt{2 - \sqrt{2}} = 0.586$. Why such a big difference while the systems remain similar, just keep adding the same mass and the same spring?

It is easy to see qualitatively why the way it is. Let us look at the scenario when we add more masses and springs into the system. As we keep adding the masses in this manner, the system's total mass becomes heavier, and it certainly could become more difficult to move; and when it moves, it would become more sluggish.

The imaginary process of adding more masses and springs into the system in a self-similar manner reminds us something that we are familiar with: a rod. Recall that in the previous chapter we have learned that a uniform rod under uniaxial loading behaves similar to a spring, with an equivalent spring constant of

$$k_{\text{equiv}} = \frac{AE}{L} \tag{4.53}$$

where A is the cross-sectional area, E is Young's modulus, and L is the length of the rod. In this sense, our mass–spring model is a lumped-parameter model for an elastic rod whose both ends are fixed. For the two-DOF model, it is similar to dividing the rod into three pieces (because of three springs) and lumping the mass into the four connecting locations, two of which are the walls. This means that the effective mass between the springs is one-half of two adjacent divisions. Working along this line of thoughts, we can correct the parameters used in the 10-DOF mode.

We start from the two-DOF model. Three springs are the results of subdividing the rod into three pieces, that is,

$$k_{\text{2DOF}} = \frac{AE}{L/3} \quad \text{or} \quad \frac{AE}{L} = \frac{k_{\text{2DOF}}}{3}$$

Each of the two masses represents a half of two adjacent pieces, which adds up to one piece. In other words, half mass of the two pieces connecting to the walls is not accounted for the masses; while the remaining mass of the rod is evenly distributed among the two pieces. Denote the total mass of the rod as m. Then,

$$m_{2DOF} = \frac{m}{3} \quad \text{or} \quad m = 3m_{2DOF}$$

Now we can work back to the 10-DOF model. In this model, we divide the rod into 11 pieces, then

$$k_{10DOF} = \frac{AE}{L/11} = 11k_{equiv} \tag{4.54}$$

and

$$m_{10DOF} = \frac{1}{10}\frac{10}{11}m = \frac{1}{11}m \tag{4.55}$$

Now it is clear what equivalent parameters should be if we are to further subdivide the rod into smaller pieces. Assume we want to construct an N-DOF system, then,

$$k_{NDOF} = (N+1)k_{equiv} \tag{4.56}$$

$$m_{NDOF} = \frac{m}{N+1} \tag{4.57}$$

We can construct systems of any number of DOFs that represent the same rod. We can expect that, as the number of DOFs increases, the lumped-parameter model would approach the continuous rod.

■ Example 4.12: Modeling to Continuum Limit

We use the natural modes and natural frequencies to verify our expectation: as the number of DOFs increases, the lumped-parameter models would approach a limit. Specifically, we want to observe the first and the fifth modes of these models while we keep increasing the number of DOFs in the system, N. Assume the mass of the rod is m.

□ Solution:

The equations of motion for the case of N-DOF are

$$[M]\{\ddot{x}\} + [K]\{x\} = \{0\} \tag{a}$$

where

$$[K] = (N+1)k_{equiv} \begin{bmatrix} 2 & -1 & 0 & \cdots & 0 \\ -1 & 2 & -1 & \cdots & 0 \\ 0 & -1 & 2 & \cdots & 0 \\ \vdots & \vdots & \vdots & \ddots & \vdots \\ 0 & 0 & 0 & \cdots & 2 \end{bmatrix} \tag{b}$$

$$[M] = \frac{m}{N+1} \begin{bmatrix} 1 & 0 & 0 & \cdots & 0 \\ 0 & 1 & 0 & \cdots & 0 \\ 0 & 0 & 1 & \cdots & 0 \\ \vdots & \vdots & \vdots & \ddots & \vdots \\ 0 & 0 & 0 & \cdots & 1 \end{bmatrix} \tag{c}$$

where $[K]$ and $[M]$ are both $N \times N$ matrices. We will explore the mode shapes of this ever growing system numerically using MATLAB.

□ Exploring the Solution with MATLAB

We can define the following nondimensionalization parameters, for $i = 1, 2, \ldots,$

$$\overline{\omega}_i = \frac{\omega_i}{\sqrt{k_{equiv}/m}} \qquad \overline{x}_0 = \frac{x_i}{X_0} \qquad \overline{t} = \sqrt{\frac{k_{equiv}}{m}} t \qquad \text{(d)}$$

where X_0 is an imagined displacement amplitude. Then, the equations of motion become

$$[\overline{M}]\{\ddot{\overline{x}}\} + [\overline{K}]\{\overline{x}\} = \{0\} \qquad \text{(e)}$$

where

$$[\overline{K}] = (N+1)^2 \begin{bmatrix} 2 & -1 & 0 & \cdots & 0 \\ -1 & 2 & -1 & \cdots & 0 \\ 0 & -1 & 2 & \cdots & 0 \\ \vdots & \vdots & \vdots & \ddots & \vdots \\ 0 & 0 & 0 & \cdots & 2 \end{bmatrix} \qquad [\overline{M}] = \begin{bmatrix} 1 & 0 & 0 & \cdots & 0 \\ 0 & 1 & 0 & \cdots & 0 \\ 0 & 0 & 1 & \cdots & 0 \\ \vdots & \vdots & \vdots & \ddots & \vdots \\ 0 & 0 & 0 & \cdots & 1 \end{bmatrix} \qquad \text{(f)}$$

The following MATLAB code calculates the natural frequencies and mode shapes for an N-DOF system.

```
% System Setup
N = 150;               % Number of DOF's
k=2*eye(N,N);
m=eye(N,N);
for i=1:N-1
    k(i,i+1)=-1;
    k(i+1,i)=-1;
end;
k = (N+1)^2 * k;

% Calculate Eigenvalues and Eigenvectors
[V,lambda]=eig(k,m);
omega(1)=sqrt(lambda(1,1));
omega(2)=sqrt(lambda(5,5));
mode1 = [0, V(:,1)', 0];       % Pad two ends of each mode shape with 0
mode5 = [0, V(:,5)', 0];

% Plotting the mode shapes. Need to construct an array for x-axis
x=0:1/(N+1):1;
subplot(2,1,1);
plot(x,mode1);
ylabel('Magnitude');
xlabel('x');
subplot(2,1,2);
plot(x,mode5);
ylabel('Magnitude');
xlabel('x');
```

Table 4.1   Natural Frequencies for Modes 1 and 5

DOF	ω_1	ω_5
5	3.1058	11.5911
10	3.1309	14.4069
15	3.1365	15.0847
20	3.1387	15.3443
30	3.1402	15.5405
50	3.1411	15.6459
70	3.1413	15.6759
100	3.1415	15.6921
150	3.1415	15.7009
200	3.1416	15.7040
300	3.1416	15.7062
500	3.1416	15.7073
700	3.1416	15.7076
1000	3.1416	15.7078

Most of the structure in the code is almost the same as the one in the previous example. The only difference is that an adjustable size parameter N is used. When plotting the mode shapes, we would like the mode shape to extend from wall to wall. Hence, the eigenvectors are padded with zeros to represent the end walls when they are assigned to mode1 and mode5.

Table 4.1 shows the values of the first and fifth natural frequencies obtained by models of different number of DOFs. Figures 4.18 and 4.19 show the mode shapes of Modes 1 and 5, respectively.

From Table 4.1, we can see that, for Mode 1, the natural frequency reaches a limiting value when the number of DOFs is in the order of ∼100; whereas for Mode 5, it needs ∼1000 DOFs to reach the same level of stability. For the mode shapes, as the number of DOFs increases, the shape becomes smoother. It takes far less DOFs to reach a smooth mode shape. Note that in the plots for the mode shapes, with different numbers of DOFs, the amplitudes are different. This is due to the normalization performed by MATLAB: it is normalized by the *second norm*,

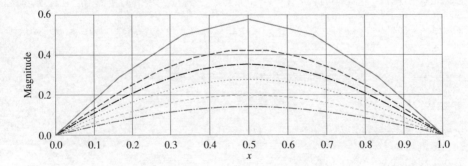

Figure 4.18   Mode shapes for Mode 1 obtained via lumped-parameter models of different number of DOFs. Mode shapes of amplitudes from large to small correspond to the number of DOF = 5, 10, 15, 25, 50, and 100. The amplitude differences are due to normalization process

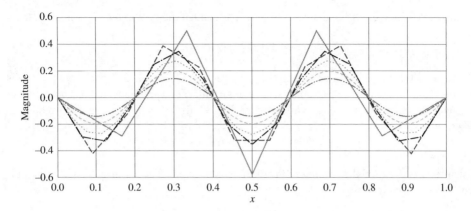

Figure 4.19 Mode shapes for Mode 5 obtained via lumped-parameter models of different number of DOFs. Mode shapes of amplitudes from large to small correspond to the number of DOF = 5, 10, 15, 25, 50, 100. The amplitude differences are due to normalization process

that is, the square root of the sum of squares of the individual elements; as the number of DOFs increases, the number of elements increases. But, for mode shapes, what matters is the shape, not the amplitude. If needed, we can renormalize them by the *infinite norm*, which is the absolute maximum value among elements.

4.5 Eigenvalues and Eigenvectors

We have mentioned in the previous section that the problem of finding the natural frequencies is called an *eigenvalue problem* in the matrix theory. We shall review this part of the matrix theory in this section.

4.5.1 Standard Eigenvalue Problem

In the context of our discussion, we shall limit an eigenvalue problem to the one that is associated with a square matrix, say $[A]$. An $n \times n$ square matrix represents a linear transformation in the n-dimension vector space. If there exists a nontrivial vector $\{X\}$ such that the transformed vector $[A]\{X\}$ is parallel to the original vector $\{X\}$, but not necessarily the same size, say, scaled by a scalar λ, that is,

$$[A]\{X\} = \lambda\{X\} \qquad (4.58)$$

then $\{X\}$ is called an *eigenvector* of the matrix $[A]$ and the corresponding scalar λ is called the *eigenvalue*.

Equation (4.58) can be rearranged as a set of linear equations as

$$([A] - \lambda[I])\{X\} = \mathbf{0} \qquad (4.59)$$

where $[I]$ is the *identity matrix*, a diagonal matrix with all elements being 1. It is also called the *unit matrix*. Equation (4.59) has nontrivial solutions *if and only if* the determinant of the square matrix on the left-hand side in front of $\{X\}$, called the *system matrix*, vanishes, that is,

$$\det([A] - \lambda[I]) = 0 \tag{4.60}$$

Equation (4.60) is an algebraic equation of order n, called the *characteristic equation*. It has n roots. Consequently, an $n \times n$ matrix has n eigenvalues; each eigenvalue has an associated eigenvector. Furthermore, for a symmetric matrix $[A]$, these eigenvectors are normal to each other, that is,

$$\{X^{(i)}\}^T\{X^{(j)}\} = 0 \quad \text{unless} \quad i = j$$

This property is called the *orthogonality* of the eigenvectors. The eigenvector $\{X^{(i)}\}$ is said to have been *normalized* if

$$\{X^{(i)}\}^T\{X^{(i)}\} = 1$$

There are cases when the characteristic equation has repeated roots. In such a case, there is an infinite number of possible eigenvectors; and it is always possible to choose a set of vectors that are normal to each other as the eigenvectors.

Because the eigenvectors are normal to each other, they are also *independent*. This makes the n eigenvectors a *complete set*, meaning that any vector in the n-dimensional space can be expressed as a linear combination of these eigenvectors. As three mutually perpendicular unit vectors in a Cartesian coordinate system, any vector in the Cartesian space is expressible as a linear combination of these three unit vectors.

4.5.2 Generalized Eigenvalue Problem

Our equation that leads to the finding of natural frequencies, eqn. (4.35), is repeated as the following:

$$\left([K] - \omega^2[M]\right)\{X\} = 0 \tag{4.61}$$

Comparing eqn. (4.61) with the standard eigenvalue problem in eqn. (4.58), we can see that we have a matrix $[M]$ in place of the identity matrix. Although in some of our examples of two-DOF systems, we might be able to normalize the $[M]$ matrix into an identity matrix, this is generally not the case. In such cases, the problem is called a *generalized eigenvalue problem*.

The process of finding the eigenvalues and eigenvectors for such a system remains the same as the standard eigenvalue problem. However, the properties of the eigenvectors are slightly different. The eigenvectors are normal to each other *with respect to* either $[M]$ or $[K]$ matrix; that is, unless $i = j$,

$$\{X^{(i)}\}^T[M]\{X^{(j)}\} = 0 \quad \text{and} \quad \{X^{(i)}\}^T[K]\{X^{(j)}\} = 0 \tag{4.62}$$

The orthogonality of eigenvectors requires that matrices $[M]$ and $[K]$ be symmetric. For a linear system, the symmetry is generally satisfied, at least fundamentally, although in appearance the symmetry can be easily "destroyed." For example, if we divide eqn. (4.2) by 2 yet maintain eqn. (4.1) unchanged, we will end up with apparently nonsymmetric $[M]$ and $[K]$ matrices. It can be shown that if we choose the generalized coordinates such that they all are measured from an equilibrium configuration, both $[M]$ and $[K]$ will be symmetric.

4.6   Coupling, Decoupling, and Principal Coordinates

So far, we have not considered any system having damping. When the damping presents, solving a set of coupled second-order differential equations becomes complicated and beyond our mathematical preparation. However, with the understanding of the eigenvalue problems, we would be able to analyze some damped systems.

4.6.1   Types of Coupling

Typically, for a multi-DOF system, each equation of motion involves several generalized coordinates. For example, in the two-mass system we considered at the beginning of this chapter, the equation of motion for x_1, eqn. (4.1), involves a kx_2 term, and the x_2-equation, eqn. (4.2), involves kx_1. Such two equations are said to be *coupled*. Correspondingly, the system is said to be *coupled*, too.

   In matrix notation, these coupling terms appear as off-diagonal elements in $[M]$ and $[K]$ matrices. If any off-diagonal element appears in $[K]$ matrix, the system is *statically coupled*; if any off-diagonal element appears in $[M]$ matrix, the system is *dynamically coupled*. Many systems are both statically and dynamically coupled.

   Imagine, for a moment, an undamped system: if there are no off-diagonal elements in both $[M]$, $[K]$, then the equations of motion for the system would appear as a set of uncoupled equations, one for each generalized coordinate. We can then treat each equation (or generalized coordinate) as a single-DOF system. In single-DOF systems, we should be able to handle damping.

   Well, this is not a fantasy. This is possible because of the orthogonality of the natural modes. We can bring this idea to work.

4.6.2   Principal Coordinates

Assume that a system is described by the following matrix equation of motion

$$[M]\{\ddot{x}\} + [C]\{\dot{x}\} + [K]\{x\} = \{F\} \tag{4.63}$$

where, for generality, we have included a forcing term $\{F\}$ on the right-hand side. Further assume that we have solved the eigenvalue problem for the system as described by the generalized coordinates we had chosen initially, have found the mode shapes, and have assembled the mode shapes into the modal matrix $[V]$ as in eqn. (4.41). Then, we define a new set of generalized coordinates $\{y\}$ such that

$$\{x\} = [V]\{y\} \tag{4.64}$$

Noting that $[V]$ does not change with time, the matrix equation of motion would become

$$[M][V]\{\ddot{y}\} + [C][V]\{\dot{y}\} + [K][V]\{y\} = \{F\} \tag{4.65}$$

Left-multiplying both sides by $[V]^T$ gives

$$[V]^T[M][V]\{\ddot{y}\} + [V]^T[C][V]\{\dot{y}\} + [V]^T[K][V]\{y\} = [V]^T\{F\} \tag{4.66}$$

Now, because of the orthogonality of mode shapes in eqn. (4.62), matrix products $[V]^T[M][V]$ and $[V]^T[K][V]$ are diagonal matrices, which are denoted as

$$[M_d] = [V]^T[M][V] \quad \text{and} \quad [K_d] = [V]^T[K][V] \tag{4.67}$$

where a subscript d is used to signify that the matrix is diagonal.

If the system has no damping ($[C] = 0$), we have decoupled the equations as

$$[M_d]\{\ddot{y}\} + [K_d]\{y\} = [V]^T\{F\} \tag{4.68}$$

If the damping presents but can be expressed as (or approximated by) the following:

$$[C] = c_M[M] + c_K[K] \tag{4.69}$$

that is, if the damping matrix $[C]$ is proportional to either $[M]$ or $[K]$ matrices or their linear combinations, we can easily see that the damping terms are also decoupled, as the matrix equation of motion becomes

$$[M_d]\{\ddot{y}\} + (c_M[M_d] + c_K[K_d])\{\dot{y}\} + [K_d]\{y\} = [V]^T\{f\} \tag{4.70}$$

Damping of the form in eqn. (4.69) is called the *proportional damping* or the *Rayleigh damping*.

If the initial conditions are given as

$$\{x(0)\} = \{x_0\} \quad \{\dot{x}(0)\} = \{v_0\}$$

substituting eqn. (4.64) into these initial conditions, the initial conditions for the new set of generalized coordinates $\{y\}$ can be written as

$$\{y(0)\} = [V]^{-1}\{x_0\} \quad \{\dot{y}(0)\} = [V]^{-1}\{v_0\} \tag{4.71}$$

Now that all the equations have been decoupled and initial conditions have been transformed, we can solve them individually, and merrily, as a set of single-DOF systems. Once the solutions for $\{y\}$ are obtained, eqn. (4.64) can be used to reexpress the solutions back in the original generalized coordinates $\{x\}$.

The remaining question is how to obtain $\{y\}$. A simple way is to inverse the model matrix according to eqn. (4.64),

$$\{y\} = [V]^{-1}\{x\} \tag{4.72}$$

In reality, we rarely need to know this explicitly. But, conceptually, it is important to know that this set of generalized coordinates $\{y\}$ is called the *principal coordinates*, as it represents the system's mode shapes. It is also called the *modal coordinates*. The decoupled equations of motion are also called the *modal equations of motion*.

■ Example 4.13: Decoupling of Undamped System

In this example, we demonstrate the process of decoupling the equations of motion of the undamped mass–spring system that we have analyzed at the beginning of this chapter, and it was used for free-vibration analysis in Example 4.7.

□ Solution:

The equations of motion, given earlier in eqn. (4.5), are repeated here for convenience:

$$\begin{bmatrix} m & 0 \\ 0 & m \end{bmatrix} \begin{Bmatrix} \ddot{x}_1 \\ \ddot{x}_2 \end{Bmatrix} + \begin{bmatrix} 2k & -k \\ -k & 2k \end{bmatrix} \begin{Bmatrix} x_1 \\ x_2 \end{Bmatrix} = \begin{Bmatrix} F_1 \\ F_2 \end{Bmatrix} \tag{a}$$

where, for generality, we have added the forcing term on the right-hand side. The mode shapes have been found in eqns. (4.32) and (4.33). The modal matrix and its inverse can be written as

$$[V] = [V]^T = \begin{bmatrix} 1 & 1 \\ 1 & -1 \end{bmatrix} \qquad [V]^{-1} = \frac{1}{2} \begin{bmatrix} 1 & 1 \\ 1 & -1 \end{bmatrix} \tag{b}$$

We define, according to eqn. (4.64), a new set of generalized coordinates $\{y\}$ such that $\{y\} = [V]\{x\}$. Then,

$$[M_d] = [V]^T[M][V] = \begin{bmatrix} 1 & 1 \\ 1 & -1 \end{bmatrix} \begin{bmatrix} m & 0 \\ 0 & m \end{bmatrix} \begin{bmatrix} 1 & 1 \\ 1 & -1 \end{bmatrix} = \begin{bmatrix} 2m & 0 \\ 0 & 2m \end{bmatrix} \tag{c}$$

$$[K_d] = [V]^T[K][V] = \begin{bmatrix} 1 & 1 \\ 1 & -1 \end{bmatrix} \begin{bmatrix} 2k & -k \\ -k & 2k \end{bmatrix} \begin{bmatrix} 1 & 1 \\ 1 & -1 \end{bmatrix} = \begin{bmatrix} 2k & 0 \\ 0 & 6k \end{bmatrix} \tag{d}$$

Following eqn. (4.66), we can write the matrix equation of motion in the new generalized coordinates as

$$\begin{bmatrix} 2m & 0 \\ 0 & 2m \end{bmatrix} \begin{Bmatrix} \ddot{y}_1 \\ \ddot{y}_2 \end{Bmatrix} + \begin{bmatrix} 2k & 0 \\ 0 & 6k \end{bmatrix} \begin{Bmatrix} y_1 \\ y_2 \end{Bmatrix} = \begin{bmatrix} 1 & 1 \\ 1 & -1 \end{bmatrix} \begin{Bmatrix} F_1 \\ F_2 \end{Bmatrix} = \begin{Bmatrix} F_1 + F_2 \\ F_1 - F_2 \end{Bmatrix} \tag{e}$$

and we have completely decoupled the equations of motion.

Let us verify, just for those curious among us. We write the original equations as

$$m\ddot{x}_1 + 2kx_1 - kx_2 = F_1 \tag{f}$$

$$m\ddot{x}_2 + 2kx_2 - kx_1 = F_2 \tag{g}$$

Adding and subtracting with each other, respectively, give

$$m(\ddot{x}_1 + \ddot{x}_1) + k(x_1 + x_2) = F_1 + F_2 \tag{h}$$

$$m(\ddot{x}_1 - \ddot{x}_1) + 3k(x_1 - x_2) = F_1 - F_2 \tag{i}$$

Recall the expression for $\{y\}$ in eqn. (4.72)

$$\{y\} = [V]^{-1}\{x\} = \frac{1}{2} \begin{bmatrix} 1 & 1 \\ 1 & -1 \end{bmatrix} \begin{Bmatrix} x_1 \\ x_2 \end{Bmatrix} = \begin{Bmatrix} \frac{1}{2}(x_1 + x_2) \\ \frac{1}{2}(x_1 - x_2) \end{Bmatrix} \tag{j}$$

Substituting the definition for y_1 and y_2 above into eqns. (h) and (i) gives

$$2m\ddot{y}_1 + 2ky_1 = F_1 + F_2 \tag{k}$$

$$2m\ddot{y}_2 + 6ky_2 = F_1 - F_2 \tag{l}$$

We can see that they are exactly the same as those in matrix form in eqn. (e).

4.6.3   Decoupling Method for Free-Vibration Analysis

For free-vibration analyses using the decoupling method, after transforming the equations of motion into a decoupled set in terms of the principal coordinates, the initial conditions are similarly transformed, as shown in eqn. (4.71). Afterward, each decoupled equation is solved separately. Finally, eqn. (4.64) is used to transform the solutions back to the original set of generalized coordinates.

In the following, we demonstrate, through examples, the process of solving free-vibration problems using the decoupling method.

■ **Example 4.14:   Free Vibration of Undamped System via Decoupling Method**

Use the decoupling method to analyze Example 4.7, which is a free vibration of the system in Example 4.13, with initial conditions $x(0) = \Delta$ and $\dot{x}(0) = 0$.

□ **Solution:**

From Example 4.13, the modal matrix for the system described by x_1 and x_2 is

$$[V] = \begin{bmatrix} 1 & 1 \\ 1 & -1 \end{bmatrix} \quad \text{and} \quad [V]^{-1} = \frac{1}{2}\begin{bmatrix} 1 & 1 \\ 1 & -1 \end{bmatrix} \tag{a}$$

The principal coordinates are

$$\{y\} = \left\{ \begin{array}{c} \frac{1}{2}(x_1 + x_2) \\ \frac{1}{2}(x_1 - x_2) \end{array} \right\} \tag{b}$$

and the equations of motion in terms of the principal coordinates are

$$\begin{bmatrix} 2m & 0 \\ 0 & 2m \end{bmatrix} \left\{ \begin{array}{c} \ddot{y}_1 \\ \ddot{y}_2 \end{array} \right\} + \begin{bmatrix} 2k & 0 \\ 0 & 6k \end{bmatrix} \left\{ \begin{array}{c} y_1 \\ y_2 \end{array} \right\} = \left\{ \begin{array}{c} 0 \\ 0 \end{array} \right\} \tag{c}$$

where the right-hand side has been set back to zero because this example deals with free vibrations of the system.

The initial conditions given in the original set of generalized coordinates are

$$\{x(0)\} = \left\{ \begin{array}{c} \Delta \\ 0 \end{array} \right\} \qquad \{\dot{x}(0)\} = \mathbf{0} \tag{d}$$

which can be expressed in the principal coordinates according to eqn. (4.71) as

$$\{y(0)\} = \frac{1}{2}\begin{bmatrix} 1 & 1 \\ 1 & -1 \end{bmatrix} \left\{ \begin{array}{c} \Delta \\ 0 \end{array} \right\} = \frac{\Delta}{2}\left\{ \begin{array}{c} 1 \\ 1 \end{array} \right\} \qquad \{\dot{y}(0)\} = \mathbf{0} \tag{e}$$

Therefore, the original coupled two-DOF problem becomes the following two decoupled single-DOF problems:

$$m\ddot{y}_1 + ky_1 = 0 \qquad \text{with initial conditions} \quad y_1(0) = \frac{1}{2}\Delta, \quad \dot{y}_1(0) = 0 \tag{f}$$

$$m\ddot{y}_2 + 3ky_2 = 0 \qquad \text{with initial conditions} \quad y_2(0) = \frac{1}{2}\Delta, \quad \dot{y}_2(0) = 0 \tag{g}$$

The solutions to these two problems are almost trivial, and we directly write

$$y_1(t) = \frac{\Delta}{2} \cos \sqrt{\frac{k}{m}} t \tag{h}$$

$$y_2(t) = \frac{\Delta}{2} \cos \sqrt{\frac{3k}{m}} t \tag{i}$$

Finally, in terms of the original set of generalized coordinates x_1 and x_2, according to eqn. (4.64)

$$\left\{ \begin{matrix} x_1(t) \\ x_2(t) \end{matrix} \right\} = [V]\{y(t)\} = \begin{bmatrix} 1 & 1 \\ 1 & -1 \end{bmatrix} \left\{ \begin{matrix} \frac{\Delta}{2} \cos \sqrt{\frac{k}{m}} t \\ \frac{\Delta}{2} \cos \sqrt{\frac{3k}{m}} t \end{matrix} \right\} \tag{j}$$

or, in expanded form,

$$x_1(t) = \frac{\Delta}{2} \left(\cos \sqrt{\frac{k}{m}} t + \cos \sqrt{\frac{3k}{m}} t \right) \tag{k}$$

$$x_2(t) = \frac{\Delta}{2} \left(\cos \sqrt{\frac{k}{m}} t - \cos \sqrt{\frac{3k}{m}} t \right) \tag{l}$$

which are identical to the solution we found in Example 4.7.

■ Example 4.15: Response of Damped System Subjected to Initial Displacement

In this example, we demonstrate the decoupling method for a damped system. We consider again the system of two masses connected by three springs. In addition, there are three dashpots in the system: two dashpots, of dashpot constant c_1, connect the masses to their respective nearby walls and the third dashpot, of dashpot constant c_2, connects the two masses, as shown in Fig. 4.20. Initially the system is at rest, and the mass on the left is subjected to a displacement of Δ and then released. Find the subsequent motion.

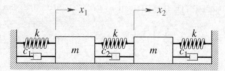

Figure 4.20   Two-mass mass–spring–dashpot system

□ **Solution:**

Without taking a digression, we directly write the equations of motion for the system as

$$\begin{bmatrix} m & 0 \\ 0 & m \end{bmatrix} \left\{ \begin{matrix} \ddot{x}_1 \\ \ddot{x}_2 \end{matrix} \right\} + \begin{bmatrix} c_1 + c_2 & -c_2 \\ -c_2 & c_1 + c_2 \end{bmatrix} \left\{ \begin{matrix} \dot{x}_1 \\ \dot{x}_2 \end{matrix} \right\} + \begin{bmatrix} 2k & -k \\ -k & 2k \end{bmatrix} \left\{ \begin{matrix} x_1 \\ x_2 \end{matrix} \right\} = \left\{ \begin{matrix} 0 \\ 0 \end{matrix} \right\} \tag{a}$$

The verification of these equations of motion is left for the readers.

Note that adding the damping does not alter the natural frequencies and mode shapes of the system. In fact, it is required to remove the damping terms in order to perform the modal analysis. Therefore, the decoupling process is the same as in Example 4.13. The natural frequencies have been found in eqn. (4.22) as

$$\omega_1 = \sqrt{\frac{k}{m}} \qquad \omega_2 = \sqrt{\frac{3k}{m}} \tag{b}$$

and the modal matrix and its inverse remain as

$$[V] = \begin{bmatrix} 1 & 1 \\ 1 & -1 \end{bmatrix} \qquad \text{and} \qquad [V]^{-1} = \frac{1}{2}\begin{bmatrix} 1 & 1 \\ 1 & -1 \end{bmatrix} \tag{c}$$

and the principal coordinates are

$$\{y\} = \left\{ \begin{array}{c} \frac{1}{2}(x_1 + x_2) \\ \frac{1}{2}(x_1 - x_2) \end{array} \right\} \tag{d}$$

In this example, the damping matrix $[C]$ can be written as

$$[C] = \begin{bmatrix} c_1 - c_2 & 0 \\ 0 & c_1 - c_2 \end{bmatrix} + \begin{bmatrix} 2c_2 & -c_2 \\ -c_2 & 2c_2 \end{bmatrix} \tag{e}$$

where the first component is proportional to the $[M]$ matrix and the second component is proportional to the $[K]$ matrix. This shows that the system has a proportional damping. We can use this proportionality to write the diagonalized damping matrix $[C_d]$, based on the diagonalized $[M_d]$ and $[K_d]$ matrices in eqns. (c) and (d) in Example 4.13,

$$[C_d] = \begin{bmatrix} 2(c_1 - c_2) & 0 \\ 0 & 2(c_1 - c_2) \end{bmatrix} + \begin{bmatrix} 2c_2 & 0 \\ 0 & 6c_2 \end{bmatrix} = \begin{bmatrix} 2c_1 & 0 \\ 0 & 2c_1 + 4c_2 \end{bmatrix}$$

Then the equations of motion in terms of principal coordinates are

$$\begin{bmatrix} 2m & 0 \\ 0 & 2m \end{bmatrix}\left\{ \begin{array}{c} \ddot{y}_1 \\ \ddot{y}_2 \end{array} \right\} + \begin{bmatrix} 2c_1 & 0 \\ 0 & 2c_1 + 4c_2 \end{bmatrix}\left\{ \begin{array}{c} \dot{y}_1 \\ \dot{y}_2 \end{array} \right\} + \begin{bmatrix} 2k & 0 \\ 0 & 6k \end{bmatrix}\left\{ \begin{array}{c} y_1 \\ y_2 \end{array} \right\} = \left\{ \begin{array}{c} 0 \\ 0 \end{array} \right\} \tag{f}$$

The initial conditions in terms of the principal coordinates are, according to eqn. (4.71),

$$\{y(0)\} = \frac{\Delta}{2}\left\{ \begin{array}{c} 1 \\ 1 \end{array} \right\} \qquad \{\dot{y}(0)\} = \{0\} \tag{g}$$

The two decoupled systems can be written as

$$2m\ddot{y}_1 + 2c_1\dot{y}_1 + 2ky_1 = 0 \qquad \text{with} \quad y_1(0) = \frac{1}{2}\Delta, \quad \dot{y}_1(0) = 0 \tag{h}$$

$$2m\ddot{y}_2 + (2c_1 + 4c_2)\dot{y}_2 + 6ky_2 = 0 \qquad \text{with} \quad y_2(0) = \frac{1}{2}\Delta, \quad \dot{y}_2(0) = 0 \tag{i}$$

Alternatively, the decoupled equations can be written in normalized form as

$$\ddot{y}_1 + 2\zeta_1\omega_1\dot{y}_1 + \omega_1^2 y_1 = 0 \tag{j}$$

$$\ddot{y}_2 + 2\zeta_2\omega_2\dot{y}_2 + \omega_2^2 y_2 = 0 \tag{k}$$

where

$$\omega_1 = \sqrt{\frac{k}{m}} \qquad \zeta_1 = \frac{c_1}{2m\omega_1} = \frac{c_1}{2\sqrt{km}} \tag{l}$$

$$\omega_2 = \sqrt{\frac{3k}{m}} \qquad \zeta_2 = \frac{c_1 + 2c_2}{2m\omega_2} = \frac{c_1 + 2c_2}{2\sqrt{3km}} \tag{m}$$

It is no coincidence that the natural frequencies of the two decoupled equations are the same as natural frequencies of the original system. Natural frequencies are intrinsic characteristics of the system that do not change when we change our way of observing the system.

Assuming the system is underdamped in both modes, the two single-DOF systems can be easily solved. Here, instead of going into details, we adapt the solution we obtained in eqn. (f) in Example 2.2, in which an underdamped single-DOF system is subjected to an initial displacement of $-\Delta$, and the solution is given in eqn. (f) in that example. Thus,

$$y_1(t) = \frac{\Delta}{2} \frac{e^{-\zeta_1\omega_1 t}}{\sqrt{1 - \zeta_1^2}} \cos\left(\omega_{1d}t - \phi_1\right) \tag{n}$$

$$y_2(t) = \frac{\Delta}{2} \frac{e^{-\zeta_2\omega_2 t}}{\sqrt{1 - \zeta_2^2}} \cos\left(\omega_{2d}t - \phi_2\right) \tag{o}$$

where

$$\omega_{1d} = \sqrt{1 - \zeta_1^2}\omega_1 \qquad \text{and} \qquad \omega_{2d} = \sqrt{1 - \zeta_2^2}\omega_2 \tag{p}$$

are the damped natural frequencies for Modes 1 and 2; and

$$\phi_1 = \tan^{-1}\frac{\zeta_1}{\sqrt{1 - \zeta_1^2}} \qquad \phi_2 = \tan^{-1}\frac{\zeta_2}{\sqrt{1 - \zeta_2^2}} \tag{q}$$

Converting back to the original set of generalized coordinates according to eqn. (4.64) gives

$$x_1(t) = \frac{\Delta}{2}\left[\frac{e^{-\zeta_1\omega_1 t}}{\sqrt{1 - \zeta_1^2}} \cos\left(\omega_{1d}t - \phi_1\right) + \frac{e^{-\zeta_2\omega_2 t}}{\sqrt{1 - \zeta_2^2}} \cos\left(\omega_{2d}t - \phi_2\right)\right] \tag{r}$$

$$x_2(t) = \frac{\Delta}{2}\left[\frac{e^{-\zeta_1\omega_1 t}}{\sqrt{1 - \zeta_1^2}} \cos\left(\omega_{1d}t - \phi_1\right) - \frac{e^{-\zeta_2\omega_2 t}}{\sqrt{1 - \zeta_2^2}} \cos\left(\omega_{2d}t - \phi_2\right)\right] \tag{s}$$

□ **Discussion:**

In many cases, the damping being proportional such as shown in eqn. (e) may not be obvious in the first sight, unless we squint really hard until it hurts our eyes. A brute-force approach is much more straightforward but with only slightly more work: we simply go ahead to carry out the matrix multiplication of $[V]^T[C][V]$. If it turns out to be a diagonal matrix, then it is a proportional damping. But the end result is still useful. Doing so gives

$$[V]^T[C][V] = \begin{bmatrix} 1 & 1 \\ 1 & -1 \end{bmatrix} \begin{bmatrix} c_1 + c_2 & -c_2 \\ -c_2 & c_1 + c_2 \end{bmatrix} \begin{bmatrix} 1 & 1 \\ 1 & -1 \end{bmatrix} = \begin{bmatrix} 2c_1 & 0 \\ 0 & 2c_1 + 4c_2 \end{bmatrix}$$

☐ **Exploring the Solution with MATLAB**

We define the following nondimensionalization parameters: for $i = 1, 2$,

$$\bar{x}_i = \frac{x_i}{\Delta} \qquad \bar{t} = \omega_1 t \qquad \omega_2 t = \sqrt{3}\bar{t} \tag{t}$$

The nondimensionalized response of the masses can be written as

$$\bar{x}_1(\bar{t}) = \frac{1}{2} \left[\frac{e^{-\zeta_1 \bar{t}}}{\sqrt{1 - \zeta_1^2}} \cos\left(\sqrt{1 - \zeta_1^2}\,\bar{t} - \phi_1\right) + \frac{e^{-\sqrt{3}\zeta_2 \bar{t}}}{\sqrt{1 - \zeta_2^2}} \cos\left(\sqrt{3(1 - \zeta_2^2)}\,\bar{t} - \phi_2\right) \right] \tag{u}$$

$$\bar{x}_2(\bar{t}) = \frac{1}{2} \left[\frac{e^{-\zeta_1 \bar{t}}}{\sqrt{1 - \zeta_1^2}} \cos\left(\sqrt{1 - \zeta_1^2}\,\bar{t} - \phi_1\right) - \frac{e^{-\sqrt{3}\zeta_2 \bar{t}}}{\sqrt{1 - \zeta_2^2}} \cos\left(\sqrt{3(1 - \zeta_2^2)}\,\bar{t} - \phi_2\right) \right] \tag{v}$$

In exploring the solution, it might be more helpful to think of the relation between the two dashpots. It is noted that ζ_1 is entirely due to dashpot constant c_1, but ζ_2 is due to the contribution of both c_1 and c_2. Keeping ζ_1, we can write ζ_2 as

$$\zeta_2 = \frac{c_1 + 2c_2}{m\omega_2} = \frac{c_1 + 2c_2}{\sqrt{3}m\omega_1} = \frac{\zeta_1}{\sqrt{3}}\left(1 + 2\frac{c_2}{c_1}\right) = \frac{1 + 2r_c}{\sqrt{3}}\zeta_1 \tag{w}$$

where

$$r_c = \frac{c_2}{c_1}$$

The following MATLAB code implements the solution and plots the responses:

```
% System Setup
zeta1 = 0.1;              % Damping ratio for Mode 1
rc = 0;                   % Ratio of two dashpots

% Other constants
zeta2 = zeta1 * (1+2*rc)/sqrt(3);
omega_d1 = sqrt(1-zeta1^2);
omega_d2 = sqrt(1-zeta2^2);
phi1 = atan(zeta1/omega_d1);
phi2 = atan(zeta2/omega_d2);

t=[0:.02:40];
y1 = exp(-zeta1*t)/omega_d1 .* cos(omega_d1*t-phi1);
y2 = exp(-zeta2*sqrt(3)*t)/omega_d2 .* cos(sqrt(3)*omega_d2*t-phi2);
x1 = .5*( y1 + y2 );
x2 = .5*( y1 - y2 );
```

```
plot(t,x1,'r',t,x2,'--b');
ylabel('x_{1,2}(t)/\Delta');
xlabel('\omega_1 t');
legend('x_1(t)','x_2(t)');
```

Figures 4.21 through 4.23 show the responses of the system at three different combinations of damping ratios: $\zeta_1 = 0.1$ and $r_c = 0$ $\left(\zeta_2 = 0.1/\sqrt{3}\right)$, $\zeta_1 = 0.001$ and $r_c = 100$ $\left(\zeta_2 = 0.2/\sqrt{3}\right)$, and $\zeta_1 = 0.1$ and $r_c = 1$ $\left(\zeta_2 = 0.3/\sqrt{3}\right)$.

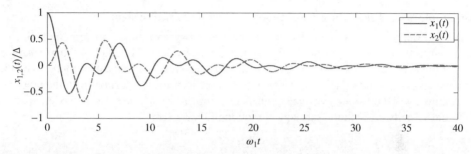

Figure 4.21 Responses of damped two-mass system subjected to initial displacement Δ on the left mass, for $\zeta_1 = 0.1$ and $r_c = 0$ $\left(\zeta_2 = 0.1/\sqrt{3}\right)$

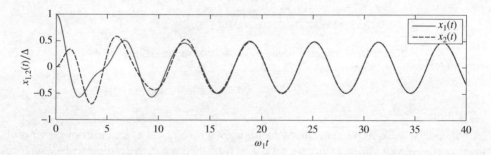

Figure 4.22 Responses of damped two-mass system subjected to initial displacement Δ on the left mass, for $\zeta_1 = 0.001$ and $r_c = 100$ $\left(\zeta_2 = 0.2/\sqrt{3}\right)$

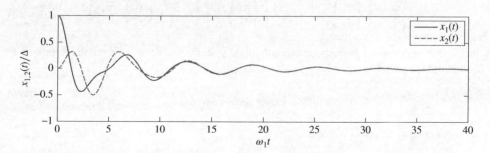

Figure 4.23 Responses of damped two-mass system subjected to initial displacement Δ on the left mass, for $\zeta_1 = 0.1$ and $r_c = 1$ $\left(\zeta_2 = 0.3/\sqrt{3}\right)$

Among the three cases, when $\zeta_1 = 0.001$ (Fig. 4.22), that is, when damping between the masses and the walls is extremely small, the motion quickly reduces to purely the first mode and the magnitude does not decrease although there is a dashpot between the two masses, producing $\zeta_2 = 0.2/\sqrt{3}$. This is because there is no relative motion between the two masses in Mode 1; hence, the dashpot c_2 is not active as soon as Mode 2 dies down.

In this example, with the presence of damping, Mode 2 only makes a brief presence in the system response. Whenever there is a dashpot, Mode 2 is always damped: $\zeta_2 \geq \zeta_1/\sqrt{3}$. This explains why in these plots Mode 2 is quickly suppressed and Mode 1 dominates.

■ Example 4.16: Free Vibration of Damped Unconstrained System

In this example, we explore a damped unconstrained system subject to an initial velocity. An undamped two-mass unconstrained system has been studied in Example 4.5 for its natural modes and in Example 4.10 for free vibrations. In this example, we add a dashpot between the two masses, as shown in Fig. 4.24. We want to explore a curiosity question: with the presence of damping, will the motion eventually comes to a stop? At $t = 0$, the mass on the left is given an initial velocity v_0. Find the subsequent motion of the system.

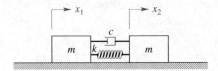

Figure 4.24   Unconstrained two-mass mass–spring–dashpot system

□ Solution:

If we use x_1 and x_2 as the generalized coordinates, where x_1 and x_2 are the displacements of the two masses with respect to ground, measured from their respective positions in a reference configuration in which the system is in equilibrium and the spring is unstretched, then the equations of motion are

$$\begin{bmatrix} m & 0 \\ 0 & m \end{bmatrix} \begin{Bmatrix} \ddot{x}_1 \\ \ddot{x}_2 \end{Bmatrix} + \begin{bmatrix} c & -c \\ -c & c \end{bmatrix} \begin{Bmatrix} \dot{x}_1 \\ \dot{x}_2 \end{Bmatrix} + \begin{bmatrix} k & -k \\ -k & k \end{bmatrix} \begin{Bmatrix} x_1 \\ x_2 \end{Bmatrix} = \{0\} \tag{a}$$

The natural modes of the corresponding undamped system have been analyzed in Example 4.5. The natural frequencies are

$$\omega_1 = 0 \qquad \omega_2 = \sqrt{\frac{2k}{m}} \tag{b}$$

The modal matrix and its inverse are

$$[V] = \begin{bmatrix} 1 & 1 \\ 1 & -1 \end{bmatrix} \qquad [V]^{-1} = \frac{1}{2} \begin{bmatrix} 1 & 1 \\ 1 & -1 \end{bmatrix} \tag{c}$$

Thus, we can proceed to diagonalize the matrices as

$$[M_d] = [V]^T[M][V] = \begin{bmatrix} 1 & 1 \\ 1 & -1 \end{bmatrix}\begin{bmatrix} m & 0 \\ 0 & m \end{bmatrix}\begin{bmatrix} 1 & 1 \\ 1 & -1 \end{bmatrix} = \begin{bmatrix} 2m & 0 \\ 0 & 2m \end{bmatrix} \tag{d}$$

$$[C_d] = [V]^T[C][V] = \begin{bmatrix} 1 & 1 \\ 1 & -1 \end{bmatrix}\begin{bmatrix} c & -c \\ -c & c \end{bmatrix}\begin{bmatrix} 1 & 1 \\ 1 & -1 \end{bmatrix} = \begin{bmatrix} 0 & 0 \\ 0 & 4c \end{bmatrix} \tag{e}$$

$$[K_d] = [V]^T[M][V] = \begin{bmatrix} 1 & 1 \\ 1 & -1 \end{bmatrix}\begin{bmatrix} k & -k \\ -k & k \end{bmatrix}\begin{bmatrix} 1 & 1 \\ 1 & -1 \end{bmatrix} = \begin{bmatrix} 0 & 0 \\ 0 & 4k \end{bmatrix} \tag{f}$$

The initial conditions in terms of the original set of generalized coordinates are

$$\{x(0)\} = 0 \qquad \{\dot{x}(0)\} = \begin{Bmatrix} v_0 \\ 0 \end{Bmatrix} \tag{g}$$

which can be transformed into the initial conditions for $\{y\}$, according to eqn. (4.71),

$$\{y(0)\} = 0 \qquad \{\dot{y}(0)\} = [V]^{-1}\{\dot{x}(0)\} = \frac{1}{2}\begin{bmatrix} 1 & 1 \\ 1 & -1 \end{bmatrix}\begin{Bmatrix} v_0 \\ 0 \end{Bmatrix} = \frac{1}{2}\begin{Bmatrix} v_0 \\ v_0 \end{Bmatrix} \tag{h}$$

The result of diagonalization indicates that the system has a proportional damping. The resulting equations of motion in terms of principal coordinates are

$$\begin{bmatrix} 2m & 0 \\ 0 & 2m \end{bmatrix}\begin{Bmatrix} \ddot{y}_1 \\ \ddot{y}_2 \end{Bmatrix} + \begin{bmatrix} 0 & 0 \\ 0 & 4c \end{bmatrix}\begin{Bmatrix} \dot{y}_1 \\ \dot{y}_2 \end{Bmatrix} + \begin{bmatrix} 0 & 0 \\ 0 & 4k \end{bmatrix}\begin{Bmatrix} y_1 \\ y_2 \end{Bmatrix} = \{0\} \tag{i}$$

or

$$m\ddot{y}_1 = 0 \tag{j}$$

$$m\ddot{y}_2 + 2c\dot{y}_2 + 2ky_2 = 0 \tag{k}$$

Equation (j) shows the acceleration $\ddot{y}_1 = 0$, which can be integrated twice to give

$$y_1(t) = C_1 t + C_2 \tag{l}$$

Applying the initial conditions in eqn. (h) gives

$$y_1(t) = \frac{1}{2}v_0 t \tag{m}$$

Equation (k) can be written as

$$\ddot{y}_2 + 2\zeta_2\omega_2\dot{y}_2 + \omega_2^2 y_2 = 0 \tag{n}$$

Using the initial conditions in eqns. (h), and the free-vibration response for an underdamped single-DOF system in eqn. (2.56),

$$y_2(t) = \frac{v_0}{2}\frac{e^{-\zeta_2\omega_2 t}}{\omega_{2d}}\sin\omega_{2d}t \tag{o}$$

where

$$\omega_{2d} = \sqrt{1 - \zeta_2^2}\,\omega_2 \tag{p}$$

Finally, we can convert the solution back to the original set of generalized coordinates according to $\{x\} = [V]\{y\}$. Then,

$$x_1(t) = y_1(t) + y_2(t) = \frac{v_0}{2}\left[t + \frac{e^{-\zeta_2\omega_2 t}}{\omega_{2d}}\sin\omega_{2d}t\right] \tag{q}$$

$$x_2(t) = y_1(t) - y_2(t) = \frac{v_0}{2}\left[t - \frac{e^{-\zeta_2\omega_2 t}}{\omega_{2d}}\sin\omega_{2d}t\right] \tag{r}$$

□ Exploring the Solution with MATLAB

Similar to Example 4.10, we introduce the following nondimensionalization parameters:

$$\bar{x}_i = \frac{x_i}{v_0/\omega_2} \qquad \bar{t} = \omega_2 t \qquad \bar{\omega}_d = \frac{\omega_{2d}}{\omega_2} \tag{s}$$

Then the nondimensionalized responses can be written as

$$\bar{x}_1(\bar{t}) = \frac{1}{2}\left(\bar{t} + \frac{e^{-\zeta_2\bar{t}}}{\bar{\omega}_d}\sin\bar{\omega}_d\bar{t}\right) \tag{t}$$

$$\bar{x}_2(\bar{t}) = \frac{1}{2}\left(\bar{t} - \frac{e^{-\zeta_2\bar{t}}}{\bar{\omega}_d}\sin\bar{\omega}_d\bar{t}\right) \tag{u}$$

The following MATLAB code plots the nondimensionalized responses of the two masses, which are shown in Fig. 4.25 for the case $\zeta_2 = 0.1$.

```
zeta = 0.1;
omega_d = sqrt(1-zeta^2);
t=[0:.01:20];
x1 = .5* (t + exp(-zeta*t) .* sin(omega_d*t)/omega_d);
x2 = .5* (t - exp(-zeta*t) .* sin(omega_d*t)/omega_d);
plot(t,x1,'r',t,x2,'--b');
ylabel('x_{1,2}(t)/(v_0 /\omega_2)');
xlabel('\omega_2 t');
legend('x_1(t)','x_2(t)');
```

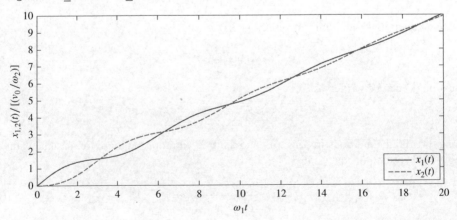

Figure 4.25 Responses of unconstrained damped two-mass system subjected to initial velocity v_0 on the left mass at damping ratio $\zeta_2 = 0.1$

We can verify the correctness of the results by checking the initial conditions: both $\bar{x}$ start from 0, and one has a zero slope and the other has a nonzero slope.

We observe that, the system does not seem to be slowing down, indicating that it will never stop. But, the pulsating relative vibration between the two masses is diminishing as the time progresses.

This brings up an interesting concept of *modal energy*: the total energy distributed into different modes. In this case, the rigid-body mode has its allocation, and since it moves at a constant speed ($\dot{y}_1 = v_0/2$), this energy is conserved. Mode 2 has its allocation, but the damping eventually consumes its entire allocation so the pulsating motion diminishes.

We can calculate the energy allocation to each mode as the following. At time $t = 0$, the total energy in the system is the kinetic energy imparted into the left mass, that is, $T(0) = \frac{1}{2}mv_0^2$. At the end, Mode 2 dies down, and both masses have velocity of $\frac{1}{2}v_0$, which means the entire system has a kinetic energy $T(\infty) = \frac{1}{4}mv_0^2$. This indicates that half of the total energy is allocated to Mode 1 and the remaining half goes into Mode 2.

4.7 Forced Vibrations I: Steady-State Responses

In this section, we focus on the steady-state responses of multi-DOF systems to harmonic excitations. Responses to nonharmonic periodic excitations can be analyzed by using Fourier expansion method in the same way as discussed in the single-DOF systems in Chapter 2, once we know how to analyze the harmonic excitations.

We assume that all harmonic excitations are of the same frequency Ω. For the situations in which the harmonic excitations in different parts of the system are in different frequencies, the superposition principle can be used.

We have seen in single-DOF systems that the steady-state response to a harmonic excitation is the simplest analysis among the three types of vibration analyses. This remains true here. Even damped systems can be analyzed with ease, with the help of using complex variables. Taking damping into account, the general equations of motion can be written as, in matrix form,

$$[M]\{\ddot{x}\} + [C]\{\dot{x}\} + [K]\{x\} = \{F_0\}e^{i\Omega t} \tag{4.73}$$

The physical nature dictates that, in the long term, which we called technically the steady state, the system will submit to the driving frequency. Thus, we assume the solution of the form $\{X\}e^{i\Omega t}$. Substituting the solution into eqn. (4.73) gives

$$[M]\{X\}(-\Omega^2)e^{i\Omega t} + [C]\{X\}(i\Omega)e^{i\Omega t} + [K]\{X\}e^{i\Omega t} = \{F_0\}e^{i\Omega t} \tag{4.74}$$

Since $e^{i\Omega t}$ is not always zero, we can cancel it from both sides, giving

$$\left(-\Omega^2[M] + i\Omega[C] + [K]\right)\{X\} = \{F_0\} \tag{4.75}$$

This equation can be solved as before; and the resulting $\{X\}$ is generally complex.

Physical systems can only be excited with real excitations, such as $\{F_0\}\cos\Omega t$ or $\{F_0\}\sin\Omega t$. They correspond to the real and imaginary parts, respectively, of the complex excitation $\{F_0\}e^{i\Omega t}$. Consequently, only either the real or the imaginary part of the complex solution $\{x(t)\}$ will be the solution to the physical system.

For instance, when X_i is a complex quantity, it can be written as

$$X_i = |X_i|e^{i\phi_i} \tag{4.76}$$

where $|X_i|$ is the *magnitude* and ϕ_i is the *phase angle* in complex variables terminologies. The corresponding complex solution is expressible as

$$x_i(t) = X_i e^{i\Omega t} = |X_i|e^{i(\Omega t + \phi_i)}$$

The final physical solution would be either $|X_i| \cos(\Omega t + \phi_i)$ or $|X_i| \sin(\Omega t + \phi_i)$, depending on the function form of the loading. In either case, the amplitude of the physical solution is $|X_i|$, and the phase lag is $-\phi_i$, that is

$$|x_i(t)| = |X_i| \qquad \text{phase lag} = -\phi_i = -\arg(X_i) \tag{4.77}$$

For this reason, X_i is also called the *complex amplitude*.

■ Example 4.17: Steady State of Damped Two-Mass System

We continue to consider the system with damping in Example 4.15. Assume that a force $F_0 \sin \Omega t$ is acting on the left mass. Find the steady-state response of the system.

□ Solution:

Using complex notation, we write the excitation on the left mass as $F_0 e^{i\Omega t}$. This means that, at the end, we will take the imaginary part of the solution $\{x(t)\}$, which is assumed to be

$$\{x(t)\} = \{X\}e^{i\Omega t} \tag{a}$$

We expand eqn. (4.75) for this particular system as

$$\begin{bmatrix} 2k - \Omega^2 m + i(c_1 + c_2)\Omega & -(k + ic_2) \\ -(k + ic_2) & 2k - \Omega^2 m + i(c_1 + c_2)\Omega \end{bmatrix} \begin{Bmatrix} X_1 \\ X_2 \end{Bmatrix} = \begin{Bmatrix} F_0 \\ 0 \end{Bmatrix} \tag{b}$$

which can be readily solved using Cramer's rule to give

$$X_1 = \frac{F_0[2k - \Omega^2 m + i(c_1 + c_2)\Omega]}{[2k - \Omega^2 m + i(c_1 + c_2)\Omega]^2 - (k + ic_2\Omega)^2} \tag{c}$$

$$X_2 = \frac{F_0(k + ic_2\Omega)}{[2k - \Omega^2 m + i(c_1 + c_2)\Omega]^2 - (k + ic_2\Omega)^2} \tag{d}$$

Then, the solution to the original problem is

$$x_1(t) = \Im\left\{X_1 e^{i\Omega t}\right\} \tag{e}$$

$$x_2(t) = \Im\left\{X_2 e^{i\Omega t}\right\} \tag{f}$$

Writing out the explicit expressions for the imaginary parts of the above expressions would be rather involved. We shall leave MATLAB to handle the complex math.

□ Exploring the Solution with MATLAB

In this portion of the solution, we shall use MATLAB to explore the complex solution as given in eqns. (e) and (f). We would like to plot the amplitude and the phase angles, which are all embedded in the complex amplitudes given in eqns. (c) and (d).

Here we use a new way of nondimensionalization because using the modal damping ratio, as it is done in Example 4.15, actually makes the expressions look more complicated. Furthermore, for steady-state analyses, the modal analysis is not a required step. Thus, in this case, we choose some nominal parameters to nondimensionalize the solution. Taking the inspiration from single-DOF systems, we introduce the nominal frequency ω_0 and nominal (pseudo) damping ratios ξ_1 and ξ_2:

$$\omega_0 = \sqrt{\frac{k}{m}} \qquad 2\xi_1\omega_1 = \frac{c_1}{m} \qquad 2\xi_2\omega_2 = \frac{c_2}{m} \tag{g}$$

in addition to the ones we used earlier:

$$r = \frac{\Omega}{\omega_0}$$

In fact, as seen in the modal analysis for the undamped system at the beginning of the chapter, ω_0 as defined is actually the system's first natural frequency and ξ_1 is the actual damping ratio for Mode 1. The physical meaning of ξ_2 is not very clear, but we know that it is unrelated to ξ_1 and can be arbitrarily changed. We can find out the relation between the pseudo damping ratio ξ_2 and the real damping ratio ζ_2 if needed. But, at this stage, the priority is to keep the nondimensionalized expressions as close to the original as possible.

With these nondimensionalization parameters, the nondimensionalized complex amplitudes are as follows:

$$\frac{X_1}{F_0/k} = \frac{2 - r^2 + 2ir(\xi_1 + \xi_2)}{[2 - r^2 + 2ir(\xi_1 + \xi_2)]^2 - (1 + 2ir\xi_2)^2} \tag{h}$$

$$\frac{X_2}{F_0/k} = \frac{1 + ir\xi_2}{[2 - r^2 + 2ir(\xi_1 + \xi_2))]^2 - (1 + 2ir\xi_2)^2} \tag{i}$$

The following MATLAB code plots both the amplitude and phase angle of x_1 and x_2:

```
xi_1 = 0.1;
xi_2 = 0.2;


r=[0:.003:5];
deno = (2-r .^2 +j*2*r* (xi_1+xi_2)) .^2 - (1+2*j*r*xi_2) .^2;
x1 = (2-r .^2 +2*j* (xi_1+xi_2)) ./ deno;
x2 = (1+j*2*r*xi_2) ./ deno;

subplot (2,2,1);
plot (r,abs (x1));
axis ([0 3 0 4]);
ylabel ('|x_1/ (F_0/k) |');
xlabel ('r');

subplot (2,2,2);
plot (r,abs (x2));
```

```
axis([0 3 0 4]);
ylabel('|x_2/(F_0/k)|');
xlabel('r');

subplot(2,2,3);
plot(r,-angle(x1));
axis([0 3 -pi pi]);
ylabel('\phi_1');
xlabel('r');

subplot(2,2,4);
plot(r,-angle(x2));
axis([0 3 -pi pi]);
ylabel('\phi_2');
xlabel('r');
```

In the above MATLAB code, the unit for imaginary number $i = \sqrt{-1}$ is j. MATLAB also recognizes i for the same. However, caution is advised because i and j are also often used as indices in an integer context, such as in loops. For a complex variable, MATLAB functions abs() and angle() return its magnitude and phase angle, respectively.

The responses are shown in Figure 4.26 for the case $\xi_1 = 0$, 0.025, 0.05, 0.1, 0.2, and 0.5 and $\xi_2 = 0$; in Figure 4.27 for the case $\xi_1 = 0$ and $\xi_2 = 0$, 0.025, 0.05, 0.1, 0.2, and 0.5; and in Figure 4.28 for the case $\xi_1 = \xi_2 = 0$, 0.025, 0.05, 0.1, 0.2, and 0.5. In each of these figures, top two subfigures show the amplitude and the bottom two subfigures show the phase angle.

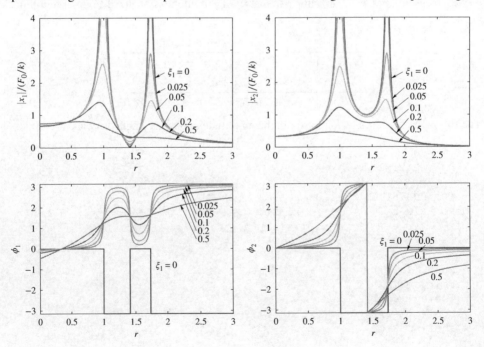

Figure 4.26 Steady-state response spectra for damped two-DOF system when ξ_1 value ranges from 0 to 0.5 and $\xi_2 = 0$

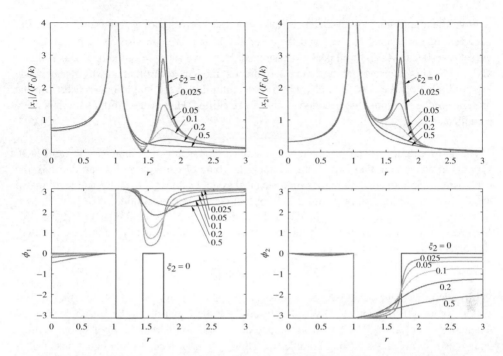

Figure 4.27 Steady-state response spectra for damped two-DOF system when $\xi_1 = 0$, and ξ_2 value ranges from 0 to 0.5

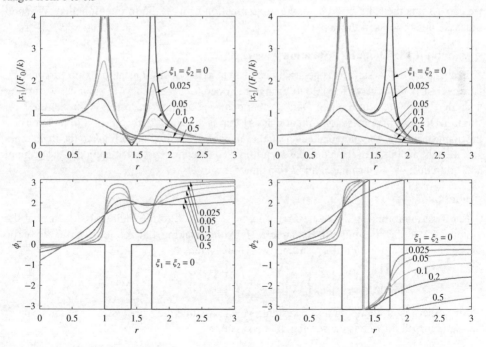

Figure 4.28 Steady-state response spectra for damped two-DOF system when $\xi_1 = \xi_2$, and their values range from 0 to 0.5

From the amplitude plots, we can see the two peaks corresponding to the two natural frequencies: one is located at $r = 1$ and the other at $r = \sqrt{3}$. We can also observe that even if we set $\xi_2 = 0$ (Fig. 4.26), both peaks are damped. But if we set $\xi_1 = 0$, the first peak remains undamped. This suggests that the damping of the first mode is controlled only by ξ_1; but Mode 2 is by both ξ_1 and ξ_2. We can also observe that, similar to the single-DOF case, with the presence of the damping, the peak locations will be shifted. However, shifting pattern is more complicated.

———————— · ————————

The above example illustrates the more general case of steady-state analysis for multi-DOF systems in which damping is present. We took the route of using complex variables. To the end, to get back to the real quantities for a physical system subjected to a physical loading, we need to take either the real or the imaginary part of the complex solution.

It is possible to stay entirely in the real realm. The process is similar to the case for single-DOF systems. We start with the assumption of the solution to be comprised of both sine and cosine components, such as

$$\{x(t)\} = \{X_c\} \cos \Omega t + \{X_s\} \sin \Omega t$$

Substituting this solution into the equations of motion and then matching the coefficients for $\sin \Omega t$ and $\cos \Omega t$, respectively, lead to a set of algebraic equations to be solved, giving the coefficients $\{X_c\}$ and $\{X_s\}$. The disadvantage of this method is that the number of resulting equations is doubled. Even for a humble two-DOF system, it leads to four equations. This is likely beyond the feasibility of hand calculation.

However, if the system has no damping, this method works just fine. Without the damping, we can assume the solution to be in the same function form as the forcing function. We shall illustrate this process in the next example.

■ Example 4.18: Dynamic Vibration Absorber

In a single-DOF system, it is generally advised to avoid operations at its natural frequency. However, it is not always feasible to change the operating frequency. In such cases, turning a single-DOF system into a two-DOF system could be a solution. In this example, the base single-DOF system consists of a main mass M that is attached to a wall (or a foundation) by the main spring K. We want to attach a smaller mass m via a smaller spring k to the main mass such that the vibration of the main mass is minimized when the system is excited at the original natural frequency. Determine m and k to achieve this goal.

□ Solution:

Let us assume the forcing to be $F(t) = F_0 \sin \Omega t$ applied to the main mass. Note that the system as described is similar to Example 1.21. Without taking a digression, we directly write the equations of motion for the resulting two-DOF system as

$$\begin{bmatrix} M & 0 \\ 0 & m \end{bmatrix} \begin{Bmatrix} \ddot{x}_1 \\ \ddot{x}_2 \end{Bmatrix} + \begin{bmatrix} K+k & -k \\ -k & k \end{bmatrix} \begin{Bmatrix} x_1 \\ x_2 \end{Bmatrix} = \begin{Bmatrix} F_0 \\ 0 \end{Bmatrix} \sin \Omega t \qquad \text{(a)}$$

Since there is no damping in the system, we can assume the steady-state solution to be of the same function form as the forcing, that is, assume

$$\{x(t)\} = \begin{Bmatrix} X_1 \\ X_2 \end{Bmatrix} \sin \Omega t \qquad \text{(b)}$$

Substituting this solution into the equations of motion gives

$$-\Omega^2 \begin{bmatrix} M & 0 \\ 0 & m \end{bmatrix} \begin{Bmatrix} X_1 \\ X_2 \end{Bmatrix} \sin \Omega t + \begin{bmatrix} K+k & -k \\ -k & k \end{bmatrix} \begin{Bmatrix} X_1 \\ X_2 \end{Bmatrix} \sin \Omega t = \begin{Bmatrix} F_0 \\ 0 \end{Bmatrix} \sin \Omega t \qquad (c)$$

Since the equations must hold for all times, dropping the common factor $\sin \Omega t$ from both sides gives

$$\begin{bmatrix} K+k-M\Omega^2 & -k \\ -k & k-m\Omega^2 \end{bmatrix} \begin{Bmatrix} X_1 \\ X_2 \end{Bmatrix} = \begin{Bmatrix} F_0 \\ 0 \end{Bmatrix} \qquad (d)$$

Using Cramer's rule, the solutions are

$$X_1 = \frac{F_0(k-m\Omega^2)}{(K+k-M\Omega^2)(k-m\Omega^2) - k^2} \qquad (e)$$

$$X_2 = \frac{F_0 k}{(K+k-M\Omega^2)(k-m\Omega^2) - k^2} \qquad (f)$$

To minimize the motion of the main mass (X_1) at its original natural frequency, that is, when $\Omega = \sqrt{K/M}$, we find that it is actually possible to eliminate the motion altogether, by setting $k - m\Omega^2 = 0$, or,

$$\frac{k}{m} = \frac{K}{M} \qquad (g)$$

that is, the natural frequency of the attached system alone is the same as that of the main system. Under this condition, the amplitude of the attached mass is

$$X_2 = -\frac{F_0}{k} \qquad (h)$$

that is, the amplitude of the attached mass equals the static deflection due to a constant force F_0 applied directly onto the auxiliary spring.

Equations (g) and (h) form the basis for the design of the vibration absorber: choose a spring constant k first based on the allowable amplitude of the auxiliary mass, and then choose a mass based on the natural frequency of the main system such that the auxiliary system's natural frequency matches that of the main system.

□ **Exploring the Solution with MATLAB**

We define the following nondimensionalization parameters:

$$\omega_0 = \sqrt{\frac{k}{m}} = \sqrt{\frac{K}{M}}, \qquad r = \frac{\Omega}{\omega_0} \qquad \mu = \frac{m}{M} = \frac{k}{K} \qquad (i)$$

where ω_0 is a nominal natural frequency, as it is not the natural frequency of the two-DOF system. But it is the natural frequency of the two single-DOF systems. Then,

$$\frac{X_1}{F_0/K} = \frac{1-r^2}{(1+\mu-r^2)(1-r^2) - \mu} \qquad (j)$$

$$\frac{X_2}{F_0/K} = \frac{1}{(1+\mu-r^2)(1-r^2) - \mu} \qquad (k)$$

The following MATLAB code plots the amplitude of the steady-state response of the two masses.

```
mu = 0.5;
r=0:.013:4;
deno = (1+mu-r.^2).*(1-r.^2)-mu;
x1 = (1-r.^2) ./ deno;
x2 = 1 ./ deno;

subplot(1,2,1)
plot(r,abs(x1));
axis([0 3 0 4]);
ylabel('X_1/(F/K)');
xlabel('r');

subplot(1,2,2);
plot(r,abs(x2));
axis([0 4 0 4]);
ylabel('X_2/(F/K)');
xlabel('r');
```

The responses are plotted in Fig. 4.29 for $\mu = 0.1$, 0.2, 0.3, and 0.5. As a verification, we observe that for all cases the amplitude X_1 is exactly 0 at $r = 1$.

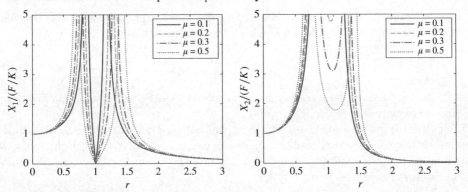

Figure 4.29 Response spectra of dynamic vibration absorber for mass ratios $\mu = m/M = 0.1, 0.2, 0.3$, and 0.5

We observe that the new system has two natural frequencies that reside on each side of the original natural frequency ($r = 1$), where both $|X_1|$ and $|X_2|$ approach infinity. Slightly off that frequency, both $|X_1|$ and $|X_2|$ increase rapidly; the larger the μ, the larger $|X_1|$, but $|X_2|$ becomes smaller.

Let us take a closer look at the feasibility of this vibration absorption mechanism, especially, the width of the "usable frequency range."

Natural Frequencies

Obviously, the modified system has two natural frequencies. The natural frequencies can be found by setting the denominator to zero, which gives

$$(1 + \mu - r^2)(1 - r^2) - \mu = 1 - (2 + \mu)r^2 + r^4 = 0 \tag{1}$$

which has the following two roots:

$$r_{1,2}^2 = 1 + \frac{\mu \pm \sqrt{\mu(4 + \mu)}}{2} \tag{m}$$

Usable Range

Let us say that the usable range is defined such that the dynamic magnification for the main mass, $|X_1/(F/K)|$, is not more than 1. This range can be found as the distance between two solutions nearest to $r = 1$

$$\left| \frac{1 - r^2}{(1 + \mu - r^2)(1 - r^2) - \mu} \right| = 1 \tag{n}$$

or

$$|1 - r^2| = |1 - (2 + \mu)r^2 + r^4| \tag{o}$$

This can be solved by simply squaring both sides. However, that would raise the order of the equation and make it more difficult to solve. We would prefer to remove the absolute sign by making the following observations of Fig. 4.29. First, both roots of interest fall within a range limited by the two natural frequencies; one root resides on each side of $r = 1$. Second, within this range, X_2 does not cross the abscissa. Comparing eqns. (j) and (k), this means that the right-hand side of eqn. (o) does not change its sign within this range; and since $[1 - (2 + \mu)r^2 + r^4] < 0$ at $r = 1$, this holds for the entire range. Denote the two roots as R_1 and R_2.

Assuming $R_1 < 1$, eqn. (o) becomes

$$1 - R_1^2 = -[1 - (2 + \mu)R_1^2 + R_1^4]$$

which can be solved for

$$R_1^2 = \frac{3 + \mu - \sqrt{1 + 6\mu + \mu^2}}{2} \tag{p}$$

Assuming $R_2 > 1$, eqn. (o) becomes

$$-(1 - R_2^2) = -[1 - (2 + \mu)R_2^2 + R_2^4]$$

which can be solved for

$$R_2^2 = 1 + \mu \tag{q}$$

The following MATLAB code can be used to plot the natural frequencies and the operating ranges as the mass ratio μ changes, which is shown in Fig. 4.30.

```
mu=0:.01:1;
r1=sqrt(1+.5*(mu-sqrt(mu.*(4+mu))));
r2=sqrt(1+.5*(mu+sqrt(mu.*(4+mu))));
R1=sqrt(.5*(3+mu-sqrt(1+6*mu+mu.^2)));
R2=sqrt(1+mu);
plot(R1,mu,'r', r1,mu,'--b',R2,mu,'r', r2,mu,'--b');
axis([0 2 0 1]);
ylabel('\mu');
xlabel('r');
legend('Operational range','Natural frequencies',4);
```

We can observe from this figure that the usable range increases as μ increases, which also separate the two natural frequencies further apart.

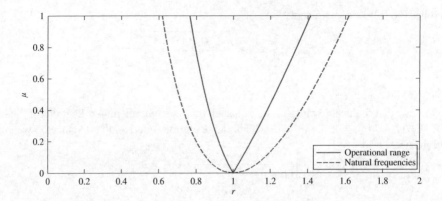

Figure 4.30 Natural frequencies and usable range of dynamic vibration absorber

4.8 Forced Vibrations II: Transient Responses

In this section, we study the motion immediately after a time-varying load is applied to the system. As we have seen earlier, this is the most challenging type of vibration analyses.

In general, we seek the total solution for the following set of equations of motion:

$$[M]\{\ddot{x}\} + [C]\{\dot{x}\} + [K]\{x\} = \{F(t)\} \tag{4.78}$$

4.8.1 Direct Analytical Method

According to the theory of differential equations, the general solution to eqn. (4.78) consists of a particular solution plus the general solution to the corresponding homogeneous problem, called the *homogeneous solution*.

For undamped systems ($\{C\} = 0$), we have found the homogeneous solution in eqn. (4.40). Hence, the general solution can be written as

$$\{x(t)\} = \{x_p(t)\} + [V]\{s(t)\} \tag{4.79}$$

If a particular solution $\{x_p(t)\}$ can be found, the initial conditions can then be used to determine the constants a_i and b_i in $\{s(t)\}$ as following: setting $t = 0$ to $\{x(t)\}$ in eqn. (4.79) and its derivative,

$$\{x(0)\} = \{x_p(0)\} + [V]\{a\} \qquad \{\dot{x}(0)\} = \{\dot{x}_p(0)\} + [V]\{\beta\} \tag{4.80}$$

where $\beta_i = \omega_i b_i$, and in the event $\omega_0 = 0$, $\beta_1 = b_1$. They can be solved to give

$$\{a\} = [V]^{-1}\left(\{x(0)\} - \{x_p(0)\}\right) \tag{4.81}$$

$$\{\beta\} = [V]^{-1}\left(\{\dot{x}(0)\} - \{\dot{x}_p(0)\}\right) \tag{4.82}$$

As with single-DOF systems, the main challenge is to find the particular solution $\{x_p(t)\}$. This method is also called the *modal summation method* or the *modal superposition method*. It is limited to systems without damping because the above-mentioned homogeneous solution is limited to systems without damping. The basic idea behind the decomposition method as discussed for the single-DOF systems is still applicable to multi-DOF systems.

■ **Example 4.19: Step Loading on Undamped Two-Mass System**

Consider the two-mass system we had studied earlier in Fig. 4.1, whose free-vibration problem has been analyzed in Example 4.7. Assume the mass on the left is subjected to a step loading: $F_1(t) = F_0$ when $t \geq 0$ and 0 otherwise, $F_2(t) = 0$. The system is initially at rest. Find the system's transient response.

□ **Solution:**

The equations of motion for the system can be written as, for $t \geq 0$,

$$\begin{bmatrix} m & 0 \\ 0 & m \end{bmatrix} \begin{Bmatrix} \ddot{x}_1 \\ \ddot{x}_2 \end{Bmatrix} + \begin{bmatrix} 2k & -k \\ -k & 2k \end{bmatrix} \begin{Bmatrix} x_1 \\ x_2 \end{Bmatrix} = \begin{Bmatrix} F_0 \\ 0 \end{Bmatrix} \tag{a}$$

The natural modes of the system have been found earlier in Example 4.7. Here, we directly quote modal analysis results from Example 4.7. The natural frequencies are

$$\omega_1 = \sqrt{\frac{k}{m}} \qquad \omega_2 = \sqrt{\frac{3k}{m}} \tag{b}$$

and the modal matrix and its inversion are

$$[V] = \begin{bmatrix} 1 & 1 \\ 1 & -1 \end{bmatrix} \qquad [V]^{-1} = \frac{1}{2} \begin{bmatrix} 1 & 1 \\ 1 & -1 \end{bmatrix} \tag{c}$$

Since the loading is a constant force, we assume the particular solution as

$$\{x_p\} = \begin{Bmatrix} X_1 \\ X_2 \end{Bmatrix} \tag{d}$$

Substituting this particular solution into eqn. (a) gives

$$\begin{bmatrix} 2k & -k \\ -k & 2k \end{bmatrix} \begin{Bmatrix} X_1 \\ X_2 \end{Bmatrix} = \begin{Bmatrix} F_0 \\ 0 \end{Bmatrix} \tag{e}$$

which can be solved as

$$\begin{Bmatrix} X_1 \\ X_2 \end{Bmatrix} = \frac{F_0}{3k} \begin{Bmatrix} 2 \\ 1 \end{Bmatrix} \tag{f}$$

Thus, the total solution can be assumed as

$$\{x(t)\} = \frac{F_0}{3k} \begin{Bmatrix} 2 \\ 1 \end{Bmatrix} + \begin{bmatrix} 1 & 1 \\ 1 & -1 \end{bmatrix} \begin{Bmatrix} a_1 \cos \omega_1 t + b_1 \sin \omega_1 t \\ a_2 \cos \omega_2 t + b_2 \sin \omega_2 t \end{Bmatrix} \tag{g}$$

The initial conditions are the trivial initial conditions, that is,

$$\{x(0)\} = 0 \qquad \{\dot{x}(0)\} = 0 \tag{h}$$

Evaluating the initial conditions gives

$$\frac{F_0}{3k} \begin{Bmatrix} 2 \\ 1 \end{Bmatrix} + \begin{bmatrix} 1 & 1 \\ 1 & -1 \end{bmatrix} \begin{Bmatrix} a_1 \\ a_2 \end{Bmatrix} = 0 \qquad \begin{bmatrix} 1 & 1 \\ 1 & -1 \end{bmatrix} \begin{Bmatrix} \omega_1 b_1 \\ \omega_2 b_2 \end{Bmatrix} = 0 \tag{i}$$

which can be solved to give, according to eqns. (4.81) and (4.82), $\{\boldsymbol{\beta}\} = \{\boldsymbol{b}\} = 0$ and

$$\begin{Bmatrix} a_1 \\ a_2 \end{Bmatrix} = -\frac{1}{2}\begin{bmatrix} 1 & 1 \\ 1 & -1 \end{bmatrix}\frac{F_0}{3k}\begin{Bmatrix} 2 \\ 1 \end{Bmatrix} = -\frac{F_0}{6k}\begin{Bmatrix} 3 \\ 1 \end{Bmatrix} \tag{j}$$

Finally, the solution is

$$\{x(t)\} = \frac{F_0}{3k}\begin{Bmatrix} 2 \\ 1 \end{Bmatrix} - \frac{F_0}{6k}\begin{bmatrix} 1 & 1 \\ 1 & -1 \end{bmatrix}\begin{Bmatrix} 3\cos\omega_1 t \\ \cos\omega_2 t \end{Bmatrix} \tag{k}$$

or

$$x_1(t) = \frac{F_0}{k}\left[\frac{2}{3} - \frac{1}{2}\cos\sqrt{\frac{k}{m}}t - \frac{1}{6}\cos\sqrt{\frac{3k}{m}}t\right] \tag{l}$$

$$x_2(t) = \frac{F_0}{k}\left[\frac{1}{3} - \frac{1}{2}\cos\sqrt{\frac{k}{m}}t + \frac{1}{6}\cos\sqrt{\frac{3k}{m}}t\right] \tag{m}$$

□ **Exploring the Solution with MATLAB**

We define the following nondimensionalization parameters: for $i = 1, 2$,

$$\bar{x}_i = \frac{x_i}{F_0/k} \qquad \bar{t} = \omega_1 t \tag{n}$$

Then, the solutions in eqns. (l) and (m) can be nondimensionalized to

$$\bar{x}_1(\bar{t}) = \frac{2}{3} - \frac{1}{2}\cos\bar{t} - \frac{1}{6}\cos\sqrt{3}\bar{t} \tag{o}$$

$$\bar{x}_2(\bar{t}) = \frac{1}{3} - \frac{1}{2}\cos\bar{t} + \frac{1}{6}\cos\sqrt{3}\bar{t} \tag{p}$$

The following MATLAB code can be used to plot the system responses, which are shown in Fig. 4.31.

```
t = [0:.01:40];
x1 = 2/3 - .5*cos(t) - 1/6*cos(sqrt(3)*t);
x2 = 1/3 - .5*cos(t) + 1/6*cos(sqrt(3)*t);
plot(t,x1,'r',t,x2,'--b');
xlabel('\omega_1 t');
ylabel('x_{1,2}(t)/(F_0/k)');
legend( 'x_1(t)', 'x_2(t)' );
```

From Fig. 4.31, we verify that both masses start with a zero displacement and a zero slope, indicating zero initial velocities. However, we shall be brief about making further observations, as system's response to a step loading is essentially the same as system's free vibration to initial displacement conditions, which we have examined in Example 4.7.

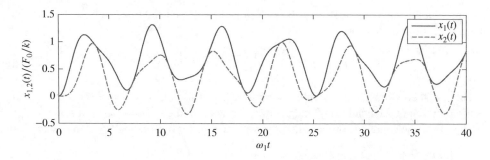

Figure 4.31   Transient responses of two-mass mass–spring–dashpot system initially at rest due to step loading on the left mass

4.8.2   Decoupling Method

The basic idea behind the decoupling method has been discussed in Section 4.6. Once decoupled, we end up with a set of single-DOF systems, and all the solution methods we have learned in Chapter 2 can be used. Furthermore, we can use and adapt known results for single-DOF systems.

An important advantage of this method is that it can analyze systems with proportional damping. Here, we use an example to illustrate the solution procedure.

■ **Example 4.20: Step Loading on Damped Two-Mass System**

In this example, we consider the damped two-mass system shown in Fig. 4.20, whose free-vibration problem has been studied in Example 4.15. Assume that the mass on the left is subjected to a step loading: when $t \geq 0$, $F_1(t) = F_0$ and $F_2(t) = 0$. The system is initially at rest. Find the system's transient response.

□ **Solution:**

The equations of motion for the system have been given in eqn. (a) in Example 4.15 for free vibrations. With the step loading as described, the right-hand side of eqn. (a) becomes

$$\{F\} = \begin{Bmatrix} F_0 \\ 0 \end{Bmatrix} \tag{a}$$

First, we conduct the modal analysis of the system. The modal analysis is independent of the damping and has been done in Section 4.2. Quoting from Example 4.15, the natural frequencies are

$$\omega_1 = \sqrt{\frac{k}{m}} \qquad \omega_2 = \sqrt{\frac{3k}{m}} \tag{b}$$

and the modal matrix and its inversion are

$$[V] = \begin{bmatrix} 1 & 1 \\ 1 & -1 \end{bmatrix} \qquad [V]^{-1} = \frac{1}{2}\begin{bmatrix} 1 & 1 \\ 1 & -1 \end{bmatrix} \tag{c}$$

In the next, we decouple the equations of motion by introducing a new set of generalized coordinates $\{y\}$ such that $\{x\} = [V]\{y\}$. This has mostly been done in Example 4.15. After having found the diagonal matrices $[M_d]$, $[K_d]$, and $[C_d]$, we need to convert the forcing into the new set of principal coordinates, that is,

$$[V]^T\{F\} = \begin{bmatrix} 1 & 1 \\ 1 & -1 \end{bmatrix} \begin{Bmatrix} F_0 \\ 0 \end{Bmatrix} = \begin{Bmatrix} F_0 \\ F_0 \end{Bmatrix} \tag{d}$$

The initial conditions can be similarly transformed. But in this example, the initial conditions are trivial initial conditions, and the transformed initial conditions are still trivial.

Therefore, the equations of motion of the system in terms of principal coordinates are

$$\begin{bmatrix} 2m & 0 \\ 0 & 2m \end{bmatrix} \begin{Bmatrix} \ddot{y}_1 \\ \ddot{y}_2 \end{Bmatrix} + \begin{bmatrix} 2c_1 & 0 \\ 0 & 2c_1 + 4c_2 \end{bmatrix} \begin{Bmatrix} \dot{y}_1 \\ \dot{y}_2 \end{Bmatrix} + \begin{bmatrix} 2k & 0 \\ 0 & 6k \end{bmatrix} \begin{Bmatrix} y_1 \\ y_2 \end{Bmatrix} = \begin{Bmatrix} F_0 \\ F_0 \end{Bmatrix} \tag{e}$$

Expanding and dividing both sides of both equations by $2m$,

$$\ddot{y}_1 + 2\zeta_1 \omega_1 \dot{y}_1 + \omega_1^2 y_1 = \frac{1}{2} f_0 \tag{f}$$

$$\ddot{y}_2 + 2\zeta_2 \omega_2 \dot{y}_2 + \omega_2^2 y_2 = \frac{1}{2} f_0 \tag{g}$$

where $f_0 = F_0/m$ and

$$\omega_1 = \sqrt{\frac{k}{m}} \qquad \zeta_1 = \frac{c_1}{2m\omega_1} \qquad \omega_{1d} = \sqrt{1 - \zeta_1^2}\,\omega_1 \tag{h}$$

$$\omega_2 = \sqrt{\frac{3k}{m}} \qquad \zeta_2 = \frac{c_1 + 2c_2}{2m\omega_2} \qquad \omega_{2d} = \sqrt{1 - \zeta_2^2}\,\omega_2 \tag{i}$$

They are now a set of completely decoupled equations of motion. We can treat them as two transient vibration problems of single-DOF systems and solve individually. Each of the system is subjected to a step loading, with trivial initial conditions. This problem has been solved in Example 2.10. Without repeating the solution process, we adapt the solutions in Example 2.10. This gives

$$y_1(t) = \frac{f_0}{2\omega_1^2} \left[1 - \frac{e^{-\zeta_1 \omega_1 t}}{\sqrt{1 - \zeta_1^2}} \cos(\omega_{1d} t - \phi_1) \right] \tag{j}$$

$$y_2(t) = \frac{f_0}{2\omega_2^2} \left[1 - \frac{e^{-\zeta_2 \omega_2 t}}{\sqrt{1 - \zeta_2^2}} \cos(\omega_{2d} t - \phi_2) \right] \tag{k}$$

where

$$\phi_1 = \tan^{-1} \frac{\zeta_1}{\sqrt{1 - \zeta_1^2}} \qquad \phi_2 = \tan^{-1} \frac{\zeta_2}{\sqrt{1 - \zeta_2^2}} \tag{l}$$

Finally, we convert the solutions back to the original set of generalized coordinates according to eqn. (4.72),

$$\{x\} = \begin{Bmatrix} x_1 \\ x_2 \end{Bmatrix} = [V]\{y\} = \begin{bmatrix} 1 & 1 \\ 1 & -1 \end{bmatrix} \begin{Bmatrix} y_1 \\ y_2 \end{Bmatrix} \tag{m}$$

Thus, the solution can be written as

$$x_1(t) = \frac{f_0}{\omega_1^2}\left[\frac{2}{3} - \frac{1}{2}\frac{e^{-\zeta_1\omega_1 t}}{\sqrt{1-\zeta_1^2}}\cos(\omega_{1d}t - \phi_1) - \frac{1}{6}\frac{e^{-\zeta_2\omega_2 t}}{\sqrt{1-\zeta_2^2}}\cos(\omega_{2d}t - \phi_2)\right] \quad (n)$$

$$x_2(t) = \frac{f_0}{\omega_1^2}\left[\frac{1}{3} - \frac{1}{2}\frac{e^{-\zeta_1\omega_1 t}}{\sqrt{1-\zeta_1^2}}\cos(\omega_{1d}t - \phi_1) + \frac{1}{6}\frac{e^{-\zeta_2\omega_2 t}}{\sqrt{1-\zeta_2^2}}\cos(\omega_{2d}t - \phi_2)\right] \quad (o)$$

We can see that when there is no damping, the solution would be identical to the solution in Example 4.19.

□ **Exploring the Solution with MATLAB**

We define the following nondimensionalization parameters: for $i = 1, 2$,

$$\bar{x}_i = \frac{x_i}{f_0/\omega_1^2} = \frac{x_i}{F_0/k} \qquad \bar{t} = \omega_1 t \quad (p)$$

and

$$\bar{\omega}_{d1} = \sqrt{1-\zeta_1^2} \qquad \bar{\omega}_{d2} = \sqrt{1-\zeta_2^2}\frac{\omega_2}{\omega_1} = \sqrt{3(1-\zeta_2^2)} \quad (q)$$

Then, the solutions in eqns. (n) and (o) can be nondimensionalized as

$$\bar{x}_1(\bar{t}) = \frac{2}{3} - \frac{1}{2}\frac{e^{-\zeta_1\bar{t}}}{\sqrt{1-\zeta_1^2}}\cos\left(\bar{\omega}_{d1}\bar{t} - \phi_1\right) - \frac{1}{6}\frac{e^{-\sqrt{3}\zeta_2\bar{t}}}{\sqrt{1-\zeta_2^2}}\cos\left(\bar{\omega}_{d2}\bar{t} - \phi_2\right) \quad (r)$$

$$\bar{x}_2(\bar{t}) = \frac{1}{3} - \frac{1}{2}\frac{e^{-\zeta_1\bar{t}}}{\sqrt{1-\zeta_1^2}}\cos\left(\bar{\omega}_{d1}\bar{t} - \phi_1\right) + \frac{1}{6}\frac{e^{-\sqrt{3}\zeta_2\bar{t}}}{\sqrt{1-\zeta_2^2}}\cos\left(\bar{\omega}_{d2}\bar{t} - \phi_2\right) \quad (s)$$

The following MATLAB code can be used to plot the system responses, which are shown in Figures 4.33 through 4.34 for three different combinations of damping ratios: $\zeta_1 = 0.1$ and $\zeta_2 = 0.1/\sqrt{3}$ ($c_2 = 0$); $\zeta_1 = 0$ and $\zeta_2 = 0.2/\sqrt{3}$ ($c_1 = 0$); and $\zeta_1 = 0.1$ and $\zeta_2 = 0.3/\sqrt{3}$ ($c_1 = c_2$).

```
zeta1=0.1;
zeta2=0.1/sqrt(3);

deno1=sqrt(1-zeta1^2);
deno2=sqrt(1-zeta2^2);
phi1=atan(zeta1/deno1);
phi2=atan(zeta2/deno2);

t = [0:.01:40];
x1 = 2/3 - .5*exp(-zeta1*t) .* cos (deno1*t-phi1)/deno1 ...
     - 1/6*exp(-sqrt(3)*zeta2*t).* cos(deno1*sqrt(3)*t-phi2)/deno2;
x2 = 1/3 - .5*exp(-zeta1*t) .* cos (deno1*t-phi1)/deno1 ...
     + 1/6*exp(-sqrt(3)*zeta2*t).* cos(deno1*sqrt(3)*t-phi2)/deno2;
```

```
plot(t,x1,'r',t,x2,'--b');
xlabel('\omega_1 t');
ylabel('x_{1,2}(t)/(F_0/k)');
legend( 'x_1(t)', 'x_2(t)' );
```

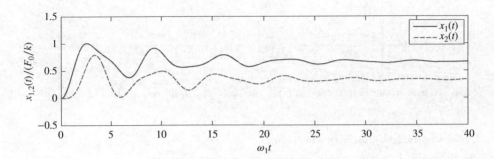

Figure 4.32 Transient responses of two-mass mass–spring–dashpot system due to step loading on the left mass at $\zeta_1 = 0.1$, $\zeta_2 = 0.1/\sqrt{3}$ $(c_2 = 0)$

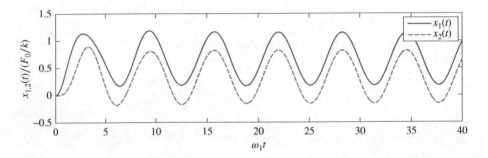

Figure 4.33 Transient responses of two-mass mass–spring–dashpot system due to step loading on the left mass at $\zeta_1 = 0$, $\zeta_2 = 0.2/\sqrt{3}$ $(c_1 = 0)$

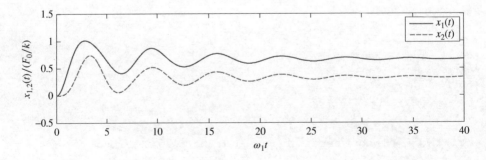

Figure 4.34 Transient responses of two-mass mass–spring–dashpot system due to step loading on the left mass at $\zeta_1 = 0.1$, $\zeta_2 = 0.3/\sqrt{3}$ $(c_1 = c_2)$

■ Example 4.21: Step Loading on Damped Unconstrained System

Consider a damped unconstrained system: two masses are connected by a spring and a dashpot, as shown in Fig. 4.24. A free-vibration problem of this system has been studied in Example 4.16. In this example, the mass on the left is subjected to a constant force $F_1(t) = F_0$ at $t = 0$. Find the subsequent transient response of the system.

□ Solution:

We continue to use x_1 and x_2 as defined in Example 4.16 as the complete and independent set of generalized coordinates. We can directly quote some results from Example 4.16: the natural frequencies are

$$\omega_1 = 0 \qquad \omega_2 = \sqrt{\frac{2k}{m}} \tag{a}$$

The modal matrix and its inversion are

$$[V] = \begin{bmatrix} 1 & 1 \\ 1 & -1 \end{bmatrix} \qquad [V]^{-1} = \frac{1}{2} \begin{bmatrix} 1 & 1 \\ 1 & -1 \end{bmatrix} \tag{b}$$

The diagonalized characteristic matrices are

$$[M_d] = \begin{bmatrix} 2m & 0 \\ 0 & 2m \end{bmatrix} \qquad [C_d] = \begin{bmatrix} 0 & 0 \\ 0 & 4c \end{bmatrix} \qquad [K_d] = \begin{bmatrix} 0 & 0 \\ 0 & 4k \end{bmatrix} \tag{c}$$

In this example, we need to transform the forcing as

$$[V]^T \{F\} = \begin{bmatrix} 1 & 1 \\ 1 & -1 \end{bmatrix} \begin{Bmatrix} F_0 \\ 0 \end{Bmatrix} = \begin{Bmatrix} F_0 \\ F_0 \end{Bmatrix} \tag{d}$$

The equations of motion in terms of the principal coordinates y_1 and y_2 are

$$\begin{bmatrix} 2m & 0 \\ 0 & 2m \end{bmatrix} \begin{Bmatrix} \ddot{y}_1 \\ \ddot{y}_2 \end{Bmatrix} + \begin{bmatrix} 0 & 0 \\ 0 & 4c \end{bmatrix} \begin{Bmatrix} \dot{y}_1 \\ \dot{y}_2 \end{Bmatrix} + \begin{bmatrix} 0 & 0 \\ 0 & 4k \end{bmatrix} \begin{Bmatrix} y_1 \\ y_2 \end{Bmatrix} = \begin{Bmatrix} F_0 \\ F_0 \end{Bmatrix} \tag{e}$$

or

$$m\ddot{y}_1 = \frac{1}{2}F_0 \tag{f}$$

$$m\ddot{y}_2 + 2c\dot{y}_2 + 2ky_2 = \frac{1}{2}F_0 \tag{g}$$

The y_1-equation, eqn. (f), describes a motion with a constant acceleration. The solution for the trivial initial conditions is

$$y_1(t) = \frac{F_0}{4m}t^2 \tag{h}$$

The y_2-equation, eqn. (g), represents a single-DOF system subjected to step loading. We directly adapt the result in Example 2.10 as

$$y_2(t) = \frac{F_0}{4k} \left[1 - \frac{e^{-\zeta_2 \omega_2 t}}{\sqrt{1 - \zeta_2^2}} \cos(\omega_{2d} t - \phi_2) \right] \tag{i}$$

where

$$\omega_2 = \sqrt{\frac{2k}{m}} \qquad \zeta_2 = \frac{2c}{m\omega_2} \qquad \omega_{2d} = \sqrt{1-\zeta_2^2}\,\omega_2 \qquad \phi_2 = \tan^{-1}\frac{\zeta_2}{\sqrt{1-\zeta_2^2}} \tag{j}$$

Converting the above solutions back to the original set of generalized coordinates gives

$$x_1(t) = y_1 + y_2 = \frac{F_0}{4m}t^2 + \frac{F_0}{4k}\left[1 - \frac{e^{-\zeta_2\omega_2 t}}{\sqrt{1-\zeta_2^2}}\cos(\omega_{2d}t - \phi_2)\right] \tag{k}$$

$$x_2(t) = y_1 - y_2 = \frac{F_0}{4m}t^2 - \frac{F_0}{4k}\left[1 - \frac{e^{-\zeta_2\omega_2 t}}{\sqrt{1-\zeta_2^2}}\cos(\omega_{2d}t - \phi_2)\right] \tag{l}$$

□ **Exploring the Solution with MATLAB**

We define the following nondimensionalization parameters: for $i = 1, 2$,

$$\bar{x}_i = \frac{x_i}{F_0/k} \qquad \bar{t} = \omega_2 t \tag{m}$$

Then the nondimensionalized responses can be written as

$$\bar{x}_1(\bar{t}) = \frac{1}{4}\left[\bar{t}^2 + 1 - \frac{e^{-\zeta_2\bar{t}}}{\sqrt{1-\zeta_2^2}}\cos\left(\sqrt{1-\zeta_2^2}\,\bar{t} - \phi_2\right)\right] \tag{n}$$

$$\bar{x}_2(\bar{t}) = \frac{1}{4}\left[\bar{t}^2 - 1 + \frac{e^{-\zeta_2\bar{t}}}{\sqrt{1-\zeta_2^2}}\cos\left(\sqrt{1-\zeta_2^2}\,\bar{t} - \phi_2\right)\right] \tag{o}$$

The following MATLAB code can be used to plot the response curve.

```
zeta2=0.01;
deno2=sqrt(1-zeta2^2);
phi2=atan(zeta2/deno2);
t = [0:.01:15];
x1 = .25*( t.^2 +1 - exp(-zeta2*t) .* cos (deno2 * t-phi2)/deno2 );
x2 = .25*( t.^2 -1 + exp(-zeta2*t) .* cos (deno2 * t-phi2)/deno2 );
y2 = .25*( 1 - exp(-zeta2*t) .* cos (deno2 * t-phi2)/deno2 );
plot(t,x1,'r',t,x2,'--g',t,y2,':m');
xlabel('\omega_1 t');
ylabel('x_{1,2}(t)/(F_0/k)');
legend( 'x_1(t)', 'x_2(t)' );
```

The response is plotted in Fig. 4.35. The motion is dominated by the constant acceleration rigid-body motion. The relative motion is in rather small scale. The response for y_2 is plotted separately to show the relative motions between the two masses.

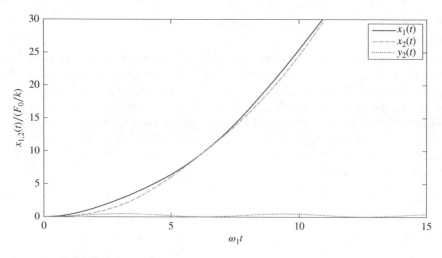

Figure 4.35 Responses of unconstrained two-mass mass–spring–dashpot system subjected to step loading on the left mass

■ Example 4.22: Free-Falling Slinky

A few years ago, a video of free-falling Slinky generated an intense public interest on the Internet. The phenomenon is called the *Slinky levitation*: a Slinky is held at its top vertically in equilibrium; then it is released. It is observed that the bottom of the Slinky remains stationary in the mid-air until the top hits it. This defies a common conception of a "free fall." The video was accompanying a research paper explaining this curious phenomenon that was published in, and featured on the cover of, December 2012 issue of the *American Journal of Physics*. A few video frame grabs are shown in Fig. 4.36. In this example, we use MatLab to model the falling Slinky.

Figure 4.36 Frames from video demonstration of free-falling Slinky. Notice that bottom of Slinky does not move in the first four frames. (Reproduced with permission from Cross & Wheatland (2012). © 2012 American Association of Physics Teachers.)

□ **Solution:**

Modeling

Slinky is a long soft spring. We assume that it has uniformly distributed mass along its wire of N coils. We model each coil as a mass and a spring; all the masses are identical; and all springs are identical. We assume that each spring has an unstretched length of l_0. Therefore, in our model for the Slinky, we have N mass–spring subsystems attached in series. We number the masses and springs from top down. This way, spring i connects to mass $i - 1$ above it (or ceiling) and mass i beneath it. The top spring is attached to the ceiling (or a steady hand). The Slinky is allowed to reach its equilibrium in this way. Then, at $t = 0$, the top spring is cut, and the Slinky starts its free fall.

Equation of Motion

The model described above is similar to the 10-DOF mass–spring system in Example 4.11. Here, we derive the equations of motion using Lagrangian dynamics first.

- *Generalized Coordinate*: We define $\{\, x_j : j = 1, 2, \ldots, N \,\}$ as a complete and independent set of generalized coordinates, where x_j is the downward displacement of mass m_j with respect to ground measured from its equilibrium position.
- *Admissible Variations*: We verify $\{\, \delta x_j : j = 1, 2, \ldots, N \,\}$ is a complete and independent set of admissible variations in this set of generalized coordinates.
- *Holonomicity*: We conclude that the system is holonomic and has N DOFs.
- *Generalized Force*: There is no nonconservative element in the system. Thus, $\Xi_j = 0$.
- *Lagrangian*: For spring i, the amount of stretch is $x_i - x_{i-1} + \Delta_i$, where Δ_i is the stretch at equilibrium. Thus,

$$\mathcal{L} = \sum_{i=1}^{N} \frac{1}{2} m_i \dot{x}_i^2 - \sum_{i=1}^{N} \frac{1}{2} k_i \left(x_i - x_{i-1} + \Delta_i \right)^2 + m_i g x_i \tag{a}$$

where the first spring ($i = 1$) is also included and $x_0 = 0$ when needed. Although it has been assumed that all masses and springs are identical, subscripts are still used for m's and k's for identification purpose.

- *Lagrange's Equation*: The Lagrange's equation for x_i is

$$\frac{d}{dt} \left(\frac{\partial \mathcal{L}}{\partial \dot{x}_i} \right) - \frac{\partial \mathcal{L}}{\partial x_i} = \Xi_{x_i} \tag{b}$$

For $i = 2, 3, \ldots, N - 1$,

$$\frac{\partial \mathcal{L}}{\partial \dot{x}_i} = m_i \dot{x}_i \qquad \frac{\partial \mathcal{L}}{\partial x_i} = -k_i \left(x_i - x_{i-1} + \Delta_i \right) + k_{i+1} \left(x_{i+1} - x_i + \Delta_{i+1} \right) + m_i g$$

Then, Lagrange's equation gives the following equation of motion:

$$m_i \ddot{x}_i - k_i x_{i-1} + (k_i + k_{i+1}) x_i - k_{i+1} x_{i+1} = m_i g - k_i \Delta_i + k_{i+1} \Delta_{i+1} \tag{c}$$

For the top spring $i = 1$, terms involving k_{i-1} and x_{i-1} vanish. There are two versions:
- Before the release, k_1 holds the entire system in equilibrium, that is,

$$m_1 \ddot{x}_1 + (k_1 + k_2) x_1 - k_2 x_2 = m_1 g - k_1 \Delta_1 + k_2 \Delta_2 \tag{d}$$

- During the flight, k_1 has been cut, giving

$$m_1\ddot{x}_1 + k_2x_1 - k_2x_2 = m_1g + k_2\Delta_2 \tag{e}$$

For the bottom mass, $i = N$, terms involving k_{i+1} and x_{i+1} vanish, that is,

$$m_N\ddot{x}_N - k_Nx_{N-1} + k_Nx_N = m_Ng - k_N\Delta_N \tag{f}$$

Equilibrium Positions

To finalize the equations of motion, we need to evaluate Δ_i. At equilibrium, the constant terms in each equation of motion vanish. This leads to a set of N linear equations. We could solve this set of equations easily using MATLAB. In fact, they can be solved analytically without much trouble, and analytical solutions are always preferred.

The last spring at the bottom connects only to one mass. Thus, from eqn. (f),

$$k_N\Delta_N = m_Ng \quad \text{or} \quad \Delta_N = \frac{m_Ng}{k_N} \tag{g}$$

We then can work back up to the top, from eqn. (c) and setting the constant terms to vanish:

$$k_i\Delta_i = m_ig + k_{i+1}\Delta_{i+1} \tag{h}$$

which is also valid for the case $i = 1$ as obtained from eqn. (d). Equation (h) is an iterative relation. Using eqn. (g) as a starting point, it can be easily worked out as

$$k_i\Delta_i = \sum_{j=i}^{N} m_jg \tag{i}$$

which is rather intuitive: spring i is stretched by the weights of all masses beneath it. In the case where all the masses are m and all springs are k, we have, for all i,

$$\Delta_i = (N - i + 1)\frac{mg}{k} \tag{j}$$

For a more intuitive illustration for the locations of the masses, we define a Cartesian coordinate system whose origin is located at the ceiling and the y-axis is pointing upward. Denote the equilibrium position of mass i as y_i^{eq}. Then, the first mass is located at $y_1^{eq} = -l_0 - \Delta_1$; the second mass is at $y_2^{eq} = y_1^{eq} - l_0 - \Delta_2$. We can deduce that, for mass i,

$$y_i^{eq} = -il_0 - \sum_{p=1}^{i}\Delta_p = -il_0 - \frac{mg}{k}\sum_{p=0}^{i}(N - p + 1) = -il_0 - \frac{img}{k}\left(N - \frac{i-1}{2}\right) \tag{k}$$

When the system moves,

$$y_i(t) = y_i^{eq} - x_i(t) = -il_0 - \frac{img}{k}\left(N - \frac{i-1}{2}\right) - x_i(t) \tag{l}$$

Equations of Motion for Slinky in Flight

We are now ready to finalize the equations of motion for the falling Slinky. For the Slinky in flight, the equation of motion for the first mass is eqn. (e). Its right-hand side does not vanish!

Comparing the right-hand side of eqn. (e) with that of eqn. (d), whose right-hand side vanishes at equilibrium, we have

$$m_1 g + k_2 \Delta_2 = k_1 \Delta_1 = Nmg \tag{m}$$

where eqn. (j) has been used. That is, the equation of motion for the first mass is

$$m_1 \ddot{x}_1 + k_2 x_1 - k_2 x_2 = Nmg \tag{n}$$

This suggests that when released, the weight of the entire Slinky acts as a step loading on to the first mass. For other masses, the right-hand sides of eqns. (e) and (f) vanish due to the equilibrium conditions. In matrix form, the set of equations of motion can be written as

$$\begin{bmatrix} m & 0 & \cdots & 0 \\ 0 & m & \cdots & 0 \\ \vdots & \vdots & \ddots & \vdots \\ 0 & 0 & \cdots & m \end{bmatrix} \begin{Bmatrix} \ddot{x}_1 \\ \ddot{x}_2 \\ \vdots \\ \ddot{x}_N \end{Bmatrix} + \begin{bmatrix} k & -k & 0 & \cdots & 0 \\ -k & 2k & -k & \cdots & 0 \\ \vdots & \vdots & \vdots & \ddots & \vdots \\ 0 & 0 & 0 & \cdots & k \end{bmatrix} \begin{Bmatrix} x_1 \\ x_2 \\ \vdots \\ x_N \end{Bmatrix} = \begin{Bmatrix} Nmg \\ 0 \\ \vdots \\ 0 \end{Bmatrix} \tag{o}$$

In writing the above equation, we have used the information that all masses are equal, denoted as m, and all springs are equal, denoted as k.

Normalizing the Equations of Motion

We shall leave much of the solution process to MATLAB. Here, we nondimensionalize the equations of motion before moving to MATLAB. From our earlier experience with the 10-DOF system in Example 4.11, we define the following nominal frequency ω_0 and use it to normalize the time:

$$\omega_0 = \sqrt{\frac{k}{m}} \qquad \bar{t} = \omega_0 t$$

This way, as we have seen in Example 4.11,

$$\frac{d}{dt} = \frac{d}{d\bar{t}} \frac{d\bar{t}}{dt} = \omega_0 \frac{d}{d\bar{t}} \qquad \frac{d^2}{dt^2} = \omega_0^2 \frac{d^2}{d\bar{t}^2}$$

Dividing both sides of eqn. (o) by k would nondimensionalize the time aspect of the equation; and dividing both sides by Nl_0 would further nondimensionalize the displacements. The nondimensionalized equations of motion become

$$\begin{bmatrix} 1 & 0 & \cdots & 0 \\ 0 & 1 & \cdots & 0 \\ \vdots & \vdots & \ddots & \vdots \\ 0 & 0 & \cdots & 1 \end{bmatrix} \begin{Bmatrix} \ddot{\bar{x}}_1 \\ \ddot{\bar{x}}_2 \\ \vdots \\ \ddot{\bar{x}}_N \end{Bmatrix} + \begin{bmatrix} 1 & -1 & 0 & \cdots & 0 \\ -1 & 2 & -1 & \cdots & 0 \\ 0 & -1 & 2 & \cdots & 0 \\ \vdots & \vdots & \vdots & \ddots & \vdots \\ 0 & 0 & 0 & \cdots & 1 \end{bmatrix} \begin{Bmatrix} \bar{x}_1 \\ \bar{x}_2 \\ \bar{x}_3 \\ \vdots \\ \bar{x}_N \end{Bmatrix} = \begin{Bmatrix} r \\ 0 \\ 0 \\ \vdots \\ 0 \end{Bmatrix} \tag{p}$$

where we continue to use a dot over a nondimensionalized variable to represent the derivative with respect to the nondimensionalized time, that is, $\dot{\bar{x}} = d\bar{x}/d\bar{t}$ and

$$\bar{x}_i = \frac{x_i}{Nl_0} \qquad r = \frac{Nmg}{kNl_0} = \frac{mg}{kl_0} \tag{q}$$

Physically, Nl_0 is the unstretched length of the entire Slinky; and r represents the ratio of the static deflection of the Slinky to its unstretched length, or the ratio for each coil of it. The above nondimensionalization process suggests that r is the only characterizing parameter for the Slinky, apart from the number of coils.

Solution via Decoupling Method

The equations of motion in eqn. (p) can be written symbolically as

$$[\overline{M}]\{\ddot{\overline{x}}\} + [\overline{K}]\{\overline{x}\} = \{\overline{F}\} \tag{r}$$

The system is undamped. We have the choices of using either the analytical method or the decoupling method to solve it. Here we choose the latter, for its broader applicability to damped systems. After decoupling, the equations of motion in terms of principal coordinates $\{\overline{\xi}\}$ are expressible as

$$[\overline{M}_d]\{\ddot{\overline{\xi}}\} + [\overline{K}_d]\{\overline{\xi}\} = \{\overline{F}_p\} \tag{s}$$

We expect that the falling Slinky has a rigid-body mode, which will be identified by a vanishing first natural frequency, along with a vanishing corresponding stiffness. In this mode, the modal equation should become, as seen in Example 4.21,

$$\overline{M}_{d1}\ddot{\overline{\xi}}_1 = \overline{F}_{p1} \tag{t}$$

With the trivial initial conditions, the solution for this mode is

$$\overline{\xi}_1(\overline{t}) = \frac{1}{2}\frac{\overline{F}_{p1}}{\overline{M}_{d1}}\overline{t}^2 \tag{u}$$

For other modes, say Mode i, the modal equation of motion is

$$\overline{M}_{di}\ddot{\overline{\xi}}_i + \overline{K}_{di}\overline{\xi}_i = \overline{F}_{pi} \tag{v}$$

and, with the trivial initial conditions, the solution is

$$\overline{\xi}_i(\overline{t}) = \frac{\overline{F}_{pi}}{\overline{K}_{di}}\left(1 - \cos\overline{\omega}_i\overline{t}\right) \tag{w}$$

□ Exploring the Solution with MATLAB

The following MATLAB code represents the complete solution process:

```
% Part 1: Problem Setup
N=100;    % Number of coils
r = 1;    % Slinky property: r = mg/kl_0

% Part 2: System Setup
m=eye(N,N);
k=2*eye(N,N);
for i=1:N-1
    k(i,i+1) = -1;
    k(i+1,i) = -1;
end;
```

```
k(1,1)=1;
k(N,N)=1;
f=zeros(N,1);
f(1,1)=r;


% Part 3: Calculate eigenvalues and eigenvectors
[V,lambda]=eig(k,m);


% Part 4: Diagonalize and transform to principal coordinates
Kd = V' * k * V;
Md = V' * m * V;
F = V' * f;


% Part 5: Setup time array and solve in principal coordinates xp.
t=[0:1:500];
xp=zeros(N,length(t));
if abs(lambda(1,1))<1.E-10    % Check if first natural frequency is zero
    xp(1,:) = F(1,1)/Md(1,1)/2* t' .* t';
else
    xp(1,:) = F(1,1)/Kd(1,1)*(1-cos( sqrt(Kd(1,1)/Md(1,1)) * t'));
end
for i=2:N;                     % For all other modes
    xp(i,:) = F(i,1)/Kd(i,i)*(1-cos( sqrt(Kd(i,i)/Md(i,i)) * t'));
end


% Part 5: Convert back to original generalized coordinates
x = V*xp;


% Part 6: Convert each coil's displacement to its position
for i=1:N
    y(i,:) = (-i -i*r*(N-(i-1)/2))/N - x(i,:);
end;
yC = sum(y)/N;     % Extra: Keep track of the mass center


% Part 7: Plotting the results, in two subplots
subplot(1,2,1);
plot(t,x);
xlabel({'\_','t'});
ylabel({'\_','x_i'}');


subplot(1,2,2);
plot(t,y);
% plot(t,yC);                  % Plot the center of mass, if needed
xlabel({'\_','t'});
ylabel({'\_','y_i'}');
```

Most of the above code should be familiar, as we have seen in Example 4.11. In the above code, xp represents the principal coordinates, which is set up as an $N \times N_t$ matrix, where N_t is the length of the time t array. This way, each column represents a time instant and each row represents the time history of a mass. Matrices x and y are also set up this way. In the middle of the code, an if condition is used to check if the first natural frequency is zero. We do not use the logical equal (==) operator to check. In numerical computations, errors are always involved and there is almost never a case when an expected result of zero would turn out to be exactly zero. We need to allow some error tolerance. In the code, we set the error tolerance as 10^{-10}. The plot function is called with matrices of different sizes: t of size $1 \times N_t$, and x or y of size $N \times N_t$, and the number of columns of x or y equals the length of t. In this case, the plot function plots all rows, each row as a curve. This way, one simple function call plots all masses' displacements or positions.

Figure 4.37 shows the Slinky's motion, both displacements and positions for all its coils, with assumed $r = 1$ and $N = 100$, for time from $\bar{t} = 0$ up to 150.

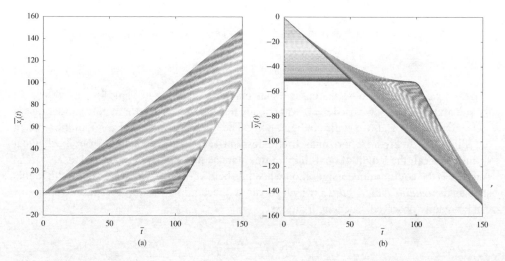

Figure 4.37   Motion of free-falling slinky, with $r = 1$ and $N = 100$, for $\bar{t}$ from 0 up to 150: (a) displacements $x_i(t)$ and (b) position $y_i(t)$

To verify the correctness of the results, we check the initial conditions. Indeed each mass starts with a zero displacement and a zero slope. In the mean time, we notice another strange behavior: the top mass, represented by the top curve in either plot, appears to be moving at a constant speed (a straight line in the figure). This defies a common conception: during the free fall, it should accelerate, and its displacement should be a function of $\bar{t}^2$.

We are curious enough to explore. So we give it a longer time to run, up to $\bar{t} = 500$, as shown in Fig. 4.38. Now this figure shows what we have expected: overall, the Slinky does fall with its displacement proportional to $\bar{t}^2$. This assures that the results are likely correct.

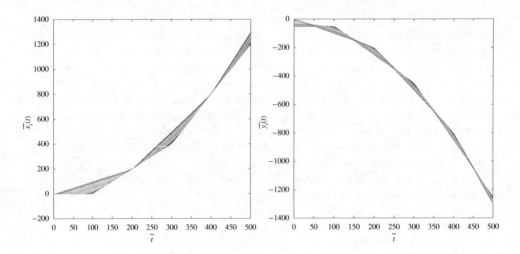

Figure 4.38   Motion of falling Slinky: same as Fig. 4.37 but for much longer time

Before we make more observations, we shall choose an appropriate value for parameter r. In the previous two plots, we used $r = 1$ without any reason. This is fine for merely validating the code. Here, we look at two static configurations. The first is the unloaded configuration: when the Slinky lies freely on a horizontal frictionless surface. To borrow a term from mechanical springs, we call the length of the Slinky in this state as the *shut length*, denoted as $L_s = Nl_0$. Another is the equilibrium configuration when it is held vertically. We denote its length in this state as the *equilibrium length* L_e, which equals $|y_N|$ according to eqn. (l). These two lengths can be readily observed. The length ratio is

$$\frac{L_e}{L_s} = \frac{Nl_0 + \frac{mg}{k}N\left(N - \frac{N-1}{2}\right)}{Nl_0} = 1 + r\frac{N+1}{2} \tag{x}$$

Thus, with $r = 1$ and $N = 100$ as we set previously, the length ratio is 50.5, which is not too far off. In the following discussions, we shall use the Slinky in the paper mentioned in the caption of Fig. 4.36: $N = 80$, $L_e = 1$ m, and $L_s = 0.06$ m, which gives $r = 0.3868$.

Figure 4.39 shows coil positions of this Slinky during an early stage of motion, up to the point where the bottom coil starts moving. In addition, a trajectory of the Slinky's centroid, which is denoted in the code as yC, is superimposed onto the trajectories of coils. It is shown that the Slinky's centroid indeed follows a smooth trajectory of a parabola. Figure 4.40 shows "snapshots" of the configurations of the Slinky in flight, for time instants from $\bar{t} = 0$ (at equilibrium) to 80 at an interval of 10. The snapshots are produced by the following MATLAB code:

```
hold on;
for ti=0:10:80;
```

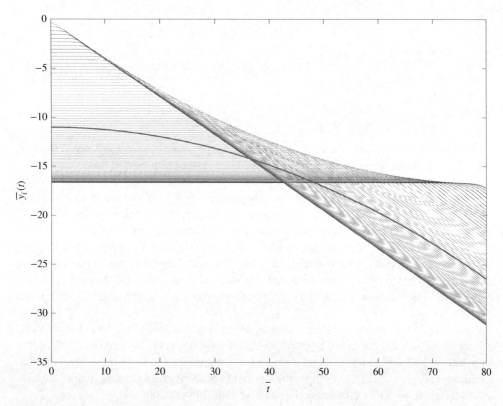

Figure 4.39 Positions of Slinky coils for case $r = 0.3868$, which represents 80-coil Slinky with equilibrium length being $16\frac{2}{3}$ times of shut length

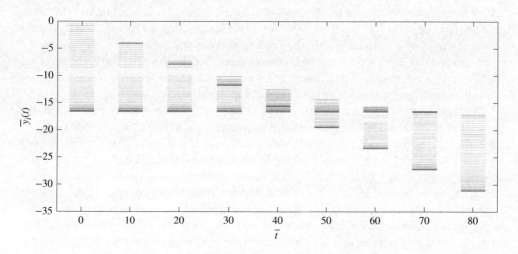

Figure 4.40 Snapshots of Slinky's configurations during free fall

```
    for i=N:-1:1
        plot([ti-2.5 ti+2.5],[y(i,ti+1) y(i,ti+1)]);
    end;
end
axis([-5 85 -35 0]);
xlabel({'\_','t'});
ylabel({'\_','y_i'}');
```

We can now start making observations. First, we see the main phenomenon that caused the Internet buzz: the bottom coil remains stationary for an extended period of time.

Second, as we mentioned earlier, the top coil appears to be moving at a constant speed, and not accelerating at all, except at the very beginning when it accelerates from zero initial velocity to the steady one. This is most clearly visible from the trajectory of the top coil in Fig. 4.39. In fact, all coils move in the same speed until something happened.

Third, the bottom coil does not move for quite a while; and when it moves, it moves at a faster speed, about twice the speed of the top coil. This speed doubling can be observed from Fig. 4.38a: it starts to move at about $\bar{t} = 100$. The displacement curves for the top coil and the bottom coil cross at about $\bar{t} = 200$. So the bottom coil takes a half of the time to catch up with the top coil.

Fourth, when the bottom coil starts to move, which occurs at around $\bar{t} = 79$ in Fig. 4.39, the Slinky's configuration is entirely flipped upside down, as seen in Fig. 4.40 at $\bar{t} = 80$. Referring back to Fig. 4.37 for the motion over a longer time, it appears that the flipping is repeated as the Slinky falls. An actual Slinky probably will not be able to achieve this flipping as the coils would tangle up as soon as the top coil gets close to the bottom coil.

As we mentioned earlier, the Slinky model is similar to the 10-mass model in Example 4.11. We recall that in single-DOF systems, the step loading of a system is essentially the same type of problem as an initial displacement free-vibration problem. The mass train in Example 4.11 was subjected to an initial displacement at the first mass, and the disturbance travels through the chain at a constant speed. In the Slinky example, the top coil is subjected to a step loading. After seeing the response for the 10-mass train in Example 4.11, it is no surprise for us to see that the bottom coil of the Slinky will not move for an extended period of time, and the top coil moves at a constant speed. The fundamental physics behind both examples is the wave phenomena. Both models are lumped-parameter models of similar continuum systems in which waves propagate. Disturbances of all sorts propagate as waves at a constant speed called the *wave speed*, which is a characteristic of a continuum.

The flight of an actual Slinky is much more complicated. Our model can only capture the essence of it. In the snapshots, we see that the top coil moves to beneath the second coil soon after the flight starts. This is not possible with a real Slinky. An actual Slinky is called, in engineering terms, a preloaded extension spring. It can only be loaded in extension; and while it is completely shut, it may have already been loaded to a certain level such that a load is needed to overcome the preload in order to extend it. In our model, we do not have a mechanism to prevent the springs from going into compression nor to prevent a coil from passing through the coils beneath. Adding such mechanisms would make the springs nonlinear. Despite these limitations, our model helps us to make some interesting observations, some are widely reported and caught people's fascination.

4.8.3   Laplace Transform Method

In general, the Laplace transform method can be used to analyze transient vibrations of general multi-DOF systems with damping. Assume that the equations of motion for a system can be written in the matrix form as

$$[M]\{\ddot{x}\} + [C]\{\dot{x}\} + [K]\{x\} = \{F(t)\} \tag{4.83}$$

The Laplace transform can be applied to the unknown $\{x(t)\}$ and its derivatives

$$\mathscr{L}[\{x(t)\}] = \{\tilde{x}(s)\} \tag{4.84}$$

$$\mathscr{L}[\{\dot{x}(t)\}] = s\{\tilde{x}(s)\} - \{x(0)\} \tag{4.85}$$

$$\mathscr{L}[\{\ddot{x}(t)\}] = s^2\{\tilde{x}(s)\} - s\{x(0)\} - \{\dot{x}(0)\} \tag{4.86}$$

where the initial conditions $\{x(0)\}$ and $\{\ddot{x}(0)\}$ have been incorporated. Then, the matrix equation of motion becomes

$$\left(s^2[M] + s[C] + [K]\right)\{\tilde{x}(s)\} = \{\tilde{F}(s)\} + (s[M] + [C])\{x(0)\} + [M]\{\ddot{x}(0)\} \tag{4.87}$$

which, despite the complicated looking, is a set of algebraic equations that can be solved by Cramer's rule. Afterward, the inverse transform can be performed for each response. The process of finding the inverse transform is exactly the same as the single-DOF system. The only difference is that the number of functions to be inverted is multiplied. The partial fraction method can be used in most situations.

In the following, we use a relatively simple example, an undamped two-mass mass–spring system, to illustrate the solution process.

■ **Example 4.23: Laplace Transform Method for Step Loading on Two-Mass Undamped System**

Rework Example 4.19 using the Laplace transform method.

□ **Solution:**

The matrix equation of motion for the system has been given in eqn. (a) in Example 4.19, which is copied here for convenience:

$$\begin{bmatrix} m & 0 \\ 0 & m \end{bmatrix}\begin{Bmatrix} \ddot{x}_1 \\ \ddot{x}_2 \end{Bmatrix} + \begin{bmatrix} 2k & -k \\ -k & 2k \end{bmatrix}\begin{Bmatrix} x_1 \\ x_2 \end{Bmatrix} = \begin{Bmatrix} F_0 \\ 0 \end{Bmatrix} \tag{a}$$

The Laplace-transformed equation, according to eqn. (4.87), can be written as

$$\begin{bmatrix} s^2m + 2k & -k \\ -k & s^2m + 2k \end{bmatrix}\begin{Bmatrix} \tilde{x}_1 \\ \tilde{x}_2 \end{Bmatrix} = \begin{Bmatrix} \frac{F_0}{s} \\ 0 \end{Bmatrix} \tag{b}$$

where the forcing is transformed according to $\mathscr{L}[F_0] = F_0/s$. The solution can be obtained using Cramer's rule as

$$\tilde{x}_1(s) = F_0\frac{ms^2 + 2k}{s[(ms^2 + 2k)^2 - k^2]} \tag{c}$$

$$\tilde{x}_2(s) = F_0\frac{k}{s[(ms^2 + 2k)^2 - k^2]} \tag{d}$$

We now use the partial fraction method to inverse those transformed solutions to the time domain. Note that a part of the denominator can be factorized as

$$(ms^2 + 2k)^2 - k^2 = (ms^2 + 3k)(ms^2 + k)$$

This factorization is inspired from the two natural frequencies of the system, denoted as

$$\omega_1 = \sqrt{k/m} \qquad \omega_2 = \sqrt{3k/m} \qquad\qquad (e)$$

Further factorizing the structure $a^2 + b^2 = (a + ib)(a - ib)$ using complex numbers, the denominator can be factorized into

$$s[(ms^2 + 2k)^2 - k^2] = m^2 s(s - i\omega_1)(s + i\omega_1)(s - i\omega_2)(s + i\omega_2) \qquad (f)$$

Then, we can write, for $\tilde{x}_1(s)$,

$$\tilde{x}_1(s) = \frac{ms^2 + 2k}{s[(ms^2 + 2k)^2 - k^2]} = \frac{c_1}{s} + \frac{c_2}{s - i\omega_1} + \frac{c_3}{s + i\omega_1} + \frac{c_4}{s - i\omega_2} + \frac{c_5}{s + i\omega_2}$$

where

$$c_1 = \left.\frac{ms^2 + 2k}{m^2(s^2 + \omega_1^2)(s^2 + \omega_2^2)}\right|_{s=0} = \frac{2}{3k} \qquad\qquad (g)$$

$$c_2 = \left.\frac{ms^2 + 2k}{m^2 s(s + i\omega_1)(s^2 + \omega_2^2)}\right|_{s=i\omega_1} = -\frac{1}{4k} \qquad\qquad (h)$$

$$c_3 = \left.\frac{ms^2 + 2k}{m^2 s(s - i\omega_1)(s^2 + \omega_2^2)}\right|_{s=-i\omega_1} = -\frac{1}{4k} \qquad\qquad (i)$$

$$c_4 = \left.\frac{ms^2 + 2k}{m^2 s(s^2 + \omega_1^2)(s + i\omega_2)}\right|_{s=i\omega_2} = -\frac{1}{12k} \qquad\qquad (j)$$

$$c_5 = \left.\frac{ms^2 + 2k}{m^2 s(s^2 + \omega_1^2)(s - i\omega_2)}\right|_{s=-i\omega_2} = -\frac{1}{12k} \qquad\qquad (k)$$

Thus,

$$x_1(t) = F_0 \left\{ \frac{2}{3k} \mathscr{L}^{-1}\left[\frac{1}{s}\right] - \frac{1}{4k}\left(\mathscr{L}^{-1}\left[\frac{1}{s - i\omega_1}\right] + \mathscr{L}^{-1}\left[\frac{1}{s + i\omega_1}\right]\right) \right.$$
$$\left. - \frac{1}{12k}\left(\mathscr{L}^{-1}\left[\frac{1}{s - i\omega_2}\right] + \mathscr{L}^{-1}\left[\frac{1}{s + i\omega_2}\right]\right) \right\}$$
$$= \frac{F_0}{k}\left[\frac{2}{3} - \frac{1}{4}\left(e^{i\omega_1 t} + e^{-i\omega_1 t}\right) - \frac{1}{12}\left(e^{i\omega_2 t} + e^{-i\omega_2 t}\right)\right]$$
$$= \frac{F_0}{k}\left[\frac{2}{3} - \frac{1}{2}\cos\omega_1 t - \frac{1}{6}\cos\omega_2 t\right] \qquad\qquad (l)$$

Similarly, for $\tilde{x}_2$, we can write

$$\tilde{x}_2(s) = \frac{k}{s[(ms^2 + 2k)^2 - k^2]} = \frac{c_6}{s} + \frac{c_7}{s - i\omega_1} + \frac{c_8}{s + i\omega_1} + \frac{c_9}{s - i\omega_2} + \frac{c_{10}}{s + i\omega_2}$$

where

$$c_6 = \left.\frac{k}{m^2(s^2 + \omega_1^2)(s^2 + \omega_2^2)}\right|_{s=0} = \frac{1}{3k} \tag{m}$$

$$c_7 = \left.\frac{k}{m^2 s(s + i\omega_1)(s^2 + \omega_2^2)}\right|_{s=i\omega_1} = -\frac{1}{4k} \tag{n}$$

$$c_8 = \left.\frac{k}{m^2 s(s - i\omega_1)(s^2 + \omega_2^2)}\right|_{s=-i\omega_1} = -\frac{1}{4k} \tag{o}$$

$$c_9 = \left.\frac{k}{m^2 s(s^2 + \omega_1^2)(s + i\omega_2)}\right|_{s=i\omega_2} = \frac{1}{12k} \tag{p}$$

$$c_{10} = \left.\frac{ms^2 + 2k}{m^2 s(s^2 + \omega_1^2)(s - i\omega_2)}\right|_{s=-i\omega_2} = \frac{1}{12k} \tag{q}$$

Thus,

$$x_2(t) = F_0 \left\{ \frac{1}{3k} \mathscr{L}^{-1}\left[\frac{1}{s}\right] - \frac{1}{4k}\left(\mathscr{L}^{-1}\left[\frac{1}{s - i\omega_1}\right] + \mathscr{L}^{-1}\left[\frac{1}{s + i\omega_1}\right]\right)\right.$$
$$\left. + \frac{1}{12k}\left(\mathscr{L}^{-1}\left[\frac{1}{s - i\omega_2}\right] + \mathscr{L}^{-1}\left[\frac{1}{s + i\omega_2}\right]\right)\right\}$$
$$= \frac{F_0}{k}\left[\frac{1}{3} - \frac{1}{4}\left(e^{i\omega_1 t} + e^{-i\omega_1 t}\right) + \frac{1}{12}\left(e^{i\omega_2 t} + e^{-i\omega_2 t}\right)\right]$$
$$= \frac{F_0}{k}\left[\frac{1}{3} - \frac{1}{2}\cos\omega_1 t + \frac{1}{6}\cos\omega_2 t\right] \tag{r}$$

The solution is exactly the same as the one we have found in Example 4.19.

4.8.4   Convolution Integral Method

Recall that, for single-DOF systems, the convolution integral method can be used for any input, resorting to numerical integration if necessary, making it the most versatile method. In this section, we explore the same idea to multi-DOF systems.

In the convolution integral method for a single-DOF system, the system's response to an unit impulse loading at $t = 0$, called the impulse response spectrum, is first obtained. Then, the system's response to a time-varying loading can be obtained by a convolution integral between the unit impulse response and the actual loading. The idea is based on the principle of superposition for linear systems. We also concluded that the system's response to the unit impulse loading at $t = 0$ is the same at a free-vibration response to a normalized initial velocity.

For an N-DOF system, the most general form of loading can be decomposed into N individual loadings, one applied to each DOF. Correspondingly, in the convolution integral method, we need to establish the system's responses due to N unit impulses, one applied to each DOF. For each impulse, there are N responses, one from each DOF. Therefore, a total of $N \times N$ impulse response spectra must be established. We arrange these impulse response spectra in an $N \times N$ square matrix $[H(t)]$, with the element at the ith row and jth column being $H_{ij}(t)$, where i is the DOF on which the impulse is applied and j is the DOF to which the response belongs. Then, in analogy to eqn. (2.128), the system's response can be written as

$$\{x(t)\} = \int_0^t [H(t-\tau)]\{F(\tau)\}d\tau = \int_0^t [H(\tau)]\{F(t-\tau)\}d\tau \qquad (4.88)$$

Unfortunately, this process is not very practical: we have to establish $H_{ij}(t)$ for every system that we deal with. This means solving a set of N^2 impulsive loading or free-vibration problems just as a preparatory step. And worse: we would have to do this for every system we deal with. This means that we cannot write a general-purpose computer code that is applicable to all systems. We really like the versatility of the Duhamel integral for single-DOF systems.

The decoupling method can lead us to that ideal. Recall that, by using the principal coordinates, we can convert a multi-DOF system into a set of uncoupled single-DOF systems. The convolution integral of eqn. (2.128) can be applied to these individual single-DOF systems. We can then convert the solutions in principal coordinates back to the original set of generalized coordinates.

Assuming that the equations of motion of the system have been so converted as

$$[M_d]\{\ddot{y}\} + [C_d]\{\dot{y}\} + [K_d]\{y\} = \{F_p\} \qquad (4.89)$$

where $\{y\} = [V]\{x\}$ is the principal coordinates, and $\{F_p\} = [V]^T\{F\}$. Note that the Duhamel integral in eqn. (2.128) is for the normalized equation of motion. Accordingly, we normalize the above equation by left-multiplying each term with $[M_d]^{-1}$. Doing so, eqn. (4.89) becomes

$$\{\ddot{y}\} + 2[\zeta][\omega]\{\dot{y}\} + [\omega][\omega]\{y\} = \{f_p\} \qquad (4.90)$$

where

$$2[\zeta][\omega] = [M_d]^{-1}[C_d] \qquad (4.91)$$

$$[\omega][\omega] = [M_d]^{-1}[K_d] \qquad (4.92)$$

$$\{f_p\} = [M_d]^{-1}\{F_p\} = [M_d]^{-1}[V]^T\{F\} \qquad (4.93)$$

Then, the matrix version of eqn. (2.128) can be written as

$$\{y(t)\} = \int_0^t [h(t-\tau)]\{f_p(\tau)\}d\tau \qquad (4.94)$$

where $[h(t)]$ is a diagonal matrix whose entry at the ith row and the ith column is, according to eqn. (2.129),

$$h_i(t) = \frac{e^{-\zeta_i \omega_i t}}{\omega_{di}} \sin \omega_{di} t \qquad (4.95)$$

where ζ_i and ω_i are the entries at the ith row and ith column of diagonal matrices $[\boldsymbol{\zeta}]$ and $[\boldsymbol{\omega}]$, respectively, and

$$\omega_{di} = \sqrt{1 - \zeta_i^2}\,\omega_i \tag{4.96}$$

Afterward, we can transform $\{\boldsymbol{y}\}$ back into $\{\boldsymbol{x}\}$ according to

$$\{\boldsymbol{x}\} = [V]^{-1}\{\boldsymbol{y}\} \tag{4.97}$$

In the above process, the core idea is still the decoupling method. By combining it with the Duhamel integral, we are now capable of handing any time-varying loading.

In the literature, there are other ways to normalize eqn. (4.89) into eqn. (4.90). The most common one is to normalize the mode shapes and, in turn, the modal matrix $[V]$, such that diagonalized $[M_d]$ is an identity matrix. But they are not significantly different from the one outlined earlier. In fact, this step is not absolutely necessary. Its primary purpose is to obtain the damping ratio for each mode. This information can be simply obtained from eqn. (4.89): $\zeta_i = C_{di}/(2\sqrt{K_{di}M_{di}})$ where C_{di}, K_{di}, and M_{di} are the ith row and the ith column entries of diagonal matrices $[C_d]$, $[K_d]$, and $[M_d]$, respectively.

■ Example 4.24: Damped Two-Mass System Subjected to Rectangular Loading

Consider the forced vibration of the damped two-mass system that we have studied earlier, shown in Fig. 4.20 in Example 4.15. In this example, the left mass is subjected to a rectangular loading of amplitude F_0 and duration t_0. The system is initially at rest. Find the transient response of the system.

□ Solution:

In this example, we will primarily use MATLAB to perform most of the analysis, including the modal analysis. To facilitate the discussions, we repeat the matrix equation of motion here:

$$\begin{bmatrix} m & 0 \\ 0 & m \end{bmatrix} \begin{Bmatrix} \ddot{x}_1 \\ \ddot{x}_2 \end{Bmatrix} + \begin{bmatrix} c_1 + c_2 & -c_2 \\ -c_2 & c_1 + c_2 \end{bmatrix} \begin{Bmatrix} \dot{x}_1 \\ \dot{x}_2 \end{Bmatrix} + \begin{bmatrix} 2k & -k \\ -k & 2k \end{bmatrix} \begin{Bmatrix} x_1 \\ x_2 \end{Bmatrix} = \begin{Bmatrix} F_1 \\ 0 \end{Bmatrix} \tag{a}$$

□ Exploring the Solution with MATLAB

Introducing the following nondimensionalization parameters:

$$\omega_0 = \sqrt{\frac{k}{m}} \qquad \bar{t} = \omega_0 t \qquad \bar{x}_i = \frac{x_i}{F_0/k} \tag{b}$$

Then

$$\dot{x} = \frac{dx}{d\bar{x}}\frac{d\bar{x}}{d\bar{t}}\frac{d\bar{t}}{dt} = \frac{F_0\omega_0}{k}\dot{\bar{x}} \qquad \ddot{x} = \frac{d\dot{x}}{dt} = \frac{d\dot{x}}{d\bar{t}}\frac{d\bar{t}}{dt} = \frac{F_0\omega_0^2}{k}\ddot{\bar{x}} \tag{c}$$

where an overdot over a nondimensionalized $\bar{x}$ represents its derivative with respect to the nondimensionalized time $\bar{t}$. Then, the nondimensionalized equations of motion are

$$\begin{bmatrix} 1 & 0 \\ 0 & 1 \end{bmatrix} \begin{Bmatrix} \ddot{\bar{x}}_1 \\ \ddot{\bar{x}}_2 \end{Bmatrix} + 2\zeta_1 \begin{bmatrix} 1 + r_c & -r_c \\ -r_c & 1 + r_c \end{bmatrix} \begin{Bmatrix} \dot{\bar{x}}_1 \\ \dot{\bar{x}}_2 \end{Bmatrix} + \begin{bmatrix} 2 & -1 \\ -1 & 2 \end{bmatrix} \begin{Bmatrix} \bar{x}_1 \\ \bar{x}_2 \end{Bmatrix} = \begin{Bmatrix} f_1 \\ 0 \end{Bmatrix} \tag{d}$$

where

$$\zeta_1 = c_1/(2m\omega_0) \qquad r_c = \frac{c_2}{c_1} \qquad f_1(\bar{t}) = F_1(t)/F_0 \tag{e}$$

The following MATLAB code performs the entire vibration analysis and plots the system's response:

```
% Part 1: Problem parameters
n = 2;                    % Number of DOF's
rc = 100;                 % Dashpot's c2/c1 ratio
zeta1 = .001;             % Damping ratio for first mode
r = .5;                   % Loading duration t0/Tn
tmax = 40;                % Maximum time of to be computed
dt = .1;                  % Time step size


% Part 2: Setup system matrices
M = eye(n,n);


C = 2*zeta1*(1+rc)*eye(n,n);
for i=1:n-1
    C(i,i+1)=-2*zeta1*rc;
    C(i+1,i)=-2*zeta1*rc;
end


K = 2*eye(n,n);
for i=1:n-1
    K(i,i+1)=-1;
    K(i+1,i)=-1;
end


% Part 3: Solve for eigenvalues and diagonalize matrices
[V,lambda] = eig(K,M);
Md = V' * M * V;
Cd = V' * C * V;
Kd = V' * K * V;


% Part 3.1: Normalize the diagonalized matrices for damping ratios
Minv = inv(Md);
Cn = Minv * Cd;
Kn = Minv * Kd;


% Part 4: Retrieve natural frequency and damping ratio for each mode
for i=1:n
    omega(i,i) = sqrt(Kn(i,i));
    zeta(i,i) = Cn(i,i)/2/omega(i,i);
end


% Part 5: Constructing the forcing
t = [0:dt:tmax];                    % Time array, used for solution
f = 1-heaviside(t-2*pi*r);          % Scalar loading form
F(1,:) = f;                         % Construct {F} in original gen. coord.
F(2,:) = zeros(1,length(t));
Fp = V' * F;                        % Convert {F} to principal coord.
Fn = Minv * Fp;                     % Normalization of {Fp}
```

```
% Part 6: Computing the response, loop over each mode
for i=1:n
    o = omega(i,i);              % Natural frequency of this mode
    z = zeta(i,i);               % Modal damping ratio
    od = sqrt(1-z*z)*o;          % Damped natural frequency
    h(i,:) = exp(-z*o*t)/od .* sin(od*t);   % Unit impulse response
    y(i,:) = conv(h(i,:),Fn(i,:)) * dt;     % Convolution sum
end;

% Part 7: Convert solution back to original generalized coordinates
x = V*y;

% Part 8: Finally, plot the results and the forcing
plot(t,x(1,1:length(t)),'r', t,x(2,1:length(t)),'b--', t,F(1,:),'g:');
xlabel('\omega_0 t');
ylabel('x_{1,2}(t)/(F_0/k), F(t)/F_0');
legend('x_1(t)','x_2(t)','F_1(t)');
```

Let us walk through the code first. Part 1 defines the adjustable parameters for the system, such as the number of DOFs, as well as parameters for the computation, such as dt for the time step size. Part 2 sets up the system matrices $[\overline{M}]$, $[\overline{C}]$, and $[\overline{K}]$ according to eqn. (d), and normalize them according to eqn. (4.90). Part 3 computes the modal matrix $[V]$ and diagonalized all system matrices. Part 4 constructs the $[\zeta]$ and $[\overline{\omega}]$ matrices. Part 5 constructs the forcing, first in the scalar form, and then the matrix $\{\overline{F}(t)\}$, of size $N \times N_t$, where N_t is the length of time array. It is then transformed to normalized modal forcing. Part 6 calculates the modal responses using convolution integral. Part 7 converts the modal responses back into the original set of generalized coordinates $\{\overline{x}\}$. Finally, Part 8 plots the results.

Before performing extensive computations, we shall validate the code first. Apart from observing the initial conditions, one additional validation is to compute a case in which the damping is set to 0, turning the system into an undamped one. For the undamped two-mass mass–spring system, a step-loading case has been analyzed in Example 4.19. We can mimic a step loading by letting the load duration t_0 to be large. Thus, we set $\zeta_1 = 0$ and $t_0/T_1 = 15$ (r_c becomes immaterial). The responses are plotted in Fig. 4.41, which, within the time period (before unloading) shown, is indeed identical to the undamped case in Fig. 4.31.

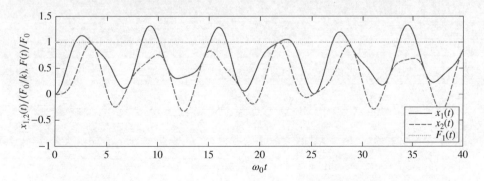

Figure 4.41 Responses of two-mass mass–spring–dashpot system to square loading on the left mass at $\zeta_1 = 0$ and $t_0/T_1 = 15$, which are identical to Fig. 4.31 when loading is step loading, within time period shown

We now can start to explore the system's parameter space. First, we set $\zeta = 0.1$ and $r_c = 1$, the system's responses for cases $t_0/T_1 = 0.5$, 1, and 2 are shown in Figs. 4.42 through 4.44. From these figures, we can see the motions dissipate rather quickly in all three cases. Recall earlier in Example 4.20, the damping ratio for Mode 2 is $\zeta_2 = (1 + 2r_c)\zeta_1/\sqrt{3} = \sqrt{3}\zeta_1$ and the motion of Mode 2 dissipates significantly quicker: after the load reduced to zero, the motion is dominated by a very weak Mode 1 motion.

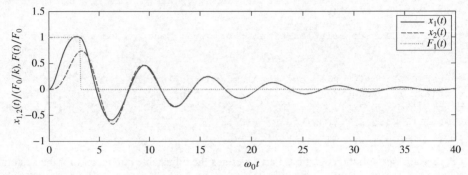

Figure 4.42 Responses of two-mass mass–spring–dashpot system to square loading on the left mass at $\zeta_1 = 0.1$, $r_c = 1$, and $t_0/T_1 = 0.5$

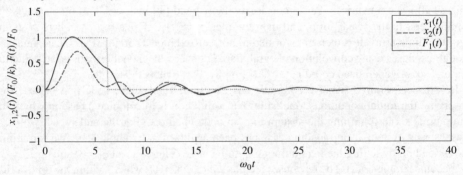

Figure 4.43 Responses of two-mass mass–spring–dashpot system to square loading on the left mass at $\zeta_1 = 0.1$, $r_c = 1$, and $t_0/T_1 = 1$

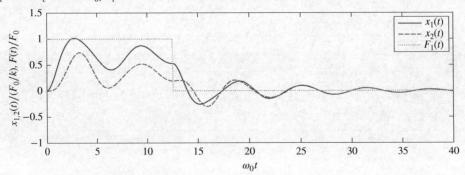

Figure 4.44 Responses of two-mass mass–spring–dashpot system to square loading on the left mass at $\zeta_1 = 0.1$, $r_c = 1$, and $t_0/T_1 = 2$

In the next, ζ_1 remains unchanged, but $r_c = 0$. This essentially removes the dashpot between the two masses. The system's responses for the same three cases are shown in Figs. 4.45 through 4.47. We notice that this set of figures does not differ noticeably from the previous set. This mainly is because, although $r_c = 0$, ζ_2 does not vanish, it just becomes smaller.

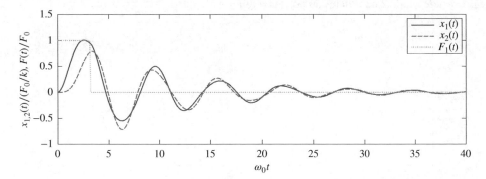

Figure 4.45   Responses of two-mass mass–spring–dashpot system to square loading on the left mass at $\zeta_1 = 0.1$, $r_c = 0$, and $t_0/T_1 = 0.5$

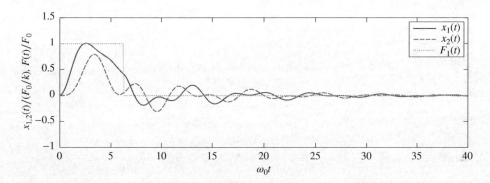

Figure 4.46   Responses of two-mass mass–spring–dashpot system to square loading on the left mass at $\zeta_1 = 0.1$, $r_c = 0$, and $t_0/T_1 = 1$

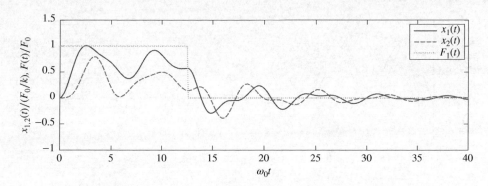

Figure 4.47   Responses of two-mass mass–spring–dashpot system to square loading on the left mass at $\zeta_1 = 0.1$, $r_c = 0$, and $t_0/T_1 = 2$

In the last set, we set $\zeta_1 = 0.001$ while $r_c = 100$. This essentially removes the damping for Mode 1, but $\zeta_2 = 0.1$. The system's responses for the same set of loadings are shown in Figs. 4.48 through 4.50. In this set of figures, the most notable difference is that the case $t_0/T_1 = 0.5$, which is dominated by Mode 1 motion, is almost undamped; while the other two cases where Mode 2 dominates do not show much difference.

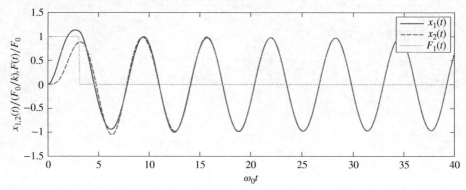

Figure 4.48 Responses of two-mass mass–spring–dashpot system to square loading on the left mass at $\zeta_1 = 0.001$, $r_c = 100$, and $t_0/T_1 = 0.5$

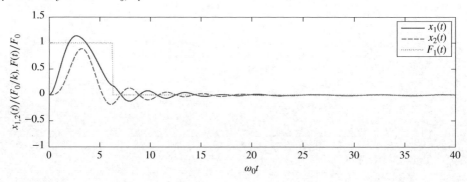

Figure 4.49 Responses of two-mass mass–spring–dashpot system to square loading on the left mass at $\zeta_1 = 0.001$, $r_c = 100$, and $t_0/T_1 = 1$

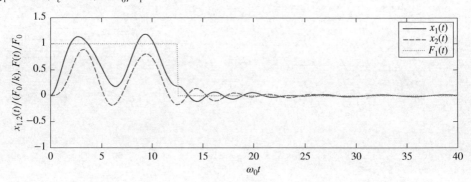

Figure 4.50 Responses of two-mass mass–spring–dashpot system to square loading on the left mass at $\zeta_1 = 0.001$, $r_c = 100$, and $t_0/T_1 = 2$

The above example demonstrates how the convolution integral method can be used. Resorting to the numerical integration, it should be as capable as the Duhamel integral for single-DOF systems. The code in the previous example can be easily adapted for use in different systems.

4.9 Chapter Summary

4.9.1 Modal Analyses

Equations of Motion: Damping should be removed for modal analysis. Thus,

$$[M]\{\ddot{x}\} + [K]\{x\} = 0 \tag{4.98}$$

Natural Frequencies: Assume $\{x\} = \{X\}e^{i\omega t}$, the equations of motion become

$$([K] - \omega^2[M])\{X\} = 0 \tag{4.99}$$

To have a nontrivial solution (so that the free vibration exists), it is required that

$$\det([K] - \omega^2[M]) = 0 \tag{4.100}$$

Equation (4.100) is an nth order algebraic equation about ω^2 where n is the number of DOFs, ω^2 is called the eigenvalue, and ω is called the natural frequency. Generally, the number of natural frequencies equals the number of DOFs. It is customary to sort the natural frequencies such that $\omega_1 < \omega_2 < \cdots < \omega_n$.

Mode Shapes: For each natural frequency, because of eqn. (4.100), the number of independent equations in eqn. (4.99) is reduced by 1. The resulting equations can be solved by treating one of X_i's as known. Often X_1 is chosen, and the solutions are expressed in the form of ratios $\left.\frac{X_i}{X_1}\right|_{\omega=\omega_j}$. The column matrix

$$\{X^{(j)}\} = \left.\left\{ \frac{X_1}{X_1} \quad \frac{X_2}{X_1} \quad \frac{X_3}{X_1} \quad \cdots \quad \frac{X_n}{X_1} \right\}^{T}\right|_{\omega=\omega_j} \tag{4.101}$$

is called the *mode shape* corresponding to natural frequency ω_j. Mathematically, it is also called an eigenvector.

4.9.2 Free Vibrations of Multi-DOF Systems

Equations of Motion: In matrix notation:

$$[M]\{\ddot{x}\} + [C]\{\dot{x}\} + [K]\{x\} = 0 \tag{4.102}$$

Undamped Systems: When $[C]$ vanishes in eqn. (4.102), the system is undamped. The general solution for eqn. (4.102) is expressible as

$$\{x(t)\} = [V]\{s(t)\} \tag{4.103}$$

where $[V]$ is the modal matrix that is made up of the mode shape column matrices,

$$[V] = \left[\{X^{(1)}\} \quad \{X^{(2)}\} \quad \cdots \quad \{X^{(n)}\} \right] \tag{4.104}$$

and

$$\{s(t)\} = \begin{Bmatrix} a_1 \cos \omega_1 t + b_1 \sin \omega_1 t \\ a_2 \cos \omega_2 t + b_2 \sin \omega_2 t \\ \vdots \\ a_n \cos \omega_n t + b_n \sin \omega_n t \end{Bmatrix} \tag{4.105}$$

The initial conditions generally include initial positions and initial velocities of all DOFs, denoted as $\{x_0\}$ and $\{v_0\}$. Denote

$$\{a\} = \begin{Bmatrix} a_1 \\ a_2 \\ \vdots \\ a_n \end{Bmatrix} \qquad \{\beta\} = \begin{Bmatrix} \omega_1 b_1 \\ \omega_2 b_2 \\ \vdots \\ \omega_n b_n \end{Bmatrix} \tag{4.106}$$

The initial conditions give the following equation:

$$\{a\} = [V]^{-1}\{x_0\} \qquad \{\beta\} = [V]^{-1}\{v_0\} \tag{4.107}$$

which can be solved for a_i and b_i, and hence the free vibration solution is obtained.

Special Case: When $\omega_1 = 0$, the system is unconstrained, and the system has a rigid-body mode, and the corresponding component in the solution is obtained in the same way except that $b_1 = \beta_1$ and $s_1(t) = a_1 + b_1 t$.

Principal Coordinates: Principal coordinates can be obtained via

$$\{y\} = [V]^{-1}\{x\} \tag{4.108}$$

Using the principal coordinates, the equations of motion become

$$[M_d]\{\ddot{y}\} + [V]^T[C][V]\{\dot{y}\} + [K_d]\{y\} = 0 \tag{4.109}$$

where

$$[M_d] = [V]^T[M][V] \qquad [K_d] = [V]^T[K][V] \tag{4.110}$$

are diagonal matrices due to the orthogonality of the mode shapes. If

$$[C] = c_m[M] + c_k[K] \tag{4.111}$$

the damping is called the proportional damping. In such cases, all equations are completely decoupled. They can be solved individually as a set of single-DOF systems.

Free Vibration via Decoupling Method: Using the principal coordinates, the system's equations of motion can be decoupled if the system is undamped or damped with proportional damping. The initial conditions in principal coordinates become

$$\{y(0)\} = [V]^{-1}\{x(0)\} \qquad \{\dot{y}(0)\} = [V]^{-1}\{\dot{x}(0)\} \tag{4.112}$$

Each mode is a single-DOF system and the solutions from Chapter 2 can be utilized.

4.9.3   Steady-State Responses of Multi-DOF Systems

Equation of Motion: In matrix and complex notation:

$$[M]\{\ddot{x}\} + [C]\{\dot{x}\} + [K]\{x\} = \{F_0\}e^{i\Omega t} \tag{4.113}$$

Solution: Using the complex notation, assume the solution to be $\{X\}e^{i\Omega t}$. The equation becomes

$$\left([K] - \Omega^2[M] + i\Omega[C]\right)\{X\} = \{F_0\} \tag{4.114}$$

This equation can be solved in the complex notation easily without much effort. If the forcing is in real sinusoidal form, the amplitude equals the modulus of the complex amplitude.

If the system is undamped, it is possible to stay entirely within the realm of real variables, by assuming the response to have the same sinusoidal form as the forcing.

4.9.4   Transient Responses of Multi-DOF Systems

Equation of Motion: In matrix notation,

$$[M]\{\ddot{x}\} + [C]\{\dot{x}\} + [K]\{x\} = \{F(t)\} \tag{4.115}$$

Analytical Method: If there is no damping, the solution is

$$\{x(t)\} = \{x_p(t)\} + [V]\{s(t)\} \tag{4.116}$$

If the particular solution can be found, then a_i and b_i in $\{s(t)\}$ can be found as

$$\{a\} = [V]^{-1}(\{x(0)\} - \{x_p(0)\}) \qquad \{\beta\} = [V]^{-1}(\{\dot{x}(0)\} - \{\dot{x}_p(0)\}) \tag{4.117}$$

Decoupling Method: Equations of motion can be decoupled by using the principal coordinates. This is only possible when the damping is proportional or undamped. The decoupled equations are

$$[M_d]\{\ddot{y}\} + [C_d]\{\dot{y}\} + [K_d]\{y\} = [V]^T\{F(t)\} \tag{4.118}$$

They can then be solved using methods suitable for single-DOF systems.

Laplace Transform Method: The Laplace-transformed equations are

$$\left(s^2[M] + s[C] + [K]\right)\{\tilde{x}\} = \{\tilde{F}(t)\} + (s[M] + [C])\{x(0)\} + [M]\{\dot{x}(0)\} \tag{4.119}$$

This can be solved for the transformed solutions via Cramer's rule. The most challenging part is finding the inverses of the transformed solutions. The good thing is: the inversion process is no more complicated than that in the single-DOF case. The only difference is that more solutions need to be inversed.

Convolution Integral Method: In matrix form, the solution can be written as

$$\{x(t)\} = \int_0^t [H(t-\tau)]\{F(\tau)\}d\tau \tag{4.120}$$

where $[H(t)]$ is an $n \times n$ matrix, whose entry at the ith row jth column, $H_{ij}(t)$, is the response of the jth DOF due to an unit impulse at the ith DOF. Finding those $H_{ij}(t)$ functions itself is solving a set of complicated transient response problems.

A more procedural approach is to decouple the equations of motion into principal coordinates, leading to eqn. (4.118), which is then normalized, resulting in the following equations of motion:

$$\{\ddot{y}\} + 2[\zeta][\omega]\{\dot{y}\} + [\omega][\omega]\{y\} = \{f\} \tag{4.121}$$

where $[\zeta]$ and $[\omega]$ are diagonal matrices. Then the following matrix-version Duhamel integral can be performed

$$\{y(t)\} = \int_0^t [h(t - \tau)]\{f(\tau)\}d\tau \tag{4.122}$$

to obtain the system's responses in terms of principal coordinates, which will then be converted back to the original set of generalized coordinates. In eqn. (4.122), $[h(t)]$ is a diagonal matrix with entries

$$h_i(t) = \frac{e^{-\zeta_i \omega_i t}}{\omega_{di}} \sin \omega_{di} t \qquad \omega_{di} = \sqrt{1 - \zeta_i^2} \omega_i \tag{4.123}$$

and ζ_i and ω_i are entries in $[\zeta]$ and $[\omega]$ matrices, respectively.

Problems

Problem 4.1:   The right wall in the mass–spring system shown in Figure 4.1 is removed. Conduct the modal analysis for the resulting system. Assume $m_1 = m_2 = m$ and $k_1 = k_2 = k$.

Problem 4.2:   Conduct the modal analysis for the mass–spring–dashpot system in Example 1.21 using both sets of generalized coordinates as defined. Assume $k_1 = k_3 = k$, $k_2 = 2k$, $c_1 = c_3 = c$, $c_2 = 2c$, $m_1 = m$, and $m_2 = 2m$.

Problem 4.3:   Conduct the modal analysis for the platform model in Example 1.26 using both sets of generalized coordinates as defined, for the following two cases: (a) $k_1 = k_2 = k$ and $c_1 = c_2 = c$; and (b) the same as case (a) except $k_2 = k - \Delta k$ where $\Delta k / k$ is small. Discuss the effects of the variation Δk on the natural frequencies and the mode shapes.

Problem 4.4:   Conduct the modal analysis for the pendulum-on-mass system described in Problem 1.6.

Problem 4.5:   Conduct the modal analysis for the mass-on-cart system described in Problem 1.10.

Problem 4.6:   Conduct the modal analysis for the pendulum-in-frame system described in Problem 1.11.

Problem 4.7:   Conduct the modal analysis for the extensible pendulum described in Problem 1.13.

Problem 4.8:   Conduct the modal analysis for the slender-rod pendulum described in Problem 1.21.

Problem 4.9:   Conduct the modal analysis for the double rigid-link pendulum described in Problem 1.22.

Problem 4.10: Conduct the modal analysis for the spring-supported beam (slender rod) described in Problem 1.25. Assume $k_1 = k_2 = k_3 = k_4 = k$ and $c_1 = c_2 = c_3 = c_4 = c$.

Problem 4.11: Conduct the modal analysis for the rod–pulley–mass system described in 1.26. Assume $k_1 = k_2 = k$, $c_1 = c_2 = c$, $M = 3m$, and $m_0 = 0$.

Problem 4.12: Conduct the modal analysis for the pivoted rod described in Problem 1.27. Assume $a = b = \frac{1}{2}L$, $k_1 = k_2 = k$, $c_1 = c_2 = c$, $m_1 = 2m$, and $m_2 = m$.

Problem 4.13: Conduct the modal analysis for the cylinder-on-cart system described in Part (a) of Problem 1.29.

Problem 4.14: Conduct the modal analysis for the connected parallel beams described in Problem 1.39. Assume $M = m_0 = m$ and $m_1 = 0$.

Problem 4.15: Conduct the modal analysis for the beam-on-carts system described in Problem 1.40. Assume $M = m$, $m_1 = 0$, and $x_0(t) = X_0 \sin \Omega t$.

Problem 4.16: Conduct the modal analysis for the pendulum-on-cylinder system described in Part (b) of Problem 1.44. Assume $m_1 = m_2 = m$, $m_3 = 0$, $r_1 = r_2 = r$, and $L = 3r$.

Problem 4.17: For the double pendulum in Example 4.8, construct two sets of initial conditions such that only motions purely in Mode 1 and Mode 2, respectively, are triggered.

Problem 4.18: The extensible pendulum described in Problem 1.13 is initially at rest. At $t = 0$, the mass is given an initial velocity of v_0 horizontally to the right. Find the subsequent motion of the system. Use MATLAB to plot the response of the system. Note that the modal analysis for the system has been conducted in Problem 4.7.

Problem 4.19: The right wall in the mass–spring system in Figure 4.1 is removed. The resulting system is initially at rest. At $t = 0$, the right mass is given an initial velocity of v_0 horizontally to the right. Assume $m_1 = m_2 = m$ and $k_1 = k_2 = k$. Find the subsequent motion of the system. Use MATLAB to plot the response of the system. Note that the modal analysis for the system has been conducted in Problem 4.1.

Problem 4.20: The slender-rod pendulum described in Problem 1.21 is initially at rest and dangles vertically. At $t = 0$, the top of the rod is given a velocity of v_0 horizontally to the right. Find the subsequent motion of the system. Use MATLAB to plot the response of the system. Note that the modal analysis for the system has been conducted in Problem 4.8.

Problem 4.21: Solve Problem 4.20 if, at $t = 0$, the bottom of the rod is given a velocity of v_0 horizontally to the right.

Problem 4.22: The connected parallel beams described in Problem 1.39 is initially at rest. The lower beam is pulled a distance Δ to the right relative to the upper beam and then released at $t = 0$. Determine the subsequent motion of the system. Use MATLAB to plot the response of the system. Note that the modal analysis for the system has been conducted in Problem 4.14.

Problem 4.23: Repeat Problem 4.22 if, at $t = 0$, the lower beam is given a velocity of v_0 horizontally to the right.

Problem 4.24: The pendulum-in-cart system described in Problem 1.11 is initially at rest. The pendulum is pulled to the right to form an angle Θ with respect to the vertical and then released at $t = 0$. Find the subsequent motion of the system. Use MATLAB to plot the response of the system for a series of values of m/M ratio: 0.2, 0.5, 1, 2, and 5. Note that the modal analysis for the system has been conducted in Problem 4.6.

Problem 4.25: Solve Problem 4.24 if, at $t = 0$, the mass of the pendulum is given a velocity of v_0 horizontally to the right.

Problem 4.26: Solve Problem 4.24 if, at $t = 0$, the cart is given a velocity of v_0 horizontally to the right.

Problem 4.27: Use the *decoupling method* to solve Problem 4.19.

Problem 4.28: Use the *decoupling method* to solve Problem 4.20.

Problem 4.29: The rod–pulley–mass system described in Problem 1.26 is initially at rest. The mass m is pulled down by a distance Δ and then suddenly released at $t = 0$. Assume $k_1 = k_2 = k$, $c_1 = c_2 = c$, $M = 3m$, and $m_0 = 0$. Determine the subsequent motion of the system. Use MATLAB to plot the response of the system for a series of values of an appropriately defined damping ratio such that all modes are undamped or underdamped. Note that the modal analysis for the system has been conducted in Problem 4.11.

Problem 4.30: The pivoted slender rod described in Problem 1.27 is initially at rest. The mass m_2 is pulled down by a distance Δ and then suddenly released at $t = 0$. Use the *decoupling method* to find the subsequent motion of the system. Assume $a = b = \frac{1}{2}L$, $k_1 = k_2 = k$, $c_1 = c_2 = c$, $m_1 = 2m$, and $m_2 = m$. Use MATLAB to plot the response of the system for a series of values of an appropriately defined damping ratio such that all modes are undamped or underdamped. Note that the modal analysis for the system has been conducted in Problem 4.12.

Problem 4.31: The right wall in the mass–spring–dashpot system in Example 1.21 is removed. The resulting system is initially at rest. Mass m_2 is pulled further to the right by the amount Δ and suddenly released at $t = 0$. Assume $k_1 = k_2 = k$, $c_1 = c_2 = c$, and $m_1 = m_2 = m$. Find the system's subsequent motion using the *decoupling method* using the generalized coordinates x_1 and x_2 as defined. Use MATLAB to plot the response of the system for a series of values of an appropriately defined damping ratio such that all modes are undamped or underdamped. Note that the modal analysis for the system has been conducted in Problem 4.1.

Problem 4.32: Solve Problem 4.31 using the generalized coordinates x_1 and y_2 as defined in Example 1.21.

Problem 4.33: Solve Problem 4.31 if, at $t = 0$, mass m_1 is given a velocity of v_0 to the right.

Problem 4.34: The mass–spring–dashpot system shown in Example 1.21 is initially at rest. At $t = 0$, mass m_1 is given an initial velocity of v_0 to the right. Assume $k_1 = k_3 = k$, $k_2 = 2k$, $c_1 = c_3 = c$, $c_2 = 2c$, $m_1 = m$, and $m_2 = 2m$. Find the system's subsequent motion using the *decoupling method*. Use the generalized coordinates x_1 and x_2 as defined. Use MATLAB to plot the response of the system for a series of values of an appropriately defined damping ratio such that all modes are undamped or underdamped. Note that the modal analysis for the system has been conducted in Problem 4.2.

Problem 4.35: Solve Problem 4.34 using the generalized coordinates x_1 and y_2 as defined in Example 1.21.

Problem 4.36: The platform model in Example 1.26 is initially at rest. At $t = 0$, the right end of the platform is lifted up by the amount Δ and then suddenly released. Assume $k_1 = k_2 = k$ and $c_1 = c_2 = c$. Find the system's subsequent motion using the *decoupling method*. Use both sets of generalized coordinates as defined. Use MATLAB to plot the response of the system for a series of values of an appropriately defined damping ratio such that all modes are undamped or underdamped. Note that the modal analysis for the system has been conducted in Problem 4.3.

Problem 4.37: The cylinder-on-cart system described in Problem 1.29 is initially at rest. The cylinder is dragged down a distance Δ along the incline and is suddenly released at $t = 0$. Use the *decoupling method* to find the subsequent motion of the system. Use MATLAB to plot the system's response for a series of values of an appropriately defined damping ratio such that all modes are undamped or underdamped, for the series of values of m/M ratio of 0.5, 1, and 2. Note that the modal analysis for the system has been conducted in Problem 4.13.

Problem 4.38: Solve Problem 4.37 if, at $t = 0$, the cart is given a velocity of v_0 horizontally to the right.

Problem 4.39: For the slender-rod pendulum described in Problem 1.21, a harmonic force $F(t) = F_0 \sin \Omega t$ is applied to the top of the rod in the horizontal direction. Find the steady-state response of the system. Use MATLAB to plot the steady-state response spectra (amplitude and phase lag versus frequency curves) for the system.

Problem 4.40: For the double rigid-link pendulum described in Problem 1.22, $F(t) = F_0 \sin \Omega t$. Find the steady-state response of the system. Use MATLAB to plot the steady-state response spectra (amplitude and phase lag versus frequency curves) for the system.

Problem 4.41: For the pendulum-on-cart system described in Example 1.20, if the specified motion $x_0(t)$ is replaced by a horizontal harmonic force $F(t) = F_0 \sin \Omega t$, find the steady-state response of the system. Use MATLAB to plot the steady-state response spectra (amplitude and phase lag versus frequency curves) of the system for a series of values of m_2/m_1 ratio of 0.2, 0.5, 1, 2, and 5.

Problem 4.42: For the connected parallel beams described in Problem 1.39, if the top beam is subjected to a harmonic horizontal force of $F_1(t) = F_0 \cos \Omega t$, while the bottom beam is subjected to a harmonic horizontal force of $F_2(t) = F_0 \sin \Omega t$, find the steady-state motion of the system. Use MATLAB to plot the steady-state response spectra (amplitude and phase lag versus frequency curves) of the system for a series of values of m_0/M ratio of 0.2, 0.5, 1, 2, and 5.

Problem 4.43: For the dynamic vibration absorber described in Example 4.18, if a dashpot is added to the axillary system, in parallel to spring k, find the dashpot coefficient c such that the amplitude of the steady-state response of the axillary system is minimized.

Problem 4.44: For the beam-on-carts system described in Problem 1.40, if the prescribed motion is $x_0(t) = X_0 \cos \Omega t$, find the steady-state motion of the system. Use MATLAB to plot the steady-state response spectra (amplitude and phase lag versus frequency curves) of the system for a series of values of m/M ratio of 0.2, 0.5, 1, 2, and 5. Produce one figure for each mass ratio, but include a series of values of an appropriately defined damping ratio encompassing the undamped, underdamped, and overdamped cases for each mode.

Problem 4.45: For the mass–spring–dashpot system shown in Example 1.21, assume $k_1 = k_3 = k$, $k_2 = 2k$, $c_1 = c_3 = c$, $c_2 = 2c$, $m_1 = m$, and $m_2 = 2m$. Mass m_1 is subjected to a harmonic force of $F(t) = F_0 \sin \Omega t$. Find the steady-state response of the system. Use MATLAB to plot the steady-state response spectra (amplitude and phase lag versus frequency curves) of the system for a series of values of an appropriately defined damping ratio encompassing the undamped, underdamped, and overdamped cases for each mode.

Problem 4.46: For the mass–spring–dashpot system shown in Example 1.21 with the right wall removed, assume $k_1 = k_2 = k$, $c_1 = c_2 = c$, and $m_1 = m_2 = m$. Mass m_1 is subjected to a harmonic force of $F_1(t) = F_1 \sin \Omega t$; while mass m_2 is subjected to another harmonic force of $F_2(t) = F_2 \cos \Omega t$. Using the generalized coordinates x_1 and y_2 as defined, find the steady-state motion of the system. Use MATLAB to plot the steady-state response spectra (amplitude and phase lag versus frequency curves) of the system for the cases $F_2/F_1 = -1$, 0, and 1. Produce one figure for each force ratio, but include a series of values of an appropriately defined damping ratio encompassing the undamped, underdamped, and overdamped cases for each mode.

Problem 4.47: For the rod–pulley–mass system described in Problem 1.26, if the mass is subjected to a harmonic vertical force of $F(t) = F_0 \cos \Omega t$, find the steady-state response of the system. Assume $k_1 = k_2 = k$, $c_1 = c_2 = c$, $M = 3m$, and $m_0 = 0$. Use MATLAB to plot the steady-state response spectra (amplitude and phase lag versus frequency curves) of the system for a series of values of m/M ratio of 0.2, 0.5, 1, 2, and 5. Produce one figure for each mass ratio, but include a series of values of an appropriately defined damping ratio encompassing the undamped, underdamped, and overdamped cases for each mode.

Problem 4.48: For the cylinder-on-cart system described in Problem 1.29, if the cart is subjected to a harmonic horizontal force of $F(t) = F_0 \cos \Omega t$, find the steady-state motion of the system. Use MATLAB to plot the steady-state response spectra (amplitude and phase lag versus frequency curves) of the system for a series of values of m/M ratio of 0.2, 0.5, 1, 2, and 5. Produce one figure for each mass ratio, but include a series of values of an appropriately defined damping ratio encompassing the undamped, underdamped, and overdamped cases for each mode.

Problem 4.49: The right wall in the mass–spring system in Figure 4.1 is removed. The left mass is subjected to a triangle-shaped periodic loading of a period T: during the first half of the period, the force increases linearly from 0 to F_0 and linearly decreases to 0 in the second half of the period. Assume $m_1 = m_2 = m$ and $k_1 = k_2 = k$. Find the steady-state response of the system. Use MATLAB to plot the steady-state response of the system for a series of values of T/T_1 ratio of 0.2, 0.5, 0.95, 1.9, and 4.9, where T_1 is the period of the system's first natural mode. (*Hint: The integral identities listed in Problem 2.31 might be useful.*)

Problem 4.50: For the double rigid-link pendulum described in Problem 1.22, $F(t)$ is a saw-tooth-shaped periodic loading of period T that increases from 0 linearly to F_0 over the period and suddenly drops to 0 at the beginning the next period. Find the steady-state response of the system. Use MATLAB to plot the steady-state response of the system for a series of values of T/T_1 ratio, of 0.2, 0.5, 0.95, 1.9, and 4.9, where T_1 is the period of the system's first natural mode. (*Hint: The integral identities listed in Problem 2.31 might be useful.*)

Problem 4.51: For the rod–pulley–mass system described in Problem 1.26, if the mass is subjected to a periodic loading of a period T, in which the force on the mass is a constant F_0 in the upward direction for the first half of a period and vanishes for the remainder of the period, find the steady-state motion of the system. Assume $k_1 = k_2 = k$, $c_1 = c_2 = c$, $M = 3m$, and $m_0 = 0$. Use MATLAB to plot the system's steady-state response for a series of values of T/T_1 ratio, of 0.2, 0.5, 1, 2, and 5, where T_1 is the period of the system's first mode, and for a series of values of an appropriately defined damping ratio encompassing the undamped, underdamped, and overdamped cases for each mode.

Problem 4.52: The right wall in the mass–spring system in Figure 4.1 is removed. The resulting system is initially at rest. Starting from $t = 0$, the mass on the right is subjected to a step loading of magnitude F_0 horizontally to the right. Find the transient response of the system using the *analytical method*. Assume $m_1 = m_2 = m$ and $k_1 = k_2 = k$. Use MATLAB to plot the system's response. Note that the modal analysis for the system has been conducted in Problem 4.1.

Problem 4.53: The double rigid-link pendulum in Problem 1.22 is initially at rest. Starting from $t = 0$, $F(t) = F_0 \sin \Omega t$. Find the system's transient response using the *analytical method*. Use MATLAB to plot the system's response for a series of values of Ω/ω_1 ratio, of 0.2, 0.5, 1, 2, and 5, where ω_1 is the system's first natural frequency. Note that the modal analysis for the system has been conducted in Problem 4.9.

Problem 4.54: The mass–spring system described in Problem 1.18 is initially at rest. Starting from $t = 0$, the mass is subjected to a constant force of F_0 for a duration of t_0, in the direction that forms an angle θ with respect to one pair of springs connected diametrically. Find the transient response of the system using the *analytical method* in conjunction with the *decomposition method*. Use MATLAB to plot the system's responses for the case $\theta = 30°$ and a series of values of t_0/T_1 ratio, of 0.2, 0.5, 1, 2, and 5, where T_1 is the period of the system's first natural mode.

Problem 4.55: The pendulum-on-cart system described in Example 1.20 is initially at rest. If the specified motion is replaced by a force $F(t) = F_0 \sin \Omega t$ in the horizontal direction, positive to the right, find the transient response of the system using the *analytical method*. Use MATLAB to plot the system's response, for a series of values for the Ω/ω_2 ratio of 0.2, 0.5, 1, 2, and 5, where ω_2 is the system's second natural frequency, and for a series of values of m_2/m_1 ratio of 0.2, 0.5, 1, 2, and 5. Note that the steady-state response of the system has been found in Problem 4.41.

Problem 4.56: The connected parallel beams described in Problem 1.39 is initially at rest. Assume $M = m_0 = m$ and $m_1 = 0$. Starting from $t = 0$, the top beam is subjected to a force $F(t)$ horizontally to the right, in a shape as shown in Fig. P2.51. Find the system's response using the *analytical method* in conjunction with the *decomposition method*. Use MATLAB to plot the system's response for a series of values of t_0/T_2 ratio, of 0.05, 0.1, 0.2, 0.5, and 1, where T_2 is the period of the system's second natural mode. Note that the modal analysis for the system has been conducted in Problem 4.14.

Problem 4.57: Solve Problem 4.52 using the *decoupling method*.

Problem 4.58: Solve Problem 4.53 using the *decoupling method*.

Problem 4.59: Solve Problem 4.55 using the *decoupling method*.

Problem 4.60: Solve Problem 4.56 using the *decoupling method* in conjunction with the *decomposition method*.

Problem 4.61: The rod–pulley–mass system described in Problem 1.26 is initially at rest. Starting from $t = 0$, the mass m is subjected to a square-shaped loading, of a magnitude F_0 and a duration t_0 in the vertical downward direction. Assume $k_1 = k_2 = k, c_1 = c_2 = c$, $M = 3m$, and $m_0 = 0$. Find the transient response of the system using the *analytical method* in conjunction with the *decomposition method*. Use MATLAB to plot the system's response for a series of values of t_0/T_1 ratio, of 0.2, 0.5, 1, 2, and 5, where T_1 is the period of the system's first natural mode, and a series of values of an appropriately defined damping ratio such that all modes are undamped or underdamped. Produce one figure for each combination of the two parameters. Note that the modal analysis for the system has been conducted in Problem 4.11.

Problem 4.62: Solve Problem 4.61 if, starting from $t = 0$, the mass m is subjected to a ramped-step force of magnitude F_0 in the vertical downward direction, with a ramping duration of t_0, as sketched in Fig. P2.40.

Problem 4.63: Solve Problem 4.61 if, starting from $t = 0$, the mass m is subjected to a triangle-shaped force of a peak value of F_0 and a duration of $2t_0$, as sketched in Fig. P2.50, in the vertical downward direction.

Problem 4.64: The mass–spring–dashpot system shown in Example 1.21 is initially at rest. Starting from $t = 0$, mass m_1 is subjected to a ramped-step force to the right, of a magnitude F_0 and a ramping duration t_0, as sketched in Fig. P2.40. Assume $k_1 = k_3 = k$,

$k_2 = 2k, c_1 = c_3 = c, c_2 = 2c, m_1 = m$, and $m_2 = 2m$. Use x_1 and x_2 as defined in Example 1.21 as the generalized coordinates, find the transient response of the system using the *decoupling method* in conjunction with the *decomposition method*. Use MATLAB to plot the system's response for a series of values of t_0/T_1 ratio of 0.2, 0.5, 1, 2, and 5, where T_1 is the period of the system's first natural mode, and for a series of values of an appropriately defined damping ratio such that all modes are undamped or underdamped. Produce one figure for each combination of the two parameters.

Problem 4.65: Solve Problem 4.64 using x_1 and y_2 as the generalized coordinates as defined in Example 1.21.

Problem 4.66: Solve Problem 4.64 if, starting from $t = 0$, mass m_1 is subjected to a triangle-shaped force to the right, of a magnitude F_0 and a total duration $2t_0$, as sketched in Fig. P2.50.

Problem 4.67: Solve Problem 4.66 using x_1 and y_2 as the generalized coordinates as defined in Example 1.21.

Problem 4.68: Solve Problem 4.54 using the *Laplace transform method*.

Problem 4.69: Solve Problem 4.55 using the *Laplace transform method*.

Problem 4.70: Solve Example 4.21 using the *Laplace transform method*.

Problem 4.71: Solve Problem 4.64 using numerical *Duhamel integral method*.

Problem 4.72: Solve Problem 4.66 using numerical *Duhamel integral method*.

Problem 4.73: The platform described in Example 1.26 is used to model the trailer of a semitrailer truck. The springs and dashpots represent tires in front and rear wheels. If the truck runs over a speed bump, both sets of wheels will run through the same bump but with a time delay for the rear set. Suppose the effect of the bump on each tire can be modeled as a force of a half-period of a sine function, of a magnitude F_0 and a duration t_0. Denote the time delay between the wheel sets encountering the speed bump as rt_0. Find the response of the system using numerical *Duhamel integral method*. Use MATLAB to plot the system's response for a series of combinations of problem's parameters: r, t_0, and appropriately defined damping ratio(s).

Problem 4.74: The platform described in Example 1.26 is used to model the trailer of a semitrailer truck. The springs and dashpots represent tires in front and rear wheels. If the truck runs over a speed bump, both sets of wheels will run through the same bump but with a time delay for the rear set. Suppose the truck is moving at a constant speed of v_0, and the bump has a profile as shown in Fig. P2.42, with a height h, and a half-span a. Other than the bump, the road surface is flat. The trailer's wheelbase (distance between the front and rear wheel sets) is $L > 2a$. During the motion, the tires never lose contact with the ground surface. Use numerical *Duhamel integral method* to find the response of the system after the front wheel set touches on the bump. Use MATLAB to plot the system's responses for a series of combinations of problem's parameters: v_0, a, h, L, and appropriately defined damping ratio(s).

Computer Projects for Many-DOF Systems

Problem 4.75: In this exercise, we model a distributed-parameter system using discrete masses and springs. Assume a uniform rod has a circular cross section whose diameter d varies according to $d = a + b \cos(2\pi x/L)$, where L is the length of the rod, a and b ($a > b$) are constants, and x is measured from a wall to which the beam's left end is fixed. A mass M is attached to the right end of the beam, as shown in Fig. P4.75. The rod has a total mass of m and Young's modulus E.

Figure P4.75   Rod of variable cross section with mass attached at free end

- Develop an $(N + 1)$-DOF spring–mass model for the system, by assigning proper values of spring constant and mass to each spring and mass. The extra (the "+1") DOF is for the mass M attached to the right end.
- Derive the equations of motion for the $(N + 1)$-DOF system.
- Nondimensionalize the equations of motion.
- Obtain the natural frequencies and mode shapes for models containing 6, 11, 51, and 101 DOFs. Use MATLAB to plot the first and the fifth mode shapes for all above models. Produce one figure for each mode and place the mode shapes from all models in the same figure.
- Assume the system is initially at rest. At $t = 0$, mass M is given an initial velocity of v_0. Use MATLAB to compute and plot the 101-DOF model's response.

Problem 4.76: In this exercise, we extend Example 4.11 to contain 201 and 501 DOFs. Assume the rod is initially at rest. At $t = 0$, the mass at the center (the 101st and 251st masses, respectively) is subjected to an impulsive force of a square shape of a magnitude F_0 and a duration t_0, which is much smaller than the period of the system's fundamental mode. Use MATLAB to compute and plot the system's response.

Problem 4.77: In this exercise, we model a distributed-parameter system by using discrete masses and springs. Assume a uniform rod has a circular cross section of radius a and length L. Its left end is attached to a wall, and the right end has a mass M attached. The material has Young's modulus E.

- Develop an $(N + 1)$-DOF spring–mass model for the system, by assigning proper values of spring constant and mass to each spring and mass. The extra (the "+1") DOF is for the mass attached to the end.
- Derive the equations of motion for the general $(N + 1)$-DOF system.
- Nondimensionalize the equations of motion.

- Obtain the natural frequencies and mode shapes for models containing 6, 11, 51, and 101 DOFs. Use MATLAB to plot the first and the fifth mode shapes for all above models. Produce one figure for each mode, and place the mode shapes from all models in the same figure.
- If a force of a square shape, of a magnitude F_0 and a time duration t_0, is applied on mass M, use MATLAB to compute and plot the 101-DOF system's response for a series of t_0/T_1 values, where T_1 is the period of the system's fundamental mode.
- Repeat the same computation and plots for a 201-DOF system. Compare the results with those of the 101-DOF system.
- Compare the results from the previous two steps with a single-DOF model in which the rod is modeled as a spring and a lumped mass.

Problem 4.78: A wave packet of a short duration is called a *pulse*, which is often used as an input signal for many probing purposes. Figure P4.78 is an example of wave packet, in the form of a force, of the following expression:

$$F(t) = F_0 \sin \frac{2\pi t}{T} \sin \frac{18\pi t}{T}$$

The signal only lasts up to time T, which is a half-period of the sine envelop. Suppose this wave packet is sent into the mass M at the end of the 101-DOF model developed in Problem 4.75. Find the resulting transient motion of the system if the system is initially at rest. Explore the system's response with different T/T_1 ratio and m/M ratio, where T_1 is the period of the system's fundamental mode.

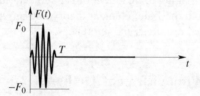

Figure P4.78   Simple wave packet of duration T

Reference

R. C. Cross & M. S. Wheatland, 2012, Modeling a falling Slinky, *Am. J. Phys.*, **80**, No. 12, pp. 1051–1060.

5

Vibration Analyses Using Finite Element Method

5.1 Objectives

The objectives of this chapter are to furnish the readers with elementary knowledge of the finite element method (FEM) and to provide guided tours on using commercial finite element software to perform vibration analyses. The simplest type of finite element is used to illustrate the relationship among the Lagrangian dynamics, the lumped-parameter modeling, and the finite element modeling. Then, a complete finite element analysis for beams is formulated and implemented using MATLAB as the programming language. The use of commercial finite element analysis software is presented by illustrated tutorials using SOLIDWORKS. This chapter is not intended as an in-depth treatment of the subject. Students are expected to gain further understanding of both the finite element theory and any particular software through more extensive and dedicated resources and references.

5.2 Introduction to Finite Element Method

The finite element method (FEM) is, in essence, the lumped-parameter modeling done systematically in a grand scale. The fundamental difference between the finite element modeling and the lumped-parameter modeling in Chapter 3 is that the lumped-parameter modeling specifies only one point of interest, while the finite element modeling typically defines many, often thousands and even hundreds of thousands, points of interest distributed throughout the structure.

In the finite element modeling, a structure is divided into *elements* of finite sizes, whence the term *finite element*. Each element is bounded by *nodes*. Treating these nodes as points of interest, all field quantities are expressed, in approximations, in terms of displacements of these nodes. For this purpose, a set of predetermined functions, called the *shape functions*, is used to interpolate the displacement across an element the same way as the Rayleigh–Ritz method used in Chapter 3. As a finite element model contains many elements, the requirement for the accuracy of the shape functions is relaxed. The accuracy is improved by using more elements such that each element only represents a small portion of the structure.

Fundamentals of Mechanical Vibrations, First Edition. Liang-Wu Cai.
© 2016 John Wiley & Sons, Ltd. Published 2016 by John Wiley & Sons, Ltd.
Companion Website: www.wiley.com/go/cai/fundamentals_mechanical_vibrations

In the following, a simple structure, a uniform rod, is used to illustrate the fundamental idea behind the FEM. The problem is first formulated following the procedure of the Lagrangian dynamics as outlined in Chapter 1. Then it is reformulated using a matrix notation of the FEM to provide a better computational structure, which is important for large-scale computations.

5.2.1   Lagrangian Dynamics Formulation of FEM Model

Consider a uniform rod whose one end is fixed and the other end is subjected to a uniaxial loading by force F. The rod has a length of L, a cross-sectional area A, a mass density ρ, and Young's modulus E. In the finite element model, the rod is divided into two elements using three nodes: one node at each end and the third node at the center, as shown in Fig. 5.1. This is equivalent to having three points of interest. In the finite element terminology, this process is called the *discretization*.

The discretization also implies that the nodal displacements, denoted as u_i and measured from their respective positions in the undeformed configuration, are used as the generalized coordinates. In the finite element modeling, u_1 is considered as a generalized coordinate and counted as a degree of freedom (DOF) initially and will be later removed in a process of *constraining* the finite element model by imposing appropriate boundary conditions. In comparison, in Lagrangian dynamics, $u_1 = 0$ is considered as a geometric constraint at the beginning.

The key step in the finite element modeling is to assume that the displacement within an element can be interpolated by a predetermined set of *shape functions*. Denote the displacement in two elements as $u^{[1]}(x)$ and $u^{[2]}(x)$, respectively. Then,

$$u^{[1]}(x,t) = N_1^{[1]}(x)u_1(t) + N_2^{[1]}(x)u_2(t) \tag{5.1}$$

$$u^{[2]}(x,t) = N_2^{[2]}(x)u_2(t) + N_3^{[2]}(x)u_3(t) \tag{5.2}$$

where $N_i^{[k]}(x)$ is the shape function for node i defined within element k. Having two nodes in each element implies a linear interpolation such that

$$N_1^{[1]}(x) = \frac{x_2 - x}{x_2 - x_1} \qquad N_2^{[1]}(x) = \frac{x - x_1}{x_2 - x_1} \tag{5.3}$$

and

$$N_2^{[2]}(x) = \frac{x_3 - x}{x_3 - x_2} \qquad N_3^{[2]}(x) = \frac{x - x_2}{x_3 - x_2} \tag{5.4}$$

The two sets of shape functions in eqns. (5.3) and (5.4) and the interpolated displacement in the two elements in eqns. (5.1) and (5.2) are sketched in Fig. 5.2. There are two shape functions for each element. Within element k, $N_i^{[k]}(x)$ evaluates to 1 at node i and 0 at the other node, and the two shape functions in the element sum to unity at any x.

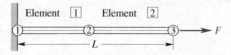

Figure 5.1   Rod under uniaxial loading modeled by two elements and three nodes

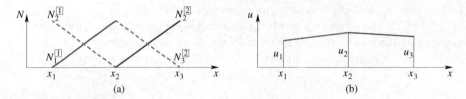

Figure 5.2 Shape functions (a) in eqns. (5.3) and (5.4) and the interpolated displacement (b) in eqns. (5.1) and (5.2) for the two elements

Furthermore, it is possible to have more than two nodes, such as three nodes, in an element. Correspondingly, there will be three shape functions defined for each element, and the displacement will be interpolated by quadratic functions.

With these preparatory setups, we now follow the procedure of the Lagrangian dynamics outlined in Chapter 1 to formulate the equation of motion for the rod.

- *Generalized Coordinates*: We define $\{u_i : i = 1, 2, 3\}$ as a complete and independent set of generalized coordinates, where u_i is the displacement of node i from its initial undeformed configuration. As mentioned earlier, u_1 is considered as a generalized coordinate at this stage.
- *Admissible Variations*: We verify that $\{\delta u_i : i = 1, 2, 3\}$ is a complete and independent set of admissible variations in this set of generalized coordinates.
- *Holonomicity*: We conclude that the system is holonomic and has three DOFs.
- *Generalized Forces*: When a force F_i acts on node i ($i = 3$ for the rod shown in Fig. 5.1, but, in general, could be any node), the resulting virtual work is

$$\delta W^{\text{n.c.}} = F_i \delta u_i \tag{5.5}$$

Thus, the corresponding generalized force is

$$\Xi_i = F_i \tag{5.6}$$

- *Potential Energy*: For a deformable structure, the potential energy is called the *strain energy*. According to the Mechanics of Materials, for a rod,

$$V = \frac{1}{2} \int_0^L EA \left(\frac{du}{dx} \right)^2 dx \tag{5.7}$$

Integrating element by element gives

$$
\begin{aligned}
V &= \frac{EA}{2} \left\{ \int_{\boxed{1}} \left[\frac{d}{dx} \left(N_1^{\boxed{1}} u_1 + N_2^{\boxed{1}} u_2 \right) \right]^2 dx + \int_{\boxed{2}} \left[\frac{d}{dx} \left(N_2^{\boxed{2}} u_2 + N_3^{\boxed{2}} u_3 \right) \right]^2 dx \right\} \\
&= \frac{EA}{2} \left[\frac{u_1^2 + u_2^2 - 2u_1 u_2}{L^{\boxed{1}}} + \frac{u_2^2 + u_3^2 - 2u_2 u_3}{L^{\boxed{2}}} \right]
\end{aligned}
\tag{5.8}
$$

where expressions for $N_i^{\boxed{k}}$ in eqns. (5.3) and (5.4) have been used.

- *Kinetic Energy*: Based on the displacement in eqns. (5.1) and (5.2), note that the shape functions are independent of t, the velocity in the rod is expressible as

$$v^{(1)}(x,t) = N_1^{(1)}(x)\dot{u}_1(t) + N_2^{(1)}(x)\dot{u}_2(t) \tag{5.9}$$

$$v^{(2)}(x,t) = N_2^{(2)}(x)\dot{u}_2(t) + N_3^{(2)}(x)\dot{u}_3(t) \tag{5.10}$$

Thus, the kinetic energy can be calculated as

$$T = \frac{\rho}{2}\left\{ \int_{(1)}\left(N_1^{(1)}\dot{u}_1 + N_2^{(1)}\dot{u}_2\right)^2 dV + \int_{(2)}\left(N_1^{(2)}\dot{u}_2 + N_2^{(2)}\dot{u}_3\right)^2 dV \right\} \tag{5.11}$$

Denote

$$m_{ij}^{(k)} = \rho \int_{(k)} N_i^{(k)}(x)N_j^{(k)}(x)dV \tag{5.12}$$

Then, the kinetic energy can be evaluated to

$$T = \frac{1}{2}\left\{ \left[m_{11}^{(1)}\dot{u}_1^2 + m_{22}^{(1)}\dot{u}_2^2 + 2m_{12}^{(1)}\dot{u}_1\dot{u}_2 \right] + \left[m_{22}^{(2)}\dot{u}_2^2 + m_{33}^{(2)}\dot{u}_3^2 + 2m_{23}^{(2)}\dot{u}_2\dot{u}_3 \right] \right\} \tag{5.13}$$

- *Lagrangian*: According to eqn. (1.3),

$$\mathcal{L} = T - V \tag{5.14}$$

- *Lagrange's Equation*: For each generalized coordinate u_i,

$$\frac{d}{dt}\left(\frac{\partial \mathcal{L}}{\partial \dot{u}_i} \right) - \frac{\partial \mathcal{L}}{\partial u_i} = \Xi_i \tag{5.15}$$

Substituting eqns. (5.14) and (5.6) into (5.15) gives

$$m_{11}^{(1)}\ddot{u}_1 + m_{12}^{(1)}\ddot{u}_2 + \frac{EA}{L^{(1)}}u_1 - \frac{EA}{L^{(1)}}u_2 = 0 \tag{5.16}$$

$$m_{12}^{(1)}\ddot{u}_1 + \left(m_{22}^{(1)} + m_{22}^{(2)} \right)\ddot{u}_2 + m_{23}^{(2)}\ddot{u}_3$$

$$- \frac{EA}{L^{(1)}}u_1 + \left(\frac{EA}{L^{(1)}} + \frac{EA}{L^{(2)}} \right)u_2 - \frac{EA}{L^{(2)}}u_3 = 0 \tag{5.17}$$

$$m_{23}^{(2)}\ddot{u}_2 + m_{33}^{(2)}\ddot{u}_3 - \frac{EA}{L^{(2)}}u_2 + \frac{EA}{L^{(2)}}u_3 = F \tag{5.18}$$

In matrix form, the above equations of motion can be written as

$$\begin{bmatrix} m_{11}^{(1)} & m_{12}^{(1)} & 0 \\ m_{12}^{(1)} & m_{22}^{(1)} + m_{22}^{(2)} & m_{23}^{(2)} \\ 0 & m_{23}^{(2)} & m_{33}^{(2)} \end{bmatrix}\begin{Bmatrix} \ddot{u}_1 \\ \ddot{u}_2 \\ \ddot{u}_3 \end{Bmatrix}$$

$$+ \begin{bmatrix} \dfrac{AE}{L^{(1)}} & -\dfrac{AE}{L^{(1)}} & 0 \\ -\dfrac{AE}{L^{(1)}} & \dfrac{AE}{L^{(1)}} + \dfrac{AE}{L^{(2)}} & -\dfrac{AE}{L^{(2)}} \\ 0 & -\dfrac{AE}{L^{(2)}} & \dfrac{AE}{L^{(2)}} \end{bmatrix}\begin{Bmatrix} u_1 \\ u_2 \\ u_3 \end{Bmatrix} = \begin{Bmatrix} 0 \\ 0 \\ F \end{Bmatrix} \tag{5.19}$$

5.2.2 *Matrix Formulation*

We now reformulate the problem using the matrix notation of the FEM. Consider a generic rod element with two nodes i and j, located at x_i and x_j, respectively, as shown in Fig. 5.3. Also, for generality, assume that a concentrated force is applied at each node, denoted as F_i and F_j, respectively.

Introducing the *element shape function* matrix $\{N^e(x)\}$ and the *element nodal displacement* matrix $\{u^e\}$, both are column matrices and are defined as

$$\{N^e(x)\} = \begin{Bmatrix} N_i(x) \\ N_j(x) \end{Bmatrix} \qquad \{u^e(t)\} = \begin{Bmatrix} u_i(t) \\ u_j(t) \end{Bmatrix} \tag{5.20}$$

Then, the displacement in the element is expressible as

$$u^e(x, t) = \{N^e(x)\}^T \{u^e(t)\} \tag{5.21}$$

Consequently, the strain and the velocity in the element are expressible as

$$\epsilon^e(x, t) = \{B^e(x)\}^T \{u^e(t)\} \qquad v^e(x, t) = \{N^e(x)\}^T \{\dot{u}^e(t)\} \tag{5.22}$$

where

$$\{B^e(x)\} = \frac{d\{N^e(x)\}}{dx} \tag{5.23}$$

is called the *strain–displacement matrix*.

The potential and kinetic energies in the element can be calculated according to

$$V^e = \frac{1}{2} \int_{V^e} E\left[\epsilon^e(x)\right]^2 dV \qquad T^e = \frac{1}{2} \int_{V^e} \rho\left[v^e(x)\right]^2 dV \tag{5.24}$$

where V^e is the volume of the element. Note that $(\epsilon^e)^2$ should be expanded as $(\epsilon^e)^T \epsilon^e$ when the scalar is expressed as a matrix product, that is,

$$(\epsilon^e)^2 = \left(\{B^e(x)\}^T \{u^e(t)\}\right)^T \{B^e(x)\}^T \{u^e(t)\}$$
$$= \{u^e(t)\}^T \{B^e(x)\}\{B^e(x)\}^T \{u^e(t)\} \tag{5.25}$$

This ensures that the resulting chain matrix products can be performed in any order. Similar form of expression can be written for $(v^e)^2$. Then, substituting eqns. (5.22) into eqns. (5.24) and noting that $\{u\}$ and $\{\dot{u}\}$ are independent of x give

$$V^e = \frac{1}{2} \{u^e\}^T [K^e]\{u^e\} \qquad T^e = \frac{1}{2} \{\dot{u}^e\}^T [M^e]\{\dot{u}^e\} \tag{5.26}$$

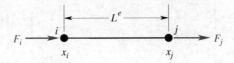

Figure 5.3   Two-node rod element

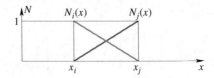

Figure 5.4 Linear shape functions for two-node element

where the material is assumed to be uniform within the element and

$$[K^e] = EA \int_{L^e} \{B^e(x)\}\{B^e(x)\}^T dx \tag{5.27}$$

$$[M^e] = \rho A \int_{L^e} \{N^e(x)\}\{N^e(x)\}^T dx \tag{5.28}$$

where L^e is the length of the element. Matrices $[K^e]$ and $[M^e]$ are called the *element stiffness matrix* and the *element mass matrix*, respectively.

The formulation up to this point is independent of any particular choice of the shape functions and the number of nodes in the element. Different shape functions can be used. For the two-node element, the two nodes define a pair of linear shape functions, similar to eqns. (5.3) and (5.4), as

$$\{N^e(x)\} = \begin{Bmatrix} N_i(x) \\ N_j(x) \end{Bmatrix} = \begin{Bmatrix} \frac{x_j-x}{x_j-x_i} \\ \frac{x-x_i}{x_j-x_i} \end{Bmatrix} \tag{5.29}$$

They are illustrated in Fig. 5.4. Then, integrating eqns. (5.27) and (5.28) gives

$$[K^e] = \frac{EA}{L^e}\begin{bmatrix} 1 & -1 \\ -1 & 1 \end{bmatrix} \qquad [M^e] = \frac{m^e}{6}\begin{bmatrix} 2 & 1 \\ 1 & 2 \end{bmatrix} \tag{5.30}$$

where $m^e = \rho A L^e$ is the mass of the element.

Next, define the *global nodal displacement* matrix as

$$\{u\} = \begin{Bmatrix} u_1 \\ u_2 \\ u_2 \end{Bmatrix} \tag{5.31}$$

The energies for each element in eqns. (5.26) are reexpressible in terms of the global nodal displacement matrix. For instance, the kinetic energy for element $\boxed{1}$ is expressible as

$$T^{\boxed{1}} = \frac{1}{2}\begin{Bmatrix} \dot{u}_1 \\ \dot{u}_2 \end{Bmatrix}^T \begin{bmatrix} m_{11}^{\boxed{1}} & m_{12}^{\boxed{1}} \\ m_{21}^{\boxed{1}} & m_{22}^{\boxed{1}} \end{bmatrix} \begin{Bmatrix} \dot{u}_1 \\ \dot{u}_2 \end{Bmatrix} = \frac{1}{2}\begin{Bmatrix} \dot{u}_1 \\ \dot{u}_2 \\ \dot{u}_3 \end{Bmatrix}^T \begin{bmatrix} m_{11}^{\boxed{1}} & m_{12}^{\boxed{1}} & 0 \\ m_{21}^{\boxed{1}} & m_{22}^{\boxed{1}} & 0 \\ 0 & 0 & 0 \end{bmatrix} \begin{Bmatrix} \dot{u}_1 \\ \dot{u}_2 \\ \dot{u}_3 \end{Bmatrix} \tag{5.32}$$

and, similarly, for element $\boxed{2}$,

$$
T^{\boxed{2}} = \frac{1}{2} \begin{Bmatrix} \ddot{u}_1 \\ \ddot{u}_2 \\ \ddot{u}_3 \end{Bmatrix}^T \begin{bmatrix} 0 & 0 & 0 \\ 0 & m_{22}^{\boxed{2}} & m_{23}^{\boxed{2}} \\ 0 & m_{32}^{\boxed{2}} & m_{33}^{\boxed{2}} \end{bmatrix} \begin{Bmatrix} \ddot{u}_1 \\ \ddot{u}_2 \\ \ddot{u}_3 \end{Bmatrix}
\tag{5.33}
$$

The total kinetic energy stored in the structure is the sum of the kinetic energies in individual elements, that is,

$$
T = T^{\boxed{1}} + T^{\boxed{2}} = \frac{1}{2} \begin{Bmatrix} \ddot{u}_1 \\ \ddot{u}_2 \\ \ddot{u}_3 \end{Bmatrix}^T \left(\begin{bmatrix} m_{11}^{\boxed{1}} & m_{12}^{\boxed{1}} & 0 \\ m_{21}^{\boxed{1}} & m_{22}^{\boxed{1}} & 0 \\ 0 & 0 & 0 \end{bmatrix} + \begin{bmatrix} 0 & 0 & 0 \\ 0 & m_{11}^{\boxed{2}} & m_{12}^{\boxed{2}} \\ 0 & m_{21}^{\boxed{2}} & m_{22}^{\boxed{2}} \end{bmatrix} \right) \begin{Bmatrix} \ddot{u}_1 \\ \ddot{u}_2 \\ \ddot{u}_3 \end{Bmatrix}
$$

$$
= \frac{1}{2} \begin{Bmatrix} \ddot{u}_1 \\ \ddot{u}_2 \\ \ddot{u}_3 \end{Bmatrix}^T \begin{bmatrix} m_{11}^{\boxed{1}} & m_{12}^{\boxed{1}} & 0 \\ m_{21}^{\boxed{1}} & m_{22}^{\boxed{1}} + m_{11}^{\boxed{2}} & m_{12}^{\boxed{2}} \\ 0 & m_{21}^{\boxed{2}} & m_{22}^{\boxed{2}} \end{bmatrix} \begin{Bmatrix} \ddot{u}_1 \\ \ddot{u}_2 \\ \ddot{u}_3 \end{Bmatrix}
\tag{5.34}
$$

In FEM terminologies, the above process is called the *assembly*. In computations, the assembly can be performed as follows: create the global matrices first of a size $N \times N$, where N is the total number of generalized coordinates, and fill the entries with zero. These matrices are called the *global mass matrix* and the *global stiffness matrix*. Then, for each element, add the values of entries in the element matrices to the global matrix entries at the positions dictated by the *global matrix index number*, which, based on the definitions in eqn. (5.31), is simply the node number. This process is repeated for all elements in the finite element model.

After the assembly, the total potential and kinetic energies are of the forms

$$
V = \frac{1}{2} \{u\}^T [K] \{u\} \qquad T = \frac{1}{2} \{\dot{u}\}^T [M] \{\dot{u}\}
\tag{5.35}
$$

In expanded form, they would give the same expressions as in eqns. (5.8) and (5.13).

The generalized force as obtained in eqn. (5.6) remains unchanged. For the element shown in Fig. 5.3, the *element nodal force* can be written as

$$
\{F^e\} = \begin{Bmatrix} F_i \\ F_j \end{Bmatrix}
\tag{5.36}
$$

When assembling the *element nodal force* matrices into the *global nodal force* matrix, all forces acting on a node shared by multiple elements sum to the externally applied force. This suggests that we can forgo writing element nodal force matrices and directly write the global nodal force matrix, according to the finite element model in Fig. 5.1 and the global displacement matrix in eqn. (5.31), as

$$
\{F\} = \begin{Bmatrix} 0 \\ 0 \\ F \end{Bmatrix}
\tag{5.37}
$$

Lagrange's equation can now be used to obtain the equation of motion for the system. In fact, Lagrange's equation can be written in the matrix form for the generalized coordinate $\{u\}$ as

$$\frac{d}{dt}\left(\frac{\partial \mathcal{L}}{\partial \{\dot{u}\}}\right) - \frac{\partial \mathcal{L}}{\partial \{u\}} = \{F\} \tag{5.38}$$

where the Lagrangian is, according to eqn. (5.35),

$$\mathcal{L} = T - V = \frac{1}{2}\{\dot{u}\}^T[M]\{\dot{u}\} - \frac{1}{2}\{u\}^T[K]\{u\} \tag{5.39}$$

The derivatives of the Lagrangian are expressible as

$$\frac{\partial \mathcal{L}}{\partial \{\dot{u}\}} = [M]\{\dot{u}\} \qquad \frac{\partial \mathcal{L}}{\partial \{u\}} = [K]\{u\} \tag{5.40}$$

Then, substituting eqns. (5.40) into eqn. (5.38) gives the following matrix form of the equations of motion for the system:

$$[M]\{\ddot{u}\} + [K]\{u\} = \{F\} \tag{5.41}$$

This indicates that, once the three matrices $[M]$, $[K]$, and $\{F\}$ are obtained through the assembly process, the matrix equation of motion for the system is determined.

In the FEM, the boundary conditions are imposed after the assembly. Imposing the boundary conditions is rather straightforward, especially if the boundary condition is expressible as $u_i = 0$, which is called a *homogeneous boundary condition*. Given this boundary condition, the u_i-equation is not needed, and all terms involving u_i in other equations vanish. In the matrix equation of motion in eqn. (5.41), for all the matrices, the ith row and the ith column are eliminated. This reduces the size of square matrices ($[M]$ and $[K]$) to $(N-1) \times (N-1)$ and the size of column matrices ($\{u\}$ and $\{F\}$) to $(N-1) \times 1$. If the problem has n boundary conditions, the final matrices would be of sizes of $(N-n) \times (N-n)$ or $(N-n) \times 1$, and, rightly, the system has $N-n$ DOFs.

If the boundary condition is not homogeneous, the process is similar but involves correcting remaining equations with the specified value of u_i.

One important characteristic of the matrix equation of motion to note is that the $[K]$ and $[M]$ matrices are symmetric. This is because the element-level matrices in eqns. (5.26) are symmetric, and both processes of assembling the global matrices and imposing the boundary conditions preserve the symmetry. The symmetry of these two system matrices means that the resulting matrix equation of motion for the system can be decoupled if the damping is proportional.

■ Example 5.1: Three-Element Model for Uniform Slender Rod

The uniform slender rod in Fig. 5.1 is divided into three elements of equal size. Follow the finite element modeling to obtain the matrix equation of motion for the system.

□ Solution:

We continue to use the two-node element, whose element stiffness and mass matrices are given in eqns. (5.26). When the rod is divided into three elements of equal size, the finite element model contains four nodes, and $L^e = L/3$ and $m^e = m/3$. Assume that the nodes are numbered from left to right, starting from 1, and denote the displacements at these nodes as u_i ($i = 1, \ldots, 4$). Then,

$$[K^e] = \frac{3AE}{L} \begin{bmatrix} 1 & -1 \\ -1 & 1 \end{bmatrix} \qquad [M^e] = \frac{m}{18} \begin{bmatrix} 2 & 1 \\ 1 & 2 \end{bmatrix} \tag{a}$$

Denote the global nodal displacement matrix as

$$\{u\} = \{u_1 \quad u_2 \quad u_2 \quad u_4\}^T \tag{b}$$

Then, the assembly of the system matrices gives

$$[M] = \frac{m}{18} \begin{bmatrix} 2 & 1 & 0 & 0 \\ 1 & 2+2 & 1 & 0 \\ 0 & 1 & 2+2 & 1 \\ 0 & 0 & 1 & 2 \end{bmatrix} = \frac{m}{18} \begin{bmatrix} 2 & 1 & 0 & 0 \\ 1 & 4 & 1 & 0 \\ 0 & 1 & 4 & 1 \\ 0 & 0 & 1 & 2 \end{bmatrix} \tag{c}$$

$$[K] = \frac{3AE}{L} \begin{bmatrix} 1 & -1 & 0 & 0 \\ -1 & 1+1 & -1 & 0 \\ 0 & -1 & 1+1 & -1 \\ 0 & 0 & -1 & 1 \end{bmatrix} = \frac{3AE}{L} \begin{bmatrix} 1 & -1 & 0 & 0 \\ -1 & 2 & -1 & 0 \\ 0 & -1 & 2 & -1 \\ 0 & 0 & -1 & 1 \end{bmatrix} \tag{d}$$

and the global nodal force matrix is

$$\{F\} = \{0 \quad 0 \quad 0 \quad F\}^T \tag{e}$$

Since the left end of the rod is fixed to the wall, the boundary condition is $u_1 = 0$. Thus, eliminating the first row and the first column of $[K]$ and $[M]$ matrices and the first row of $\{u\}$ and $\{F\}$ matrices gives the following equation of motion, according to eqn. (5.41),

$$\frac{m}{18} \begin{bmatrix} 4 & 1 & 0 \\ 1 & 4 & 1 \\ 0 & 1 & 2 \end{bmatrix} \begin{Bmatrix} \ddot{u}_2 \\ \ddot{u}_3 \\ \ddot{u}_4 \end{Bmatrix} + \frac{3AE}{L} \begin{bmatrix} 2 & -1 & 0 \\ -1 & 2 & -1 \\ 0 & -1 & 1 \end{bmatrix} \begin{Bmatrix} u_2 \\ u_3 \\ u_4 \end{Bmatrix} = \begin{Bmatrix} 0 \\ 0 \\ F \end{Bmatrix} \tag{f}$$

5.3 Finite Element Analyses of Beams

In this section, we use the finite element modeling to work out a more challenging problem: formulating and implementing the finite element analysis of a one-dimensional beam using MATLAB as the programming language. We choose to analyze a beam because we have done extensive lumped-parameter modeling with beams in Chapter 3.

5.3.1   Formulation of Beam Element

Consider a two-node one-dimensional beam element as shown in Fig. 5.5. Recall in beam analysis, the boundary conditions are specified as deflections and slopes at certain locations. This means that each node in the beam element has two generalized coordinates: one for the deflection and one for the slope. In FEM's terminology, each node is said to have two DOFs. Denote the nodes as i and j, located in a local coordinate system at $x = 0$ and L_e, respectively, where L_e is the length of the element. For generality, a concentrated force and a concentrated bending moment are applied at each node.

The element nodal displacement can be written as

$$\{u^e(t)\} = \begin{Bmatrix} u_i(t) \\ \theta_i(t) \\ u_j(t) \\ \theta_j(t) \end{Bmatrix} \tag{5.42}$$

Here, the term displacement should be understood in a generalized sense as in generalized coordinates. Having four nodal displacements, the deflection of the element can be approximated by a polynomial of four terms as

$$u = a + bx + cx^2 + dx^3 \tag{5.43}$$

Coefficients a through d are determined by the nodal displacements:

$$u|_{x=0} = u_i \qquad u'|_{x=0} = \theta_i \qquad u|_{x=L_e} = u_j \qquad u'|_{x=L_e} = \theta_j \tag{5.44}$$

Solving this set of equations, the deflection in the element becomes expressible as

$$u = u_i + \theta_i x + \left(-3\frac{u_i - u_j}{L_e^2} - \frac{2\theta_i + \theta_j}{L_e} \right) x^2 + \left(2\frac{u_i - u_j}{L_e^3} + \frac{\theta_i + \theta_j}{L_e^2} \right) x^3 \tag{5.45}$$

On the other hand, the FEM defines a set of shape functions $\{N^e(x)\}$ for each element such that the deflection in the element is expressible as

$$u(x, t) = \{N^e(x)\}^T \{u^e(t)\} \tag{5.46}$$

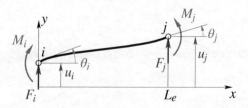

Figure 5.5   Two-node beam element

Comparing eqns. (5.45) and (5.46), the shape functions can be identified as

$$N_{u_i}(\xi) = 1 - 3\xi^2 + 2\xi^3 \tag{5.47}$$

$$N_{\theta_i}(\xi) = \left(\xi - 2\xi^2 + \xi^3\right) L_e \tag{5.48}$$

$$N_{u_j}(\xi) = 3\xi^2 - 2\xi^3 \tag{5.49}$$

$$N_{\theta_j}(\xi) = \left(-\xi^2 + \xi^3\right) L_e \tag{5.50}$$

where $\xi = x/L_e$. These four shape functions are sketched in Fig. 5.6.

The element stiffness and mass matrices can be obtained by evaluating the potential and kinetic energies in the element, that is,

$$V^e = \frac{EI}{2} \int_{L_e} [u''(x, t)]^2 dx \qquad T^e = \frac{\rho A}{2} \int_{L_e} \dot{u}^2(x, t) dx \tag{5.51}$$

where the expression for the strain energy for the beam has been used in Chapter 3, in eqn. (3.42). Substituting eqn. (5.46) into the above expressions gives

$$V^e = \frac{1}{2} \{u^e\}^T [K^e] \{u^e\} \qquad T^e = \frac{1}{2} \{\dot{u}^e\}^T [M^e] \{\dot{u}^e\} \tag{5.52}$$

where

$$[K^e] = \frac{EI}{2} \int_{L_e} \{B^e\} \{B^e\}^T dx \tag{5.53}$$

$$[M^e] = \frac{\rho A}{2} \int_{L_e} \{N^e\} \{N^e\}^T dx \tag{5.54}$$

and

$$\{B(x)\} = \frac{d^2 \{N^e(x)\}}{dx^2} \tag{5.55}$$

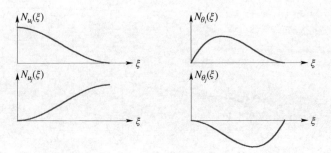

Figure 5.6 Shape functions for two-node beam element

With the shape functions in eqns. (5.47) through (5.50), it is convenient to perform the integrations in eqns. (5.53) and (5.54) in terms of variable ξ. Then,

$$\{B\} = \frac{1}{L_e^2} \begin{Bmatrix} -6 + 6\xi \\ -4 + 6\xi \\ 6 - 12\xi \\ -2 + 6\xi \end{Bmatrix} \tag{5.56}$$

and the integrations give

$$[K^e] = \frac{EI}{L_e^3} \begin{bmatrix} 12 & 6L_e & -12 & 6L_e \\ 6L_e & 4L_e^2 & -6L_e & 2L_e^2 \\ -12 & -6L_e & 12 & -6L_e \\ 6L_e & 2L_e^2 & -6L_e & 4L_e^2 \end{bmatrix} \tag{5.57}$$

$$[M^e] = \frac{m_e}{420} \begin{bmatrix} 156 & 22L_e & 54 & -13L_e \\ 22L_e & 4L_e^2 & 13L_e & -3L_e^2 \\ 54 & 13L_e & 156 & -22L_e \\ -13L_e & -3L_e^2 & -22L_e & 4L_e^2 \end{bmatrix} \tag{5.58}$$

where $m_e = \rho A L_e$ is the mass of the element.

For the nodal forces, we can construct the global nodal force matrix $\{F\}$ directly by placing applied forces and moments at appropriate rows. At this stage, the so-called *global matrix index* is important. Each node has two generalized coordinates or DOFs; and each DOF has one corresponding equation. For instance, node i has generalized coordinates u_i and θ_i. Their row numbers in the global displacement matrix are called their global matrix index numbers, which are also called the *global equation numbers* or the *global DOF numbers*. In constructing the global nodal force matrix, concentrated forces F_i and F_j are placed at rows corresponding to u_i and u_j, respectively, and the concentrated moments M_i and M_j are placed at rows corresponding to θ_i and θ_j, respectively. In addition, a beam is capable of withstanding a distributed load. Converting a distributed load along the element into nodal forces is left as an exercise for the readers.

The boundary conditions can be imposed in the same way as for a rod: by eliminating the affected rows and columns from the assembled global matrices.

■ Example 5.2: Single-Element Cantilever Beam

A uniform cantilever beam of mass m, length L, and bending rigidity EI is subjected to a concentrated force F at its free end. Follow the finite element formulation above to obtain the equation of motion for the beam represented by a single two-node beam element. Use the equation of motion to find the fundamental natural frequencies of the beam.

□ **Solution:**

Denote the cantilevered end as node 1 and the free end as node 2. Since the finite element model has only a single element, without the use of an assembly process, the equation of motion for the beam can be written based on eqns. (5.57) and (5.58) as

$$\frac{m}{420}\begin{bmatrix} 156 & 22L & 54 & -13L \\ 22L & 4L^2 & 13L & -3L^2 \\ 54 & 13L & 156 & -22L \\ -13L & -3L^2 & -22L & 4L^2 \end{bmatrix}\begin{Bmatrix} \ddot{u}_1 \\ \ddot{\theta}_1 \\ \ddot{u}_2 \\ \ddot{\theta}_2 \end{Bmatrix}$$

$$+\frac{EI}{L^3}\begin{bmatrix} 12 & 6L & -12 & 6L \\ 6L & 4L^2 & -6L & 2L^2 \\ -12 & -6L & 12 & -6L \\ 6L & 2L^2 & -6L & 4L^2 \end{bmatrix}\begin{Bmatrix} u_1 \\ \theta_1 \\ u_2 \\ \theta_2 \end{Bmatrix}=\begin{Bmatrix} 0 \\ 0 \\ 0 \\ F \end{Bmatrix} \tag{a}$$

The boundary conditions for the cantilever beam include $u_1 = 0$ and $\theta_1 = 0$, which correspond to global matrix indices of 1 and 2, respectively. Thus, eliminating the first two rows and first two columns of the matrices gives

$$\frac{m}{420}\begin{bmatrix} 156 & -22L \\ -22L & 4L^2 \end{bmatrix}\begin{Bmatrix} \ddot{u}_2 \\ \ddot{\theta}_2 \end{Bmatrix}+\frac{EI}{L^3}\begin{bmatrix} 12 & -6L \\ -6L & 4L^2 \end{bmatrix}\begin{Bmatrix} u_2 \\ \theta_2 \end{Bmatrix}=\begin{Bmatrix} 0 \\ F \end{Bmatrix} \tag{b}$$

To obtain the natural frequencies of the beam, assume the solution of the form

$$\{u\}=\begin{Bmatrix} u_2 \\ \theta_2 \end{Bmatrix}=\begin{Bmatrix} U_2 \\ \Theta_2 \end{Bmatrix}e^{i\omega t}=\{U\}e^{i\omega t} \tag{c}$$

Substituting this solution into the corresponding homogeneous equation of motion gives

$$\begin{bmatrix} 12\frac{EI}{L^3}-\frac{156}{420}m\omega^2 & -6\frac{EI}{L^2}+\frac{22}{420}mL\omega^2 \\ -6\frac{EI}{L^2}+\frac{22}{420}mL\omega^2 & 4\frac{EI}{L}-\frac{4}{420}mL^2\omega^2 \end{bmatrix}\begin{Bmatrix} U_2 \\ \Theta_2 \end{Bmatrix}=0 \tag{d}$$

For nontrivial solutions, the determinant of the system matrix must vanish, giving

$$\left(\frac{12EI}{L^3}-\frac{156}{420}m\omega^2\right)\left(\frac{4EI}{L}-\frac{4}{420}mL^2\omega^2\right)-\left(-\frac{6EI}{L^2}+\frac{22}{420}mL\omega^2\right)^2=0 \tag{e}$$

which can be solved to give

$$\omega^2=12\left(51\pm8\sqrt{39}\right)\frac{EI}{mL^3} \tag{f}$$

The fundamental natural frequency is the lowest natural frequency. Taking the negative sign,

$$\omega_1^2\approx12.48\frac{EI}{mL^3} \tag{g}$$

□ **Discussion:**

In the lumped-parameter modeling, the equivalent spring constant and the equivalent mass for a cantilever beam have been found in Examples 3.2 and 3.8, respectively, as

$$k_{\text{equiv}}=\frac{3EI}{L^3}\qquad m_{\text{equiv}}=\frac{33}{140}m \tag{h}$$

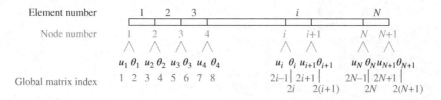

Figure 5.7 Numbering schemes for finite element model for one-dimensional beam

which give the beam's fundamental natural frequency as

$$\omega_{LPM}^2 = \frac{140}{11}\frac{EI}{mL^3} \approx 12.73\frac{EI}{mL^3} \tag{i}$$

The two values are rather close, but both are approximate. In essence, both the lumped-parameter modeling and the finite element modeling in this example are based on the Rayleigh–Ritz method. The reason for the difference is that they use different shape functions.

5.3.2   Implementation Using MATLAB

The finite element method is a computational method. To see the complete process of the finite element modeling, the formulation for the one-dimensional beam is now implemented using MATLAB as a computational code.

To organize the computation systematically, numbering schemes must be established first. The most important number in a finite element model is the node number, from which the global matrix index is derived. These numbering schemes are collectively called the *topology* of the finite element model.

Assume that the beam is divided into N elements. All topological entities are numbered consecutively from left to right, starting from 1. The numbering schemes for elements, nodes, and the global matrix indices, as well as the relations among these numbers, are illustrated in Fig. 5.7. The MATLAB code is listed as follows:

```
% MATLAB code for 1-D Beam FEM Analysis -- (c)2015 Liang-Wu Cai
%
% Part 1: Structural and material properties, using SI units
E = 207E9;        % Young's modulus, Pa
rho = 7900;       % Mass density, kg/m^(-3)
d = 0.025;        % Beam diameter, m
L = 0.5;          % Total beam length, m

I = 1/64*pi*d^4; % Moment of area
A = pi*d^2/4;    % Cross-sectional area
m = rho*A*L;     % Total mass

% Part 2: FEM model details
Ne = 40;    % Number of elements
BC =[1,2]; % Boundary conditions: listing by DOFs in ascending order
```

```
% Part 3: FEM  modeling
% 3.1: Element mass and stiffness matrices
Me = @(l,m) m/420*[156, 22*l, 54, -13*l; 22*l, 4*l^2, 13*l, -3*l^2;...
    54, 13*l, 156, -22*l; -13*l, -3*l^3, -22*l, 4*l^2];
Ke = @(l,EI) EI/l^3*[12, 6*l, -12, 6*l; 6*l, 4*l^2, -6*l, 2*l^2;...
    -12, -6*l, 12, -6*l; 6*l, 2*l^2, -6*l, 4*l^2];

Ntotal = 2*(Ne+1);        % Size of global matrix = total # of DOF's
M=zeros(Ntotal,Ntotal);   % Declare global stiffness and mass matrices
K=zeros(Ntotal,Ntotal);

% 3.2 Assembling global M & K matrices. Element i has nodes i & i+1
for i=1:Ne;
    me = Me(L/Ne,m/Ne);
    ke = Ke(L/Ne,E*I);
    M(2*i-1:2*(i+1),2*i-1:2*(i+1)) = M(2*i-1:2*(i+1),2*i-1:2*(i+1))+me;
    K(2*i-1:2*(i+1),2*i-1:2*(i+1)) = K(2*i-1:2*(i+1),2*i-1:2*(i+1))+ke;
end;

% 3.3 Imposing boundary conditions by eleminiating rows and columns
for i=1:length(BC);
    M(BC(length(BC)-i+1),:)=[];
    M(:,BC(length(BC)-i+1))=[];
    K(BC(length(BC)-i+1),:)=[];
    K(:,BC(length(BC)-i+1))=[];
end;

% Part 4: Solve for eigenvalues and convert to natural frequency
[V D] = eig(K,M);
for i=1:2*(Ne+1)-length(BC);
    omega(i) = sqrt(D(i,i));
end;
min(omega)
```

In the above code, Part 1 collects the physical and geometrical parameters of the problem. Some of these parameters are given and some are derived. The particular values in the above code reflect the situation in the examples to follow.

Part 2 specifies the parameters for the finite element model: Ne is the number of elements; and array BC stores the boundary conditions. This simplistic finite element analysis code makes the following assumptions: all elements are of equal size, and all boundary conditions are homogeneous. Because all boundary conditions are homogeneous, knowing only their global matrix indices suffices. They are listed in BC in an ascending order. These restrictions are not due to the formulation. They are purposely designed to produce a compact code so that we can focus on conceptually important steps, rather than mundane details, in the computational process.

Part 3 is the core of the finite element modeling. Part 3.1 implements element matrices in eqns. (5.57) and (5.58) as MATLAB anonymous functions; Part 3.2 assembles the global stiffness and mass matrices; and Part 3.3 imposes the boundary conditions. When imposing the boundary conditions, eliminating an affected row or column is done by assigning a blank matrix [] to the entire row or column.

Part 4 calls MATLAB's built-in function eig to compute the eigenvalues and convert the eigenvalues into an array of natural frequencies. The last line ends without a semicolon. This way, at the end of the computation, the fundamental natural frequency of the finite element model will be displayed.

In the following, numerical examples are used to demonstrate how to adapt the code above to suite the specifics of a given finite element analysis problem.

■ **Example 5.3: Multiple-Element Cantilever Beam**

Use the code above to determine the fundamental natural frequency of a cantilever beam. The beam is made of steel of a mass density 7900 kg/m³ and a Young's modulus 207 GPa. The beam has a circular cross section of diameter 0.025 m and a length of 0.5 m.

□ **Exploring the Solution with MATLAB**

The structural and material properties of this cantilever beam and its boundary conditions have already been incorporated into the code. Thus, the code can be run without any modification. The code is run for a different number of elements Ne. The resulting fundamental natural frequency of the beam is tabulated as follows:

Number of Elements	Fundamental Natural Frequency (rad/s)
1	452.0873
2	449.2017
3	449.4021
4	449.5881
5	449.6996
10	449.8785
20	449.9299

The convergence of the fundamental natural frequency toward a steady value as the number of elements increases provides a preliminary validation of the beam FEM code. Comparing with the FEM result using 20 elements, the simplistic lump-parameter modeling gives an estimate of the natural frequency, according to eqn. (i) in Examples 5.2, as

$$\omega_{\text{LPM}} = \sqrt{\frac{140}{11} \frac{EI}{mL^3}} = 456.5405 \text{ rad/s} \tag{a}$$

whose error is approximately 1.5%!

■ **Example 5.4: Simply-Supported Beam**

A simply-supported beam has the same geometric and physical parameters as in Example 5.3. Use the FEM to determine its fundamental natural frequency.

□ **Exploring the Solution with MATLAB**

This beam differs from the one in Example 5.3 only in its boundary conditions. Referring to the numbering schemes in Fig. 5.7, for an FEM model of N elements, the last node is $N + 1$. Thus, the boundary conditions are $u_1 = 0$ and $u_{N+1} = 0$. Their corresponding global matrix indices are 1 and $2N + 1$, respectively. Thus, the only change in the code is the following line for specifying the boundary conditions:

```
BC =[1,2*Ne+1];   % Boundary conditions: listing by DOFs in ascending order
```

The computation results are listed as the following:

Number of Elements	Fundamental Natural Frequency (rad/s)
1	1487.5
2	1268.1
5	1261.9
10	1262.6
20	1262.9

Recall that the equivalent spring constant and the equivalent mass for the simply-supported beam have been found in Examples 3.3 and 3.9, respectively,

$$k_{\text{equiv}} = \frac{48EI}{L^3} \qquad m_{\text{equiv}} = \frac{17}{35}m \tag{a}$$

which give, using the numerical values,

$$\omega_{\text{LPM}} = \sqrt{\frac{48EI}{L^3}\frac{35}{17m}} = 1272.2 \text{ rad/s} \tag{b}$$

Again, treating the finite element analysis result obtained by using 20 elements as the accurate one, the error in the lump-parameter modeling is approximately 0.75%.

■ **Example 5.5: Guided-Cantilever Beam**

A guided-cantilever beam has the same geometrical and physical parameters as in Example 5.3. Use the FEM to determine its fundamental natural frequency.

□ **Exploring the Solution with MATLAB**

This beam differs from the one in Example 5.3 only in the boundary conditions. Assume that the left end is cantilevered and the right end is guided, which means the slope vanishes.

Thus, the only change in the code is the following line for the boundary conditions:

```
BC =[1,2,2*(Ne+1)]; % Boundary conditions: listing by DOFs in ascending order
```

The computation results are listed as the following:

Number of Elements	Fundamental Natural Frequency (rad/s)
1	727.3854
2	716.7331
5	715.5804
10	715.7137
20	715.7642

Recall that the equivalent spring constant and the equivalent mass for the guided-cantilever beam have been analyzed in Example 3.24 when the micro-electro-mechanical systems (MEMS) cantilever beam is used for in-plane applications. From that example,

$$k_{equiv} = \frac{12EI}{L^3} \qquad m_{equiv} = \frac{13}{35}m \tag{a}$$

which gives, using the numerical values,

$$\omega_{LPM} = \sqrt{\frac{12EI}{L^3}\frac{35}{13m}} = 727.3854 \text{ rad/s} \tag{b}$$

Again, using the finite element analysis result obtained by using 20 elements as a reference, the error in the lump-parameter modeling is approximately 1.62%.

■ Example 5.6: Tapered Beam

A cantilever beam made of a wooden plank has a rectangular cross section and a uniform thickness of 0.5 in. Its width varies linearly from 5 in. at the cantilevered end to 1 in. at the free end, as shown in Fig. 5.8. The beam has a length of 2 ft. The material has Young's modulus of 1.15 Mpsi and a specific weight of 43.2 lbf/ft³. Use the FEM to find its fundamental natural frequency.

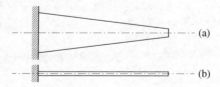

Figure 5.8 Tapered beam with linearly varying width: (a) top view and (b) side view

□ Exploring the Solution with MATLAB

In the US ft-lb-s unit system, the unit for a mass is derived as its weight divided by the gravitational acceleration, of 32.2 ft/s^2, and is called the *slug*. The given physical and geometric parameters are converted to the standard units as the following:

$$\text{Beam thickness} = 0.5 \text{ in.} \times \frac{1 \text{ ft}}{12 \text{ in.}} = 0.04167 \text{ ft} \tag{a}$$

$$\text{Smallest width} = 1 \text{ in.} \times \frac{1 \text{ ft}}{12 \text{ in.}} = 0.0833 \text{ ft} \tag{b}$$

$$\text{Largest width} = 5 \text{ in.} \times \frac{1 \text{ ft}}{12 \text{ in.}} = 0.4167 \text{ ft} \tag{c}$$

$$\text{Mass density} = \frac{43.2 \text{ lbf/ft}^3}{32.2 \text{ ft/sec}^2} = 1.342 \text{ slug/ft}^3 \tag{d}$$

$$\text{Young's modulus} = 1.15 \times 10^6 \text{ lbf/in}^2 \times \left(\frac{12 \text{ in.}}{1 \text{ ft}} \right)^2 = 1.656 \times 10^8 \text{ lbf/ft}^2 \tag{e}$$

Furthermore, the cross-sectional area and the moment of area of the beam are

$$A = w(x)h \qquad I = \frac{1}{12}w(x)h^3 \tag{f}$$

where $w(x)$ is the width and h is the thickness. Because of the linear variation of the width, both A and I vary linearly. The essence of the finite element modeling is to use small enough elements such that those parameters can be assumed to be constant within each element. In this example, the width at the center of each element is used as the width for the entire element. Two parts of the code need to be modified, as listed in the following:

```
% Part 1: Structural and material properties, using US ft-lbf-sec units
E = 1.15E6 *12^2;        % Young's modulus, lbf/ft^2
rho = 43.2/32.2;         % mass density, slug/ft^3
L = 2;                   % length, ft
h = 0.5/12;              % thickness, ft
w = @(x) (5-2*x)/12;     % width, in ft, as a funtion of x


I = @(w) 1/12*w*h^3;     % Moment of area, as function of width
A = @(w) w*h;            % Cross-sectional area, as function of width

...


% 3.2 Assembly of global M & K matrices. Element i has nodes i & i+1
for i=1:Ne;
    x = (i-.5)*L/Ne;           % x coordinate for center of each element;
    ke = Ke(L/Ne,E*I(w(x)));
    me = Me(L/Ne,rho*A(w(x))*L/Ne);
    ...
end;
```

In Part 1, the unit conversions in eqns. (a) through (e) are performed, and a set of anonymous functions is defined to calculate the beam width w, the cross-sectional area A, and the moment of area I of the beam element. These anonymous functions are called in Part 3.2 to calculate the element mass and stiffness matrices. The code is run with a different number of elements, and the resulting fundamental natural frequency is listed as follows:

Number of Elements	Fundamental Natural Frequency (rad/s)
2	158.5171
5	176.3010
10	179.2892
20	180.0631
30	180.2079
40	180.2587
50	180.2823

■ Example 5.7: MEMS Cantilever Beam

Use the FEM to determine the fundamental natural frequency for both in-plane and out-of-plane applications of the cantilever beam designed for MEMS applications in Example 3.24, which is shown in Fig. 5.9. Geometric parameters are listed in Table 5.1. The material is silicon wafer of mass density 2300 kg/m^3 and Young's modulus 130 GPa.

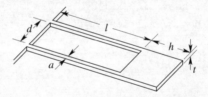

Figure 5.9   MEMS cantilever beam model

□ Exploring the Solution with MATLAB
In-Plane Applications

For in-plane applications, the beam can be viewed as comprising of two segments: one segment of length l; and the remainder, the plate portion, as the second segment.

Table 5.1   Parameters for the MEMS cantilever beam example

Name	Symbol	Value
Leg length	l	400 μm
Plate length	h	100 μm
Leg width	a	8 μm
Leg spacing	d	64 μm
Layer thickness	t	8 μm

The two segments differ in both the moment of area and the cross-sectional area. The first segment is made up of two guided-cantilever beams of length l, which requires the slope to vanish at $x = l$, which is 4/5 of the total length. To ensure that there is a node at this location, the finite element model will have to contain multiples of five elements. The code needs to be modified in two places, similar to the case of tapered beam in Example 5.6.

```
% Part 1: Structural and material properties, using SI units
E = 130E9;        % Young's modulus, Pa
rho = 2300;       % mass density, kg/m^(-3)
h = 100E-6;       % plate length, m
a = 8E-6;         % leg thickness, m
t = 8E-6;         % wafer thickness, m
d = 64E-6;        % leg spacing, m
l = 400E-6;       % leg length, m

I1 = 2/12*t*a^3;
I2 = 1/12*t*(d+2*a)^3;
A1 = 2*a*t;
A2 = (d+2*a)*t;
I = @(x) I1 + (I2-I1)*heaviside(x-1);
md = @(x) rho*A1 + rho*(A2-A1)*heaviside(x-1);   % linear mass density

% Part 2: FEM modeling details
...
BC =[1,2,2*(Ne*4/5+1)];
...
% 3.2 Assembly of global M and K matrices. Ele-
ment i has nodes i and i+1
for i=1:Ne;
    x = (i-.5)*(1+h)/Ne;
    ke = Ke( (1+h)/Ne, E*I(x));
    me = Me( (1+h)/Ne, md(x)*(1+h)/Ne);
    ...
end;
```

In the above, the step changes in the cross-sectional area and the moment of area are expressed as MATLAB's heaviside step function; and md calculates the linear mass density. The code is run through a different number of elements in multiples of 5 to ensure that the edge of the plate falls on a node. The resulting fundamental natural frequency is as follows:

Number of Elements	Fundamental Natural Frequency (rad/s)
5	294,925
10	294,947
20	294,957
30	294,958
40	294,947
50	294,979

Out-of-Plane Applications

For out-of-plane applications, the beam can still be viewed as comprising of two segments. The only difference from the case of in-plane applications is the direction of bending and, in turn, the expressions for the moment of area. Thus, only the following two lines are changed from the code used for the in-plane applications:

```
I1 = 2/12*a*t^3;
I2 = 1/12*(d+2*a)*t^3;
...
BC = [1,2];
```

The code is run through a different number of elements in multiples of 5 to ensure that the edge of the plate falls on a node. The resulting fundamental natural frequency is as follows:

Number of Elements	Fundamental Natural Frequency (rad/s)
5	132,079
10	132,123
20	132,135
30	132,140
40	132,155
50	132,095

□ Discussions:

With given numerical values for the geometry and the material, the fundamental natural frequencies are, according to eqns. (h) and (j) in Example 3.24,

$$\omega_{\text{LPM, in-plane}} = 2.9521 \times 10^5 \text{rad/s} \qquad (a)$$

$$\omega_{\text{LPM, out-of-plane}} = 1.5420 \times 10^5 \text{rad/s} \qquad (b)$$

The lumped-parameter modeling result for the in-plane applications is very close to the finite element modeling results, with errors in the level of 0.08% based on the 50-element result. On the other hand, the lumped-parameter modeling for out-of-plane applications gives a noticeably higher fundamental natural frequency, with errors in the level of 16.7%.

5.3.3   Generalization: Large-Scale Finite Element Simulations

The code in the previous section is compact yet anatomically complete. It serves as a basis for us to explore the workings of large-scale finite element analyses. In this section, we explore how the code can be augmented to accommodate a variety of problems, and how the code can guide us through the process of using commercial large-scale finite element analysis software.

5.3.3.1   From a Code Developer's Perspective

Part 1 of the code is about the geometric and physical properties of a structure. An essential level of versatility is to allow the user to supply the data at the time of execution, instead of directly modifying the code. An elementary mechanism is via input data files, in such cases, the code needs to be able to read the data from a file and interpret the data. A more advanced mechanism is to collect common engineering materials as a built-in library while accepting the user-supplied data.

Part 2 of the code, in conjunction with Fig. 5.7, is about the finite element model topology. In the early years of the FEM, this part is done by the user manually compiling lists of nodal coordinates and element nodal compositions; and then the code reads and interprets the data. This could be an extremely laborious and error-prone process. With the advances of computer graphics and display technology in the 1980s and 1990s, this part is now essentially taken over by computer-aided modeling and automatic mesh generation. Recent commercial finite element software integrates this part into the computer-aided design and engineering systems.

Part 3.1 is where different types of element models, such as the rod and beam elements studied in this chapter, are implemented. For different types of structural members, the physics behind the element formulation may be very different. Also, finite element formulation could be based on different theories, such as variational formulation that is based on a theorem in the theory of elasticity and the weighted residual method that is based on minimizing errors. A versatile finite element code typically provides an array of types of elements for users to choose from.

Part 3.2 assembles the system matrices. For large-scale computations, efficiently storing these matrices in the computer memory becomes important: how to minimize the memory requirement while taking advantages of the symmetry of system matrices, as well as optimizing the numbering schemes.

Part 3.3 considers the boundary conditions and loadings of the structure. More general non-homogeneous boundary conditions should be allowed. More complex boundary conditions can also be modeled as special type of elements.

Part 4 solves the assembled matrix equation of motion, based on the given initial conditions. In the code in the previous section, calling MATLAB's `eig` function to perform one type of vibration analysis serves the purpose of demonstrating the complete finite element analysis without getting too involved with computational algorithms and the associated issues. Other types of vibration analyses can be implemented here and would be required to address the algorithms and the associated issues. As the number of DOFs of the problem grows larger, many other computational issues emerge, and strategies to address those issues are developed and improved. Solving the resulting matrix equation system is by itself a topic of endeavor, which in fact was the focal point for many research and developmental efforts in the 1960s and up to 1980s, and the collimation of those efforts forms the technological backbone of MATLAB software itself.

A finite element analysis solves for the nodal displacements for a structure. Stresses, strains, and many other parameters of interest, such as effective stresses and factors of safety, can be obtained in secondary computations, generally called the *post-processing*. In the old days, computed results are printed out numerically. The results of modern finite element analysis are displayed graphically.

5.3.3.2   From an End-User's Perspective

A number of commercial software packages have been available for large-scale simulations since the early years of the development of the FEM in the 1970s. With the advances in computer technology seen in the past decades, large-scale computations become possible with personal computers. Despite the different software implementations, the procedure of performing a finite element simulation is conceptually the same and can be summarized as the following steps:

- *Geometric Modeling*: A geometric model is first constructed using geometric entities such as points, lines, and surfaces. A typical geometric modeling process works as follows: draw a sketch on a plane first since we work most comfortably on planar geometries. Then a shape-forming method, such as extrusion or rotation, is used to convert the planar sketch into a three-dimensional object.
- *Finite Element Modeling*: The geometric model is divided into finite elements. This process is called *meshing*. The finite element model is an approximation of the geometric model because errors could be introduced in the meshing process. In theory, the finite element modeling also includes the development of element-level matrices and the assemblage of global matrices. However, for most users of commercial software, this part of modeling is rather invisible.
- *Constraining and Loading*: The finite element simulation is finding stresses and deformations of a structure under certain loading. The finite element model will need to be constrained in space and loaded by the specified loading. In theory, constraints and loads can only be applied to nodes and elements in the finite element model. However, most commercial software allow constraints and loads be applied to geometric models and then meshed around the locations of these constraints and loads. This is a much preferred method because when the model is remeshed and the constraints and loads remain unchanged. In most commercial finite software, constraining and loading of the model are performed before the geometric model is meshed into the finite element model.
- *Solution Process*: This is the central step of the finite element simulation. Ironically, it does not require much work of the user beyond pressing a button. But, it still takes most of computer time to perform the computation.
- *Post-Processing*: Useful information is extracted from the displacement solution and displayed, plotted, or exported to be further processed by other programs.

However, as a general rule: the computational results should not be accepted blindly: the computation needs to be validated before the results could be accepted. Validation of computational results has always been a daunting task if we are given a problem that no one knows the answer. One approach to this is to construct simpler analytical models to obtain approximate solutions and to compare the approximate solutions to the simulation results. The lump-parameter modeling we have done in Chapter 3 provides a way to produce such simplified analytical models.

5.3.4   Damping Models in Finite Element Modeling

Recall that when the matrix equation of motion for a system is written as

$$[M]\{\ddot{u}\} + [C]\{\dot{u}\} + [K]\{u\} = \{F\} \tag{5.59}$$

the damping is called the proportional damping or Rayleigh damping if

$$[C] = \alpha[M] + \beta[K] \tag{5.60}$$

where α and β are constants. The decoupling method discussed in Chapter 4, which is more often called the *modal superposition* method in which the decoupled coordinates are called the *modal coordinates* in the finite element modeling, can accommodate damping only if the damping is proportional.

Using modal coordinates, one more damping model, called the *modal damping*, can also be included. Assume that, for mode i, the principal coordinate is y_i and the corresponding decoupled equation of motion, according to eqn. (5.60), is

$$m_i\ddot{y}_i + (\alpha m_i + \beta k_i + 2m_i\zeta_i\omega_i)\dot{y}_i + k_i y_i = F_i(t) \tag{5.61}$$

where ζ_i-term is the added *modal damping* term. Without the α- and β-damping terms, eqn. (5.61) would be identical to that of a single-DOF system. Dividing both sides of eqn. (5.61) by m_i gives

$$\ddot{y}_i + 2\left(\zeta_i + \frac{\alpha}{2\omega_i} + \frac{\omega_i}{2}\beta\right)\omega_i\dot{y}_i + \omega_i^2 y_i = f_i(t) \tag{5.62}$$

Thus, the effective damping ratio for mode i is

$$\zeta_{\text{eff}} = \zeta_i + \frac{\alpha}{2\omega_i} + \frac{\omega_i}{2}\beta \tag{5.63}$$

These three damping models behave differently when the frequency changes, as illustrated in Fig. 5.10. They are considered as essential damping models that have been incorporated in most commercial finite element software, allowing user to specify the particular model, or a combination of them, and the associated parameters to be used in the finite element modeling.

Readers might recall that, in Chapter 3, another damping model, called the *structural damping model*, has been praised as the most plausible damping model for structural components. The structural damping model assumes a complex stiffness for a structure

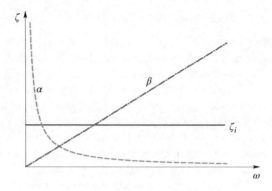

Figure 5.10 Effective damping ratio due to different damping models

component. In doing so, the amount of computer memory required to store the finite element model and the time needed for the computation would also be doubled. Historically, striving to fit as large a problem as possible into the limit amount of available computer memory has been an important line of technical development of the finite element method. To the end, most commercial finite element software takes a compromised route of using only the above-mentioned three damping models.

5.4 Vibration Analyses Using SOLIDWORKS

In this section, a series of large-scale finite element simulations of vibrations of a cantilever beam designed for MEMS applications is performed using commercial FEM software SOLIDWORKS. The MEMS cantilever beam is shown in Fig. 5.9, with parameters given in Table 5.1 and in Example 5.7. It has been modeled by the lumped-parameter modeling first in Example 3.24 and then by the FEM using the beam element in this chapter in Example 5.7.

These large-scale finite element simulation examples using commercial software serve two purposes. First, they provide step-by-step tutorials on using a leading software package and open the door for students to explore more real-life vibration problems. Second, they demonstrate that the overall steps listed Section 5.3.3 remain true in any finite element analysis software such that students who are exposed to one software package at school should be well prepared for the new simulation environment without much difficulty.

SOLIDWORKS started out in the early 1990s and soon became a leader in solid modeling. Before SOLIDWORKS, in the 1980s, a company called *Structural Research & Analysis Corporation* developed a finite element software package COSMOS/M as an alternative to more expensive commercial FEM software. It later developed COSMOSWORKS to provide an interface for models constructed in SOLIDWORKS to be analyzed by COSMOS/M. In the early 2000s, both companies were acquired by Dassault Systèmes of France. Since then, the finite element analysis capabilities of COSMOS/M are consolidated into *SOLIDWORKS Simulation* as an add-in package for SOLIDWORKS.

5.4.1 Introduction to SOLIDWORKS Simulation

SOLIDWORKS is centered around the typical workflow of computer-aided engineering. It offers extensive tools for producing three-dimensional geometric models. A typical workflow includes stages of parts construction, assembly of parts, and engineering drawings. In *SOLIDWORKS Simulation*, a finite element analysis is called a Study, of the model just constructed in SOLIDWORKS. In addition to static structural finite element analysis capabilities, *SOLIDWORKS Simulation* offers a complete set of the vibration analysis tools for modal, steady-state, and transient vibration analyses. It also offers some advanced vibration analyses such as random vibration and response spectrum analyses. Readers might recall that a response spectrum has been computed in Example 2.17 in Chapter 2.

When SOLIDWORKS is launched, its user interface is rather simplistic, or bland, as shown in Fig. 5.11. More important components of the user interface appear only after a task is started. For the first time using SOLIDWORKS, the *SOLIDWORKS Simulation* package must be turned on, by clicking on the small triangle beside the Options (check-list) icon on the Menu bar, and then select Add-in..., as shown in Fig. 5.12. In the add-in configuration interface, check the appropriate boxes for *SOLIDWORKS Simulation*, as shown in Fig. 5.13: the left checkbox is for the current session, and the right is for all future sessions. It is advisable to check both boxes.

To start constructing a new part, click on the New (blank file) icon on the Menu bar and then select Part, as shown in Fig. 5.14. The complete SOLIDWORKS graphical user interface now appears, as shown in Fig. 5.15. In this interface, the most often used items are the Command Manager, the Feature/Property Manager, and the Heads-Up View Toolbar. The interfaces of

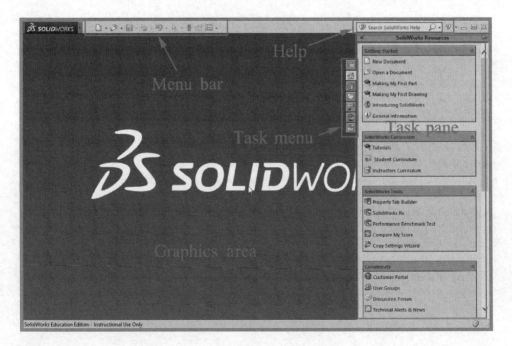

Figure 5.11   User interface when SOLIDWORKS starts

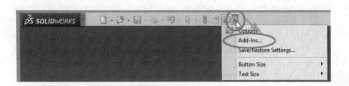

Figure 5.12   Accessing the add-in interface

Figure 5.13   Turning on *SOLIDWORKS Simulation* add-in

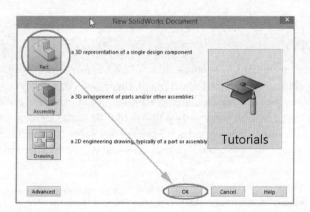

Figure 5.14   User interface when SOLIDWORKS starts

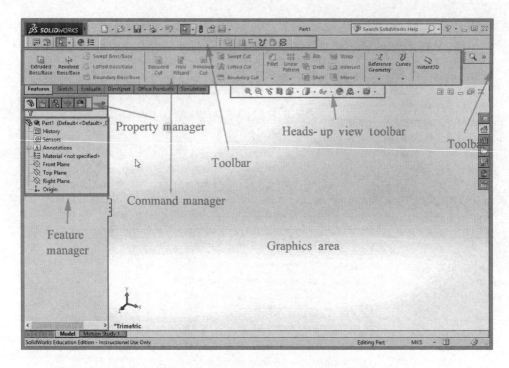

Figure 5.15   Main SolidWorks graphical user interface (GUI)

SolidWorks are highly configurable, and the Toolbar may be scattered around in pieces. Also note that the Task Pane has been minimized in order to maximize the Graphics Area. Most items on the user interface appear as graphic icons. When the cursor is hovered over an icon, an explanatory text appears in the tip box.

5.4.2   Static Analysis

Before performing any vibration analysis, it is advisable to perform a static analysis first, because a static problem is simpler to analyze and its results can tell a lot about the structure and the modeling process. For example, performing the static analysis can check whether parts in the model are connected correctly and whether the finite element model is properly constrained; the static stress field provides a guidance for mesh refinement in areas of higher stress; and the static deformation could hint about how the system will deform when the system is in motion.

■ **Example 5.8:   Static Structural Analysis Using SolidWorks**

Use SolidWorks to determine the effective spring constants of the MEMS cantilever beam in Fig. 5.9 for in-plane and out-of-plane applications. Use geometric and physical parameters in Example 5.7. The Poisson ratio for the material is 0.25. The center of the plate is the point of interest, where a concentrated force, of magnitude of 1 μN in both in-plane and out-of-plane directions, is applied. Compare the results with the lumped-parameter modeling results in Chapter 3.

□ Preparatory Setups

Before constructing the model, we shall set up an appropriate unit system and prepare the material data to be used for the model.

To set up a unit system, click on the Options icon on the Menu bar to open the options interface. Open the Document Properties tab and click on Units. SOLIDWORKS has built-in a few sets of standard unit systems. In this example, the dimensions in Table 5.1 are given in micrometers, or *microns* in SOLIDWORKS, we shall use micron for lengths. We will build our customized unit system: click on Custom and enter the selections shown in Fig. 5.16.

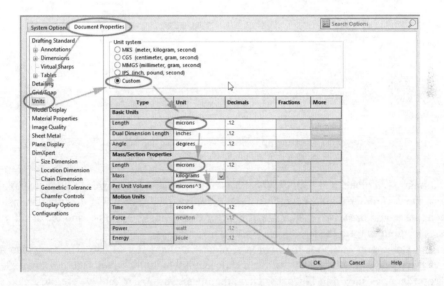

Figure 5.16   Setting up custom unit system in SOLIDWORKS

The next is to enter the material properties data. SOLIDWORKS has built-in an extensive set of engineering material data; but it unlikely has the exact property values of our particular material. We will set up our own material. Click on Edit Materials icon on the Toolbar, the material editing interface opens, as shown in Fig. 5.17. On the left pane, right-click anywhere and select New Library in the context menu to create a new library, and name it MyMaterial-Library. Right-click on MyMaterialLibrary and select New Category and name it Silicons. In a similar manner, add a New Material and name it Silicon Wafer. Then, on the right pane, enter the values for the relevant properties, as shown in Fig. 5.17. Then click Close to close the material editing interface.

□ Creating Geometric Model

We are now ready to create the geometric model for the MEMS beam. Here is the out-line of our modeling strategy. We will first create a sketch of the beam's cross section on a

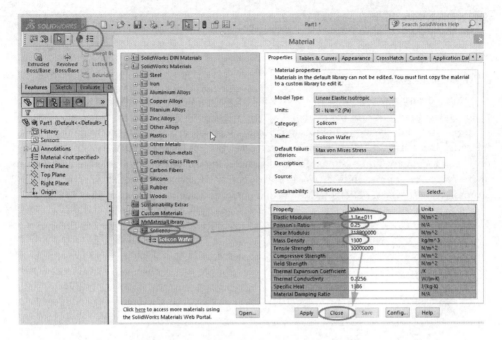

Figure 5.17   Setting up "Silicon Wafer" material in SOLIDWORKS

two-dimensional plane, and then use the three-dimensional shape-forming tool called extrude to turn the two-dimensional sketch into a three-dimensional model. On the two-dimensional sketch, we would draw two rectangles: the larger one forms the outline, and the smaller one creates a cutout to form the two legs.

In the Command Manager for Features, click Extruded Boss/Base. The Graphics Ares shows three perpendicular planes for us to choose, on which the sketch is to be drawn. Select the Top Plane, as shown in Fig. 5.18. Then, the planes disappear, and we are presented with the direct view of the top plane, with two red arrows indicating the location of the origin.

The Command Manager automatically switches to Sketch tab. (If not, click on the Sketch tab below the Command Manager window.) Select the rectangle drawing tool with two diagonally placed red dots, called Corner Rectangle, to draw the first rectangle around the origin, as shown in Fig. 5.19. At this stage, the size is not important.

Use the Smart Dimension tool, click on a vertical edge of the rectangle and pull it to the left, and click again to place the dimensioning mark. On the pop-up window, enter 80 and click on the green check mark to complete dimensioning the side length of the rectangle. Repeat the same process to dimension a horizontal side, the distance between the origin and a vertical side, and the distance between the origin and a horizontal side. The completely dimensioned first triangle is shown in Fig. 5.20.

We now draw the second rectangle. Use the rectangle tool again, move the cursor to the left vertical side of the first rectangle near its top-left corner, ensuring a T-junction sign appears, signaling that the cursor coincides with the side, as shown in Fig. 5.21. Click to locate the first corner of the second rectangle. Drag the cursor to the appropriate location for the diagonal and click to finalize the second rectangle, as shown in Fig. 5.22.

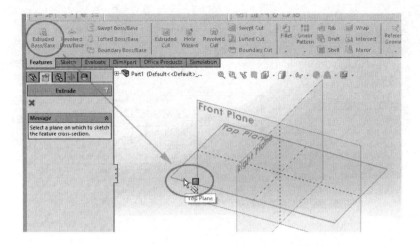

Figure 5.18   Selecting Top Plane to create sketch for extrusion

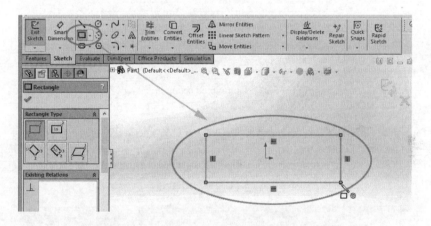

Figure 5.19   Creating the first rectangle around the origin

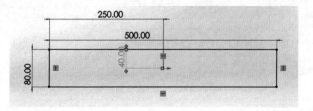

Figure 5.20   Completely dimensioned first rectangle

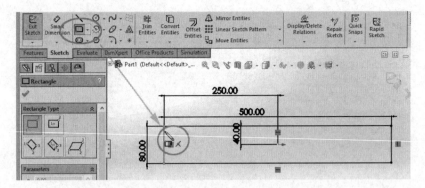

Figure 5.21 Starting the second rectangle, ensure that starting point coincides with the side of first rectangle

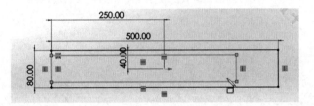

Figure 5.22 Completed second rectangle

Before dimensioning the second rectangle, we want to remove the line segment between the two legs. Use Trim Entities tool with Power Trim option. Click near the line segment targeted for removal. As the cursor moves around, the line segment is eliminated, as shown in Fig. 5.23.

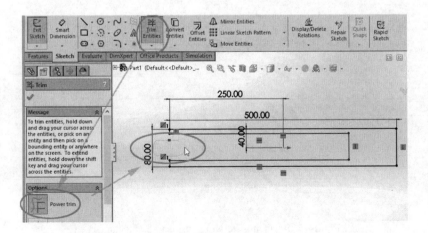

Figure 5.23 Trimming off overlapped side of two rectangles

Then, use the Smart Dimension tool to dimension the leg widths and the distance between the origin and the edge of the plate, as shown in Fig. 5.24. This completes the sketch of the outline of the MEMS beam. Click Exit Sketch to complete the sketch.

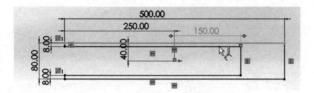

Figure 5.24   Completed two-dimensional sketch of MEMS beam

SOLIDWORKS automatically switches to the extrusion stage. Enter the extrusion thickness of 8 in the extrude Property Manager window as shown in Fig. 5.25, then click the green check mark. The part is extruded. We might have to click the Zoom to Fit button on the Heads-Up View Toolbar to see the completed part, as shown in Fig. 5.26.

Figure 5.25   Setting thickness for extrusion

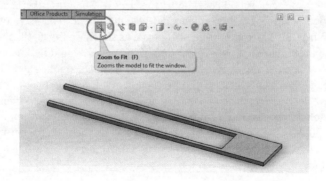

Figure 5.26   Extruded three-dimensional MEMS beam model

As the last step of solid modeling, we assign the material for the MEMS beam. On the Feature Manager window, right-click on Material <not specified> and select Edit Material, as shown in Fig. 5.27. Then, on the material editing interface, open the folder MyMaterialLibrary → Silicon and select the Silicon Wafer we defined earlier. Then click Apply and Close, as shown in Fig. 5.28.

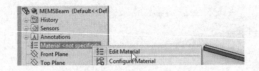

Figure 5.27   Editing the material for MEMS beam model

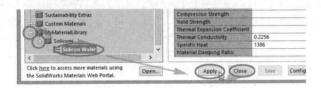

Figure 5.28   Assigning "Silicon Wafer" as material for MEMS beam model

□ Preparing Geometric Features for Finite Element Analysis

The center of the plate is the point of interest, where we would apply a concentrated force, and measure the resulting displacements. It is possible to place a concentrated force at the node of a finite element mesh. However, there is no guarantee that a node will be located exactly at the center of the plate. A more elegant approach is to place a geometric feature there before meshing. Thus, we want to add such a feature while we are still at the geometric modeling stage. A mechanism that can be used is to add a so-called *split line*: a curve that splits a surface into two portions, and in the process, the two portions as well as the split line become identifiable geometric entities. Here, we add a split line to cordon off a quarter of the top surface of the plate, along its symmetry lines.

Click on the top surface of the beam model to select it. Then, use the Line tool from the Command Manager window for the Sketch tab. Move the cursor to the edge of the plate near its center line. When the T-junction symbol appears, as shown in Fig. 5.29, click to place the first point. Move the cursor along the center line of the plate. A guiding line appears. Follow this guiding line. When the cursor is near the center of the plate, another guiding line appears in the perpendicular direction. Click on the intersection of the guiding lines to place the second point. Then, move the cursor in the length direction along the guiding line. When the T-junction appears, double-click to place the third point and end the multisegment line. The process is shown in Fig. 5.30. Use the Smart Dimension tool to specify the distance between the second point of the split line and the edge of the plate, as shown in Fig. 5.31. Click Exit Sketch to finish the split line.

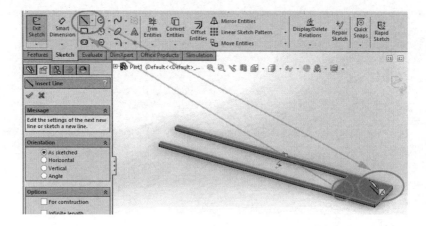

Figure 5.29   Starting the split line on the top surface

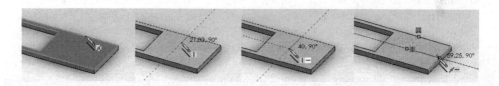

Figure 5.30   Drawing two-segment split line on the top surface

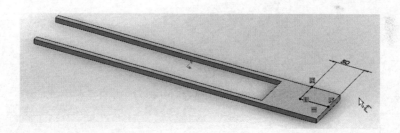

Figure 5.31   Finished and dimensioned split line

Then, on the Command Manager window for the Features tab, select Curves → Split Line. The two-segment line just created is automatically selected for the first entry of the selection, the source of splitting. Select the Type of Split as Projection and select the top surface of the beam as the second entry in Selections, shown as Face <1>, as the surface to be split. Click on the green check mark, the top surface is now split into two portions, and the center of the plate can now be identified as the junction of the two segments, as shown in Fig. 5.32.

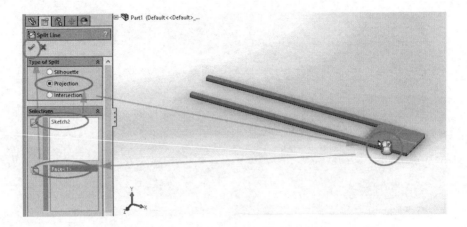

Figure 5.32 Splitting top surface by split line just created

SOLIDWORKS has a mechanism called the *sensors* that allow the user to collect a variety of data from particular locations on the part/assembly. This is exactly the concept of *points of interest*. Since the center of the plate is the point of interest, we want to embed a sensor there. In the feature tree in the Feature Manager, right-click on Sensors and select Add Sensor..., as shown in Fig. 5.33. In the sensor Property Manager window, select Simulation Data for the Sensor Type and Workflow Sensitive for the Data Quality. This allows the sensor data to be defined later as we run through different finite element studies. In the Graphics Area, select the junction point, and Vertex <1> appears in the Properties window, as shown in Fig. 5.34. Click the green check mark to complete the sensor definition.

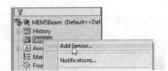

Figure 5.33 Adding sensor to MEMS beam model

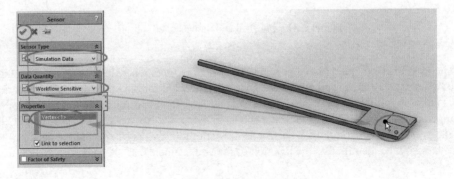

Figure 5.34 Specifying sensor properties

□ Creating a Static Finite Element Study

Click on the Simulation tab under the Command Manager window, we now move into SOLIDWORKS' realm of finite element simulations. Pull down the Study Advisor and select New Study, as shown in Fig. 5.35. In the Feature Manager window, select Static. Click on the green check mark to OK the selection, as shown in Fig. 5.36.

Figure 5.35 Starting finite element simulation study in SOLIDWORKS

Figure 5.36 Creating Static study

Now, the Feature Manager window shows the tree view of the finite element study called Static 1, as shown in Fig. 5.37. It can be renamed if so desired. The first item listed is the solid model. The green check mark signifies that the material has been assigned. The Connections will be needed if the structure contains multiple parts, which need to be connected. The MEMS beam has only one part, so this feature is irrelevant. The Fixtures are for constraints. The Command Manager window also shows icons for similar tasks in the same order. This means that the features can be accessed through either interface.

□ Restraining and Loading Finite Element Model

Right-click on the Fixtures in Static 1 feature tree in the Feature Manager, select Fixed Geometry..., which means cantilever support: both deflection and slope are set to vanish. To select the ends of the two legs, the view may have to be rotated. This can be done by changing the view using Heads-Up View Tools or by using the rotation tool: right-click on anywhere in the Graphics Area and select Zoom/Pan/Rotate → Rotate View, as shown in Fig. 5.38. Select the end faces of the two legs. The Selections window shows two Faces, as shown in Fig. 5.39. Click the green check mark to finalize the selection.

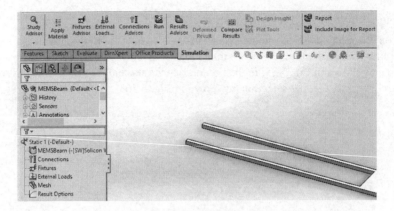

Figure 5.37   Feature Manager and Command Manager windows showing steps in finite element Static Study

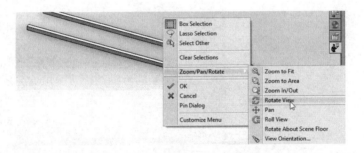

Figure 5.38   Accessing view control from context menu of the Graphics Area

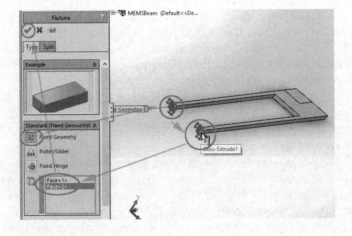

Figure 5.39   Selecting end faces of two legs for Fixed Geometry restraint

To apply a concentrated force at the point of interest, on the Static 1 feature tree, right-click on External Loads and select Force. The Feature Manager becomes Property Manager for the force, as shown in Fig. 5.40. In the Graphics Area, click at the center of the plate when a green dot appears, indicating the item to be selected is a point, to select. The selection box in the Property Manager window shows a Vertex <1>. Back to the Graphics Area, select an edge that runs along the thickness direction. The Edge <1> shows in the corresponding selection box. Enter 1E-6 as the Force and click the green check mark.

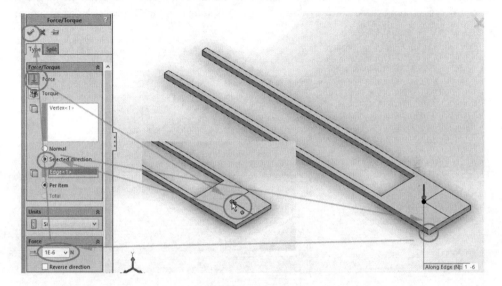

Figure 5.40   Adding concentrated force in thickness direction

Repeat the same process to apply another concentrated force at the same location but along the lateral direction. Alternatively, the concentrated force created above can be copied, by dragging the force in the feature tree and drop it onto the External Loads folder, and then modified for the direction.

The completely constrained and loaded model is shown in Fig. 5.41.

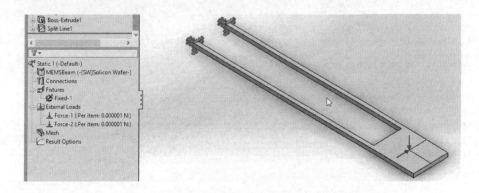

Figure 5.41   Completely loaded and constrained MEMS beam model

☐ **Meshing the Geometric Model**

To create a mesh, right-click on the Mesh in the Static 1 feature tree (or from the Run menu in the Command Manager window) and select Create Mesh... In mesh Property Manager, there are a few selections, as shown in Fig. 5.42. The Mesh Density specifies the size of elements, and Mesh Parameters allow two types of meshes: the Standard mesh and Curvature based mesh. In general, the latter is suitable for parts with different curvatures. As the first meshing exercise, we do not change any setting and simply click the green check mark to create the mesh. When it is done, the mesh is shown in the Graphics Area, as shown in Fig. 5.43.

Figure 5.42   Specifying details in meshing

Figure 5.43   Completed first trial mesh

Although a mesh can be created almost painlessly, the quality of the mesh is crucially important for the quality of finite element simulation results. There are rules of thumb for a good mesh. For example, element sizes should not differ too greatly, the aspect ratios should not be too large, and smaller sizes should be used in areas of higher stress levels. SOLIDWORKS has a set of mesh refinement tools for creating better meshes. Here, we demonstrate one of them. One of the rules of thumb for a good mesh is that a finer mesh should be used in areas having higher stress levels. We anticipate that there might be stress concentrations at the roots of the two legs, and we would use a finer mesh there.

For this, right-click on Mesh and select Apply Mesh Control... In the Property Manager for Mesh Control, as shown in Fig. 5.44, the first is the Selected Entities to apply the mesh control. We select the two edges at the plate–leg junctions. In Mesh Density, slide the slider to a fine size and set a/b to 1.2. This parameter determines the graduation from finer to coarser mesh, and a smaller value means a more gradual change. Click the green check mark to close the Mesh Density. While using a mesh control for mesh refinement, a coarser overall mesh can be used, which could significantly reduce the total number of elements and nodes. The final mesh is shown in Fig. 5.45. Users are strongly encouraged to explore different meshing options and see the difference in the mesh quality.

Figure 5.44   Details in Mesh Control

□ **Solution Process**

Once we are satisfied with the mesh, we can run the finite element simulation, by clicking on the Run button in the Command Window, as shown in Fig. 5.46.

□ **Post-Processing**

When the solution process is completed, SOLIDWORKS automatically produces some common post-processing results, such as the stress, displacement, and strain plots. Users can define more post-processing results, by right-clicking on the Results folder.

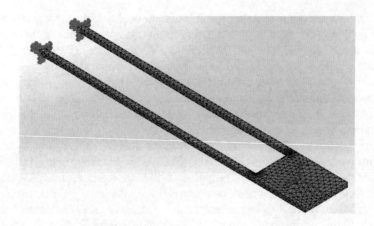

Figure 5.45   Final mesh to be used for Static study

Figure 5.46   Clicking on Run button to start Static study

To see the displacement plot, double-click on the Displacement1 in the Results folder. The displacement color plot is shown the Graphics Window. To see the displacements at particular locations, we need to first toggle off the Deformed Results in the Command Manager and then select Plot Tools → Probe, as shown in Fig. 5.47. Clicking on a location on the model, a label showing the location's coordinates and the displacement value is attached to the location, as shown in Fig. 5.48. Meanwhile, numerical values are also tabulated in the Property Manager, which can be subsequently saved to a file.

Figure 5.47   Selecting Probe tool to probe displacements at any location in model

To modify the post-processing results SOLIDWORKS produced automatically, right-click on the Displacement1 and select Edit Definition... to open up its definition in the Property Manager, as shown in Fig. 5.49. In this case, we want to show the y-component

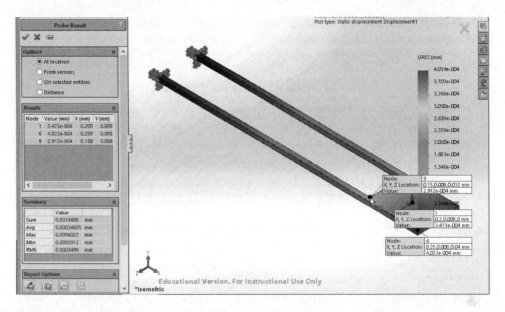

Figure 5.48   Probed results using Probe tool on the displacement plot

Figure 5.49   Editing properties of displacement plot

Figure 5.50   Duplicating a displacement plot

(out-of-plane) displacement. Select UY: Y Displacement component and micron as unit, click on the green check mark to complete the editing. We can also duplicate the plot by dragging-and-dropping. Use a slow double-click to change its name to Displacement2, as shown in Fig. 5.50, and edit its properties to show UZ: Z Displacement component.

Since we have embedded a sensor at the point of interest, we do not have to locate and select the point of interest. From the Command Manager, select Plot Tools → List Selected.

In the Property Manager, select From sensors in the Options. The sensor reading is listed in the Results section. The displacements in both y- and z-directions are shown in Fig. 5.51.

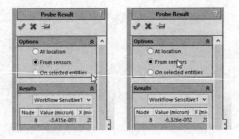

Figure 5.51   Displacement results from sensor by using From sensors option

□ **Discussions:**

From the sensor data, the deflection at the point of interest is $0.3415\,\mu m$ in the y-direction (out-of-plane), and $0.06326\,\mu m$ in the z-direction (in-plane). Given the applied force of $1\,\mu N$ in each direction, they give the effective spring constants as

$$k_{\text{in-plane}} = \frac{1\,\mu N}{0.06326\,\mu m} = 15.805 \text{ N/m} \tag{a}$$

$$k_{\text{out-of-plane}} = \frac{1\,\mu N}{0.3415\,\mu m} = 2.9283 \text{ N/m} \tag{b}$$

In Example 3.24, the results of lumped-parameter modeling for both in-plane and out-of-plane applications are, using the numerical values of the present example,

$$k_{\text{LPM, in-plane}} = \frac{2Ea^3t}{l^3} = 16.640 \text{ N/m} \tag{c}$$

$$k_{\text{LPM, out-of-plane}} = \frac{Eat^3}{2l^3} = 4.1600 \text{ N/m} \tag{d}$$

The difference for in-plane applications is small but is rather large for out-of-plane applications. This is because the lumped-parameter modeling considers only the legs as cantilever beams. This is equivalent to making the ends of the legs as the point of interest. In the finite element model, the force is applied at the center of the plate. In light of this difference, if $l + h/2$ is used in place of l in eqn. (d), the equivalent spring constant would be

$$k_{\text{LPM, out-of-plane}} = \frac{Eat^3}{2(l + h/2)^3} = 2.9217 \text{ N/m} \tag{e}$$

which is much closer to the finite element modeling result. This comparison serves as a rudimentary validation of the finite element modeling.

If we are not interested in the stress and strain results, they can be turned off by editing the Result Options. While this is turned off, some other post-processing tasks must be defined before the solution process. Otherwise, by default, SOLIDWORKS will attempt to compute a stress and a strain post-processing tasks, and errors would occur. With predefined post-processing tasks, SOLIDWORKS computes only the specified post-processing task.

In the solution process, an important aspect has not been explored: settings for the solution process, which can be accessed through Study Advisor → Study Properties. Most of these properties require advanced knowledge of the finite element modeling. As a novice user, simply using the default setting generally suffices.

Throughout the process, a Quick Tip window anticipating the user's needs and giving guidance pops up from time to time, as shown in Fig. 5.52. Clicking on the links will point the user to the appropriate menu locations. This could be useful for a new user. Once the user is familiar with the process and the SOLIDWORKS interfaces, it can be turned off by clicking on the green question mark at the lower-right corner of the main window.

Figure 5.52   Example of Quick Tip pop-up windows in SOLIDWORKS

5.4.3   Modal Analysis

The modal analysis finds the natural frequencies and mode shapes of a multi-DOF system. For a system of N DOFs, there are N natural frequencies. For the systems modeled by the FEM, the number of DOF equals the total number of notes multiplied by the number of DOFs for each node. For a finite element model containing a large number of DOFs, it is neither practical nor necessary to find all the modes. On the other hand, finding the modes is extremely helpful in understanding the system's dynamics and could be useful for other vibration analyses. In SOLIDWORKS, a modal analysis is called a Frequency Study, and both harmonic and transient vibration analyses require a modal analysis to be performed.

■ **Example 5.9: Modal Analysis of MEMS Cantilever Beam**

Conduct a model analysis for the MEMS cantilever beam using SOLIDWORKS. Limit the modal analysis to finding the first 10 natural modes.

□ **Creating a Frequency Study**

Creating a new finite element study of the same part is simply by selecting the New Study under the Study Advisor in the Command manager and then selecting Frequency. SOLIDWORKS automatically names this study as Frequency 1, which can be changed if desired. The Feature Manager shows the feature tree for the Frequency 1, almost identical to that for Static 1, which is now minimized into a tab at the bottom of the main window.

□ Restraining the Model

We can reuse some features in the previous studies. Click on the Static 1 tab at the bottom of the window to reactivate the Static 1 study. Drag the Fixtures folder from the feature tree of the Static 1 and drop it onto the Frequency 1 tab, as shown in Fig. 5.53.

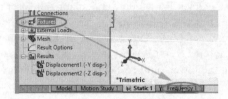

Figure 5.53   Copying Fixtures from Static Study to Frequency Study

Modal analysis is performed on unloaded system. Thus, no external load is to be applied. SOLIDWORKS does provide an interface for specifying the external loads. Those are intended for prestressed structures.

□ Meshing the Model

The same drag-and-drop copying process can be repeated for the Mesh, which copies only the mesh controls. The mesh itself still needs to be created. This can be done by right-click on Mesh and then select Create Mesh...

□ Solution Process

To specify first 10 natural modes to be computed, click Study Advisor → Study Properties to edit the properties of this study, as shown in Fig. 5.54. Among the settings, the first is the number of modes to be computed. Change it from default 5 to 10. For the Solver, use the Automatic. Then, hit the Run button to start the solution process.

□ Post-Processing

Once the solution process is completed, SOLIDWORKS automatically produces the mode shape plots for the number of modes specified in Study Properties, and the mode shape for Mode 1 is shown in the Graphics Area.

To view a mode shape in comparison with the undeformed shape, right-click on the post-processing object and select Edit Definition.... In the Property Manager, under the Settings tab, check the box in front of the Superimpose model on the deformed shape and set the Transparency to approximately 0.5, as shown in Fig. 5.55. Click the green check mark to complete the change. Repeat this setting for all mode shape plots.

Up to four plots can be viewed simultaneously. Click on Compare Results in the Command Manager and select mode shapes 2 through 5 for comparison, as shown in Fig. 5.56. The mode shape plots for Modes 2 through 5 are shown, as in Fig. 5.57.

Figure 5.54 Settings for Frequency Study

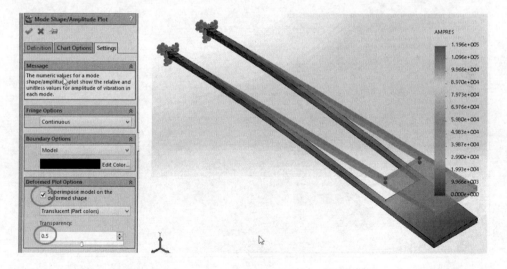

Figure 5.55 Settings for mode shape plot for Mode 1

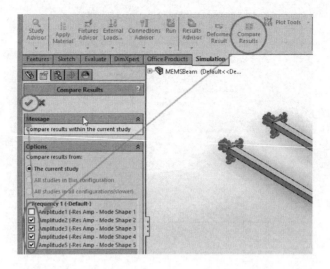

Figure 5.56 Selecting mode shapes 2 through 5 for comparison

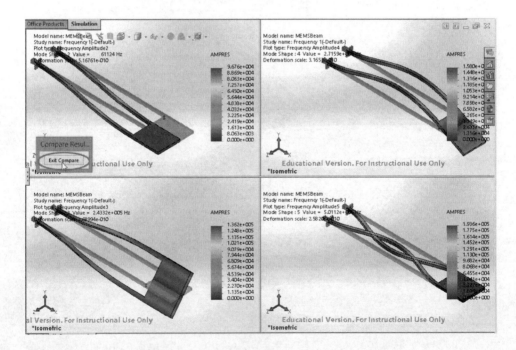

Figure 5.57 Comparing mode shapes for Modes 2 through 5

To obtain numerical values of the natural frequencies, right-click the Results and select List Resonant Frequencies... The 10 natural frequencies are tabulated, as shown in Fig. 5.58.

 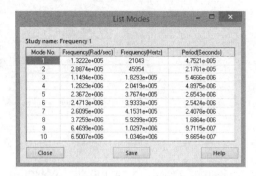

Figure 5.58   Listing of natural frequencies

□ **Discussions:**

The SOLIDWORKS finite element simulation gives the first two natural frequencies as

$$\omega_1 = 1.3222 \times 10^5 \text{ rad/s} \qquad \omega_2 = 2.8874 \times 10^5 \text{ rad/s} \qquad \text{(a)}$$

Recall our earlier finite element modeling using beam element in Example 5.7, which gives the following fundamental natural frequencies for the 50-element model:

$$\omega_{\text{out-of-plane}} = 132095 \text{ rad/s} \qquad \omega_{\text{in-plane}} = 294979 \text{ rad/s} \qquad \text{(b)}$$

while the lumped-parameter modeling in Example 3.24 gives, using the numerical values for the present example,

$$\omega_{\text{LPM, out-of-plane}} = 1.5420 \times 10^5 \text{ rad/s} \qquad \omega_{\text{LPM, in-plane}} = 2.9521 \times 10^5 \text{ rad/s} \qquad \text{(c)}$$

The two finite element results for the out-of-plane fundamental natural frequency match well, with an error of 0.08%, but they differ notably from the lumped-parameter modeling. The source of this discrepancy has been discussed in Example 5.8. For in-plane applications, the beam element result matches well with the lumped-parameter modeling but differs slightly from SOLIDWORKS's. The reason for this discrepancy is that in the beam element modeling, a boundary condition of zero slope is applied at the end of legs, making the two legs guided-cantilever. This makes the beam element model closer to the lumped-parameter model.

SOLIDWORKS frequency analysis also provides another important parameter called *mass participation factor* for the natural modes. This parameter provides a quantitative measure of the relative importance of each mode. To view these data, right-click on the Results and select List Mass Participation. The results are shown in Fig. 5.59.

5.4.4   Harmonic Vibration Analysis

A harmonic analysis refers to a steady-state vibration analysis in which the forcing is a harmonic function of a single frequency. Typically, such an analysis is aimed at producing a

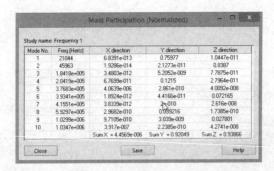

Figure 5.59   Listing of mass participation factors for all first 10 modes

steady-state response spectrum for the system over a range of forcing frequency. This is a Harmonic option in Linear Dynamic study in SOLIDWORKS.

■ Example 5.10: Harmonic Response of MEMS Cantilever Beam

Use SOLIDWORKS to obtain the harmonic response of the MEMS cantilever beam at the center of the plate when a concentrate sinusoidal force having both in-plane and out-of-plane components of a magnitude 1 μN is applied at the center of the plate.

□ Creating Harmonic Response Study

Select the New Study under the Study Advisor in the Command Manager and then select Linear Dynamic in the Type and Harmonic in Options, as shown in Fig. 5.60. The new study is presented with a similar feature tree, and all previous studies are minimized.

□ Restraining and Loading the Model

We can reuse the constraints and external loads from the past studies, using the drag-and-drop process as described earlier. However, it is interesting to note that sometimes an entire folder cannot be copied while individual items can.

For harmonic analysis, we need to consider system damping. SOLIDWORKS offers all three damping models as discussed in Section 5.3. Here, we use Modal Damping and set the damping ratio for all the first 15 modes to 0.01, as shown in Fig. 5.61. Note that 15 here is the default number of modes to be included in the solution process.

□ Meshing the Model

The mesh controls can be copied from a previous study.

□ Solution Process

To set up for the harmonic analysis, open the Study Properties, as shown in Fig. 5.62. The Frequency Options tab is for setting up the modal analysis. SOLIDWORKS performs a modal

Figure 5.60 Stating Harmonic option of Linear Dynamic Study

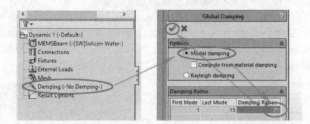

Figure 5.61 Setting damping property for Harmonic study

analysis before the harmonic analysis. Although we have already performed a separate Frequency Study in Example 5.9, there is no way to reuse the results from that study. The options under this tab are the same as those for the Frequency 1 study. A minor difference is the default number of modes being 15. We will use the default value. The deciding factor is the *mass participation factor* obtained from the modal analysis in Example 5.9. The Harmonic Options tab is for the harmonic analysis. We want to set the Upper limit of the harmonic analysis to be 50 kHz. Then, hit the Run to start the solution process.

□ **Post-Processing**

After the solution process is completed, SOLIDWORKS computes the stress and the displacement plots. By default, the plot shown in the Graphics Area is for the final step. Other steps can be displayed by editing the definitions for those plots.

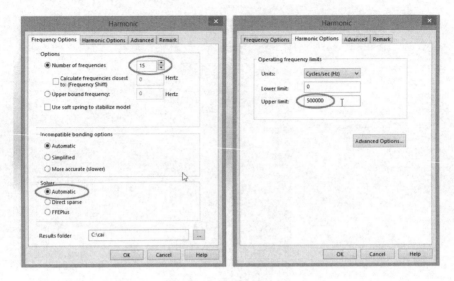

Figure 5.62 Other settings for Harmonic study

To produce the spectrum for the displacement at the point of interest, where we have embedded sensors, right-click the Result Options and select Edit/Define... In the Property Manager, under Save Results, select For specified solution steps, as shown in Fig. 5.63. This opens up more options, including the access to the sensor data. To use all computed steps, we set Increment in the Solution Steps - Set 1 to 1. In the Locations for Graphs, select Workflow Sensitive1, which was the name of the sensor SOLIDWORKS assigned during the solid modeling stage. Click on the green check mark to finish the setting.

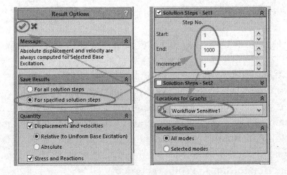

Figure 5.63 Retrieving data from sensor for producing response spectrum

Then, right-click the Results folder and select Define Response Graph... In the Property Manager, the Predefined locations is automatically selected, implying that the sensor data to be used. In the Y axis section, select Displacement, UY: Y Displacement, and micron; and in Options, select Amplitude, as shown in Fig. 5.64. Click the green check mark, a response spectrum for the *y*-displacement at the point of interest appears, as shown in Fig. 5.65.

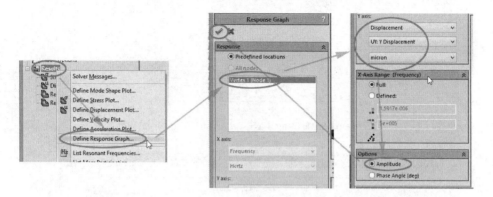

Figure 5.64   Defining response spectrum graphs for Harmonic study

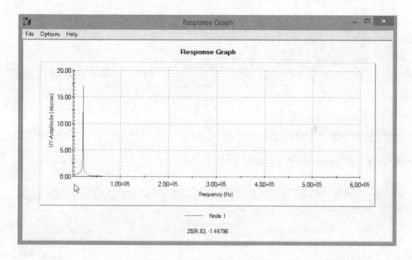

Figure 5.65   Harmonic response spectrum for the *y*-displacement at plate's center

Only one peak is prominently shown in Fig. 5.64. This is because the first peak is dominant. A response spectrum is often plotted in logarithmic scale. On the plot, select Options → Properties. In the property editing interface, open the Axes tab, select Y axis, and check the box in front of IsLogarhimic, as shown in Fig. 5.66. The modified response spectrum is replotted in semi-log scale, as shown in Fig. 5.67. Now, two peaks are visible, matching the natural frequencies of the beam.

In a similar process, or alternatively, using a drag-and-drop copying process, other spectra can be defined: the *z*-displacement and the phase angles. The response spectrum for the *z*-displacement is shown in Fig. 5.68.

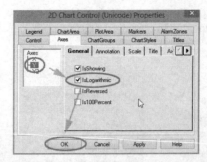

Figure 5.66   Editing y-axis properties in spectrum plot

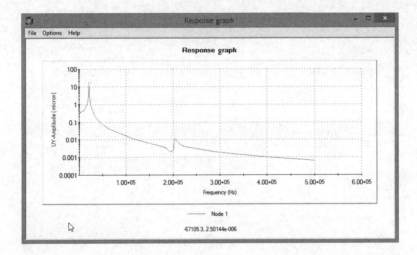

Figure 5.67   Harmonic response spectrum for the y-displacement in semi-log plot

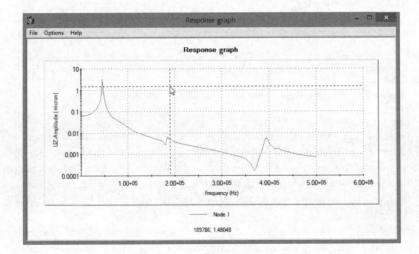

Figure 5.68   Harmonic response spectrum for the z-displacement at plate center

□ **Discussions:**

SOLIDWORKS uses the *modal superposition method* for the harmonic analysis. This is the reason why a modal analysis is performed prior to the harmonic analysis. It is unfortunate that SOLIDWORKS cannot use the results from a separate modal analysis. However, it is still advisable to perform a separate modal analysis, as it can provide much useful information about the structure. For example, from the mass participation data, we would not be surprised to see that the linear plot of the response spectra shows only one peak.

For solution settings, only the lower and upper limits of the frequency are set. By default, SOLIDWORKS uses fine steps around each natural frequency. Detailed step controls can be accessed through the Advance Options inside the Harmonics Options in the Study Properties. Interested readers are encouraged to explore those options.

5.4.5   Transient Vibration Analysis

A transient vibration analysis calculates the system's response due to a time-varying loading starting from the time when the load is applied. SOLIDWORKS uses the *modal superposition method* for this type of analysis and called it a Modal Time History study, which is an option under the Linear Dynamic study.

■ **Example 5.11: Transient Vibration Analysis of MEMS Cantilever Beam**

The MEMS cantilever beam is subjected to a time-varying force applied to the center of the plate. The same amount of force is applied in both in-plane and out-of-plane directions. The time history of the force in each direction is shown in Fig. 5.69. Use SOLIDWORKS to find the response of the beam when it is initially at rest.

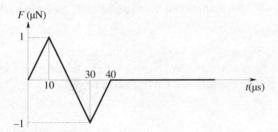

Figure 5.69   Time-varying loading to be applied to center of plate

□ **Creating a Modal Time History Study**

Select Study Advisor → New Study and then select Linear Dynamic for the Type section and Modal Time History in the Options section. When the green check mark is clicked, a new study becomes active, and all previous studies are minimized.

□ **Restraining and Loading the Model**

The Fixtures and External Loads can be copied from the past studies, as before.

We copy only one of the two forces for now. Open the Property Manager for the copied force, scroll down to Variation with time section, check the Curve option, and then click on Edit... This opens up the interface for specifying the time-varying curve. The total force is determined by the force value as shown in the force's Property Manager, which is 1 μN in this example, multiplied by the value of this time-varying curve, which is dimensionless. On the Time curve interface, for the Shape, select Used Defined. In the Curve Data section, enter the times and values as shown in Fig. 5.70. To add a new data point, double-click on any cell of the Point column. When finished, click OK to close the Time curve window and then click the green check mark in the force's Property Manager.

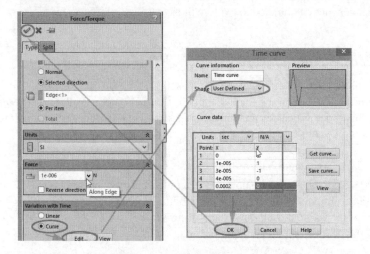

Figure 5.70   Setting time-varying loading for Transient Study

Then, this force is drag-and-drop copied into the External Loads folder. This way, the time-varying curve is copied. Then, open the copied force to change its direction.

The damping can also be copied from the Harmonic Study.

□ **Solution Process**

There are few options that need to be specified: the End time and the Time increment, as shown in Fig. 5.71. For a transient vibration analysis, when selecting a time increment size, two factors must be taken into account: the system's fundamental natural frequency and the loading's fundamental frequency. The rule of thumb is to use no more than 1/20 of the system's natural period, and no more than 1/5 of the loading's period, whichever is the smaller. Thus, performing a modal analysis before any transient vibration analysis is crucially important.

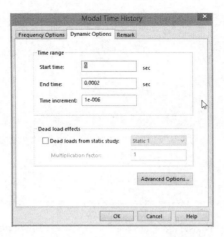

Figure 5.71   Setting for the Modal Time History study

In this example, we set the end time as 200 μs and the time increment as 1 μs. When finished, click on the green check mark to finalize the setting. Then, click the Run button.

□ **Post-Processing**

The post-processing is rather similar to the harmonic analysis. To plot a time history response of the displacement at the point of interest, define a Define Response Graph... and set up in a similar manner. The resulting displacement responses for the y- and z-components are shown in Figs. 5.72 and 5.73.

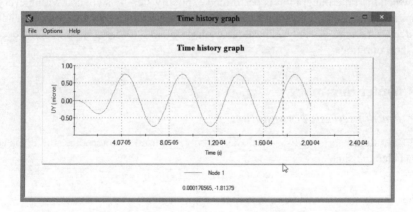

Figure 5.72   Transient displacement at center of plate in the y-direction

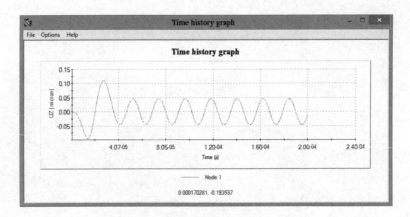

Figure 5.73   Transient displacement at center of plate in the z-direction

☐ **Discussions:**

There are surprisingly little settings to be set up for a transient analysis. The main reason is the use of the modal superposition method as the solution method.

This example does not apply any initial conditions. For most transient problems, it is assumed that the system start "from rest." SOLIDWORKS allows simple forms of initial conditions be specified, such as displacements, velocities, or accelerations at specified geometric entities, much like specifying boundary conditions. Initial Condition is not automatically added to the feature tree when a transient study is created. It can be added from the Command Manager. Readers are encouraged to explore this topic on their own.

SOLIDWORKS offers another dynamic transient analysis capability under the Nonlinear study, with Dynamic option. This is commonly called the *explicit dynamics* in finite element modeling, in which the time integration is performed explicitly step by step, and within every step, iterations might be used to accommodate nonlinearities arising from both geometric and material sources. This type of analysis is suitable for problem with large deformations occurring during short durations.

5.5   Chapter Summary

5.5.1   Finite Element Formulation

In the FEM, a predetermined set of shape functions is used to interpolate the displacement within an element such as

$$u(x, t) = \{N(x)\}^T \{u^e(t)\} \tag{5.64}$$

The strain in the element can be related to the displacement as

$$\epsilon(x, t) = \{B(x)\}^T \{u^e(t)\} \tag{5.65}$$

where $\{B(x)\}$ is called the strain–displacement matrix and is typically related to the shape function $\{N(x)\}$. Then, at the element level, the mass and stiffness matrices are

$$[M^e] = \int_{V^e} \{N\}\{N\}^T dV^e \qquad [K^e] = \int_{V^e} \{B\}\{B\}^T dV^e \tag{5.66}$$

When the global displacement matrix $\{u\}$ is defined, the corresponding locations of each DOF in the system are determined, and the element mass and stiffness matrices can be assembled by adding the entries to entries in accordance with their global matrix index. This process is called the assembly in the finite element method.

For concentrated load applied to the nodes, they can be directly assembled into the global load matrix. For a distributed load $p(x)$ along the element, the nodal force can be calculated as

$$\{F^e\} = \int_{V^e} p(x)\,\{N(x)\}\,dV^e \tag{5.67}$$

After the global matrices are assembled, the global matrix equation of motion will be

$$[M]\{\ddot{u}\} + [K]\{u\} = \{F\} \tag{5.68}$$

The boundary conditions can be imposed after the assembly. When a boundary condition is homogeneous, the corresponding row and column of the global matrices are eliminated. For a nonhomogeneous boundary condition, the remaining rows on the right-hand side forcing matrix $\{F\}$ need to be corrected.

5.5.2   Using Commercial Finite Element Analysis Software

For all commercial finite element analysis software, the process of performing a finite element analysis can be divided into the following steps:

- Geometric modeling;
- Finite element meshing to turn geometric model into a finite element model;
- Constraining and loading the finite element model;
- Solution; and
- Post-processing.

One important aspect of performing numerical analysis is to validate the results. Only after careful validations the finite element analysis results can be usable.

Problems

Problem 5.1: In the two-node rod element in Fig. 5.3, a linearly varying distributed pressure in the axial direction is applied along the element, with values p_i at node i and p_j at node j. Derive the expression for the element nodal force matrix due to this pressure.

Problem 5.2: In the two-node beam element in Fig. 5.5, a linearly varying distributed pressure perpendicular to the beam is applied along the element, with values p_i at node i and p_j at node j. Derive the expression for the element nodal force matrix due to this pressure.

Problem 5.3: Develop a finite element analysis code using MATLAB for computing the fundamental natural frequency of a general one-dimensional rod. The code should print out the fundamental natural frequency of the rod.

Problem 5.4: Use the finite element code developed in Problem 5.3 to determine the fundamental natural frequencies of a uniform rod, with one end fixed and one end free. Assume the rod is made of steel, of mass density 7900 kg/m^3, Young's modulus 207 GPa, length 0.5 m, and a diameter of 0.05 m. Compare the results from using 1, 2, 5, 10, and 20 elements. Also, compare the finite element modeling results with the lumped-parameter modeling results: the equivalent spring constant in eqn. (3.16) and the equivalent mass in eqn. (3.25).

Problem 5.5: Use the finite element code developed in Problem 5.3 to determine the fundamental natural frequencies of a tapered rod in Problem 3.3. Assume the rod is made of steel, of mass density 7900 kg/m^3, Young's modulus 207 GPa, length 0.5 m, and the diameter varies linearly, from 0.15 m at the root to 0.025 m at the free end. Compare the results from using 1, 2, 5, 10, and 20 elements. Also, compare the finite element modeling results with the lumped-parameter modeling results in Problem 3.3.

Problem 5.6: Use the finite element beam code to determine the fundamental natural frequency of the doubly-overhung simply-supported beam in Problem 3.4. The beam is made of steel, of mass density 7900 kg/m^3 and Young's modulus 207 GPa, and has an overall length of 0.5 m and a diameter of 0.05 m. Compare the results from using 4, 8, 20, 40, and 120 elements. Also, compare the finite element modeling results with the lumped-parameter modeling results in Problem 3.4.

Problem 5.7: Use the finite element beam code to determine the fundamental natural frequency of leaf-spring model in Problem 3.8. The beam is made of steel, of mass density 7900 kg/m^3 and Young's modulus 207 GPa, and has an overall length of 0.5 m. Each strip has a rectangular cross section of thickness 0.01 m and width 0.06 m. Compare the results from using 4, 8, 20, 40, and 120 elements. Also, compare the finite element modeling results with the lumped-parameter modeling results in Problem 3.8.

Problem 5.8: Use the finite element beam code to determine the fundamental natural frequency of the beam with built-in end as described in Problem 3.10. The beam is made of steel, of mass density 7900 kg/m^3 and Young's modulus 207 GPa, and has a nominal length of $L = 0.5$ m. The cross section is a square of a side length of 0.05 m. Compare the results from using 9, 18, 45, 90, and 180 elements. Also, compare the finite element modeling results with the lumped-parameter modeling results in Problem 3.10.

Problem 5.9: Use the finite element beam code to determine the fundamental natural frequency of cantilever–simply-supported beam in Problem 3.12. The beam is made of steel, of mass density 7900 kg/m^3 and Young's modulus 207 GPa, and has an overall length of 0.5 m and a diameter of 0.05 m. Compare the results from using 1, 2, 5, 10, and 20 elements. Also, compare the finite element modeling results with the lumped-parameter modeling results in Problem 3.12.

Problem 5.10: Use the finite element beam code to determine the fundamental natural frequency of simply-supported beam with extra roller support described in Problem 3.14. The beam is made of wooden plank, whose physical properties are the same as those in

Example 5.6. It has an overall length of 5 ft, a width of 6 in., and a thickness of 1 in. Compare the results from using 2, 4, 10, 20, and 40 elements. Also, compare the finite element modeling results with the lumped-parameter modeling results in Problem 3.14.

Problem 5.11: Formulate the element stiffness and mass matrices for a one-dimensional two-node *frame element*, which behaves as a beam when loaded in the lateral direction, and behaves like a rod when loaded in the axial direction. (*Hint: Element matrices can be obtained by the finite element assembly process.*)

Problem 5.12: Formulate the element stiffness and mass matrices for two-dimensional *frame element*, which is the same as the one-dimensional frame element except that the ends are located in two-dimensional space *OXY*. The formulation may utilize the results from Problem 5.11 by introducing a local coordinate system *Oxy* that aligns its *x*-axis with the element, as shown in Fig. P5.12. (*Hint: Use a coordinate transformation. Energies are scalar quantities and hence coordinate-invariant: their different expressions in different coordinate systems should equal to each other.*)

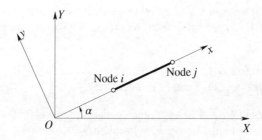

Figure P5.12   Local and global coordinate systems for the frame element

Problem 5.13: Formulate the element nodal force matrix of the two-node two-dimensional frame element, considering the contributions of externally applied forces at nodes in any direction, externally applied bending moments at nodes, and linearly distributed pressures along the length of the element in any direction. (*Hint: Results from Problems 5.1 and 5.2 can be adapted.*)

Problem 5.14: Based on the formulations in Problems 5.11 and 5.12, write a finite element code using MATLAB to compute the fundamental natural frequency of a general two-dimensional frame structure.

Problem 5.15: Use the two-dimensional frame element code developed in Problem 5.14 to determine the fundamental natural frequencies of the L-beam in Problem 3.24. The beam is made of steel, of mass density 7900 kg/m^3 and Young's modulus 207 GPa, and has square cross section of a side length of 0.01 m and $a = 0.5$ m. Compare results from using different numbers of elements.

Problem 5.16: Compare the finite element analysis results for the L-beam in Problem 5.15 with the lumped-parameter modeling results in Problems 3.24 and 3.25. Explain the differences in the three natural frequencies: two from the lumped-parameter modeling and one from finite element modeling. With this hindsight, discuss the best way to redo the lumped-parameter modeling such that the natural frequency would be close to the finite element modeling result.

Problem 5.17: Use the two-dimensional frame element code developed in Problem 5.14 to determine the fundamental natural frequencies of the U-beam in Problem 3.26. The beam is made of steel, of mass density 7900 kg/m^3 and Young's modulus 207 GPa, and has a square cross section of a side length of 0.01 m and $a = 0.5$ m. Compare results from using a different number of elements and a series of values of b/a ratio: 0.2, 0.5, 1, 2, and 5.

Problem 5.18: Use SOLIDWORKS or any other general-purpose finite element software to model the uniform L-shaped beam in Problem 3.24. The beam is made of steel, of mass density 7900 kg/m^3 and Young's modulus 207 GPa, and has a circular cross section of a diameter of 0.05 m, $a = 0.5$ m, and $b = 0.3$ m.

Problem 5.19: Use SOLIDWORKS or any other general-purpose finite element software to model the folded beam design for MEMS applications in Problem 3.54.

Problem 5.20: Use SOLIDWORKS or any other general-purpose finite element software to model the uniform H-shaped beam in Problem 3.55. The beams are assumed to be welded together.

Problem 5.21: Use SOLIDWORKS or any other general-purpose finite element software to model the uniform T-shaped beam in Problem 3.57.

Problem 5.22: Use SOLIDWORKS or any other general-purpose finite element software to model the uniform triangular hardness in Problem 3.59.

Appendix A

Review of Newtonian Dynamics

The study of Newtonian dynamics is typically split into two major topics: kinematics and kinetics. *Kinematics* studies the geometries of the motion, without regard to what causes the motion or how the motion is achieved or sustained. *Kinetics* relates the geometries of the motion to their causes and consequences. As a brief review, here we focus on motions in two-dimensional spaces.

A.1 Kinematics

A.1.1 Kinematics of a Point or a Particle

The study of kinematics starts with defining a *position vector*, denoted as r, to locate the point of interest in the space. As the time progresses, the position changes along its path, called the *trajectory*. The change in the position vector is called the *displacement* vector. This change leads to the definitions of the *velocity* vector: the time derivative of the position vector. Similarly, the change in the velocity vector leads to the definition of the *acceleration* vector: the time derivative of the velocity, that is,

$$v = \frac{dr}{dt} \equiv \dot{r} \qquad a = \frac{dv}{dt} \equiv \dot{v} = \ddot{r} \tag{A.1}$$

where it is customary to denote the time derivative by an overdot.

Mathematically describing these vectors must be aided by using a coordinate system, which defines an origin and a set of mutually perpendicular directions. Three coordinate systems are commonly used in kinematics, namely, the Cartesian, the polar, and the path coordinate systems.

In the Cartesian coordinate system, represented by its coordinates (x, y, z), the three respective directions are represented by unit vectors i, j, and k, arranged to follow the so-called *right-hand rule*, as shown in Fig. A.1. In this coordinate system, the position, velocity, and

Fundamentals of Mechanical Vibrations, First Edition. Liang-Wu Cai.
© 2016 John Wiley & Sons, Ltd. Published 2016 by John Wiley & Sons, Ltd.
Companion Website: www.wiley.com/go/cai/fundamentals_mechanical_vibrations

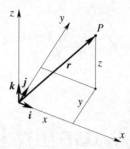

Figure A.1 A position vector in Cartesian coordinate system

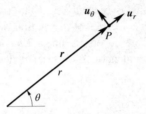

Figure A.2 A position vector in polar coordinate system

acceleration vectors are expressible as

$$r = xi + yj + zk \qquad\qquad (A.2)$$

$$v = \dot{x}i + \dot{y}j + \dot{z}k \qquad\qquad (A.3)$$

$$a = \ddot{x}i + \ddot{y}j + \ddot{z}k \qquad\qquad (A.4)$$

In the polar coordinate system for a two-dimensional space, represented by coordinates (r, θ), the unit vectors are defined as the following: u_r is radially pointing from the origin outward, toward the point of interest, and u_θ is formed by rotating u_r counterclockwise by 90°, as shown in Fig. A.2. This way, different points in space have different unit vectors. In this coordinate system, position, velocity, and acceleration vectors are expressible as

$$r = r u_r \qquad\qquad (A.5)$$

$$v = \dot{r} u_r + r\dot{\theta} u_\theta \qquad\qquad (A.6)$$

$$a = \left(\ddot{r} - r\dot{\theta}^2\right) u_r + \left(r\ddot{\theta} + 2\dot{r}\dot{\theta}\right) u_\theta \qquad\qquad (A.7)$$

In three-dimensional space, the z-coordinate is added and is called the *cylindrical coordinate system*.

The third coordinate system is the *path coordinate system*, which is sometimes called the *curvilinear coordinate system*. The two unit directions are defined based on the particle's trajectory at the point of interest. The *tangential direction* unit vector u_T is defined at the point of interest pointing tangentially to the direction of motion; the *normal direction* unit vector u_N

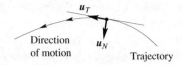

Figure A.3   Path coordinate system

is perpendicular to the tangential direction, pointing to the concave side of the trajectory, as shown in Fig. A.3. This coordinate system actually does not have a way to describe the position vector, as it does not define coordinates. The velocity and acceleration vectors in this coordinate system are expressible as

$$v = v u_T \tag{A.8}$$

$$a = \dot{v} u_T + \frac{v^2}{\rho} u_N \tag{A.9}$$

where $v = |v|$ is called the *speed* and ρ is the *radius of curvature* of the trajectory at the point of interest.

When the position needs to be described, another coordinate system has to be used. When using a Cartesian coordinate system, and when the trajectory is given by a function $y(x)$, the radius of curvature can be calculated according to

$$\rho = \frac{\left(1 + y'^2\right)^{3/2}}{|y''|} \tag{A.10}$$

When using a polar coordinate system, and when the trajectory is given by a function $r(\theta)$, the radius of curvature is given by

$$\rho = \frac{\left(r^2 + r'^2\right)^{3/2}}{|r^2 + 2r'^2 - rr''|} \tag{A.11}$$

A.1.2   Relative Motions

Sometimes it is more convenient to describe a motion using a moving coordinate system. But Newton's second law relates the force to the motion described in an *inertial reference frame*, that is, a coordinate system that is fixed in space or moving at a constant velocity. A motion described in an inertial coordinate system is called the *absolute motion*; while a motion described in a moving coordinate system is called the *relative motion* with respect to that moving reference frame. In this context, a reference frame and a coordinate system are basically the same thing.

Suppose two reference frames have been defined: *OXYZ* is an inertial reference frame, and *oxyz* is a moving reference frame. When the intermediate reference frame *oxyz* is translating but not rotating, the absolute motion of point *A* is given as

$$v_A = v_o + v_{A/o} \tag{A.12}$$

$$a_A = a_o + a_{A/o} \tag{A.13}$$

where v_o and a_o are the absolute motion of point o, the origin of the $oxyz$ reference frame; and $v_{A/o}$ and $a_{A/o}$ are the relative motion of A with respect to $oxyz$.

If reference frame $oxyz$ is also rotating, denote the angular velocity and angular acceleration of reference frame $oxyz$ as ω and α, respectively. The absolute motion of point A is given as

$$v_A = v_o + v_{A/o} + \omega \times r_{A/o} \tag{A.14}$$

$$a_A = a_o + a_{A/o} + \alpha \times r_{A/o} + 2\omega \times v_{A/o} + \omega \times \omega \times r_{A/o} \tag{A.15}$$

where $r_{A/o}$ is the position vector of point A within the $oxyz$ coordinate system.

In eqns. (A.12) through (A.15), reference frame $OXYZ$ in fact could be moving, too. In that case, the only modification to the above description would be changing the "absolute motion" to "relative motion with respect to $OXYZ$ reference frame."

A.1.3   Kinematics of a Rigid Body

The general motion of a rigid body can be decomposed into a translation plus a rotation. Based on this, for any two points A and B located on the same rigid body, the following relations hold

$$v_A = v_B + \omega \times r_{A/B} \tag{A.16}$$

$$a_A = a_B + r_{A/B} \times \alpha + \omega \times \omega \times r_{A/B} \tag{A.17}$$

where $r_{A/B}$ is a vector starting from B pointing to A, and ω and α are the angular velocity and the angular acceleration of the rigid body. These two equations are a special case of eqns. (A.14) and (A.15) with o replaced by B, for $v_{A/B} = 0$ and $a_{A/B} = 0$, because they are located on the same rigid body.

For velocity analysis of a rigid body, a useful concept is the *instantaneous center of rotation*. It can be located by drawing lines perpendicular to the velocity vectors of two points on the rigid body, without needing to know their magnitudes. The intersection of these two lines is the instantaneous center of rotation, as shown in Fig. A.4. Once this point is located, the velocity at any point on the rigid body can be calculated as if the rigid body is rotating about that point. However, this concept is not applicable for acceleration analysis.

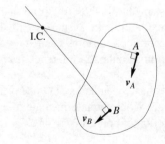

Figure A.4   The instantaneous center of rotation (I.C.) identified by knowing the directions of velocities of two points A and B on the rigid body

A.2   Kinetics

A.2.1   Newton–Euler Equations

For a particle, the Newton's second law states that

$$F = ma \tag{A.18}$$

where F is the sum of all forces acting on the particle, m is its mass, and a is its *absolute* acceleration. This is a vector equation and can be decomposed into different perpendicular directions, depending on the coordinate system in use.

This relation also hold for a system of particles, in which case, F is the sum of all external forces acting on the system, m is the total mass, and a is the acceleration of the center of the mass, called the *centroid*. The centroid of a system of particles is located at

$$r_C = \frac{\sum m_i r_i}{\sum m_i} \tag{A.19}$$

where r_C is the position vector for the centroid, and r_i and m_i are the position vector and mass, respectively, of individual particles.

A rigid body can be viewed as a system of particles. Hence, eqn. (A.18) also holds, and summations in eqn. (A.19) become integrals over the entire body.

However, particles in a rigid body are constrained in a particular way. Instead of applying Newton's second law to individual particles, a new concept is devised to characterize the rotational motion of the rigid body. This is called the *mass moment of inertia*. In general, in a three-dimensional space, the moment of inertia is a tensor of the second rank. If we limit our considerations to planar motions in the xy-plane, all rotations are about axes parallel to the z-axis; the moment of inertia becomes a scalar and is defined as

$$I_A = \int_V r^2 dm \tag{A.20}$$

where I_A is the moment of inertia about an axis parallel to the z-axis and passing through point A, r is the distance of any point in the rigid body to the axis, and the integration is performed over the entire volume V of the rigid body. Often, in planar motions, I_A is called the *moment of inertia about point A* for short. The moment of inertia about the centroid is also called the *centroidal moment of inertia*.

Many textbooks have tabulated the centroidal moments of inertia for rigid bodies of common shapes. Often, finding the centroidal moment of inertia is simply a matter of table lookup. The moment of inertia of a rigid body about any point A other than the centroid can be calculated via the following *parallel axes theorem*:

$$I_A = I_C + m d_{AC}^2 \tag{A.21}$$

where d_{AC} is the distance between the centroid C and point A.

The rotation of a rigid body is governed by the following *Euler equation*:

$$M_A = I_A \alpha + m a_A d \tag{A.22}$$

where α is the angular acceleration of the rigid body, M_A is the moment about point A, m is the mass, $a_A = |a_A|$, a_A is the acceleration of point A, and d is the distance from the centroid to

a_A (the point-to-line distance). The second term on the right-hand side of eqn. (A.22) can be dropped if the point under consideration is either a point with no acceleration (such as a fixed point, such that $a_A = 0$), or the centroid (such that $d = 0$). Denote such a special point as P, then,

$$M_P = I_P \alpha \tag{A.23}$$

A.2.2   Energy Principles

The most general form of the energy principle states that the work done by all nonconservative forces in a system equals the change in the total mechanical energy in the system, that is, between two states, denoted as 1 and 2, of motion,

$$U_{1-2}^{\text{n.c.}} = E_2 - E_1 \tag{A.24}$$

where $U_{1-2}^{\text{n.c.}}$ is the total work done by all nonconservative forces during the process, E is the total mechanical energy defined as the sum of the kinetic energy T and the potential energy V of the system, that is

$$E = T + V \tag{A.25}$$

For mechanical systems, nonconservative forces typically include externally applied forces, frictional forces, and so on. Forces due to the gravity and linear springs are conservative. When there is no nonconservative force in the system, we have the *principle of conservation of mechanical energy: $E_1 = E_2$ if $U_{1-2}^{\text{n.c.}} = 0$.*

The kinetic energy of a particle of mass m moving at speed v can be written as

$$T = \frac{1}{2}mv^2 \tag{A.26}$$

The kinetic energy of a rigid body, also of mass m, can be written as

$$T = \frac{1}{2}mv_C^2 + \frac{1}{2}I_C\omega^2 \tag{A.27}$$

or

$$T = \frac{1}{2}I_O\omega^2 \tag{A.28}$$

where v_C is the speed of the centroid, ω is the angular velocity of the rigid body, and I_C and I_O are moments of inertia of the rigid body about its centroid C and about a fixed point O, respectively. Note that eqn. (A.27) is generally applicable, whereas eqn. (A.28) is applicable only when O is a fixed point on the body.

The gravitational potential energy due to gravity for a particle or a rigid body, of mass m, can be uniformly written as

$$V = mgh \tag{A.29}$$

where h is the vertical elevation of the particle or the centroid of a rigid body from a fixed reference point, also called a *datum*.

The potential energy in a linear spring can be written as

$$V = \frac{1}{2}k\Delta^2 \tag{A.30}$$

where Δ is the amount of stretch in the spring and k is the spring constant.

A.2.3   Momentum Principles

The Newton's second law in eqn. (A.18) can be integrated over a period of time, from t_1 to t_2, into the following form:

$$\int_{t_1}^{t_2} F dt = p_2 - p_1 \tag{A.31}$$

where the integral on the left-hand side is called the *linear impulse* and

$$p = mv = \sum m_i p_i \tag{A.32}$$

is called the *linear momentum*: the first version for a single particle and the second version for a set of particles. Equation (A.31) represents the *principle of linear momentum*: the linear impulse equals the change in the linear moment of the system. If the linear impulse vanishes, we have the *conservation of linear momentum*.

   This principle is useful when the time duration $t_2 - t_1$ is short while there are significant changes in the motion. Such an event is often called an *impact* or a *shock*. In such cases, the impact forces can be so large that many other forces such as gravitational force become comparatively small and in turn negligible. If those impact forces can be treated as internal forces, we can approximate a situation as one in which the linear momentum of the system is conserved.

   For rotational motions, a similar equation can be obtained

$$\int_{t_1}^{t_2} r \times F dt = H_2 - H_1 \tag{A.33}$$

where quantity on the left-hand side is called the *angular impulse* and

$$H = r \times mv = \sum r_i \times m_i v_i \tag{A.34}$$

is called the *angular momentum*. Equation (A.33) represents the *principle of angular momentum*: the angular impulse equals the change in the angular moment in the system. For a planar motion, the above relation can be written as

$$\int_{t_1}^{t_2} M_A dt = H_{A2} - H_{A1} \tag{A.35}$$

where

$$H_A = I_A \omega \tag{A.36}$$

This principle is useful when the force always points to a central point. In such a case, the angular impulse about the force center vanishes, we have the *conservation of angular momentum*.

Appendix B

A Primer on MATLAB

MATLAB is a mathematical software developed by MathWorks, Inc. of Natick, Massachusetts. It is specialized in matrix computations. In fact, the name MATLAB is an abbreviation for Matrix Laboratory. Its development has incorporated a vast trove of extensively researched algorithms dealing with linear equation systems and many other matrix computations by the research community in the 1970s through 1980s. It is a very powerful tool for many matrix- and linear algebra-related problems that many of us probably have never heard of. Over the years, it has found extensive use in many engineering fields, as well as in scientific research endeavors.

B.1 Matrix Computations

B.1.1 Commands and Statements

MATLAB typically is used in the *interactive mode*, also known as the *interpretative mode*; that is, after the user has entered a command, MATLAB immediately interprets the command into computer's language, executes it, and feeds the results, or the error messages in case of an error, back to the user. Its user interface comprises a number of windows. The one with the >> prompt is the *command window*. This is the window in which most of interactions occur.

A command is a set of instructions to complete a specific task, such as invoking a build-in function, assigning value(s) to a variable, or simply displaying a variable, and sometimes can be much more involved. In the command window, a command ends when the user hits the enter (return) key; the interpretation, execution, and feedback start in sequence. There are two exceptions:

- If the last three characters of a command line are three dots (. . .), MATLAB will continue to read the next line of command while gobbling up the three dots.
- If a command starts with one of the flow control keywords (such as if, see Section B.1.5), MATLAB needs to read the entire flow control logic in order to perform the specified task: the command will end with an end keyword.

Fundamentals of Mechanical Vibrations, First Edition. Liang-Wu Cai.
© 2016 John Wiley & Sons, Ltd. Published 2016 by John Wiley & Sons, Ltd.
Companion Website: www.wiley.com/go/cai/fundamentals_mechanical_vibrations

A statement is almost the same as a command, but it is a smaller instruction unit that performs one single task. Multiple statements can be made into a block to be executed as a single command.

After a command is executed, MATLAB displays the answer in the command window. To suppress the answer display, a command can be followed by a trailing semicolon. The following example executes several commands in sequence:

```
>> theta = 0: pi()/4: pi()

theta =

        0      0.7854      1.5708      2.3562      3.1416

>> sin(theta)

ans =

        0      0.7071      1.0000      0.7071      0.0000

>> cos(theta);
>>
```

The first command is to construct a row matrix of named theta (see Section B.1.2 *Matrix Generation* for details on matrix generation). After the execution, it displays the full matrix it just constructed. The second command instructs MATLAB to call its build-in sine function (sin) to operate on matrix theta. As expected, it calculates the sine of individual theta values and constructs a matrix having the same size as theta to store the result. As the result is not assigned to any variable, it displays the prompt ans as the stand-in name, which stands for the *answer*. The third command computes the cosine (cos) of theta. But the command line has a trailing semicolon, and the display of the answerer is suppressed.

B.1.2 Matrix Generation

MATLAB treats every user-defined variable as a matrix. Defining a matrix is very simple, and there are numerous ways.

The most straightforward way of defining a matrix is putting all the elements inside a pair of square bracket:

```
A=[1, 2, 3; 4, 5, 6]
```

where a comma separates elements in a row and a semicolon separates rows. The above command creates the following matrix:

$$A = \begin{bmatrix} 1 & 2 & 3 \\ 4 & 5 & 6 \end{bmatrix}$$

With such explicit assignments, the size of the matrix does not need to be declared in advance. The commas between the elements in a row can be omitted, as long as there are spaces or tabbed spaces in between, such as

```
>> A=[1 2 3; 4 5 6]
```

If we put either of the above statements defining matrix A into MATLAB's command window, it produces the following response:

```
>> A =

    1       2       3
    4       5       6
```

A matrix can also be generated by an implicit loop

```
t=[0:0.1:2]
```

which produces a row matrix whose elements' values start from 0, and increase by 0.1 for each successive element, until a value of 2 is exceeded. At the end, in this example, it takes 21 iterations to reach the value of 2. Thus, this statement produces a 1×21 matrix. In this form, the square brackets can be omitted.

Several build-in functions also generate matrices at user-specified sizes:

- ones(n), ones(n,m): a matrix with all elements having a value of 1
- zeros(n), zeros(n,m): a matrix with all elements having a value of 0
- eye(n), eye(n,m): an *identity matrix*, that is, all diagonal elements having a value of 1, and all others 0
- rand(n), rand(n,m): a matrix with all elements filled with different random numbers.

In the above functions, the form having a single parameter n produces a square matrix of size n × n; and the form having two parameters n and m produces a rectangular matrix of size n × m.

B.1.3 Accessing Matrix Elements and Submatrices

A particular matrix element can be referred to by its row and column numbers. The row and column numbers start from 1. For example, for A matrix defined above, its elements can be accessed as the following:

```
>> A(2,3)

ans =

     6

>> A(1,2)

ans =

     2

>>
```

If a matrix is a row matrix (having only one row) or a column matrix (having only one column), the elements can be referred to by just one index. For example,

```
>> t(20)

ans =

    1.9000

>>
```

The single-index access in fact works with any matrix, but the way the index is defined is not intuitive and may cause confusion. It counts the elements column by column. The count starts from the first column down to its rows and continues onto the second column, and so on. For a matrix of the size m × n, its ith row jth column entries has a single index as (j-1) × n + i. For beginners, this method is not recommended.

An entire column of a matrix can be referred to by using a colon as its row number. Thus, A(:,2) would produce the second column of A matrix.

```
>> A(:,2)

ans =

    2
    5
```

Similarly, using a colon as the column number produces an entire row:

```
>> A(2,:)

ans =

    4    5    6
```

On the other hand, two matrices of compatible sizes can be concatenated to form a new larger matrix. If we joint two A matrices together, there is a horizontal concatenation as [A A] and a vertical concatenation [A; A].

```
>> [A A]

ans =

    1    2    3    1    2    3
    4    5    6    4    5    6

>> [A;A]

ans =
```

```
        1       2       3
        4       5       6
        1       2       3
        4       5       6


>>
```

B.1.4   Operators and Elementary Functions

B.1.4.1   Operators

MATLAB uses common operators as in most modern programming languages. The following is a list of often-used operators in MATLAB:

- Assignment operator: =
- Arithmetic operators: +, -, *, \, /, ∧; and the dot-preceded variety: .+, .-, .*, .\, ./, .∧
- Relational operators: >, >=, <, <=, ==, ~=
- Logical operators: &, |, ~, xor.

The meanings of these operators should be familiar to the readers who are assumed to have done elementary computer programming, at least such as Microsoft's Excel or Visual Basic. But, as every variable is treated as a matrix in MATLAB, there are some important distinctions that we shall discuss as follows.

- *Dot-Preceded Operators*, also known as *Element-Wise Operators*: A dot can precede immediately (with no space in between) most of arithmetic operators and change the meaning of the operator to an element-by-element operation. For example, A * B and A .* B are completely different operations. The former produces the matrix product $[A][B]$; the latter produces a matrix whose element at the ith row and the jth column is $A_{ij}B_{ij}$.
- *Size Matching*: The matrices to be operated on by an operator, called the *operands*, must have sizes conforming with the operator's definition. For example, the operands for the + and − operators must be of the same size. For the * operator, the number of columns in the first matrix must equal to the number of rows of the second matrix.
- *Scalar Context*: A 1×1 matrix is treated as a scalar. When it is operated on either side of a matrix, an element-by-element operation is carried out.
- *Division*: Matrix division is not mathematically defined. Being focused on linear equation systems, MATLAB actually has two division operators defined. The regular one is called the *right division*, such as A/B, which calculates $[B][A]^{-1}$. Another form is to have the slash flipped, such as B\A, which is called the *left division*, and calculates $[A]^{-1}[B]$. For a linear equation systems written as $[A]\{x\} = \{B\}$, the solution can be simply written as x = B\A. Note that the size requirements for B in the two division operators are different.
- *Matrix Transposition*: A' is the transposition of A, mathematically $[A]^T$.
- Both relational and logical operators operate in the element-by-element manner.

Being a mathematical software, MATLAB has a rather extensive set of elementary mathematical functions defined.

B.1.4.2   Trigonometric and Hyperbolic Trigonometric Functions and Their Inverses

Function	Inverse	Meaning
cos	acos	Cosine
cosh	acosh	Hyperbolic cosine
cot	acot	Cotangent
coth	acoth	Hyperbolic cotangent
csc	acsc	Cosecant
csch	acsch	Hyperbolic cosecant
sec	asec	Secant
sech	asech	Hyperbolic secant
sin	asin	Sine
sinh	asinh	Hyperbolic sine
tan	atan	Tangent
	atan2	Two-argument four-quadrant inverse tangent
tanh	atanh	Hyperbolic tangent

B.1.4.3   Exponential, Logarithms, and Other Functions

Function	Meaning
exp	Exponential
log	Natural logarithm (base e)
log10	Common logarithm (base 10)
sqrt	Square root
floor	Round down to the previous integer
ceil	Round up to the next integer
round	Round toward the nearest integer
mod	Modulus after division
rem	Remainder after division

B.1.4.4   Complex Variable Functions

Function	Meaning
abs	Absolute value
angle	Phase angle
conj	Complex conjugate
i or j	Imaginary unit
imag	Imaginary part
real	Real part

B.1.5   Flow Controls

Similar to any programming language, MatLab has a set of the so-called *flow control* devices for altering or alternating the sequences of commands to be executed, based on conditions that are evaluated during the computation.

B.1.5.1   The `if` Condition

There are several variations, with rather straightforward logic. The basic one is

```
if condition
  ...
end
```

Here, the `condition` is a statement that returns an answer of logical `true` or `false`; the `...` represents a block of statements to be executed if the `condition` is `true`. The next is a more complete logic:

```
if condition
  ...
else
  ...
end
```

It executes two alternate blocks of statements based on the `condition`. The last one is more flexible yet complicated, with multiple conditions to be tested:

```
if condition
  ...
elseif condition
  ...
else
  ...
end
```

In this last variation, the `elseif` clause can appear multiple times, each with a different `condition`, but the `else` clause can only appear no more than once. Attention is called to the logical condition here because multiple conditions have to be tested. The actual condition to execute the final `else` statement block is all of the previous conditions being `false`.

B.1.5.2   The `switch` Condition

The `switch` condition is very similar to the last version of the `if` condition.

```
switch variable
  case value1
    ...
  case value2
    ...
  otherwise
    ...
end
```

The `switch` differs from the multicondition `if` in that the branching is based on the value of one single `variable`: it executes different statement blocks based on the different values of that switch variable.

B.1.5.3   The `for` Loop

A typical `for` loop is for executing a set of statements multiple times, with changed states. Its typical structure is the following:

```
for index = startvalue:interval:endvalue
    ...
end
```

During the loop, the `index` value starts from `startvale`, and MatLab performs the computation contained in the statement block. Then, the index value is increased by the `interval`, and it repeats the computation of the statement block. This process is repeated until the `index` value exceeds the `endvale`. Typically, the `index` is an integer. If the `interval` is 1, which is also typical, it can be omitted: `index = startvalue:endvalue`.

B.1.5.4   The `while` Loop

The `while` loop looks simpler: it keeps executing the statement block as long as the `condition` is `true`, or, until the `condition` becomes `false`.

```
while condition
    ...
end
```

But a word of caution is necessary here. The `while` loop has a very high risk of turning the loop into an endless loop, called as an *infinite loop*.[1] It is important to ensure that changing the `condition`, either implicitly or explicitly, is a part of the statement block. Also, implicitly changing the condition is dangerous because the `condition` could become unpredictable.

Let us use an example to demonstrate the use of flow controls. In this example, we define an *identity matrix* the hard way[2]: we will explicitly assign a value to each element: if an element is located on the diagonal of the matrix, it has a value of 1; otherwise, it has a value of 0. The code using the `for` loop is shown below:

```
N = 4;
for i=1:N
     for j=1:N
          if i==j
               A(i,j)=1;
          else
               A(i,j)=0;
          end
     end
end
```

After the execution, we can type in A in the command window to display the newly constructed matrix:

```
>> A

A =

     1     0     0     0
     0     1     0     0
     0     0     1     0
     0     0     0     1

>>
```

The same process can also be implemented using the `while` loop. The corresponding code is as the following:

```
N = 4;
i=1;
j=1;
while i<=N
     while j<=N
          if i==j
```

---

[1] Apple Inc.'s headquarter has the address of "1 Infinite Loop."
[2] The easy way would be to use MATLAB's build-in function `eye`.

```
            A(i,j)=1;
        else
            A(i,j)=0;
        end
        j = j+1;
    end
    i = i+1;
end
```

This code is a little longer because we have to take care of initializing the indices i and j at the beginning and advancing them during the loop.

B.1.6   M-Files, Scripts, and Functions

Very often, we need to execute a series of commands or statements in order to perform a complicated task. In such cases, we can input the commands in the command window by hands, one by one. But, it is much more convenient to save the command sequence into a text file. This way, the saved commands can be edited, corrected (called *debugged*), and re-used in the future. Moreover, if we save such a file with a file-name extension .m in the current *working directory*, MATLAB will recognize the file name (without the extension) as a command just like the build-in ones. Such a file is called an m-file for the sake of its extension. There are two types of m-files. One is simply a script; and the other is a user-defined *function*.

A script m-file simply contains the commands you would enter in the command window, and MATLAB executes it in the same way as it would in the command window. Writing a script is just like writing a code for any other programming language. In order to insert explanatory texts to aid the reading and understanding of the code, comments can be added into the script. MATLAB treats everything after a % symbol as a comment and simply ignores it. A good programming practice is to add plenty of comments to aid the reading.

A function m-file requires a specific structure. A function allows its argument to be specified at the time of invocation, just like the build-in sin and cos functions. A function m-file starts with a function declaration line and is followed by the body of the function. The function deceleration line must start with the keyword function. It is followed by the output variable for the result to be returned; and then is followed by the assignment operator, the function name, and the argument(s) placed inside a pair of parentheses. The following is an example of a function m-file that defines a function called sqrtsin to calculate the square root of the sine function.

```
% Function that calculates the square root of sine
function a = sqrtsin( theta )
a = sqrt(sin(theta));
```

In this function we have one value to return, which is called a. Correspondingly, inside the function body, the calculation result is assigned to a. When the function is called with an argument, the argument is assigned to theta and the calculation is performed according to the script in the function body. This function definition can be saved as `sqrtsin.m` in the working directory, and it then can be used just like a build-in function.

A function can have multiple arguments or no argument at all and can have multiple outputs or no output at all. The rules are as the following:

- If a function has no argument, the parentheses after the function name can be omitted.
- If a function has multiple arguments, all the arguments must be listed inside the parentheses and separated by commas. For example, in the above code, (`theta`) would be replaced by (`theta1, theta2`).
- If a function has no output, "`a =`" in the function declaration line can be omitted.
- If a function has multiple outputs, then each output needs to be assigned a variable name and placed in a square bracket. For example, `a` and `b` are the outputs, "`a =`" in the function declaration line above will need to be replaced by "`[a b] =`" or "`[a, b] =`." In fact, `a` and `b` are forced into a matrix to be returned. Therefore, we need to watch out for the size compatibility in case `a` and `b` are matrices.
- The specific output variable names and argument names are not visible outside the m-file. In other words, when calling a function, we do not need to worry about what are the names for the input arguments and output variables. What matters is the form they are organized, such as the sequence.

Based on the above rules, we define another function with multiple inputs and multiple outputs. This function calculates the square roots of both sine and cosine, and we put both `omega` and `t` as the arguments. We call this function `sqrtsc`. The function definition is the following:

```
% Function that calculates the square root of both sine and cosine of the
% product of the two imput parameters
function [a b] = sqrtsc( omega, t )
a = sqrt( sin(omega*t) );
b = sqrt( cos(omega*t) );
```

Now, we can test these two functions while practicing running an m-file script. We save the above function definition in a file named `sqrtsc.m` (must have the same file name as the function name) and put the following script in a text file named `testfunc.m`:

```
% testfunc.m: Testing function definitions
t = [0:.5:3];
omega = 3.;
res = sqrtsin( omega*t );
[res_sin res_cos] = sqrtsc( omega, t );
```

The following is what happened in the command window:

```
>> testfunc
>> t
```

```
t =

    0     0.5000    1.0000    1.5000    2.0000    2.5000    3.0000

>> res

res =

   Columns 1 through 6

    0     0.9987    0.3757    0 + 0.9887i    0 + 0.5286i    0.9685

   Column 7

    0.6420

>> res_sin

  res_sin =

   Columns 1 through 6

    0     0.9987    0.3757    0 + 0.9887i    0 + 0.5286i    0.9685

   Column 7

    0.6420

>> res_cos

res_cos =

   Columns 1 through 6

    1.0000    0.2660    0 + 0.9950i    0 + 0.4591i    0.9799    0.5888

   Column 7

    0 + 0.9545i

>>
```

Here is what has just happened. First, since every line (except the function deceleration lines) in the function definitions and in the script file is ended with a semicolon, the execution of the script does not print out anything. This is the preferred behavior of a script or a function, as we would only want to see selected results, proactively selected by us, the users. When we enter a command t, it displays the stored values for variable t, which has been defined as a row matrix having seven columns in the first line in the script. Similarly, we enter the variable names `res`, `res_sin`, and `res_cos` as commands and MATLAB displays the corresponding values.

Among these results, two things to note. As all these variables are row matrices with seven columns, the command window is not able to accommodate all seven columns, it breaks them down into groups of six. Second, when sine and cosine functions return negative values, their square roots become imaginary numbers. MATLAB handles complex number effortlessly. It uses `i` as the unit for imaginary numbers. We can also use `j` in commands and in scripts, because in some fields such as electrical engineering, *j* is also a commonly used symbol for the unit for imaginary numbers.

When the function body is simple, MATLAB provides another mechanism to define functions on the fly: the *anonymous functions*. An anonymous function does not need to have a name and does not need to be saved into a separate file. It is simply a statement in a script. For example, the `sqrtsin` function we defined earlier can be implemented as an anonymous function in the following script:

```
% Anonymous function for calculating the square root of sine
sqrtsc = @(t) sqrt(sin(t));

% rest of the script invoking the anonymous function
theta = [0.: pi()/4: pi()];
sqrtsc(theta)
```

In the above script, `= @(t)` is a construct that declares an anonymous function having t as its argument. To the left of the construct, `sqrtsc` is called the *function handle*, which in a strict sense is a pointer to the function but in reality can be used in the subsequent script just like the function's name. To the right of the construct is the statement for the function body. Because the last statement does not end with a semicolon, the execution of the above script returns the following:

```
ans =

        0     0.8409    1.0000    0.8409    0.0000

>>
```

B.1.7   Linear Algebra

As mentioned earlier, MATLAB was founded for solving problems related to matrices and linear algebra. It has a rich and extensive library of functions that handle an extremely large variety of matrix-related computational tasks. In our most elementary use of MATLAB, some useful functions are listed in Table B.1.

Table B.1   Elementary MATLAB Functions for Linear Algebra

dot	Dot-product of two vectors
cross	Cross-product of two vectors
transpose	Transpose, the function form of operator'
inv	Inverse of a matrix
det	Determinant of a matrix
norm	Matrix or vector norm
rank	Rank of a matrix
length	Size of the larger dimension of a matrix
size	Size (numbers of row and column) of matrix
cond	Condition number of a matrix
trace	Trace of a matrix
sum	Sum elements of a matrix

B.1.7.1   Solving Linear Equation System

As mentioned earlier, solving a linear equation system $[A]\{x\} = \{b\}$ in MATLAB is rather simple and convenient. With properly defined matrices A and b, we can simply write x = A\b. Of course, when we studied the linear algebra, we know that there are many different methods of solving such a linear equation system; and MATLAB has corresponding functions for all those methods. But this elementary way of solving a linear equation system suffices in this course.

B.1.7.2   Eigenvalues and Eigenvectors

In matrix theory, a *matrix eigenvalue problem* is finding the eigenvalue λ, which is a scalar quantity, and its corresponding eigenvector $\{V\}$, which is a vector or a column matrix, for a square matrix $[A]$ such that

$$[A]\{V\} = \lambda\{V\} \tag{B.1}$$

For an $n \times n$ nonsingular matrix, there are n eigenvalues and n corresponding eigenvectors. Solving a matrix eigenvalue problem can be done by a build-in function eig in two forms, assuming matrix A has been appropriately defined:

- lambda = eig(A) where lambda is a column matrix that stores all the eigenvalues
- [V,D] = eig(A) where V is square matrix of the same size as A, which stores the eigen-vector, in which each eigenvector takes up one column; and D is a diagonal matrix of the same size that stores the corresponding eigenvalues.

The *generalized matrix eigenvalue problem* is finding the eigenvalue λ and its corresponding eigenvector $\{V\}$ for a pair of square matrices $[A]$ and $[B]$ such that

$$[A]\{V\} = \lambda[B]\{V\} \tag{B.2}$$

Such a problem still has n eigenvalue and eigenvector pairs if both $[A]$ and $[B]$ are $n \times n$ square nonsingular matrices. They can also be found using the two forms of the eig function as earlier, but it takes both A and B, presumably having been appropriately defined, as the arguments.

B.2   Plotting

Using MATLAB to plot curves is one of the main purposes we introduce MATLAB in this course. Plotting a curve used to be a rather intimidating task, with many mundane details. MATLAB makes it simple by automating many of those mundane details. Also, MATLAB offers extensive ways of presenting a set of data.

B.2.1   Two-Dimensional Curve Plots

The main two-dimensional curve plotting command is plot. It has several forms:

- plot(x,y): This function plots a curve with x as the abscissa (the horizontal axis) and y as the ordinate (the vertical axis). The results will be shown in the *figure window*. It can handle different sizes of x and y matrices as the following:
  - The most straightforward is when both x and y are vectors of the same size. This is what we expect from a simple plotting command.
  - If both x and y are matrices of the same size: each column of y is plotted against the corresponding column of x, producing a set of curves.
  - If one of x and y is a matrix and the other is a vector: the length of the vector must be equal to one of the dimensions of the matrix. Then, multiple curves will be plotted. For example, if x is vector, and it has the same size as the number of rows in y, then each column of y will be plotted against x.
- plot(x,y, LineSpec): In this case, the additional argument LineSpec defines the attributes of the curve. The attributes are as the following:
  - *Line Style*: the choices are - for solid line; -- for dashed line; : for dotted line; and .- for dot–dashed line.
  - *Marker*, which places a symbol at each data point on the curve. The common choices are o: a circle; +: plus sign; *: asterisk; .: dot; s: square; d: diamond; ∧: upward-pointing triangle; and v: downward-pointing triangle.
  - *Color*: the common choices are y: yellow; m: magenta; c: cyan; r: red; g: green; b: blue; and k: black.

  These attributes, at most one from each category, can be arranged in any order but must be placed within a pair of single quotes. For example, '--ok' means: dashed line, with circle marker, and in black color.
- plot(x1,y1,LineSpec1,x2,y2,LineSpec2,...): In this case, the x, y, Line-Spec triplet appears multiple times. It has the same effect as multiple commands.

Apart from plotting the curve, we need a few more details to complete a figure, such as naming the axes and modifying the limits to the axes. The following are a few functions that can be used to fix those details:

- xlabel('name'): Put the name on the abscissa (the horizontal axis).
- ylabel('name'): Put the name on the ordinate (the vertical axis).
- axis([xmin xmax ymin ymax]): Set the limits of both axes.
- legend('text1', 'text2', ...): Place a legend box in the figure. In the box, each curve is represented by a short segment of the line and is followed by the text.

- `hold on|off`: When turned on, subsequent `plot` commands will plot on the same figure. This way, multiple `plot` commands produce multiple curves. When `off`, each `plot` command erases the figure space. This way, multiple `plot` commands will leave only the last one on display.
- `box on|off`: Turn axes border box on or off.
- `grid on|off`: Turn the grids on or off.
- `subplot(n,m,p)`: Divide the current figure window into n × m subplots and direct all the subsequent plotting commands' outputs to the pth subplot. The subplots count by the row. For example, `subplot(2,2,2)` will place the plot onto the upper right quadrant in a 2 × 2 subdivision.

Typically, these functions are called after plotting the curve. MATLAB works in a way such that the user does not need to predeclare anything. When it comes to plotting a curve, a `plot` command needs to be issued so that MATLAB has information to create an object representing the curve. Only after this curve object has been created, we will be able to modify it by calling these functions. The following code example plots two curves using many of the ingredients discussed earlier:

```
theta = [0:5:360];
y1 = sin(theta*pi/180) - .25 * sin(theta*pi/180) .* sin(theta*pi/180);
y2 = cos(theta*pi/180) - .25 * cos(theta*pi/180) .* cos(theta*pi/180);
plot(theta,y1,':+r',theta,y2,'-.^b');
axis([0,360,-2,1]);
legend('sin\theta-sin^2\theta/4', 'cos\theta-cos^2\theta/4');
xlabel('\theta');
ylabel('y1, y2');
```

The execution results are shown in Fig. B.1. In the above code, `pi` is a build-in constant for π. The angle `theta` is originally defined in degrees, and it is converted into radians for

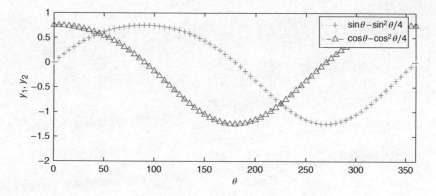

Figure B.1 A figure with two curves and a legend box. Dotted with plus signs: $y_1 = \sin\theta - \frac{1}{4}\sin^2\theta$; Dot–dashed with triangles: $y_2 = \cos\theta - \frac{1}{4}\cos^2\theta$

computation. In the legend box, θ is produced by \theta. This is the way MATLAB handles Greek letters and other symbols.

In a computer, curves are plotted using straight line segments to connect adjacent data points. In order to produce a smooth-looking curve, a sufficient number of data points should be used, unless the number of data point itself has some significance (such as each data point being experimentally gathered). Too few data points would make the curve looking zigzagged. Typically, a minimum of 200 points across the width direction is needed, and 500 points would be better. In the event that a curve has sharp peaks, significantly more data points would be desirable.

B.2.2 Three-Dimensional Curve Plots

A three-dimensional curve is typically defined by one or more parameters, mathematically known as a *parametric curve*. The main function to use is plot3. Its synopsis is essentially the same as plot. After the curve has been plotted, those curve-modifying functions still work. The following example plots a three-dimensional helix defined as:

$$x = \sin(t) \qquad y = \cos(t) \qquad z = t \tag{B.3}$$

where t is a parameter. The plot is shown in Fig. B.2, which is produced by the following script:

```
t = [0:5:720];
x = sin(t*pi/90);
y = cos(t*pi/90);
plot3(x,y,t,'r');
axis([-1.5,1.5,-1.5,1.5,0,720]);
```

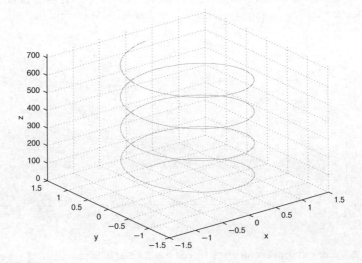

Figure B.2 Three-dimensional curve plot for the helix defined in eqn. (B.3)

```
grid on;
xlabel('x');
ylabel('y');
zlabel('z');
```

B.2.3   Three-Dimensional Surface Plots

Another type of three-dimensional data representation is often used in scientific computations: a three-dimensional surface whose elevation is a function of its location on a two-dimensional base plane, that is, mathematically, $z = f(x, y)$. MATLAB has two functions for producing such surface plots: `surf(x,y,z,C)` and `surfc(x,y,z,C)`. The difference is that the latter also plots a set of contours on the base plane, which is presumed to be the xy-plane. In the argument list, x, y, and z are matrices of the same size, and they specify the x-, y-, and z-coordinates of every point on the surface. The optional argument C is another matrix of the same size, based on which MATLAB colors the surface. Let us use an example to explain how these arguments are specified.

We imagine a curve of cosine function on the xz-plane for one half of a period, that is, x varies from 0 to π and z varies from -1 to 1. The curve is rotated about the z-axis, forming an axisymmetric surface defined by the equation

$$z = \cos \sqrt{x^2 + y^2} \tag{B.4}$$

The following code plots this surface:

```
xvec = [-180:10:180];
[x,y]=meshgrid(xvec,xvec);
z = cos( sqrt( x .* x + y .* y) * pi/180);
subplot(1,2,1);
surf(x,y,z,z);
colorbar;
axis([-200,200,-200,200,-4,2]);
xlabel('x');
ylabel('y');
zlabel('z');
subplot(1,2,2);
surfc(x,y,z,gradient(z));
colorbar;
axis([-200,200,-200,200,-4,2]);
grid on;
xlabel('x');
ylabel('y');
zlabel('z');
```

To plot the surface, we need to lay out a rectangular, and preferably square, grid on the xy-plane and then evaluate the function on the grid points. In the above code, we first create a vector

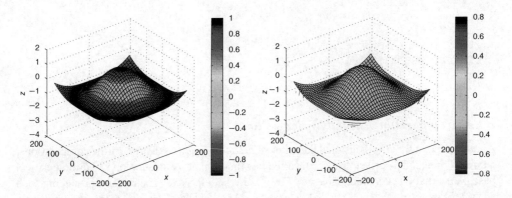

Figure B.3   Surface plot for the surface defined in eqn. (B.4)

xvec that specifies the grid points along the x-axis. We can see that this is a 1×37 row matrix. We can similarly define one for the y-axis. We skip this step as we want a square grid and use the one for the x-axis on the y-axis. Then MATLAB function meshgrid is used to convert two vectors into two grid matrices x and y, of a size 37×37. These matrices hold values for x- and y- coordinates, respectively, for the grid. Having the grid's x- and y-coordinates, z value is then evaluated accordingly. This completes the generation of the data set.

Then, this set of data is plotted twice. The first is a straightforward plot, in which the color is based on the surface height (the z-value). In the second time, the color is based on the surface gradient, for which the function gradient is called, and the plotting command is surfc. Note that command subplot is used to place two subfigures in the same figure. The resulting plots are shown in Fig. B.3.

There are many other ways to alter the curve and surface objects in MATLAB, such as colors to be used on the surface, line styles, and thicknesses of curves. But those are more advanced topics that we shall refrain from divulging deeper.

_____ · _____

So far we have touched upon only a tiny fraction of MATLAB's capabilities. A good thing of MATLAB is that it comes with a very detailed and extensive online help, with many examples. The users are encouraged to look into those information to further explore the vast capabilities of MATLAB.

Appendix C

Tables of Laplace Transform

Laplace transform is an integral transform that is often applied to transform a time-domain function, of real argument $t \geq 0$, into a function of complex parameter s. This is often denoted as $\mathscr{L}\{f(t)\} = \tilde{f}(s)$. The definition is the following:

$$\mathscr{L}\{f(t)\} = \int_0^\infty e^{-st} f(t) dt \tag{C.1}$$

where $\tilde{f}(s)$ is called the transformed function of $f(t)$.

C.1 Properties of Laplace Transform

Properties are the rules that govern all functions that can be transformed. General properties of the Laplace transform is summarized in Table C.1.

In Table C.1, $f(t)$ and $g(t)$ are any transformable functions, a and b are constants, $H(t)$ is the *Heaviside step function*, defined as

$$H(t) = \begin{cases} 1 & \text{when } t \geq 0 \\ 0 & \text{when } t < 0 \end{cases} \tag{C.2}$$

C.2 Function Transformations

Table C.2 lists the Laplace transform pairs for functions that are often used in vibration analyses. The readers are also advised to look up other sources for more comprehensive listing of functions.

In Table C.2, n is an integer, $\delta(t)$ is the *unit impulse function*, defined as

$$\delta(t) = \begin{cases} \infty & \text{when } t = 0 \\ 0 & \text{otherwise} \end{cases} \quad \text{and} \quad \int_{-\epsilon}^{\epsilon} \delta(t)\,dt = 1 \tag{C.3}$$

where ϵ is an arbitrary infinitesimal value.

Fundamentals of Mechanical Vibrations, First Edition. Liang-Wu Cai.
© 2016 John Wiley & Sons, Ltd. Published 2016 by John Wiley & Sons, Ltd.
Companion Website: www.wiley.com/go/cai/fundamentals_mechanical_vibrations

Table C.1   Properties of Laplace Transform

	Time domain	s-Domain
Linearity	$af(t) + bg(t)$	$a\tilde{f}(s) + b\tilde{g}(s)$
Differentiation	$f'(t)$	$s\tilde{f}(s) - f(0)$
Second differentiation	$f''(t)$	$s^2\tilde{f}(s) - sf(0) - f'(0)$
Time shifting	$f(t - a)H(t - a)$	$e^{-as}\tilde{f}(s)$
Integration	$\int_0^t f(\tau)d\tau$	$\dfrac{\tilde{f}(s)}{s}$
Convolution	$\int_0^t f(\tau)g(t - \tau)d\tau$	$\tilde{f}(s)\tilde{g}(s)$
s-Domain differentiation	$tf(t)$	$-\tilde{f}'(s)$
s-Domain integration	$\dfrac{f(t)}{t}$	$\int_s^\infty \tilde{f}(\sigma)d\sigma$

Table C.2   Laplace Transform of Common Functions

Function	Time Domain	s-Domain
Unit impulse	$\delta(t)$	1
Unit step	$H(t)$	$\dfrac{1}{s}$
Delayed unit step	$H(t - a)$	$\dfrac{e^{-as}}{s}$
Ramp	$tH(t)$	$\dfrac{1}{s^2}$
nth Power	$t^n H(t)$	$\dfrac{n!}{s^{n+1}}$
Exponential decay	$e^{-at}H(t)$	$\dfrac{1}{s + a}$
Two-sided exponential decay	$e^{-a|t|}H(t)$	$\dfrac{2a}{a^2 - s^2}$
Sine	$\sin(at)H(t)$	$\dfrac{a}{s^2 + a^2}$
Cosine	$\cos(at)H(t)$	$\dfrac{s}{s^2 + a^2}$
Hyperbolic sine	$\sinh(at)H(t)$	$\dfrac{a}{s^2 - a^2}$
Hyperbolic cosine	$\cosh(at)H(t)$	$\dfrac{s}{s^2 - a^2}$
Exponentially decaying sine	$e^{-bt}\sin(at)H(t)$	$\dfrac{a}{(s + b)^2 + a^2}$
Exponentially decaying cosine	$e^{-bt}\cos(at)H(t)$	$\dfrac{s + b}{(s + b)^2 + a^2}$

Index

Fundamentals of Mechanical Vibrations, First Edition. Liang-Wu Cai.
© 2016 John Wiley & Sons, Ltd. Published 2016 by John Wiley & Sons, Ltd.
Companion Website: www.wiley.com/go/cai/fundamentals_mechanical_vibrations

Printed and bound by CPI Group (UK) Ltd, Croydon, CR0 4YY